Industrial Exploitation of MICROORGANISMS

Industrial Exploitation of MICROORGANISMS

Editors

D.K. Maheshwari

R.C. Dubey

R. Saravanamuthu

I.K. International Publishing House Pvt. Ltd.

NEW DELHI • BANGALORE

Published by

I.K. International Publishing House Pvt. Ltd.
S-25, Green Park Extension
Uphaar Cinema Market
New Delhi – 110 016 (India)
E-mail: info@ikinternational.com

ISBN 978-93-80026-53-4

Published by Krishan Makhijani for I.K. International Publishing House Pvt. Ltd., S-25, Green Park Extension, Uphaar Cinema Market, New Delhi – 110 016 and Printed by Rekha Printers Pvt. Ltd., Okhla Industrial Area, Phase-II, New Delhi – 110 020.

PREFACE

Microorganisms have been exploited in multifarious ways since the dawn of human civilization. The ancient humans developed techniques of producing alcohols, acids, beverages, fermented foods, etc. in different countries of the world. For example, yeasts have been exploited throughout the world in several countries for production of different kinds of preparation from kitchen to industry. Many yeast plasmid- and chromosome-based vectors have been constructed for biotechnological applications in industry. Still it is also a major tool for biotechnological applications. Considering the importance of microorganisms, demand of industrial application of microorganisms got increased gradually. Nowadays thousands of antibiotics have been produced through microorganisms during the last 80 years from the discovery of the first antibiotic and many more have been commercialized. In recent years the researchers have developed the techniques to produce products through introducing the desired genes of interest in fungal/bacterial/plant/animal cells. Therefore, prokaryotic and eukaryotic cells are being used as hosts for introduction of recombinant vector(s) containing genes of desired functions for production of valuable products.

Though the direct use of microorganisms is known since the beginning of 16th century but much emphasis has been given on industrial microbiology in recent years for improvement of human and animal health. Among these the lactic acid bacteria are considered and selected as 'Probiotics'. Some non-lactic acid bacteria are also considered as Probiotics. In addition, different species of *Spirulina* have also been exploited in food industry and *Spirulina*-based products have now been commercialized.

However, due to expanding population day-by-day global food shortage has alarmed the world population. Many varieties of mushrooms have been explored and utilized in multifarious ways since vedic period. Recently, mushrooms are looked up on for the presence of industrially important bioactive molecules, for example anticancerous compound in Shitaki mushroom. In addition, certain higher fungi hold much promise due to the presence of medicinal principles like digestive enzymes, stress reliever, antiviral effects, anti-tumour activity, strengthening health resisting disease, etc. Certain mushrooms like *Cordyceps sinensis* has traditionally been used for treatment of human diseases like chronic bronchitis, insomnia, hypertension, pneumonia, tuberculosis, anaemia, etc.

There is resurgence in the search of new antibiotics in clinically important bacteria, especially in actinomycetes. There are many companies throughout the world that are devoted to commercialization of bioactive compounds from actinomycetes. Therefore, prospects of bioactive compounds in Pharma industries are also increasing day-by-day for large-scale production and commercialization of bioactive compounds. Use of bacterial consortium with or without yeast as Probiotics has revolutionized the area of industrial microbiology because of combating the health-related problems of humans. Several bacterial, fungal and viral preparations in different forms are being formulated and marketed because of being eco-friendly and non-hazardous to plants and animals.

Commercial production of highly valued microbial enzymes is gaining attention in recent years due to its great role in many areas of industry, for example, microbial proteases, lipases, cellulases, laccases, phytases, etc. Besides, production of 'nanoparticles' through microorganisms has attracted the attention of researchers throughout the world. The nanoparticles can play a key role in many technologies of the future. One of the key aspects is the development of reliable experimental protocols for the synthesis of nanoparticles over a range of chemical compositions, sizes and high nanodispersity. Several microorganisms are known to produce inorganic materials intra- or extracellularly, for example, nanoparticles of silver, gold, magnetite, siliceous materials, calcium carbonate, metal sulphides, etc. Such microorganisms can be a boon for industries in future. The ability of pharmaco-kinetic behaviour, of different molecules, their toxicity in mammals and the possibility for their future application in many areas of medical sciences need to be explored in future.

All the articles included in this book are devoted to industrial exploitation of potential microorganisms which would be very helpful to the undergraduate and postgraduate students, researchers and teachers of microbiology and biotechnology in India and abroad.

D.K. Maheshwari
R.C. Dubey
R. Saravanamuthu

EDITORS

D.K. Maheshwari started his teaching carrier as a lecturer in the department of Botany, D.A.V (PG) College, Muzaffarnagar and thereafter served as Reader in the Department of Microbiology, Barkatullah University, Bhopal. He joined Gurukul Kangri University, Haridwar as Professor in the year 1990 and served as Dean, Faculty of Life Science (2004-2006). Prof. Maheshwari is an active member of several scientific bodies of international repute and in the board of panels of various academic and administrative bodies including UPSC and NAAC of Government of India. He has been elected as Editor of the *Journal of Indian Botanical Society* (1999-2002) and he was awardee of young scientist "Prof. Y.S. Murthy Medal" for his outstanding contribution.

As a young scientist, he was selected under UNESCO programmes and worked in Biological Research Center. Szeged (Hungary) in the year 1983-84 and also visited Czechoslovakia, Belgium, Holland, Germany, Japan and S. Korea. He was visiting Professor in Science University of Tokyo, in 1993 and 1998, and Guest Professor, in the University of Ulm (Germany). Prof. Maheshwari visited S. Korea in the year 2000, 2003 and 2006 under bilateral scientific exchange programmes sponsored by Indian National Science Academy (INSA), New Delhi.

Prof. Maheshwari has many research papers in leading peer-reviewed journals both Indian as well as foreign. More than 30 candidates have been awarded Doctorate (Ph.D.) degree under him. He has completed 10 major research projects sponsored by various funding agencies. Prof. Maheshwari is the editor of *Microbes: Agriculture, Industry and Environment* (2000) and *Innovative Approaches in Microbiology* (2002). He is co-author of *Practical Microbiology* (2001) and *A Textbook* of Microbiology and member Editorial Board, Korean J. Agric, Chem. and Biotechnology.

He is a Fellow of the Indian Botanical Society and Indian Phytopathological Society and Life Member of Indian Science Congress, Phytopathological Society of India, Botanical Society and Association of Microbiologists in India. He is nominated as Convenor "Science forum" from Uttarakhand Council for Science and Technology, Govt. of India, Dehradun.

His field of interest is soil microbiology and microbial biotechnology having expertise in biological control of soil-borne plant pathogens, mycorrhizae, Rhizosphere microbiology and botanical pesticides, besides enzymology and heavy metal toxicity. He has published more than 75 research and review papers in the national and international journals of repute. He has produced 14 Ph.Ds and is supervising several research students on varied aspects of microbiology. He has handled research projects sponsored by UGC, UCOST and CSIR.

He has authored three books viz., *Practical Microbiology* (2004), *A Text Book of Biotechnology* (4th edn., 2006) and *A Text Book of Microbiology* (2nd edn., 2005). He is also co-editor of *Himalayan Microbial Diversity* (1997); *Microbes: Agriculture, Industry and Environment* (2000) and *Innovative Approaches in Microbiology* (2002).

He was the Organizing Secretary of the National Seminar on Bioinoculants for Holistic Sustainable Rural Development jointly organized by the Department of Botany & Microbiology and DDU state Institute for Rural Development (U.P. Govt.) (1998). He was the *Co-coordinator* of the Workshop on *Molecular Biology and Applied Microbiology*, and Hands-on-Training on *Microbial Fermentation and Inoculants Production* (2008).

He is the Life Member and Fellow of the Indian Botanical Society; Indian Phytopathological Society and the International Society for Conservation of Natural Resources. He was Associate Editor of the *Journal of the Indian Botanical Society* (2000-2004), and at present he is in the Editorial Board of the *Journal of Environmental Biology and Conservation.*

R. Saravanamuthu did his postgraduate degree at the St. Joseph's College, Tiruchirappalli (TN), M. Phil. at the Centre of Advanced Study in Botany, University of Madras (Chennai) and Ph.D. at the Centre of Advanced Study in Botany, Banaras Hindu University, Varanasi. He has more than thirty years of teaching and research experience in the Department of Botany, A.V.C. College (Autonomous), Mannampandal, Tamil Nadu.

Dr. Saravanamuthu has successfully guided eight Ph.D. scholars and published number of research articles in the reputed national and international journals. He has contributed about ten chapters of research interest in different edited books, and also has been the editor of books, journals and proceedings.

Dr. Saravanamuthu is the Life Member and Fellow of the Indian Botanical Society and member of several professional bodies and academic societies. He has served as the member of the Board of Management of the Tamil Nadu Agricultural University.

CONTRIBUTORS

A.K. Gade, Department of Biotechnology, SGB Amravati University, Amravati–444 602, (India)

A. Mehta, Microbial Technology Laboratory, Department of Botany, Dr. H.S.Gour University, Sagar–470 003, (India)

A. Panneerselvam, Department of Botany and Microbiology, A.V.V.M Sri Pushpam College (Aut) Poondi, Thanjavur–613 503, (India)

A. Satlewal, Department of Microbiology, G.B. Pant University of Agriculture & Technology, Pantnagar–263 145, (India)

D. Dhanasekaran, Department of Microbiology, School of Life Sciences, Bharathidasan University, Tiruchirappalli-620 024 Tamil Nadu, (India)

G. Chithra, Department of Biotechnology, Bharath College of Science and Management, Thanjavur–613 005, (India)

G.K. Garg, Department of Molecular Biology & Genetic Engineering, G.B. Pant University of Agriculture & Technology, Pantnagar–263 145, (India)

J. Vetriselvi, Department of Microbilogy, S.T.E.T. Womens's College, Sundarakkottai, Mannargudi–614 001, (India)

K.K. Gupta, Department of Botany and Microbiology, Gurukul Kangri University, Haridwar–249 404, Uttarakhand, (India)

K. Sowndhararajan, Department of Botany, Bharathiar University, Coimbatore–641 046, (India)

M.K. Rai, Department of Biotechnology, SGB Amravati University, Amravati–444 602, (India)

M.L. Bari, National Food Research Institute, 2-1-12, Kannondai, Tsukuba, 3058642, (Japan)

M. Radhakrishnan, Department of Microbiology, Sri Sankara Arts & Science College, Kanchipuram–631 561, (India)

N. Kango, Applied Microbiology and Biotechnology, Dr. Hari Singh Gour Vishwavidyalaya, Sagar–470 003, (India)

N. Thajuddin, Department of Microbiology Bharathidasan University Tiruchirappalli–620 024, (India)

P. Dwivedi, Department of Biotechnology, Indian Institute of Technology Roorkee Roorkee–247 667, (India)

P. Gupta, School of Biotechnology Guru Ghasidas University Bilaspur–495 009, (India)

P.C. Jain, Applied Microbiology and Biotechnology, Dr. Hari Singh Gour Vishwavidyalaya, Sagar–470 003, (India)

R. Balagurunathan, Department of Microbiology, Sri Sankara Arts & Science College, Kanchipuram-631 561, (India)

R.C. Dubey, Department of Botany and Microbiology, Gurukul Kangri University Haridwar–249 404, Uttarakhand, (India)

R.C. Kuhad, Lignocellulose Biotechnology Laboratory, Department of Microbiology, University of Delhi South Campus, Benito Juarez Road, New Delhi–110 021, (India)

R.C. Ray, Central Tuber Crops Research Institute (Regional Center), Bhubaneswar–751 019, Orissa, (India)

R. Chandrasekaran, Department of Botany and Microbiology, A.V.V.M. Sri Pushpam College (Autonomous), Poondi, Thanjavur, T.N., (India)

R. Goel, Department of Microbiology, G.B. Pant University of Agriculture & Technology, Pantnagar–263 145, (India)

R. Gothalwal, School of Biotechnology Guru Ghasidas University Bilaspur–495 009, (India)

R. Jain, Applied Microbiology and Biotechnology, Dr. Hari Singh Gour Vishwavidyalaya, Sagar–470 003, (India)

R.P. Singh, Department of Biotechnology Indian Institute of Technology Roorkee Roorkee–247 667, (India)

R.R. Pandey, Department of Life Science, Manipur University, Imphal, (India)

R. Rajendran, Department of Microbiology PSG College of Arts and Science Civil Aerodrome PO, Coimbatore–641 014, (India)

R. Saravanamuthu, Department of Botany, A.V.C College (Autonomous), Mannampandal, Mayiladuthurai–609 005. Nagapattinam, Tamil Nadu, (India)

S. Balamurugan, Department of Botany, A.V.C College (Autonomous), Mannampandal, Mayiladuthurai, Nagapattinam–609 005, (India)

S. Dhiva, AGAPPE Diagnostics Ltd., AGAPPE Hills, Pattimatom, Ernakulam Dist.–683 562, (India)

S. Isobe, National Food Research Institute, 2-1-12, Kannondai, Tsukuba–3058642, (Japan)

S. Kuhar, Lignocellulose Biotechnology Laboratory, Department of Microbiology, University of Delhi South Campus, Benito Juarez Road, New Delhi–110 021, (India)

S. Manian, Department of Botany, Bharathiar University, Coimbatore–641 046, (India)

T. Satyanarayana, Department of Microbiology, University of Delhi South Campus, Benito Juarez Road, New Delhi–110 021, (India)

V. Konasani, Department of Microbiology, University of Delhi South Campus, Benito Juarez Road, New Delhi–110 021, (India)

V. Dhivaharan, Department of Microbilogy, S.T.E.T. Womens's College, Sundarakkottai, Mannargudi–614 001, (India)

Vivekanand, Department of Biotechnology, Indian Institute of Technology Roorkee, Roorkee–247 667, (India)

CONTENTS

Chapter 1

Biotechnological Potential and Industrial Applications of Yeast

J. Vetriselvi, R. Saravanamuthu* and V. Dhivaharan

Department of Microbiology, S.T.E.T. Women's College, Sundarakkottai, Mannargudi – 614 001, Tamil Nadu, India

**Department of Botany, A.V.C. College, Mayiladuthurai – 609 305, Tamil Nadu, India*

** E-mail: rs_avcc@rediffmail.com*

Yeast is one of the eukaryotic microorganisms, which has wide applications in various spheres and is being used since very long time. Its application in the brewries and bakeries is well known. The history of application of yeast in industries is very long. This organism is exploited in the production of ethanol, beers, wine and bread. Besides, it is also a source of microbial protein (SCP), and directly used as food, feed and food supplements. Hence, it is also referred as 'nutritional yeast'. Its use as probiotics is not uncommon. Yeast is also a major source of enzymes such as invertase, β-galactosidase, alcohol dehydrogenase, glyceraldehydes 3-phosphate dehydrogenase, hexokinase, etc. The processed yeast products (yeast extract) are used as food additives or flavours. Some strains of yeast find place in bioremediation. Yeast was the first choice among eukaryotic microorganisms for the application of recombinant DNA technology. It is the best organism suited for gene manipulation and transformation experiment. Thus, yeast has been used as industrial process organism, as protein source and also as tool in gene manipulation and transformation technologies.

INTRODUCTION

Yeasts are a group of eukaryotic microorganisms classified in the kingdom of fungi. Approximately, 1,500 species of yeasts have been described (Kurtzman, 2006), most of which reproduce asexually by budding, although a few cases also reproduce by binary fission. Yeasts are cosmopolitan and widely distributed in almost all natural habitats. The most well-known and commercially significant yeasts are the related species and strains of *Saccharomyces cerevisiae*. These organisms have long been utilized to ferment the sugars of rice, wheat, barley and corn to produce alcoholic beverages, and in the baking industry to expand or raise the dough. *S. cerevisiae* is commonly used as baker's yeast and also for some types of fermentation. Yeast is often taken as a vitamin supplement because it has 50% protein and also a rich source of B vitamins, niacin and folic acid. During the last two decades, *S. cerevisiae* has been the model system of the molecular genetic research because of the basic cellular mechanisms of replication, recombination and cellular metabolism, which are generally considered between yeasts and larger eukaryotes, including mammals. Yeasts have recently been used to generate electricity in microbial fuel cells (Walker, 1998) and to produce ethanol for the biofuel industry.

HISTORY

The word "yeasts" comes from the old English language "gist"/"gyst" ultimately from the Indo-European root "yes" meaning boil, form or bobble. Yeasts are microbes, which are probably one of the earliest domesticated organisms. People have used yeasts for fermentation and baking since a long time. Archaeologist digging in Egyptian ruins found early girding stones and baking chambers for yeast mediated manufacturing bread as well as drawings depict of the 4,000 years old bakeries and breweries. In 1680, the Dutch naturalist Antony Van Leeuwenhoek was the first person to observe the yeast microscopically, but at that time he did not consider them to be living organisms but rather described as globular structure. In 1857, French microbiologist Louis Pasteur proved and reported in his paper entitled memorie *sur la fermentation alcoholique* "that alcoholic fermentation was conducted by living yeast and not by a chemical catalyst". Pasteur showed that by bubbling oxygen into the yeast broth, cell growth could be increased, but the fermentation inhibited an observation later called the "Pasteur effect". Thus, yeast is one of the earliest or the first known microorganisms.

TAXONOMY

Yeasts do not form a specific taxonomic or phylogenetic grouping. At present it is estimated that only 1% of all the yeast species have been described (Kurtzman and Piskur 2006). The term "yeast" is often taken as a synonym for *S. cerevisiae* (Kurtzman 1994), however, the phylogenetic diversity of yeast is shown by their placement in both divisions, Ascomycota and Basidiomycota. The budding yeast (true yeast) is classified in the Saccharomycetales.

YEASTS AS A MODEL EUKARYOTE

The yeasts species *S. cerevisiae* has been a very important model organism in modern cell biology research and is the most thoroughly researched eukaryotic microorganism. Although yeast has greater genetic complexity than bacteria, containing 3.5 times more than *Escherichia coli* cells, they share many of the technical advantages that permitted rapid progress in the molecular genetics of prokaryotes and viruses. Some of the properties that make yeast particularly suitable for biological studies include rapid growth, dispersed cells, the ease of replica plating and mutant isolation, a well-defined genetic system, and the most important aspect is a highly versatile DNA transformation system. Unlike many other microorganisms, *S. cerevisiae* is visible with numerous markers. Being a non-pathogenic organism, yeast can be handled with little precautions. Large quantities of normal baker's yeast are commercially available and can provide a cheap source for biochemical studies. Contrary to most of the other microorganisms, strains of *S. cerevisiae* have both a stable haploid and diploid state. Thus, recessive mutation can be conveniently induced and the manifested haploid strains can be isolated, and complementation test can be carried out in diploid strains. The development of DNA transformation has made the yeast accessible to gene cloning and genetic engineering techniques.

GROWTH AND NUTRITION

Yeasts use organic compounds as the source of energy and do not require light to grow, hence they are chemoautotrophs. The main source of carbon is obtained from hexose such as glucose or disaccharides such as sucrose and maltose and pentose sugars such as fructose. Some species can metabolize alcohols and organic acids. Species of yeasts are aerobic, i.e., require oxygen for aerobic cellular respiration (obligate aerobes) or anaerobic, which also have aerobic methods of energy production (facultative anaerobes). Yeast is ubiquitous and is most frequently isolated from sugar rich samples. Some strains are found in association with soil and insects (Slavikova and Vadkertiova, 2003).

REPRODUCTION

Yeasts have sexual and asexual reproduction cycle, however, the most common mode of vegetative growth in yeast takes place asexually by budding or fission (Balasubramanian *et al.,* 2004). A small bud or daughter cell is formed on the parent cell. The nucleus of the parent cell splits into daughter nucleus and migrates into the daughter cell. The bud continues to grow until it separates from the parent cell, forming a new cell. The haploid cells will generally die under high stress conditions. However, under the same conditions, diploid cells can undergo sporulation, entering sexual reproduction (meiosis) and reducing a variety of haploid spores, which can go onto mate (conjugate), and regenerate the diploid daughter cells.

INDUSTRIAL AND BIOTECHNOLOGICAL POTENTIAL OF YEASTS

Yeasts are directly and indirectly very useful microorganisms to mankind. Species or strains of *Saccharomyces* are widely employed by human beings in (i) baking industry – *S. cerevisiae,* (ii) distilleries industry – *S. cerevisiae* and *S. diastaticus*, (iii) wine industry – *S. ellipsoideous,* and (iv) the production of alcoholic beverages – *S. cerevisiae* and *S. carlsbergengis*.

The useful physiological properties of yeasts have led to their use in the field of biotechnology. Fermentation of sugars by yeasts is the oldest and largest application, which is recognized as old technology. Many types of yeasts are used for making many foods, such as baker's yeast in bread making and brewer's yeast in the process of fermentation leading to the production of wine and xylitol (Sreenivas Rao *et al.*, 2004).

Production of Ethanol

S. cerevisiae is the main producer of ethanol. It is used as a major tool for the production of ethanol since the discovery of fermentation process by Louis Pasteur. During 1980, fermentation of broth was discovered in sugar solution by the addition of yeast extracts obtained by its grinding. The ability of yeast to convert sugar into ethanol has been harnessed by the biotechnological industries, which has various uses. The process starts with milling sugar cane, sweet corn or cheap cereal grains and then adding dilute sulphuric acid, and fungal alpha amylase enzymes, which enhances the breakdown of the starch into simple sugars. A glucoamylase is then added to simple sugars. After this, yeasts are added to convert the simple sugar to ethanol, which is then distilled off to obtain ethanol up to 96% in concentration. After the products are harvested, these yeasts are used as microbial source of protein and animal feed supplements. Molasses, cheese, whey, sulphite liquor and gas oil fractions are used as main fermentation media components. But these media usually need some growth stimulating nutrients such as biotin, ammonium ions, sulphate, etc.

S. cerevisiae have been genetically engineered to ferment xylose-one of the major fermented sugars present in cellulosic biomasses, such as crop residues, paper wastes, and wood chips (Dujon, 1996). This development proves that ethanol can be potentially and efficiently produced from inexpensive agricultural waste. This become an alternative source of ethanol fuel which could be competitively produced as a low priced fuel alternative to gasoline fuels.

Beer

Dan Emil Christian Hansen classifies beer brewers yeasts into top fermenting yeast and bottom fermenting yeast. Top fermenting yeasts are so called because they form foam at the top of the wort during fermentation. They can produce high quantity of alcohol concentration and prefer relatively high temperatures producing frutifier, sweetener and ale type beers, e.g., *S. cerevisiae.* Bottom fermenting yeasts are used to produce lager type beers. Bottom fermenting yeasts ferment more sugars, leaving a crisper taste, and grow well at low temperature, e.g., *S. pastorianus.* In industrial brewing, to ensure purity of strain, a clean sample of yeast is stored at refrigeration temperature. After a certain number of fermentative cycles, full-scale propagation is made from this laboratory sample. Typically, it is grown up in about 3 or 4 stages using sterile brewing wort and oxygen.

Root Beer and Sodas

Root beer is also produced by the methods used for the production of beer with only variation in carbonation process—the active phase of yeast is stopped sooner producing only trace amount of alcohol (consumed by all ages), and a significant amount of sugar is left in the drink. Low calorie or nearly sugar free root beer and soda can be produced in such a way by using small amount of sugar and also replacing the majority with artificial sweeteners.

Distilled Beverages

It is a beverage that contains ethanol, which has been purified by distillation. Carbohydrate containing plant material is fermented by yeast, producing a dilute solution of ethanol. Alcoholic beverages such as whiskey and rum are prepared by distilling this dilute solution of ethanol. Components other than ethanol are collected in the condensate including water, esters and other alcohols, which account for the flavour of the beverages.

Wine

Yeast is used in wine making where it converts the sugars present in grape juice or *must* into alcohol. Yeast is normally already present on the grapes, often visible as a powdery film (also known as bloom or blush) on their exterior. The fermentation can be done with this indigenous (or wild) yeast, however, this may lead to unpredictable results depending on the exact type of yeast species that is present. For this reason a pure culture is generally added to the *must*, which predominantly enhances the process of fermentation as it proceeds. Most of the widely used wine yeast include *S. cerevisiae,* however, not all strains of the species are suitable for wine-making (Gonzalez *et al.*, 2001). Different strains of *S. cerevisiae* differ in their physiological and fermentative properties. Therefore, the actual strain of yeast should be selected that exhibits a direct impact on the finished wine (Dunn *et al.*, 2005). Significant researches have been undertaken for the development of novel wine yeast strains that produce atypical flavours, profiles or increased complexity in wines (McBryde *et al.*, 2006). The growth of yeast species such as *Zygosaccharomyces* and *Brettanomyces* in wine can result in wine faults and subsequent spoilage (Loureiro and Malfeito Ferreria, 2003). *Brettanomyces* produces an array of metabolites when growing in wine, some of which are volatile phenolic compounds. Altogether these compounds are often referred as "*Brettanomyces character*" and are often described as antiseptic or "barnyard" type aromas. *Brettanomyces* is a significant contributor to wine faults in the wine industry.

Baking

S. cerevisiae is used in baking as a leavening agent, where it converts the fermentable sugars present in the dough into carbon dioxide. This causes the dough to expand or raise the carbon dioxide formed as pockets or air bubbles. When the dough is baked and carbon dioxide pockets remain, the baked product becomes soft and spongy in texture. The use of potatoes, water from potato boiling, eggs or sugars in bread dough accelerates the growth of yeast. Salt fats and butter slow down the growth of the yeast. The majority of the yeasts used in baking are the same species, which are commonly used in alcoholic fermentation. *Saccharomyces exiguus* (also known as *S. minor*) is a wild yeast found on plants, fruits and grains that is occasionally used for baking. Today there are several retailers of baker's yeasts; one of the best known is Fleishmann's yeast, which was developed in 1868. During World War II, Fleischmann's developed granulated active dry yeasts, which did not require refrigeration and had a longer shelf life than fresh yeast.

Yeast as SCP—Food, Feed and other supplements

Since ancient times a number of microorganisms have been used as a part of diet. Fermented yeast (*Saccharomyces* sp) was recovered as a leavening agent for bread as early as 2500 B.C. (Fleet, 2001). Importance of mass production of microorganisms as a direct source of microbial protein was realized during World War I in Germany, and consequently baker's yeast (*S. cerevisiae*) was produced in an aerated molasses medium with ammonium salts. During the World War II (1939-1945) the aerobic yeast, e.g., (*Candida utilis*) was produced for food and feed in Germany. The dried cells of microorganisms including algae, bacteria, actinomycetes and fungi, which are used as food and feed, are collectively called "microbial protein". The major components of substrates are the raw materials, which contain sugars (sugar cane, sugar beet and their processed products), starch (grains, potato and their by-products) and lignocelluloses from woody plants and herbs having residues with nitrogen and phosphorus contents and other raw materials.

Yeasts are used in the houses of million of families throughout the world everyday to prepare food and beverages. Their ability and usefulness in the conversion of variety of raw materials, low in protein, to nutritious products have increased the efforts to develop their use as food. Although at the beginning the technology was used for making the yeast supplements during the World War II in Germany, the field became much more active in the 1960s, coinciding with the recognition of increasing global protein shortages. Currently yeasts are used more often as animal feed rather than as human food. For e.g., *Kluveromyces marxianus* is used to ferment whey waste from cheese making and results in the production of SCP for animal feed as well as some alcohol with the added value of reduction of pollution by the whey—one high yeast content product that has been available for many years for human consumption in the UK.

Nutritional Supplements

Its use, as a nutritional supplement, is popular with vegans and the health conscious people, where it is often referred as "nutritional yeast". It is the deactivated yeast that is usually *S. cerevisiae* used in food supplements. It is an excellent source of protein and vitamins, especially the B complex vitamins, whose functions are related to metabolism as well as other minerals and cofactors required for growth. It is also naturally low in fat and sodium. Nutritional yeast has nutty, cheesy and creamy flavours which makes it popular as an ingredient in cheese substitutes. It is often used by vegans in place of parmesan cheese. Another popular aspect is its use as topping for popcorn. They are offered along with salt or cayenne pepper as a popcorn condiment. It is sold in the form of flakes, or as a yellow powder similar in texture to cornmeal and can be found in most of the natural food stores. In Australia it is sometimes sold as "savory yeast flakes". Though "nutritional yeast" is usually referred to commercial products, prisoners are fed with "home grown" yeast to prevent vitamin deficiency.

Probiotics

In some probiotic preparations, the yeast, *S. boulardii,* is used as a supplement to maintain and restore the natural flora in the gastrointestinal tract. *S. boulardii* has been shown to reduce the symptoms of acute diarrhoea in children (Centina Sauri and Sierra,1994; Kurugol and Koturoglu, 2005), prevent reinfection of *Clostridium difficle* (Oswal *et al.*, 2002), reduce bowel movement in diarrhoea predominant IBS patients (Maupas *et al.*, 1983) and reduce the incidence of antibiotic (MaFarland *et al.*, 1995), traveller's (Kollaritcsh *et al.*, 1989) and HIV/AIDS associated diarrhoea (Saint *et al.*, 1995).

Kombucha

Yeast has synergistics effect when used with acetic acid bacteria in the preparation of kombucha, a fermented sweetened tea. Species of yeast found in the tea can vary and may include *Brettanomyces bruxellensis, Candida stellata, Schizosaccharomyces pombe, Torulapora delbrueckii* and *Zygosaccharomyces baili* (Teoh *et al.*, 2004).

Yeast constituents

Yeast is a major source of enzymes, which have applications in industry, and is a biochemical component of research.

Invertase: This enzyme is obtained from strains of *S. cerevisiae* grown aerobically on sucrose. It is available in the market as an invertase rich, dried yeast preparation or as concentrated syrup.

β-glycosidase: It is obtained from *S. fragilis, C. pseudotropicalis* or *C. utilis* grown on cheese whey. It is used in the manufacture of ice cream (to remove lactose) and to prepare lactose free milk products.

Other enzymes: Alcohol dehydrogenase, glyceraldehyde-3-phosphate dehydrogenase, hexokinase and many other enzymes are also obtained from *S. cerevisiae.* All these enzymes are useful in biochemical research.

Cofactor: Cofactor viz., co enzymes A, oxidized and reduced NAD, NADP and ATP and glutathione are also obtained from *S. cerevisiae*. They also have application in understanding the biochemical reaction and research activities.

Nucleic acid: Nucleic acid, particularly RNA, is obtained from *C. utilis* and *Saccharomyces* sp. RNA is used as a raw material for the production of derivatives for biochemical compounds. RNA is also used in the preparation of flavour compounds (guanisine and inosine 5 monophosphate).

Yeast extract

It is the common name for various forms of processed yeast products that are used as food additives or flavours. They have free glutamic acid such as MSG (Mono Sodium Glutamate). The general method for making yeast extract is to add hypertonic solution to a suspension of yeast. This leads to the cells shivering up. This triggers autolysis, where the yeasts digestive enzymes break their own proteins down into simpler compounds, and a process of self-destruction occurs. Yeast autolysates are used in various countries of the world for various purposes.

Bioremediation

Some yeast can find potential application in the field of bioremediation. One such yeast, *Yarwinia lipolytica,* is known to degrade palm oil mill effluent, TNT (an explosive material) (Jain *et al.*, 2004) and other hydrocarbons such as alkanes, fatty acids, fats and oils. It has been experimentally proved that these compounds are effectively degraded by the species of yeast.

Yeast vector as a biological tool

Generally, yeast is a unicellular eukaryotic fungus containing a small well characterized genome unlike plant or animal cells. It is a non-pathogenic fungus and has a fast growth rate. Most of its gene contain intrones, which are spliced during purification of mature mRNA. It appears that introns found in yeast contain sequence for correct splicing as they are totally absent in higher eukaryotes. Moreover, yeast can carry out post-translational modification such as removal of signal sequences from a precursor polypeptide after the secretion of cells. This removal is a major advantage of yeast over the bacteria. Yeast cells contain their own plasmid (Broach *et al.*, 1991) known as '2' cm circle plasmid, which is used as vector for foreign genes. It is present in many strains in 50-100 copies per cell. The foreign DNA taken up by yeast integrates by a specific crossing over, such as origin of replication (*ori*). Yeast DNA linked with foreign DNA has been demonstrated to transform very efficiently (about 104 cells) (Hall and Linder, 1992). The transforming cells replicate autonomously within the cell. DNA sequences of such property known as Ars (Autonomously replicating sequence) fragment plasmid have been constructed by using Ars fragments when joined at centromeric sequences (Cen region of about 6-10 kb DNA) to ensure the replication and stability in yeast. Cen regions of yeast plasmid are useful fragments because they confer stability (Glover, 1984). A wide range of vectors is available to meet various requirements for insertion, deletion, alteration and expression of genes in yeast. Most of the plasmids used for yeast studies are shuttle vectors, which contain sequences permitting them to be selected and propagated in *E. coli*, thus allowing for convenient amplification and subsequent alteration *in vitro*. The most common yeast vectors originated from pBR322 and contain an origin of replication (*ori*), promoting high copy-number maintenance in *E. coli*, and the selectable antibiotic markers, the β-lactamase gene, *bla* (or Amp^R), and sometime to tetracycline-resistance gene, *tet* or (Tet^R), conferring resistance to ampicillin and tetracycline respectively (Brown and Tuite, 1998). In addition, all yeast vectors contain markers that allow selection of transform ants containing the desired plasmid. The most commonly used yeast markers include *URA3*, *HIS3*, *LEU2*, *TRP1* and *LYS2*, which complement specific autotrophic mutations in yeast, such as *ura3-52*, *his3-Δ1*, *leu2-Δ1*, *trp1-Δ1* and *lys2-201*. These complement able yeast mutations have been chosen because of their low-reversion rate. Also, the *URA3*, *HIS3*, *LEU2* and *TRP1* yeast markers can complement specific *E. coli* autotrophic mutations (Table 1).

Table 1. Components of common yeast plasmid vectors

	Yap	Yap	Yap	Yap
Plasmid				
E. coli genes or segments				
ori, bla; tet	+	+	+	+
Yeast genes or segments				
URA3; HIS3; LEU2; TRP1; LYS2; etc.	+	+	+	+
leu2-d	0	+	+	0
2 μm; 2 μm-*ori REP3;*	0	+	0	0
ARS1; ARS2; ARS3; etc.	0	0	+	+
CEN3; CEN4; CEN11; etc.	0	0	0	+
Host (yeast) markers				
ura3-52; his3-Δ1; leu2-Δ1; trp1-Δ1; lys2-201; etc.	+	+	+	+
Stability	++	+	±	+

Although there are numerous kinds of yeast shuttle vectors, those used currently can be broadly classified in either of following three types of (Table 1) integrative vectors, YIp; autonomously replicating high copy-number vectors, YEp; or autonomously replicating low copy-number vectors and YCp.

Yeast Artificial Chromosome (YACs)

It is the most compact yeast vectors that are used to clone large DNA fragments in yeast cells. They are linear in shape and comprise telomers, centomers and ars; hence called minichromosome. There are two telomers, one on either side of an YAC, and highly conserved and homologous to each other. Each telomere consists of multiple copies of palindromic sequence (CCC AAA). It is mainly used for the construction of genomic libraries and gene map of chromosome in human genome project.

The initial step in the molecular characterization of eukaryotic genomes generally requires cloning of large chromosomal fragments, which is usually carried out by digestion with restriction endonucleases and ligation to specially developed cloning vectors. Usually, 200 to 800 kb fragments are cloned as Yeast Artificial Chromosomes (YACs), and 100-200 kb fragments are cloned as Bacterial Artificial Chromosomes (BACs or PACs). The importance of YAC technology has been heightened by the recently developed methods for transferring YACs to cultured cells and to the germline of experimental animals (Wodicka *et al.*, 1997).

YAC cloning systems are based on yeast linear plasmids, denoted YLp, containing homologous or heterologous DNA sequences that function as telomeres (*TEL*) *in vivo*, as well as containing yeast *Ars* (origins of replication) and *CEN* (centromeres) segments (Goffeau *et al.*, 1996). Manipulating YLp linear plasmids *in vitro* is complicated by their inability to be propagated in *E. coli* (Ostergaard *et al.*, 2002). However, specially developed circular YAC vectors have been developed for amplification in *E. coli*. For example, a circular YCp vector, containing a head-to-head dimer of *Tetrahymena* or yeast telomeres, is resolved *in vivo* after yeast transformation into linear molecules with the free ends terminated by functional telomeres.

Shuttle vector

This vector is constructed by using bacterial origin of replication in a plasmid, which can be manipulated and cloned in bacteria. Finally, they are transferred into yeast cells for the possible expression of eukaryotic genes, e.g., β-lactamase, a prokaryotic gene, is expressed in yeast.

Pathogenic yeast

Some species of yeast are opportunistic pathogens, where they can cause infection in people with compromised immune system. *Cryptococcus neoformans* is a significant pathogen of immunocompromised people causing the disease termed cryptococcosis. This occurs in about 7-8 % of AIDS patient in USA and slightly smaller per cent (3-6) in Western Europe (Wheals *et al.*, 1995). The cells of the yeast are surrounded by a rigid polysaccharide capsule, which helps to prevent them from being recognized and engulfed by white blood cells in the human body.

Yeast of the genus *Candida* includes another group of opportunistic pathogens, which cause oral and vaginal infections in human, known as candidiasis. It is commonly found as a commensal yeast in the mucous membrane of humans and other warm blooded animals. The yeast cells sprout a hyphal outgrowth, which locally penetrates the mucosal membrane causing irritation and shedding of the tissues (Hurley *et al.*, 1987). *C. albicans* causes infections of the urogenital tract and the bloodstream (candidemia) (Stoyan *et al.*, 2004).

Spoilage of food

Yeasts are able to grow in foods with a low pH (5.0 or lower) and in the presence of sugars, organic acids and other easily metabolized carbon sources (Kurtzman, 2006). During their growth, yeast metabolize some food components and produce metabolic end product. This causes the physical, chemical and sensory properties of food to change and the food is spoiled (Fleet and Praphailong, 2001). The growth of yeast within the food products is often seen on their surface, as in cheese or meats, or by the fermentation of sugars in beverages, such as syrups and jams. The yeast of the Zygosaccaromyces genus has a long history as a spoilage yeast within the food industry. This is mainly due to the fact that these species can grow in the presence of high sucrose, ethanol, acetic acid, sorbic acid, benzoic acid and sulphur dioxide.

CONCLUSION

Thus, yeast is the eukaryotic microorganism, which has the long history of exploitation for the welfare of human beings. It is also a thoroughly studied industrial process and well understood microorganism with reference to fermentation technology. Its exploitation in the production of bread, alcohol and wine is well known. Yeast (*S. cerevisiae*) was consumed as food in Germany during World War I, and thus assumed great significance as a source food feed and food supplements. It is obvious from the foregoing paragraphs that yeasts are not only a source of microbial protein but also a source of various types of enzymes, essential amino acids and minerals. Moreover, it exhibits rapid growth and possesses small well-characterized genome. The genome has already been sequenced, and it was the first eukaryotic microorganism with which recombinant DNA technology was attempted. It has been established that yeasts can carryout post translational modification such as removal of a signal sequence from a precursor polypeptide after the secretion of cells. Interestingly, DNA can be cloned in yeast, which depends on the uptake of DNA by its sphaeroplast. The yeast plasmids are used as vectors for foreign genes. Thus, it is concluded that yeast is an organism with high potential of biotechnological and industrial applications.

REFERENCES

Balasubramanian. M., Bi, E. and Glotzer, M. (2004). *Curr Biol* **14 (18):** R806-18.

Broach, J.R., Jones, E.W. and Pringle, J.R. (1991). Cold Spring Harbor Laboratory Press, Cold Spring Harbor, New York. 457-464.

Brown, A.J.P. and Tuite, M.F. (1998). Methods in Microbiology, Vol. 26. Acad. Press, New York 306-309.

Centina-Sauri, G., and Sierra-Basto, G. (1994). *Ann Pediatr* **41:** 397-400.

Dujon, B. (1996). *Trends in Genetics*, **7:** 263-270.

Dunn, B., Levine, R.P. and Sherlock, G. (2005). BMC Genomics. **16-6(1):** 53.

Fleet, G.H. and Praphailong, W. (2001). AIFST, Southwood Press. 383-397.

Glover, D.M. (1984). The mechanisms of DNA manipulatiion, Chapman and Hall: London, pp. 222.

González-Techera, S., Juba New-York, F.M., Carrau, C. and Gaggero. (2001). *Letters in Applied Microbiology* **33:** 1-71.

Goffeau, A., Barrell, B.G., Bussey, H., Davis R.W., Dujon, B., Feldmann, H., Galibert, F., Hoheisel, J.D., Jacq, C., Johnston, M., Louis, E.J., Mewes, H.W., Murakami, Y., Philippsen, P., Tettelin, H. and Oliver, S.G. (1996). *Science* **274:** 546-552.

Hall, M.N. and Linder, P. (1992). *The Early Days of Yeast Genetics*. Cold Spring Harbor Laboratory Press, Cold Spring Harbor, New York, 99.

Hurley, R., de Louvois, J. and Mulhall, A. (1998). *In*: *The yeasts*, (Eds. Rose, A.H. and Harrison, J.S.) Vol. 1. Academic Press, Inc., New York.

Jain, M.R., Zinjarde, S.S., Deobagkar, D.D. and Deobagkar, D.N. (2004). *Mar Pollut Bull.* **49(9-10):** 783-8.

Kollaritsch, H., Kemsner, P., Wiedermann, G. and Scheiner, O. (1989). *Acta Paediatrica* **94**: 44-47.

Kurtzman, C.P. (1994). *Yeast.* **10(13):** 1727-40.

Kurtzman, C.P. and Piskur, J. (2006). *In:* Comparative Genomics: Using Fungi as Models. (Eds. Sunnerhagen, P. and Piskur, J.) Vol. 15. Berlin: Springer-Verlag, Berlin. pp. 29-46.

Kurtzman, C.P. (2006). *In:* Blackburn, Food spoilage microorganisms. Cambridge, England: C. (Ed. De, W) Woodhead Publishing. pp. 28-54.

Loureiro, V. and Malfeito-Ferreira, M. (2003). *Int. J. Food Microbiol* **86(1-2):** 23-50. PMID 12892920

Maupas. J., Champemont. P. and Delforge, M. (1983). *Medicine Chirurgie Digestives* **12(1):** 77-9.

McFarland, L., Surawicz, C. and Greenberg, R. (1995). *Am J Gastroenterol* **90:** 439-48.

McBryde, Colin, Gardner, Jennifer, M., de Barros Lopes, Miguel, Jiranek, and Vladimir., (2006). *Am. J. Enol. Vitic.* **57:** 423-430.

Ostergaard, S., Olsson, L. and Nielsen, J. (2000). *Microbiol. Mol. Biol. Rev.* **64:** 34-50.

Oswal, N., Sarma, P.M., Zinjarde, S.S. and Pant, A. (2002). *Bioresour Technol.* **85**(1): 25-32.

Saint-Marc, T., Blehaut, H., Musial, C. and Touraine, J. (1995). *Sem Hôsp Paris* **71:** 735-41.

Slávikovà, E. and Vadkertiová, R. (2003). *J Basic Microbiol* **43(5):** 430-6.

Sreenivas, R., Prakasham, R.S., Krishna Prasad, K., Rajesham, S., Sarma, P.N. and Venkateswar Rao, L. (2004). *Process Biochemistry* **39:** 951-956.

Stoyan, Tanja, and John Carbon, (2004). *Eukaryotic Cell* **3(5):** 1154-1163.

Teoh, Ai Leng Gillian Heard, and Julian Cox, (2004). *International Journal of Food Microbiology* **95(2):** 119-126.

Walker, G.M. (1998). Yeast Physiology and Biotechnology. John Wiley and Sons, Chichester pp. 117-119.

Wheals, A.E., Rose, A.H. and Harrison, J.S. (1995). *The Yeast,* Vol. 6, *Yeast Genetics.* Academic Press, London.

Wodicka, L., Dong, H., Mitt Mann, M., Ho, M.H. and Lockhart, D.J. (1997). *Nature Biotechnol.* **15:** 1359-1367.

Chapter 2

Probiotic Microorganisms

S. Dhiva**, R. Chandrasekaran* and R. Saravanamuthu

Dept. of Botany and Microbiology, A.V.C. College (Autonomous), Mannampandal, Mayiladuthurai–609 305, Tamil Nadu

**Dept. of Botany and Microbiology, A.V.V.M. Sri Pushpam College (Autonomous) Poondi, Thanjavur, Tamil Nadu, India*

***AGAPPE Diagnostics Ltd., AGAPPE Hills, Pattimatom, Ernakulam Distt.–683 562*

Probiotics is originally defined as "microorganisms promoting the growth of other microorganisms." Recently probiotics have been more precisely defined as "mono or mixed cultures of live microorganisms when applied to animal or man, beneficially affect the host by improving the proportion of indigenous microflora". In relation to food, it can be considered as 'viable preparation' in food or dietary supplements to improve the health of human and animals. Though an impressive number of microbial species are considered as probiotics, only those strains that are classified as Lactic Acid Bacteria (LAB) are considered important with regard to food and nutrition and hence are selected as probiotics, e.g., species of *Lactobacillus, Bifidobacterium, Pediococcus,* etc. Besides some non-lactic acid bacteria such as *Enterococcus, Bacillus, Clostridium difficle,* etc., are also considered as probiotics. Most probiotics are marketed as foodstuffs or drugs. So safety and functional aspects of probiotics is of utmost importance while selecting the organisms for use as probiotics. Probiotics helps in combating the health related problems including cancer, hypertension, urinogenital infections, enhancement of immune system, lactose imbalance, cholesterol level, ulcer, asthma, etc. Probiotic products are mostly milk-based. However, efforts are also made for the development or formulation of probiotic food using other substrates such as cereal and soya. Currently, there is a wide range of probiotic products commercially viable to consumers. Such products include animal feeds, dairy foods, infant and baby foods, fruit juice-based products, cereal-based products and pharmaceuticals.

INTRODUCTION

There is a common notion or tendency to consider most microorganisms as potentially harmful. But the number of non-pathogenic species far exceeds the pathogenic species and many of the non-pathogens are in fact useful and even essential for the continued existence of life on earth.

One such group of beneficial microorganisms is those which inhabit the gastrointestinal tract of animals. The microbiota of the human large intestine influences health and well-being. It is a home of trillions of bacteria comprising hundreds of beneficial species. A beneficial association of microorganisms on the human host was probably first suggested by Doderlein (1892) who proposed that vaginal bacteria produced lactic acid from sugars to prevent or inhibit the growth of pathogenic bacteria. Such lactic acid bacteria were also found in association with fermented milk products and were advocated for their health benefits by Metchnikoff in 1908.

It has been considered that the long lifespan and good health of the Bulgarian people is related to their intake of fermented milk products, which could prevent "fouling" in the large intestine, and as a result, if consumed regularly, would lead to a longer and healthier life (Metchnikoff, 1908). He suggested that gut microbes were detrimental rather than beneficial to human health, although he admitted that the substitution of gut microbes by yoghurt bacteria may be beneficial. This paved the idea of using microbes to promote a good health and to prevent diseases. Initially, several microbes have been used unintentionally in food production such as dairy products, fermented cereals and fermented vegetables. These friendly microorganisms are called probiotics, meaning "for life".

Elie Metchnikoff, a Russian Physiologist and Nobel Prize winner is credited for inventing probiotics in 1907. This concept was developed further through the decades, and today, especially in Europe and Japan, probiotic-focused research, product development and marketing are at an all-time high.

To understand how probiotics work, it is important to know about the microbiology and physiology of human gastrointestinal tract.

GASTROINTESTINAL TRACT (GI)

It has been estimated that there are about 10^{14} bacterial cells that make their homes in humans, especially 400 different species of bacteria are found in the gut. In fact, it even exceeds the human population. So taking this into consideration, it is not surprising that microbes have been found to play an important role in human health.

Most of these bacteria are not harmful and in fact contribute positively to normal human growth and development. But some of these bacteria can have negative effects on the host. So it is important to maintain the balance of microbes to favor the beneficial bacteria over the potentially harmful one.

The composition of gut microflora is affected by various factors:

- **Age:** The microflora of live young one is different from that of the adult.
- **Diet:** To some extent this will be responsible for the changes seen with age. The composition of diet can affect the composition of the gut microflora.
- **Environment**: The load of gut microflora may vary with the location of the host residing.
- **Stress**: Hormonal changes in the host body due to various stress can affect mucus secretion and flora composition of the gut.
- **Medication**: Repeated use of antibiotics and other chemical antimicrobial compound either as growth promoters or as therapeutic agents can change the gut microflora and allow the growth of pathogens.

ROLE OF MICROBES IN THE GI TRACT

The microbes present in the gastrointestinal tract have the potential to act in a positive, negative or neutral manner. Due to unfavorable conditions microbes are not very prevalent in the stomach or upper small intestine. However, towards the lower small intestine, they begin to attain higher populations (10^6 – 10^8/g of small intestinal contents) and in the colon they constitute about 10^{11} – 10^{12}/g of colon contents (a very large number).

The collection of microbes in the large and small intestine has a powerful influence on the host in which it resides. It brings out the complete digestion of any food components that are not digested in the small intestine, such as lactose in lactose intolerant people or fibers resistant to the enzymes they encounter in the small intestine, i.e., the probiotics which can contribute to the benefits derived from commensal lactic acid bacteria residing in the gastrointestinal tract.

Probiotics are defined as non-digestible food ingredients that beneficially affect the host by selectively stimulating the growth and/or the activity of one or a limited number of bacteria in the colon (Gibson and Roberfroid, 1995).

But there is an evidence of non-digestive microbial activities as well. Certain intestinal microbes are known to produce vitamins. The intestinal microflora are important for the maturation of the immune system, the development of normal intestinal morphology and in order to maintain a chronic and immunologically balanced inflammatory response. The microflora reinforce the barrier function of the intestinal mucosa, helping in the prevention of the attachment of pathogenic microorganisms and the entry of allergens.

Some members of the microflora may contribute to the body's requirements for certain vitamins, including biotin, pantothenic acid and vitamin B_{12}. But the alteration of the microbial flora of the intestine, as such may occur with antibiotic use, disease and aging can negatively affect its beneficial role.

PROBIOTIC BACTERIA

Early taxonomic and gut (fecal) ecology studies on lactic acid bacteria were conducted by Moro in 1900 and by Beijerinck and Cahn in 1901.

Originally probiotics is defined as 'microorganisms promoting the growth of other microorganisms' (Lilly and Stillwell, 1965). Probiotics according to present-day interpretation refers to viable microorganisms that promote or support a beneficial balance of the autochthonous microbial population of the gastrointestinal tract (GT). Such microorganisms may not necessarily be constant inhabitants of the GT, but they should have a "beneficial effect on the general and health status of man and animal (Parker, 1974; Fuller, 1989).

In recent years, probiotics have been defined more precisely as "mono or mixed cultures of live microorganisms which when applied to animal or man, beneficially affect the host by improving the proportion of the indigenous microflora (Havenaar *et al.*, 1992).

In relation to food, probiotics are considered as viable preparation in food or dietary supplements to improve the health of human and animals (Salminen *et al.*, 1998).

According to these definitions, an impressive number of microbial species and genera are considered as probiotics. However, only those strains that are classified as lactic acid bacteria (LAB) are considered of importance with regard to food and nutrition (Table 1).

Table 1. Microorganisms regarded as probiotics

	Lactobacillus species	*Bifidobacterium* species	Other lactic acid bacteria
1.	*L. acidophilus*	*B. adolescentis*	*Enterococcus faecalis*
2.	*L. amylovorus*	*B. animalis*	*Enterococcus faecium*
3.	*L. casei*	*B. bifidum*	*Lactococcus lactis*
4.	*L. crispatus*	*B. bifidum*	*Leuconostoc mesenteroides*
5.	*L. delbrueckii* sub sp. *bulgaricus*	*B. breve*	*Pediococcus acidilactic*
6.	*L. gallinarum*	*B. infantis*	*Sporolactobacillus inulinus*
7.	*L. gasseri*	*B. lactis*	*Sweptococcus thermophilus*
8.	*L. johnsonii*	*B. longum*	
9.	*L. paracasei*		
10.	*L. plantarum*		
11.	*L. reuteri*		
12.	*L. rhamnosus*		

Non-lactic acid bacteria
Bacillus cereus var *toyoi*
Escherichia coli strain *nissle*
Propionibacterium treudenreichii
Saccharomyces cerevisiae
Saccharomyces boulardii

Among the numerous intestinal microbes, those that are expected to beneficially affect the host by improving the intestinal microbial balance, and hence are selected as probiotics, include species of genera *Lactobacillus*, *Bifidobacterium* and *Enterococcus* (Fuller, 1991; Gordin and Gorhach, 1992). *Bifidobacterium* sp. that specifically inhabit the intestinal tracts of animals such as *Bifidobacterium thermophilium* and *Bifidobacterium pseudolongum* are used in animal probiotics (Abe *et al.*, 1995).

Some bacteria that do not normally inhabit the intestinal tract may also come under the category of probiotics. They are used as starters in dairy products and include mainly *Lactobacillus bulgaricus, Streptococcus thermophilus* and *Leuconostoc* and *Lactococcus* sp. However, these bacteria do not colonize the intestinal tract and their effect on intestinal microbial balance is expected to be small (Alm, 1991).

Most probiotics are marketed as foodstuffs or drugs. The food industry in United States is the country's largest manufacturing industry with revenues exceeding $500 billion/year. One key research priority of this industry is to increase the development of foods that promote health and well-being. Among these the exploiting area of functional foods and probiotics show considerable promise to expand the industry into new arena (Klaenhammer, 2000).

The economic effect of the projected U.S. functional foods market is significant, recently estimated as $134 billion (Sanders, 1998). It includes foods such as natural functional foods (cornberry juice, green tea), foods and ingredients for specified health use (FOSHU), formulas (infants and elderly), medical foods, nutraceuticals and drug foods. Within this continuum between food and drug, there are unlimited niches for the development of food systems that promote optimal health and general well-being. Today and tomorrow's challenge is to strengthen the science supporting the probiotic concept and establish direct cause and effect mechanisms that are responsible for effects on human health and nutrition (Klaenhammer, 2000).

SAFETY OF PROBIOTICS

Food products or supplements and pharmaceutical preparation containing viable probiotic strains are supplied in the market either as fermented food commodities or in lyophilized form. Consideration of the safety of probiotics is therefore, of utmost importance. It can be described in short:

(i) Fermented products are in use for centuries.
(ii) No report of probiotic pathogens.

Reports of a harmful effect of these microbes to the host are rare. However, in recent years, many species of the genera *Lactobacillus*, *Leuconostoc*, *Pediococcus*, *Enterococcus* and *Bifidobacterium* were isolated frequently from various types of infective lesions.

According to Gasser, *L. rhamnosus, L. carci, L. acidophilus, L. plantarum, L. paracasei, L. salivarius, Lactobacillus lactis* and *Leuconostoc mesenteroides* are some examples of *Lactobacillus* isolated from bacterial endocarditis. *L. rhamnosus, L. plantarum, Leuconostoc mesenteroides, Pediococcus acidilactic, Bifidobacterium eriksonii* and *B. adolescentis* have been isolated from bloodstream infections and many have been isolated from local infections. In addition, Jett *et al.* (1994) reported several virulence factors in enterococci. These reports raised much debate in recent years over the safety of probiotics.

The factors that must be addressed in the evaluation of safety of probiotics include pathogenicity, infectivity and virulence factors comprising toxicity, metabolic activity and the intrinsic properties of the microbes. So the absence of pathogenicity and infectivity is a requisite for probiotic safety. However, even these lactic acid bacteria and bifidobacteria, long considered to have no infectivity, are isolated from infections, it is unlikely that they universally possess generalized infectivity. The isolation of these organisms from infections is likely to be the result of opportunistic infections (Ishibashi and Yamazaki, 2001).

Although these organisms may invade the host body by bacterial translocation or other routes, causing bacterimia (Berg, 1992; Dietch *et al.*, 1992), for these bacteria to actually cause systemic infections from endocarditis and other infections to septicemia, both the bacterial factors and the host factors probably need to be involved. However, the assessment of such an interaction is difficult.

The causes of opportunistic infection include skin injury, chronic diseases, cancer, drug-induced abnormality and bacterial translocation. Bacterial translocation is a phenomenon caused by a diminished intestinal barrier, resulting in the passage of bacteria (or bacterial components or products) across the mucus membrane and epithelium. The bacteria are then transported through the tunica propria to the mesenteric lymph nodes (MLN) and other organs (Fig. 1). This results in bacteremia which may lead to multiple organ failure and septicemia (Berg, 1985, 1992; Van Leeuwen *et al.*, 1994).

Bacterial translocation may be promoted by various factors such as intestinal mucosa, injury, immunodeficiency in the host and an abnormal overgrowth of intestinal bacteria (Fig. 1).

Intestinal bacteria

↓

Mucosal barrier

↓

Translocation (invasion)

↓

Through MLN or portal vein

↓

Organs

↓

Infections

Fig. 1. Various factors affecting bacterial translocation.

But the assessment of the ability to cause opportunistic infection is difficult. The acute and chronic toxicity tests probably provide circumstantial evidence for these controversies.

SELECTION CRITERIA FOR PROBIOTICS

Microbial probiotics should have a beneficial effect and should not cause any harm to the host. So all the strains have to be checked prior to use in humans or animals and given GRAS (Generally Regarded As Safe) status (Table 2).

Table 2. Microbial probiotics and their safety status

Organism	Infection potential
Lactobacillus	Mainly non-pathogens; a few opportunistics reported in AIDS patients.
Lactococcus	Mainly non-pathogens.
Streptococcus	Opportunistics, only *S. thermophilus* is used in dairy products.
Enterococcus	Opportunistics, some strains exhibit antibiotic resistance.
Bacillus	Only *B. subtilis*, GRAS status, is reported in probiotics use.
Bifidobacterium	Mainly non-pathogens; some strains are isolated from human infection.
Propionibacterium	Dairy propioni bacterial group in a potential candidate of probiotics.
Saccharomyces	Mainly non-pathogens; some strains are isolated from human infection.

Source : Adapted from Chukeatirote, 2003.

Typically, safety evaluation of a proposed or accepted probiotic strain includes :

(i) ability of the cells to produce metabolites and enzymes
(ii) colonisation or adhesion
(iii) factors that influence the strain survival
(iv) interactions with host, especially in pathogenicity

As mentioned earlier, the predominant species used as probiotic agents belongs to the group of lactic acid bacteria (LAB). Due to their long history of safe use in foods, most species of LAB are considered as commensal microorganism without pathogenic potential. Apart from this an effective probiotic should: (i) have beneficial effect on the host, (ii) be non-pathogenic and non-toxic, (iii) contain large number of viable cells, (iv) be capable of surviving and metabolizing in the gut, (v) remain viable during storage and use, (vi) have good sensory properties, and (vii) be isolated from the same species as its intended host (Collins and Gibson, 1999). Criteria for selection of probiotics for target specific action are given in Table 3.

Table 3: Criteria for selection of probiotics for target specific action.

Target for probiotic action	Selection criteria
Alleviation of lactose maldigestion probiotics	High lactase producing strongly site specific symptoms.
Intestinal inflammation	Site specific adhesion properties, anti-inflammatory cytokine expression, mucosal properties to alleviate permeability disorder and gut microflora abberancy.
Alleviation of food allergy symptoms, reducing the risk proteolytic properties of atopic disease.	Adherence to small intestine, induction of local transforming growth factor β-production.
Reducing the risk of colon cancer	Target specific adhesion to distal or proximal colon, mucosal butyric acid production, competitive exclusion of inflammatory bacteria, toxin binding and promotion of non-toxigenic mucosal microflora.

IN VITRO ASSESSMENT OF PROBIOTIC STRAINS

The assessment of microbial strains for use as potential probiotics involves the assessment of resistance to gastric acidity and bile toxicity, adhesion to gut epithelial tissue, ability to colonize the gastrointestinal tract, production of antimicrobial substances and ability to modulate immune responses.

Resistance to Gastric Acidity

Before reaching the intestinal tract, probiotic bacteria must first survive transit through the stomach (Henriksson *et al.*, 1999). There, the secretion of gastric juice with a pH between 2.0 and 3.4 constitutes a primary defense mechanism against most of the ingested microorganisms. So the survival of probiotic bacteria in human gastric juice is a more accurate indication of the ability of the strains to survive passage through the stomach (Dunne *et al.*, 2001).

Bile Acid Resistance

Bile acids are synthesized in the liver from cholesterol and are secreted from the gall bladder into the duodenum in the conjugated form (500–700 ml/d) (Hoffmann *et al.*, 1983). Both conjugated and deconjugated bile acids exhibit antibacterial activity, inhibiting the growth of the microbes. So it is necessary to evaluate the ability of probiotic species to resist the effects of bile acids.

Adherence of Probiotic Species to Human Cell Lines

Preselection of probiotic strain also includes its adhesion to gut epithelial tissues and their ability to colonize the gastrointestinal tract. It is an important quality control method for assessing gut barrier effects. Adhesion has been related to shortening the duration of diarrhoea, immunogenic effects, competitive exclusion and other health effects.

Adhesion properties should be carefully monitored, including adhesion to intestinal cells and human intestinal mucus. HT 29 and CaCo-2 cells are human intestinal cell lines expressing morphologic and physiologic characteristics of normal human enterocytes and are used for the adhesion studies *in vitro* (Bernet *et al.*, 1994; Nesser *et al.*, 1989; Kerneis *et al.*, 1991; Chauviere *et al.*, 1992; Coconnier *et al.*, 1993).

Antimicrobial Activity

Lactic acid bacteria produce several metabolic products such as organic acids, fatty acids, hydrogen peroxide and diacetyl, which has antimicrobial effects (Ouwehand, 1998) on various pathogens. In addition, it produces nisin, a proteinaceous substances (bacteriocin), which also has a specific inhibitory activity against closely related species (Dodd and Gasson,1994; Jack *et al.*, 1995).

Apart from all these, viability is an important factor but not the only criterion for quality assurance. To be effective, probiotic strains must retain the functional health characteristics for which they were originally selected.

Implications for Human Health

Microbial colonization has important health implications for humans. But on rare occasions, microbes develop a pathogenic relationship with a host, and illness or death of the host can result. Negative influences on human health by colonizing or invading microbes should not be acute.

Microbial metabolites may possess genotoxic, mutagenic or carcinogenic activity and may contribute to the development of cancer over a period of long-term exposure. The effect of colonising microbes in association with human body should encourage the positive and discourage the negative activities of commensal and invading microbes and it is supposed to be the probiotic theory.

Cancer

Probiotic bacteria play a major role in a variety of health effects, and the mechanisms proposed for mediating these effects are numerous (Table 4).

Table 4: Mechanisms for probiotic functionality

- Antimicrobial activity
- Colonization resistance
- Immune effects
  - Adjuvant effect
  - Cytokine expression
  - Stimulation of phagocytosis by peripheral blood leucocytes
  - Secretary IgA
- Antimutagenic effects
- Antigenotoxic effects
- Influence on enzyme activity
- Enzyme delivery

Source: Sanders Mary Ellen, 2000.

Focused epidemiologic studies using populations consuming defined probiotic products over a long period of time suggest a protective influence of probiotic bacteria against cancer. Mechanisms whereby probiotic bacteria might influence the incidence of cancer (colon cancer) and given below.

- Enhancing host immune response
- Suppression of growth and activities of intestinal microbes that produce carcinogens and promoters by competitive colonization or production of inhibitors (short-chain fatty acids or bacteriocins).
- Binding and removal of carcinogens.
- Production of antimutagenic compounds.
- Production of butyrate to stimulate programmed cell death of abnormal cells.
- Inhibition of the conversion of bile salts to secondary bile salts.

Source: Sanders, Mary Ellen., 2000.

Hypertension

Probiotic bacteria may play a role in controlling the hypertension. The bioactive peptides resulting from the proteolytic action of probiotic bacteria on casein (milk protein) during milk fermentation may suppress the blood pressure of hypertensive individuals (Takano, 1998).

Two tripeptides, valine–proline–proline and isoleucine–proline–proline, isolated from a dairy based fermentation of milk by *Saccharomyces cerevisiae* and *Lactobacillus helveticus* functions as angiotension-1-converting enzyme inhibitors and reduce blood pressure. Cell wall fragments of *L. case* Y1T9018 (Sawada *et al.*, 1990) has antihypertensive activity by reducing systolic pressure, diastolic pressure and heart rate.

Urinogenital Infections

Urinary tract infections are caused by anaerobic Gram-negative rods, *E. coli*, *Chlamydia* and *Candida* (Reid *et al.*, 1998). Although effective therapies are available to cure these infections, they lead to various side effects. Pathogens associated with vaginal tract infections are mainly from gastrointestinal tract. But at the same time a healthy vaginal tract is associated with high populations of lactobacilli, especially hydrogen peroxide producing lactobacilli and a pH < 5.0 (Eschenbach *et al.*, 1989; Hawas *et al.*, 1996; Hillier *et al.*, 1992; Klebanott *et al.*, 1991).

So this fact may be coupled for the intestinal route of transmission of bacteria to the urinogenital tract, which even led to the theory that oral probiotics may be useful in treatment or prevention of urinogenital infections. In addition, it can even be applied as dried lactobacilli $> 10^9$CFU/dose intravaginally, which in fact provides a "natural cure" against infection, as it helps in preventing infection without any side effects.

Lactose Intolerance

Lactose intolerance is the inability to absorb lactose, a milk sugar into the digestive system. Lactose has to be split into two smaller sugars by the enzyme lactase, which is present in the lining of the small intestine. If it is not absorbed, it is fermented by the intestinal bacteria and leads to diarrhoea with excess production of gas, stomach rumbling, abdominal colic, etc. But the contribution of lactase by the bacterial cultures used for yoghurt manufacturing is thought to mediate enhanced lactose digestion. In general, yoghurt starter cultures *Streptococcus thermophilus* and *Lactobacillus delbrueckii* subsp. *bulgaricus* ($\geq 10^8$/g), effectively improve the digestion of lactose in maldigesters. But at the same time products with low probiotic cell count ($\sim 2 \times 10^6$/ml) may contribute to better lactose maldigestion when compared to products with high amount of probiotic bacteria (e.g., sweet acidophilus milk) (Payne *et al.*, 1981).

Cholesterol

Elevated levels of certain blood lipids are a risk factor for cardiovascular disease. Some strains of *L. acidophilus* can assimilate the cholesterol molecule (Gilliland *et al.*, 1985). Some probiotic bacteria have the ability to deconjugate bile acids enzymatically, increasing their rates of excretion (De Smet *et al.*, 1994). To replace the bile acids lost through excretion, cholesterol molecules are converted to bile acids. This could lead to reduction in serum cholesterol.

At the same time deconjugated bile acids may be converted to secondary bile acids by colonic microbes. These secondary bile acids are known as cancer promoters. A potential increased risk of colon cancer may outweigh any benefit of reduction of serum cholesterol levels.

Another mechanism is that 3-hydroxy-3-methyl glutaric acid (HMG) present in the fermented milk inhibits hydroxyl methyl glutaryl CoA reductance, the rate limiting enzyme in cholesterol biosynthesis.

Enhancement of Immune System

Probiotic microbes (LAB) may influence the systemic immune systems in various ways (Perdigon *et al.*, 1995; Fang *et al.*, 2000).

First of all they enhance the defensive property of intestinal mucosa whose function is like a barrier against the antigens. For example, Crohn's disease, associated with impairment of the barrier function, can be cured by including probiotic supplement in patients diet (Isolauri *et al.*, 1998).

Some strains of lactic acid bacteria and bifidobacteria are able to induce tumor necrosis factor-∝ (TNF-∝) and Interleukin-6 (IL-6) as well as stimulate other non-specific immune system. *L. casei* is also used for stimulating immune responses preventing enterobacterial infection (Matsuzaki, 1998). *Lactobacillus* GG is also used as an effective oral vaccine for rotavirus (Isolauri *et al.*, 1995).

Other Benefits

Probiotics can give a positive protection against *Helicobacter pylori* infections in the stomach, alcoholic liver diseases, small bowel bacterial overgrowth, ulcerative colitis, allergy to milk protein, juvenile chronic colitis, antioxidative effects, asthma, hepatic encephalopathy, and they can even be used as vehicle for vaccines delivery (Mercenier, 1999). The probiotic effects reported and their putative mechanisms (Sanders, 2003) are given in the Table 5.

Table 5: Effects of probiotics and their putative mechanisms.

Benefit	Function	Proposed mechanism
Digestive comfort	Irritable bowel syndrome, symptoms affecting the gastrointestinal tract in general (constipation, non-pathogenic diarrhea, distension, flatulence, cramp cirrhosis of a digestive cause)	Change in populations or activities of the intestinal microflora.
	Lactose intolerance	Delivery of microbial lactase to the small intestine.
Defense	Allergy (atopical eczema, allergy to milk, rheumatoid arthritis)	Translocation, barrier effect.
	Carcinogenicity	Changes in the populations, activity of the oral microflora or its ability to adhere to the teeth.
	Carcinogenicity, mutagenicity, tumor	Absorption of the mutagen, stimulation of the immune system, inhibition of carcinogen production by the intestinal microflora.
	Diarrhoea linked to antibiotics, diarrhea caused by Rotavirus, colitis caused by *C. difficile*, nosocomial diarrhoea.	Competitive exclusion, translocation/barrier effect, immune response promoted.
	Helicobacter pylori	Antipathogenic activity.
	Immuno modulation (immune status, vaccinal response)	Interaction with the immune cells or cell receptors leading to an increase in the phagocytic activity of the white cells, increasing IgA levels after exposure to the antigen, increasing the proliferation of the intra-epithelial leukocytes, regulating the Th_1/Th_2 ratio, induction of cytokine synthesis.
	Intestinal inflammation, ulcerative colitis, Crohn's disease, pouchins.	Immune response downregulated.
	Excessive intestine bacterial growth vaginosis, coronary infections	Antimicrobial activity competitive exclusion Antipathogenic activity competitive exclusion

Table 5: Contd.

Benefit	Function	Proposed mechanism
Others	Lowering of blood cholesterol	Deconjugation of bile acids.
	Endotoxemia combined with cirrhosis	Inhibition of the production of endotoxins by the intestinal microflora.
	Hypertension	Cellular constituents or peptides derived from fermentation acting as inhibitors of ACE (angiotension –converting enzyme)
	Renal calculi	Changes in the digestive flora influencing the breakdown of oxalate.

PROBIOTIC PRODUCTS

Most probiotic foods in the markets worldwide are milk based and very few attempts are made for development of probiotic foods using other fermentation substrates such as cereals.

Why Dairy Products

Specifically, yoghurt-like products form the largest segment by far in the market for probiotic products. Probably this may be due to the following reasons:

- Fermented dairy products such as yoghurt already have a record as being good for health.
- Consumers are familiar with the presence of viable microorganisms in the fermented products.
- Probiotics as fermentation organisms combine the positive images of both probiotics and fermentation organisms.
- All these images facilitate the utility of dairy products in daily consumption as probiotic substrates and in turn the probiotic microbes.

In addition to that, there is an important technological reason for the use of dairy products as carriers of probiotics. Many of these products have already been optimized to some extent for the survival of the fermentation organisms. Thus, the existing technology can be relatively easily adapted to guarantee sufficient survival of the added probiotic bacteria (Heller, 2001).

The importance of using milk dairy as preferred substrate for probiotic products is evident from the following information.

- ❑ **D**rink milk with food.
- ❑ **A**ged cheeses like Cheddar and Swiss are low in lactose.
- ❑ **I**ntroduce dairy slowly. Gradually, increase the amount.
- ❑ **R**educe it. Enjoy lactose–free milk and milk products.
- ❑ **Y**oghurt with live and active cultures helps to digest lactose.

Cereals as Substrates

The development of non-dairy probiotic products is a challenge to the food industry, to utilize the abundant natural resources by producing high quality functional products.

These natural resources provide dietary fiber, proteins, energy, minerals and vitamins required for health. This has attracted the attention on their use as raw materials for the development of new fermented functional foods (Table 6).

Table 6. Inhibitory activity of some fermented foods on bacteria

Fermented food	Organisms Inhibited
Maize dough	*S. flexneri* ETEC, Gram-negative bacteria
Sorghum	*Salmonella typhimurium*
Uji	*Coliforms*
Motoho	*Shigella boydii, S. typhi, E. coli*
Tempeh	*Bacillus subtilis, Klebsiella pneumonia* *Staphylococcus aureus.*
Fish silage	*Staphylococcus* sp. *Coliforms*
Milk	*Staphylococcus aureus, E. coli* *Campylobacter jejuni* *Vibrio cholerae*

The possible application of cereals or cereal constituents in functional food formulations could be summarized.

- As fermentable substrates for the growth of probiotic microorganisms, especially lactobacilli and bifidobacteria
- As dietary fiber promoting several beneficial effects
- As probiotics due to their content of specific non-digestible carbohydrates
- As encapsulation materials for probiotic in order to enhance their stability

Cereals such as malt, rice, corn, wheat, sorghum, milk, oats and soya are used as substrate for lactic acid fermentation based on the availability of nutrients such as carbohydrates, amino acids, peptides, salts, vitamins, minerals, etc. These nutrients are required for the growth of probiotic microbes also. Other than the cereals, brined olives, salted gherkins and sauerkraut contain high concentration of lactobacilli (Molin, 2001).

With the apparent market success of probiotic products, questions concerning the nature of probiotic bacteria asked by consumers of particular importance is whether foods containing probiotics provide on added value compared with traditional fermented foods containing living microorganisms and whether this value is maintained during manufacturing and is provided during the entire shelf-life of the product.

Probiotic preparation may be presented in different ways depending on the organisms receiving the supplement and the condition to be treated. Thus, they may be in the form of powders, tablets, capsules, pastes or sprays.

COMMERCIALLY AVAILABLE PROBIOTICS

Currently, there is a wide range of probiotic products commercially available to consumers. These are animal feeds, dairy foods, infant and baby foods, fruit juice based products, cereal based products, and pharmaceuticals.

But before these products are sold, the probiotic strains should be characterised to ensure their safety and effective use. Organisms other than lactic acid bacteria which are being used in probiotic preparations include *Bacillus* sp., yeasts (*Saccharomyces cerevisiae, S. boularolii*) (Green *et al.*, 1999) and filamentous fungi (*Aspergillus oryzae*). Microbes involved in the fermentation of some cereals are given in Table 7.

Table 7. Microbes involved in the fermentation of some cereals

	Substrate	Food	Microorganisms
1.	Maize	Ogi	*Corynebacterium* sp., *Clostridium* sp., *Rhodotorula* sp.
		Kenkey, Koko, Akara	*Enterobacter cloacae, Acinetobacter* sp., *L. plantarum, L. brevis, S. cerevisiae, Candida mycoderma.*
		Nansha	*Streptococcus* sp., *Lactobacillus* sp., *Candida* sp., *S. cerevisiae.*
		Uji	*Leuconostoc mesenteroides, L. clatarum.*
2.	Millet	Busa, ogi	*L. plantarum, L. delbrueckii, Saccharomyces busae.*
		Mahewu	*Streptococcus lactis, L. delbrucki, L. bulgaricus, L. acidophilus.*
3.	Cassava	Gari	*L. plantarum, Streptococcus sp.*
4.	Rice	Idli	*Leuconostoc mesenteroides, Streptococcus faecalis, Pediococcus cerevisiae.*

Preparation of some fermented products are as follows (Adapted from Reid, 1998).

Preparation of some fermented products

1. Koji (Rice based)

Polished rice

↓

Soak in water for 17 hrs at 25°C

↓

Drain excess water

↓

Steam for 70 min and cool to 35°C

↓

Inoculate with ***A. oryzae***

↓

Incubate at 27-28°/50 hrs

↓

Dry

↓

Koji (Japan)

2. Fermented Oat-meal gruel (Oat based)

Malted barley flour 2.5%
oat meal by wt.

↓

Oat meal and water

↓

Heating

↓

L. plantarum ← Starter culture cooling to 37°C 5×10^{10} CFU

↓

Fermentation pH < 4.0

↓

Cooling to 4-8°C

3. Tofu: Soya based

Soya soaked beam
in H_2 for 10-16 hrs

↓

Grind

↓

Extract milk,
sterilize it ← Starter culture *Lactobacillus acidophilus*

↓

Incubate for 4-6 hrs

↓

Soy-curd

↓

Tofu

4. Sikhae (Fermented fish)

Flat – fish
Degut, wash, cut and drain

↓

Mix with salt and leave overnight

↓ Drain

Mix with cooked millet, red pepper
powder, minced garlic and ginger

↓

Pack tightly in an earthern jar

↓

Incubate at 20°C/2 weeks
(*L. mesenteroides*, *L. plantarum*)

↓

Gajami sikhae

5. Balao-Balao (shrimp)

Live-shrimp

↓

Wash and drain

↓

Add 20% (w/w) solar salt

↓

Allow to stand for 2 hrs

↓

Drain and cut antennae

↓

Mix shrimp with cooked rice (1:4.8)

↓

Pack in glass jars and cover

↓

Ferment at tropical room temperature
(***L. mesenteriodes*** *and* ***P. cerevisiae)***

6. Yoghurt (milk-based)

Boiling of raw milk (90-95°C/20 min.)

↓

Cool to 45°C

↓

Inoculate ***L. bulgaricus***
and ***Streptococcus thermophilus***

↓

Incubate at 37°C/4 hrs.

↓

Product

↓

Stored at 4°C

7. Kefir (milk-based)

Boiling of raw milk

↓

Cooling 20-25°C

↓

Inoculation 20-25°C ← Kefir grains

↓

Fermentation 20-25°C,
18-24 hrs

↓

Separation ——→ Kefir grains

↓

Maturation and cooling 4°C

↓

Stored at 4°C

Table 8. Commercial probiotic products available to date.

Brand product	Strain	Distributor
Yakult	*L. casei* Shirota	Yakult, Tokyo Japan
Adult formula CP-1	Mixtures of *L. acidophilus, L. rhamnosus, L. plantarum, B. bitidum* and *B. longum*	Custom probiotics Glendale, CA
KE-99 LACTO	*L. casei*	Probiohealth, Bererly Hills, CA.
Trener Trio capsules	*L. acidophilus, B. bifidum* and *L. bulgarium*	Probiotics, New Zealand
Multitrionta	*L. acidophilus, B. bifidum* and *B. longum*	Seven Seas Healthcare, Hull, UK.
Enterogermina	*Bacillus subtilis*	Sanoti Winthrop, Milan, Italy.
Yoghurt	*L.bulgaricus, Streptococcus thermophilus*	Aawin, Chennai, India.
Biosubtyl	*B. subtilis*	Bio Phar. Co. Ltd., Nha Trnag, Vietnam.

Probiotics must meet several basic requirements for the development of marketable probiotic products. The most important requirements are that it should survive in sufficient numbers in the product. Its physical and genetic stability during the storage of the product should be guaranteed and all its properties essential for expressing their health benefits after consumption should be maintained during manufacture and storage of the product.

In addition it should not have adverse effects on the taste or aroma of the product and should not enhance acidification during the shelf life of the product.

All fermented dairy products containing living bacteria, probiotic products must be cooled during storage. This is necessary both to guarantee high survival rates of the probiotic organisms and to ensure sufficient stability of the product (Shah *et al.*, 1995; Roy *et al.*, 1997). Furthermore, because the intestinal tract is considered to be the natural environment of the probiotic bacteria, the oxygen content, redox potential and water activity of the medium must be considered (Dave and Shah, 1997). Commercial probiotic products available to date are shown in Table 8.

INTERACTIONS BETWEEN PROBIOTICS AND COMPONENTS OF FERMENTED FOODS

The chemical composition of the dairy product is of paramount importance for the metabolic activities of the microorganisms. Essential variables are the kind and amount of carbohydrate available, the degree of hydrolysis of milk protein (which determines the availability of essential amino acids), and the composition and degree of hydrolysis of milk lipids (which determine the availability of short-chain fatty acids in particular (Fox *et al.*, 1993; 1996). On the other hand, the proteolytic (Kunji *et al.*, 1996) and lipolytic properties of probiotics may be important for further degradation of proteins and lipids. These two properties may have considerable effects on the taste and flavor of dairy products (Fox *et al.*, 1993).

Major aspect of the production of probiotic fermented dairy products is the interaction between probiotics and starter organisms. Both synergistic and antagonist effects between different starter organisms are well established. For example, the classic yoghurt culture is characterized by a protosymbiosis between *Streptococcus thermophilus* and *L. delbrueckii* sub sp. *bulgaricus*.

This synergism is seen as an accelerated and efficient acidification of the milk and multiplication of the culture organisms and is based on cross-feeding of both organisms. Antagonism, on the other hand, is often based on the production of substances that inhibit or inactivate more or less specifically other related starter organisms or even unrelated bacteria (Heller, 2001).

Most importantly, antagonism is caused by bacteriocins, which are peptides or proteins exhibiting antibiotic properties (DeVuyst and Kandamme, 1994; Dodd and Gasson, 1994). However, antagonism to starter cultures and vice versa may be a limiting factor for combinations of starters and probiotics (Joseph *et al.*, 1998). Further antagonistic activities produced by lactic acid bacteria have been described and the substances involved are H_2O_2, benzoic acid (produced from the minor milk constituent lippuric acid), biogenic amines (formed by decarboxylation of amino acids) and lactic acids (Desma Zeaud, 1996; Lankaputhra *et al.*, 1996; Sieber *et al.*, 1995; Weber, 1996).

The intensity of the interaction between probiotics, food matrix and starter organisms depends on the physiologic state and the time that probiotics are added to the product (Heller, 2001).

DOSAGE OF PROBIOTICS

An adequate number of viable cells namely the "therapeutic minimum" has to be consumed regularly to transfer the "probiotic" effect to consumers (Viljoen, 2001). But quantity to be consumed should be more than 100 g per day of bioyoghurt containing more than 10^6 CFU/ml. Shah (2000) recommended that a dosage of minimum viable number should have 10^6 CFU/ml or per gram but recommends 10^8 CFU/g to compensate their reduction during their passage through stomach.

Yoghurt is a classic example for functional probiotic food. Yoghurt with probiotics are called bio-yoghurt. It should contain living bacterial cells. According to regulation, yoghurt should contain 2×10^6 living bacteria in 1 ml at the end of recommended storage period. The daily dose of probiotic microorganisms should reach 1×10^9 cells. The titre of bacteria in fermented drinks reaches 10^8 to 10^9/ml and decreases with storage.

Even tablets or capsules can also be used as addition to foodstuffs, that contains lyophilised cultures of bacteria. Probiotics are also available as pharmacopoeia preparation such as Linex 1.2×10^7, Mutaflor 2.5×10^9, Lactoseren 1×10^9, Yoghurt capsules 2×10^9 contain freeze-dried bacterial cells per caps correspondingly (Maijatoma and Pokrotnicks, 2006).

PROBIOTIC ORGANISMS

Probiotic organisms are mostly lactic acid bacteria (LAB), Gram-positive, sugar fermenting groups.

(i) *Lactobacillus acidophilus*

It is the best known of the friendly bacteria, important in maintaining a healthy, balanced flora in the small intestine. It enhances digestion of milk sugar (lactose) and aids in the production of vitamins and enzymes. It produces lactic acid which suppresses harmful microbes in the intestine and helps in the control of yeast overgrowth such as *Candida.*

(ii) *Lactobacillus casei*

It is highly probiotic, hardy and most potent of all *Lactobacillus* strains. It strengthens the immune response and helps with leaky gut, intestinal dysbiosis. It improves the integrity of mucosal lining and produces lactic acid. It digests a wide range of carbohydrates. It acts on tumor cells and also has a beneficial stimulating activity in normal cells.

(iii) *Lactobacillus plantarum*

It has high digestive capacity, particularly for the breakdown of proteins, because it has the ability to eliminate protein wastes from the intestine before they enter the bloodstream. It benefits the immune systems. It adheres strongly to intestinal lining. Various studies also showed that it helps with leaky gut, intestinal dysbiosis and improves integrity of mucosal lining. It produces lactic acid and acts as a natural antibiotic (as acidophilus).

(iv) *Lactobacillus salivarius*

It is effective in eliminating symptoms of bowel toxemia. It produces Vitamin B and K, enzymes and lactic acid and aids in the production of lactase. It is very active on proteins and by-products of protein purification. It is helpful during food poisoning. It has a unique ability to eradicate the dreaded *Helicobacter pylori* (*H. pylori*) known to be the major cause of peptic ulcer.

(v) *Lactobacillus bulgaricus*

It is used in yoghurt start-up culture. It produces lactic acid, and enhances the digestion of milk sugar. It also has natural antibiotic properties.

(vi) *Lactobacillus rhamnosus*

It is the most intensively studied probiotic strain and has advantages in the prevention of gastro-intestinal disorders. It is known to be effective against colon tumors.

(vii) *Lactobacillus lactis*

It acts as a natural antibiotic, which reduces the ability of pathogenic bacteria to grow and act against the bacteria *Listeria monocytogenes* which causes severe food poisoning.

(viii) *Bifidobacterium bifidum*

It is a prominent probiotic microorganism that takes up residence primarily in the lining of the colon. It prevents the colonization of invading pathogenic bacteria by attaching to the intestinal walls and crowding out and taking nutrients from these unfriendly bacteria and yeasts.

(ix) *Bifidobacterium longum*

It is found abundantly in the large intestine. It is found to be effective in reducing the frequency of gastrointestinal disorders (diarrhea, nausea, etc.) during antibiotic use. Dietary intake of *B. longum* significantly inhibits the development and growth of breast cancers.

(x) *Other lactic acid bacteria*

Other than *Lactobacillus* sp. and *Bifidobacterium* sp., there are other organisms such as *Lactococcus* sp., *Streptococcus thermophilus*, *Pediococcus* sp., *Enterococcus* and *Bacillus* sp. which are also used as probiotic organisms. In addition, *Saccharomyces* sp. a principal probiotic yeast is used to treat diarrhoea associated with antibiotics.

CONCLUSION

From the foregoing account it is clear that many of the non-pathogenic microbes are infact useful especially for good health and prevention of diseases. So we can even say that these probiotic microbes are essential for the continued existence of life on earth.

REFERENCES

Abe, F., Ishibashi, N. and Shimamura, S. (1995). *J. Dairy Sci.* **78:** 2838-2846.

Alm, L. (1991). *In:* Therapeutic Properties of fermented milks (Ed. Robinson, R.K.) Elsevier Science, Barking, England, pp. 45-64.

Beijerinck, M.W. (1901). *Arch. Neer Land des Sci. Extract. et Naturelles*, **6:** 212-243.

Berg, R.D. (1985). *Exp. Anim.* **34:** 1-16.

Berg, R.D. (1992). *In:* Probiotics, The Scientific Basis (Ed. Fuller, R.) Chapman and Hall, London, pp. 55-85.

Cahn, D.R. (1901). *Abteilung Originale*, **30:** 721-726.

Chauriere, G., Coconnier, M.H., Kerneis, S., Darteuille, M.A., Joly, B. and Servin, A.L. (1992). *FEMS Microbiol. Lett.*, **91:** 213-218.

Chukeatirote, E. (2003). *J. Sci. Tech.*, **25(2):** 275-282.

Coconnier, M.H., Bernet, M.F., Kerneis, S., Chauriere, G., Fourniat, J. and Servin, A.L. (1993). *FEMS Microbiol. Lett.*, **110:** 299-305.

Collins, D.M. and Gibson, G.R. (1999). *Am. J. Clin. Nutrn.*, **69(5):** 1052-1057.

Dare, P.I. and Shah, N.P. (1997). *Intl. Dairy J.* **7:** 537-545.

De Ruyst, L. and Vandamme, E.J. (Eds.) (1994). *In:* Bacteriocins of Lactic Acid Bacteria. Chapman and Hall, London.

Desmazeaud, M. (1996). *In:* Dairy Starter Cultures (Eds. Cogan, T.M. and Accolas, J.P.), VCH Publishers, New York, pp. 131-155.

Desmet, I., Van Hoorde, L., De Saeyer, N., Vande Woestyne, M. and Verstraete, W., (1994). *Microb. Ecol. Helth Dist.*, **7:** 315-329.

Dietch, E.A., Hempa, A.C., Specian, R.P. and Bery, R.D. (1992). *J. Trauma*, **32:** 141-147.

Dodd, H.M. and Gasson, M.J. (1994). *In:* Biotechnology of Lactic Acid Bacteria (Eds. Gasson. M.J. and de Vos, W.M.), Blackio Academic and Professional, London, pp. 211-252.

Doderlein, A., Scheidensekret, D. and Seine Bedeutung Fur Das, (1892). *Central blatt fur Bacteriologie*, **11:** 699-700 (in German).

Dunne, C., Liam O' Mohony, Lisa Murphy, Thornton, G., Morrissey, D., Gile O' Halloran, Jeeney, M., Flynn, S., Fitzgerald, G., Daly, C., Kiely, B., Sullivan, Gerald CO', Shanahan, F. and Collins, J.K. (2001). *Am. J. Clin. Nutrn.*, **73(2):** 386S-392S.

Eschenbach, D.A., Davick, P.R., Klilliams, B.C., Klebanoft, S.J., Young-Smith, K., Critchlow, C.M. and Holmes, K.K. (1989). *J. Clin. Microbiol.*, **27:** 251-256.

Fox, P.F., Law, J., McSweeney, P.L.H. and Wallance, J., *In:* Chemistry, Physics and Microbiology (Ed. Fox, P.F.) Chapman and Hall, London, **1:** 389-438.

Fox, P.F., Wallace, J.M., Morgan, S., Lynch, C.M., Niland, E.J. and Tobin, J., *Antonie Van Leewnhoek,* **70:** 175-201.

Fuller, R. (1989). *J. Appl. Bacteriol.,* **66:** 365-378.

Fuller, R. (1991). *Gut.,* **32:** 439-442.

Gasser, F. (1994). *Bull. Inst. Pasteur,* **92:** 45-67.

Gibson, G.R. and Roberfroid, M.B. (1995). *J. Nutrn.,* **125:** 1401-1412.

Gilliland, S.E., Nelson, C.R. and Maxwell, C. (1985). *Appl. Environ. Microbiol.,* **49:** 377-381.

Gordin, B.R. and Gorbach, S.C. (1992). *In:* Probiotics, the Scientific Basis (Ed. Fuller, R.) Chapman and Hall, London, pp. 355-376.

Green, D.H., Wakeley, P.R., Page, A., Barnes, A., Baccigalupi, L., Ricca, E. and Wtting, S.M., (1999). *Appl. Environ. Microbiol.,* **65:** 4288-4291.

Havenaar, R. (1992). Ten brink B. Huisint veld JHJ. *In:* The Scientific Basis (Ed. Fuller, R.) Chapman and Hall, London, 209-224.

Hawas, S.F., Hillier, S.L., Benedetti, J., Sterens, C.E., Koutsky, L.A., Wolner-Hanssen, P. and Holmes, K.K. (1996). *J. Infect. Dis.,* **174:** 1058-1063.

Heller, K.J. (2001). *Am. J. Clin. Nutrn.,* **73(2):** 374S-379S.

Hillier, S.L., Krohn, M.A., Klebanoff, S.J. and Eschenbach, D.A. (1992). *Obstet. Gynecol.,* **79:** 369-373.

Hoffman, A.F., Molino, G., Milanese, M. and Belforte, G. (1983). *J. Clin. Invest.,* **71:** 1003-1022.

Ishibashi, N. and Yamazaki, S. (2001). *Am. J. Clin. Nutrn.,* **73(2):** 465S-470S.

Jack, R.W., Tagg, J.R. and Ray, B. (1995). *Microbiol Rev.,* **59:** 171-200 **(Abstract).**

Jeft, B.D., Huycke, M.M. and Gilmore, M.S. (1994). *Clin. Microbiol. Rev.,* 462-478.

Joseph, P.J., Dave, R.I. and Shah, N.P. (1998). *Food Aust.,* **50:** 20-23.

Kerneis, S., Bilge, S.S., Fourel, V., Chauviere, G., Coconnier, M.H. and Servin, A.L. (1991). *Infect Immun.,* **59:** 4013-4018.

Klaenhammer, T.R. (2000). *J. Nutrn.,* **130:** 415S-416S.

Klebanoft, S.J., Hillier, S.L., Eschenbach, D.A. and Waltersdorph, A.M. (1991). *J. Infect. Dis.,* **164:** 94-100.

Kunji, E.R.S., Mierau, I., Hogting, A., Poolman, B. and Konings, W.N. (1996). *Antonie Van Leewenhoek,* **70:** 91-125.

Lankaputhra, W.E.V., Shah, N.P. and Britz, M.L. (1996). *Milchwissen Schaft,* **51:** 65-70.

Lilly, D.M. and Stillwell, R.H. (1965). *Sci.,* **147:** 747-748.

Maijatoma, M. and Pokrotnieks, J. (2006). *Acta Univ. Latviensis,* **710(Biology):** 117-129.

Matsuzaki, T. (1998). *Intl. J. Food Microbiol.,* **41:** 133-140.

Mercenier, A. (1999). *In: Probiotics: A Critical Review* (Ed. Tannock, G.W.), Horizon Scientific Press, Nortolk, England, pp. 113-127.

Metchni Koff, E. (1908). Prolongation of Life. Putnam, New York.

Moro, E. (1900). *Jahrbuch Fur Kinderheil Kunde,* **52:** 38-55 (in German).

Nesser, J.R., Chambaz, A., Goliard, M., Link-Amster, H. and Fryder, V.K. (1989). *Infect Immun.* **57:** 3727-3734.

Ouwehand, A.C. (1998). *In:* Lactic Acid Bacteria: Microbiology and Functional Aspects. 2nd ed. (Eds Salminens. S. and Von Wright, A.) Marcel Dekker Inc., New York, pp. 139-160.

Parker, R.B. (1974). *Anim. Nutrn. Helth.,* **29:** 4-8.

Parker, R.B. (1979). *Anim. Nutrn. Helth.,* **29:** 4-8.

Payne, D.L., Welsh, J.D., Manion, C.V., Tsegaye, A. and Herd, L.D. (1981). *Am. J. Clin. Nutrn.,* **34**: 2711-2715.

Reid, G., Bruce, A.W. and Smeianov, V. (1998). *Intl. Dairy J.,* **8:** 555-562.

Roy, P., Mainville, I. and Mondou, F. (1997). *Microecol. Ther.* **26:** 167-180.

Salminen, S., Deighton, M.A., Benno, Y. and Gorbach, S.L. (1998). *In:* Lactic acid Bacteria: Microbiology and Functional Aspects. 2nd ed. (Eds. Salminen S. and Von Wright, A.) Marcel Dekker Inc., New York, pp. 211-253.

Sanders, M.E. (1998). *Intl. Dairy J.,* **8:** 341-347.

Sanders, Mary Ellen. (2000). *J. Nutrn.,* **130:** 384S-390S.

Sawada, H., Furushiro, M., Hirai, K., Motoike, M., Watanabe, T. and Yokokura, T., (1990). *Agric. Biol. Chem.,* **54:** 3211-3219.

Shah, N.P. (2000). *British J. Nutrn.,* **84:** 3-10.

Shah, N.P., Lankaputhra, W.E.V., Britz, M.L. and Kyle, W.S.A. (1995). *Intl. Dairy J.,* **5:** 515-521.

Sieber, R., Butikofer, U. and Bosset, J.O. (1995). *Intl. Dairy J.,* **5:** 227-246.

Takano, T. (1998). *Intl. Dairy J.,* **8:** 375-381.

Van Leeuwen, P.A.M., Boermeester, M.A., Houdijk, A.P.J., Ferwerdach, C., Cuesta, M.A. and Meyer, S. (1994). *Gut,* **35(Suppl.):** S28-S34.

Viljoen, B.C. and Lourens-Hatlingh, A. (2001). *Intl. Dairy J.,* **11:** 1-17.

Weber, H. (1996). *In: Mikrobiologie der Lebensmittel: Milch and Milch Products (Microbiology Food: Milk and Dairy Products)* (Ed. Weper, H.) Behr's Verlag, Hamsburg, Germany, pp. 105-152 (in German).

Chapter 3

Spirulina : Its Role in Food Industry

S. Balamurugan[1], R. Saravanamuthu and *G. Chithra
Department of Botany, A.V.C. College (Autonomous), Mannampandal, Mayiladuthurai–609 005, Nagapattinam, Tamil Nadu
** Department of Biotechnology, Bharath College of Science and Management Thanjavur–613 005, Tamil Nadu*
[1]E-mail: sundarbala_avc@yahoo.com

Microbial sources of protein are recognized as alternative to the conventional protein sources throughout the world. Cell biomass of algae, bacteria and fungi are the microbial proteins, now called single cell protein. *Spirulina* is a photosynthetic, filamentous, spiral-shaped, multicellular and blue-green alga. Before Columbus, Mexicans (Aztecs) exploited this microorganism as human food; presently, African tribes (Kanembu) use it for the same purpose. Its chemical composition includes proteins (60-70%), carbohydrates (15-25%), essential fatty acids (18%), vitamins, minerals and pigments like carotenes, chlorophyll a and phycocyanin. It is considered as an excellent food, lacking toxicity and having corrective properties against viral attacks, anaemia, tumour growth and malnutrition. In literature the use of these micro algae has been reported as animal food supplement for the enhancement of the yellow coloration of skin and eggs yolk in poultry and flamingos, growth acceleration, sexual maturation and increase of fertility in cattle.

INTRODUCTION

Population explosion is one of the major concerns of the scientists, who attempt to ensure the food security. The population of India has touched 1000 million. It is expected to reach 1560 million by 2020 AD. Indeed there is an increase in the food production and availability of food grains. In spite of that it has been estimated that there will be a net storage of nearly about 26 million tonnes by 2020 AD. More than 75 percent of the people live in villages and their primary occupation is agriculture. One of the estimates shows that nearly about 800 million people, majority of them are rural women and children, go hungry to bed everyday. WHO reported the prevalence of several nutritional disorders among the people who live in the under developed and developing countries due to insufficient supply of protein, vitamin A, C and calcium deficiencies.

Nearly about four decades ago it was forecasted that there would be protein famine by 1980. This made the food processing industries and the food technologists to search for protein sources alternative to the conventional protein sources during the period. The traditional foods and the food collection practices prevailed among the tribes throughout would were looked into with scientific eye, and number of microorganisms were identified as potential sources of rich protein. *Spirulina,* a blue-green alga (Cyanobacterium), is one of such organisms and has been recognized as the best food for tomorrow.

SINGLE CELL PROTEIN

Microorganisms have been used as food either directly or indirectly for centuries. In most of the instances they have been used as process organisms in the production or in the enrichment of the quality of the product. Obviously, they have been used in the manufacturing of bread, beverages, milk products, etc. However, the direct use of microorganisms as food was known since the beginning of the sixteenth century. The dried cells of microorganisms such as bacteria, algae, actinomycetes and fungi have been used as food or feed. Since they are the rich source of protein, they were collectively called microbial protein. The term *single cell protein* was coined and substituted the term microbial protein by the Massachusetts Institute of Technology, USA, and introduced at the first international conference on microbial protein held in 1967.

Thus, single cell protein (SCP) is the term that refers to a monoculture of microbial cells or the total protein extracted from cell cultures, either of which can be used as a human or animal protein supplement. They are also called *food grade or feed grade single cell protein.* The single cell protein that are considered suitable for human consumption are called food grade, and the one which is suitable for consumption of animal is called feed grade.

Microalgae as a Source of SCP

Microorganisms have been processed and produced as source of proteins since World War I. Among all the microorganisms, the microalgae are the most important source of SCP as they can be grown profusely in open ponds. Species of *Anabaena, Chlorella, Calothrix, Coelastrum, Dunaliella, Nostoc, Scenedesmus* and *Spirulina* have been demonstrated as potential organisms, which could be exploited as SCP. Among these organisms species of *Chlorella, Scenedesmus* and *Spirulina* have received great attention. Use of *Spirulina* as a diet component has been recorded as early as 1521. In Mexico, people used to eat small cakes prepared from the ooze extracted from lakes. In 1521, Bernal Diaz del Castile described about the lakes, ooze cakes and their cheese like odour. Now, it has been authentically established that it constituted nothing other than *Spirulina,* a blue-green algae (Cyanobacterium). A similar observation was made by Dr. G. Clemnt in 1962, (Clement and Durand Chastel, 1972) who noticed that the people living around the Chad Lake of Africa collected the algal mass floating on the surface of the lake and took it as their daily diet. The people had good health when compared to the people who did not have such diet. The organism was later identified as *Spirulina*, which had been known for many years without the knowledge of scientific information.

The genus *Spirulina* belongs to Oscillatoriaceae of Cyanophyceae. It is a non-nitrogen fixing and a filamentous alga. It is widely distributed and has been reported from varied environments including soil, sand, marshes, brackish water, sea water and fresh water. It has also been reported to thrive in the highly alkaline waters of Taxcoco lake. Thus, the organism is capable of adaptation to diverse type of habitats and environmental conditions. Intensive researches on the various aspects of *Spirulina* revealed the nutritive value of the organism as it possessed rich protein content, essential amino acids and vitamins. Various methods of mass cultivation and processes for the extraction of desirable products have been developed.

Spirulina is nutritionally very important and has gained commercial importance. The overall superiority in the nutritional quality is due to its high protein content (60-70%) of its dry weight, low fat content and rich source of β-carotene, thiamine, riboflavin and also vitamin B_{12}. It contains relatively rich quantity of rare fatty acids, especially linolenic acid. Hence, it has been reported to have therapeutic properties. The protein efficiency ratio (PER) of *Spirulina* is generally higher than any other vegetable cereal and soybean. Thus, *Spirulina* is not only considered as a rich potential source of protein but also a rich source of vitamins and some essential amino acids. It has also been reported that the bioavailability of vitamins is good, and hence it has been rightly suggested that it could be possibly used as a health food and also as a food supplement (Anusya Devi *et al.*, 1979, Anusya Devi and Venkataraman 1983, Venkataraman, and Becker, 1986, Venkataraman, 1989, 2001, Venkataraman *et al.,* 1994).

The powdered form of *Spirulina* has been used in many ways in the preparation of food, and there are many formulations available as tablets, capsules, recipes, fermented foods, ointments, etc. Besides the high nutritional quality of *Spirulina* it also possesses therapeutic properties. Species of *Spirulina* are used in immunodiagnostics as they possess biliproteins. It has been reported that it can be successfully used for the treatment of obesity. Thus, it can also lower the cholesterol contents in the human blood. It has also been successfully demonstrated to be used in the treatment of malignant tumours and anaemia and to improve the haemoglobin levels in the human blood. It has been claimed that the species of *Spirulina* are effective in curing pancreatitis, cirrhosis and hepatitis. It can also be used as a prophylactic measure against cancer. The phycocyanins present in *Spirulina* have been found to play a major role in the recovery from renal failure caused by mercury and pharmaceutical drugs.

The nutritional quality and therapeutic value of *Spirulina* drew the attention of the nations throughout the world. The importance of the alga was realized and strategies of mass cultivation of *Spirulina* have been evolved.

In India the production technology of *S. platensis* was pioneered and extensively studied by Dr. L.V. Venkataraman and his team of workers at the Central Food Technological Research Institute (CFTRI), Mysore. His ingenious works on this particular alga motivated the researchers in different parts of the country and he encouraged the scientists and also had been a good counselor. Now-a-days private institutions such as Murugappa Chettiar Research Centre (MCRC), Chennai, Panbari Tea Company in Calcutta, Ballarpur Industries Limited, Mysore, and New Ambadi Estate, Chennai Earthrise Forms (USA) and Solarium Biotechnology (Chile) are actively engaged in the cultivation of *Spirulina* and formulation of *Spirulina* products (Ayalla and Vargas, 1987; Jourdan, 1993; Belay, 1997). Today *Spirulina* is consumed by million of people all over the world, and it has been established that there are many health benefits apart from its nutritive value.

Ultrastructure

Transmission electron microscope observations show that *Spirulina* has prokaryotic organization with capsule, pluri-stratified cell wall, photosynthetic or thylakoid lamellar system, ribosomes and fibrils of DNA region and numerous inclusions. The capsule has fibrillar structure and covers each filament thus protecting it. The regular presence of capsule around the filaments in *S. platensis* is a morphological characteristic that differentiates *S. maxima* (Balloni, *et al.,* 1980, Belay, 1997). Trichomes vary from 6 to 12 μm, and are composed of cylindrical cells. The helix diameter varies from 30 to 70 μm (Tomaselli, 1997), and the trichome length is about 500 μm, although in some cases when stirring of culture is deficient, the length of filament reaches approximately 1 mm. It is very important to explain that the helical shape of *Spirulina* in liquid culture is changed to spiral shape in solid media. These changes are due to hydration or dehydration of oligopeptides in the peptidoglycan layer (Ciferri, 1983).

The *Spirulina* is formed of four numbered layers, which are formed from the innermost to outward as, LI, LII, LIII and LIV. All these layers are very weak, except layer LII which is made up of peptidoglycan, the substance that gives rigidity to the wall (Ciferri, 1983). The LI layer contains beta 1,2-glucan, a polysaccharide not very digestible by human beings. However, the low concentration (<1%) of this layer is 12 nm in thickness, and the protein and lipo-polysaccharide nature of the LII layer are considered favourable for the easy human digestion of *Spirulina* (Balloni, *et al.*, 1980).

In this microorganism, chlorophyll a, carotenes and phycobilins, which contain phycocyanin (blue pigment), are located in the thylakoid system or photosynthetic lamellae. The inter thylakoid space is limited by the presence of electronically transparent protein gas vesicles in the cylindrical form of *Spirulina,* which give buoyancy (Ciferri, 1983).

Spirulina contains numerous characteristic peripheral inclusions associated with thylakoids. These include cyanophycin granules, polyhedral bodies, polyglucan granules, lipid and polyphosphate granules (Balloni *et al.*, 1980; Ciferri, 1983). The cyanophycin granules, or reserve granules are important due to their chemical nature of a series of pigments. The polyhederal bodies or carboxysomes mainly contain the enzyme ribulose 1, 5-diphosphate carboxylase that allows the fixation of CO_2 in the photosynthetic organisms and probably carry out a reserve function. The polyglucan granules or glycogen granules or alpha granules are glucose polymers, small, circular and widely diffused in the interthylakoidal space. The lipid granules, beta granules or osmophile granules form the reserve deposit, constituted by poly-beta-hydroxybutyrate (PHB) found only in prokaryotes. PHB acts as a carbon and energy reserve (Vincenzini, *et al.*, 1990).

Nutritional Value of *Spirulina*

Early researches demonstrated that *Spirulina* is safe for consumption. When scientists discovered that *Spirulina* grew so fast, yielding 20 times more protein per acre than soyabean, they rightly pointed out it as a food for future. *Spirulina* is one of the best vegetable sources of protein with the protein content of 70%, which is higher than any other natural food and yet even greater value is found in terms of vitamins, minerals and other unusual nutrients. *Spirulina* preparations intake of three to ten times a day delivers impressive amounts of β carotene, vitamin B12, B-complex, essential trace minerals, and γ-linolenic acid.

It is legally approved as a food or feed supplement in Europe, Japan and many other countries around the World. It has been well established by the FDA of USA in 1981 that *Spirulina* is a good source of protein and contains various vitamins and minerals, and may be legally marketed as the food supplement.

Proteins

Spirulina has a high protein concentration (60-70% of dry weight). A comparative status of protein in different foods is presented in Table 1 (Ciferri, 1983). *Spirulina* is useful in human nutrition due to high quality and quantity of its protein. The nutritive value of the protein is related to the quality, flavour of amino acid and digestibility (Dillon and Phan, 1993; Richmond, 1992) *Spirulina* contains high levels of essential amino acids, which include leucine (10.9% total amino acids), valine (7.5%) and isoleucine (6.8%) (Cohen, 1997). Denaturation of *Spirulina* protein is absorbed when the alga was heated above at neutral aqueous solutions. Hydrophobic reagents interaction during heating and hydrogen bonds formation during cooling or aggregation factors of *Spirulina* protein have also been reported (Chronakis, 2001)

Table 1: A comparative account of protein content in *Spirulina* and other sources of food (Henrikson, 1994).

Food type	Crude Protein %
Spirulina powder	65
Whole dried egg	47
Beer yeast	45
Skimmed powdered	37
Whole soyabean flour	36
Wheat germ	27
Pea nuts	26
Chicken	24
Fish	22

Phycobilins

The proteins having the highest economic potential are the biliproteins. *Spirulina* contains two biliproteins, which include c-phycocyanin and allophycocyanin. The protein fraction may contain up to 20 percent of phycocyanin, a water-soluble blue pigment (Ciferri, 1983).

Phycocyanin is already being produced commercially. It was introduced in Japan as a natural colouring agent for feed and cosmetics, and is produced at a rate of 600 kg/month. The main use of phycocyanin is as a food pigment. Phycocyanin could be utilized as a natural pigment for the food, drug, and cosmetics industries to replace the currently used synthetic pigments that are suspected to possess carcinogens. Indeed, phycocyanin from *Spirulina* has already been commercialized by Dainippon Ink and Chemicals of Japan under the name of Lina blue. The product is an odourless non-toxic blue powder with slight sweetness.

Amino Acids

The amino acid composition has been determined for *S. maxima* (Tredici *et al.*, 1988). The species contains 16 amino acids, eight of which are essential, including leucine (10.9 percent of total amino acids), valine (7.5 percent) and isoleucine (6.8 percent). The major amino acids are glutamic acid (17.4 percent) and aspartic acid (13.2 percent). It appears that *Spirulina* can store amino acids. The growth rate at low temperature is decreased due to fact that the demand for amino acids is reduced at that temperature (Ciferri, 1983). Excess nitrogen is metabolized into cyanophycin, a copolymer of amino acids that is found as granules occupying up to 18 percent of the cell volume. The polymer is mainly made up of chains of poly L-aspartic acid with arginine attached to the β-carboxyl group.

Mutant strains of *S. platensis* are induced by cultivating the species on media containing amino acid analogs (Riccardi *et al.*, 1981). In the mutant strains, which acquired resistance to analogs, it appears that the resistance was due to overproduction of the corresponding amino acids.

Vitamins

Among the nutritional constituents, *Spirulina* has relatively high provitamin A concentration (Belay, 1997) and relatively an excessive amount of β carotene (Table 2), which may be toxic. But when β carotene is ingested from the *Spirulina* or any other vegetable, it is usually harmless since it is converted into vitamin A depending on the quantity it needs (Henrikson, 1994). *Spirulina* is also a very rich source of vitamin B_{12}, hence, this organism is of great value for people who need vitamin B_{12} supplements in the treatment of pernicious anemia (Richmond, 1992; Becker, 1984; Belay, 1997).

Table 2. Vitamin content in the *Spirulina* powder (Belay, 1997)

Vitamins	mg 100 g^{-1}
Pro vitamin A	2.330.000 IU kg^{-1}
β carotene	140
Thiamin B_1	3.1
Riboflavin B_2	4.0
Niacin B_3	14.0
Vitamin B_6	0.8
Vitamin B_{12}	0.32
Vitamin K	2.2

Minerals

Spirulina is rich in minerals. The mineral content is presented in Table 3. Iron is one of the important nutrients complement, which is not appropriately absorbed. But iron in *Spirulina* is

60% better absorbed than ferrous sulphate and other complements. Hence it could be an adequate source of iron for the anaemic pregnant women (Puyfoulhoux, *et al.,* 2001)

Table 3. Mineral contents in *Spirulina* powder (Belay, 1997)

Minerals	mg 100 g^{-1}
Calcium	700
Chromium	0.28
Copper	1.2
Iron	100
Magnesium	400
Sodium	900
Zinc	200

Carbohydrates

Spirulina plantensis contains about 13.6% carbohydrates, which include glucose, rhamnose, mannose, xylose, galactose (Shekharam, *et al.,* 1987). It does not have cellulose in its cell wall, a feature that makes it appropriate and important foodstuff for people with problems of poor intestinal absorption and geriatric patients (Richmond, 1992). A new high molecular weight polysaccharide with immuno stimulatory activity has been isolated from *Spirulina.* It is called imulina. It is a highly water-soluble polysaccharide present in the range between 0.5% and 2.0% (w/w) of the dry weight of micro algae (Pugh, *et al.,* 2001).

Lipids

Typically cyanobacteria are generally poor in lipids, and *Spirulina* contains 6-13 percent lipids, half of which are fatty acids.

Mono galactosyl diacyl glycerol (MGDG), sulfo quinovosyl diacy glycerol (SQDG) and phosphatidyl glycerol (PG) are the major lipids which amount to 20-25 percent each (Petkov and Furnadzieva, 1988). Digalactosyldiacylglycerol (DGDG) contributed only 7-10 percent, while triglycerides are found as minor component (1-2 percent).

Nucleic Acid Content

One of the main concerns about the consumption of microorganism is the high nucleic acid content, which increases the level of urea and results in gout. *Spirulina* contains 2.2% to 3.2% RNA and 0.65 to 1% DNA, which represents 5% on the basis of dry weight. These values are smaller than those of other micro algae such as *Chlorella* and *Scenedesmus* (Ciferri, 1983).

Pigments

Some natural pigments are found in *Spirulina,* (Table 4) and these pigments are responsible for the characteristic colour pattern of certain species of flamingo that consumed *Cyanobacteria* in the African valley. This knowledge has promoted the use of this microorganism as the source of pigmentation for fish and eggs (Cifrri, 1983; Saxena, *et al.,* 1983; Henrickson, 1994). *Spirulina* also increases the yellowness and redness of broiler chicken due to accumulation of zeaxanthin (Toyomizu, *et al.,* 2001).

Table 4. Pigments in *Spirulina* powder (Belay, 1997)

Pigments	mg 100 g^{-1}
Carotenoids	370
Chlorophyll A	1000
Phycocyanin	14000

MASS PRODUCTION OF *SPIRULINA*

The main commercial large-scale culture of micro algae started in the early 1960s in Japan with the culture of *Chlorella*, followed by *Spirulina* in the early 1970s. The third major microalgal industry was established in Austrialia in 1986. Commercial production of *Dunaliella salina* was made as a source of β-carotene (Borowitzka, 1998).

Production process of *Spirulina* requires clonal or unialgal cultures—isolation of a single algal unit or trichome (Vonshak, 1984; Gonzalez, *et al.*, 1995; Hoshaw and Rosowski, 1979). The method begins with the determination of physical and chemical parameters of the water sample, which constitute the main ingredients for the growth medium. The *Spirulina* samples should remain under dim light or in darkness and at 20-25°C. Freezing temperature is not recommended because it favours lysis and death of the cells (Rippka, 1988).

Spirulina can be cultivated in large scale under controlled as well as natural conditions. Mass production of *Spirulina* in synthetic culture medium is uneconomical because it requires relatively high quantity of minerals which are very costly. Hence, cultivation of *Spirulina* in sewage water is more economical than in synthetic media or in clean water. In this section, the strategies involved in the mass cultivation of *Spirulina* have been dealt with. The strategies include strain selection, culture medium and mass cultivation.

Strain Selection

The major criteria used in the selection of suitable strain for commercial production are growth rate, biochemical composition and resistance to mechanical and physiological stress. A wide variety of species strains of *Spirulina* have been screened by several people in various countries. Most of such works have been done in Israel (Vonshak, 1987b). A large number of strains have been screened for desirable traits as mentioned above, and the suitability of such strains to be grown in mass culture under the prevailing condition has been studied. While the selection of suitable strains and scaling of culture are relatively easy, the harvesting and recycling process result in cross contamination. Thus, maintenance of pure strains throughout the growth season is relatively difficult. However, maintenance is relatively easy when grown and harvested within the same period.

Culture Medium

Culture medium is based on the original medium (Zarouk, 1966) and is typically composed of sodium carbonate, a source of nitrogen, phosphorus, iron and trace elements. After harvesting, the culture water is recycled continuously into the bonds throughout the growth season. Only food grade chemicals are used. The choice of chemicals and the desired concentration depend on the cost and the compatibility of chemicals to the raw water. Cost of nutrients account for about 15 to 25% of the total production cost. Extensive research has been made to develop the medium concentration and supply regime that provide optimum growth rate at minimum cost. The later is based on the laboratory and outdoor study on the requirements of algae for specific nutrients.

Mass Cultivation of *Spirulina*

There are various designs and methods available for the large scale production of *Spirulina*, both in indoor and outdoor conditions. Some of the cultivation systems have been described and summarized (Singh *et al.* 2001, Saravanamuthu, 2007), the outline of which has been highlighted in the subsequent paragraphs.

Shallow mixed pond system

This system was introduced by Oswald as early as 1960 for the production of micro algae. It is a system of open shallow race ways, and it is a type of shallow pond with meandering configuration of channels. Paddle wheels of different designs with low shearing force are used for mixing water. They are used for thorough mixing of water for aeration and to direct streaming of algal mixture. Solid deposition on hydrolic losses are eliminated by establishing a single loop configuration with babbles, which direct streaming.

There are two types of shallow pond race way systems used in many countries including India. In one type of race ways system, the lining of the channel is made of concrete, and in the second type, it is a shallow earth run tunnel line with 1 to 2 mm thick sheet of polyvinyl chloride or some other durable food grade plastic material. The sizes of commercial pond systems ranged from 1000 to 5000 m^2. Continuous stirring of culture is preferred because it maintains constant supply of nutrient and CO_2, and at the same time removes the excessive oxygen from the system.

Closed tubular system

This system was developed by John Pirt and his co-workers in 1982 for the cultivation of micro algae. This system design overcomes the problem encountered in the open shallow race way system. It is also called '*closed photo bioreactor*'. It is a system made up of several tubes laid side by side on white polythene sheet. The parallel tubes are joined by connection and form loops. The release of collected oxygen due to photosynthesis of the autotrophic algae has been facilitated by incorporating a narrow tube with each connection. Comparative studies revealed that the system is more protective per unit culture volume than the shallow open race ways. This system also has the advantage of increasing the efficiency of CO_2, and the contamination is also minimized.

Experience revealed that there had been two major limitations in the use of closed tubular system which include self-shadowing and increased heat in the closed system.

Flat plate bioreactor

This system was conceptualized and explained by Samson and Leduy (1985) and Romos de Ortega and Roux (1986). This system was reported by Tridici *et al.* (1991), and their experience made them to introduce a system of bioreactor made up of commercially available panels of transparent sheets. They are partitioned to form narrow internal channels. The panel reactors can be set up with different orientations. This system provides maximum exposure of solar energy and prevents build up of oxygen.

Glass photo bioreactors

Richmond (1986, 1995) described a well-designed, a new photo bioreactor for mass cultivation of *Spirulina platensis*. This new glass photo bioreactor is made up of series of individual glass reactor. Each reactor is measures 70 cm high, 90 cm long and 2.6 or 1.3 cm wide.

Dome type photo bioreactor

One of the most recent types of photo bioreactors is the dome type photo bioreactor. It is also called '*Bio dome*'. The system was designed and developed by Hirabyashi and Guevarra (1999) in Japan. The problem and difficulties experienced in the existing systems such as contamination, self-shadowing, high light intensity, heating up of algal mixture, increasing temperature, algal density, poor mixing up of gases, mounting oxygen tension, etc., have been taken care of and rightly addressed. This system has been optimized to provide maximum production and yield of the desirable micro algae under cultivation. Some of the facilities are as follows:

(i) The dome-shaped material design facilitates the penetration of sufficient quantum of light and efficient utilization of light energy during photosynthesis by the algae and maintain the algal cells in high density.
(ii) A mixing mechanism provided with this system facilitates an effective turbulent flow of air (oxygen or CO_2).
(iii) A cooling system, a sprinkler, sprinkles water over the outer surface of the dome and acts as temperature control device.
(iv) Washing and cleaning of the system have been made easy.

The production process has nine stages, which include (i) *Filtration and Cleaning* – a nylon filter at the entrance of the water filters that cleans the water; (ii) *Pre-concentration* – the algal biomass is washed and salt content is reduced; (iii) *Concentration* – the sewage water is removed at its maximum level (located among the filaments); (iv) *Neutralization* – the algal biomass is neutralized with the addition of acid solution (v) *Disintegration* – trichomes of microalga is broken down by using a grinder (vi) *Dehydration* – water content is removed by spray drying (this operation has great economic importance since it involves about 20-30% of the production cost) (Ayala and Laing, 1990) (vii) *Packing* – done in sealed plastic bags to avoid hygroscopic action on the dry *Spirulina* (viii) *Storage* – stored in cardboard boxes, and in fresh, dry, dim, pest-free and clean storeroom in preventing the pigments of *spirulina* from deteriorating (Ayala, 1998) and (ix) *Quality control* – as a food source it includes microbiological standard tests, chemical composition test etc. (Belay, 1997).

USES OF *SPIRULINA*

Human Consumption

Spirulina is the major source of protein supply to native tribes in South America and Africa. The reintroduction of *Spirulina* as a health food for human consumption in the late 1970s and in the beginning of 1980s was associated with many controversial clients. Nevertheless, it has been well established and marketed as a food supplement and health food. It is available in the market as *Spirulina* tablets and powders.

Feed and Feed Additives

Intensification of poultry and aquaculture is increasing the demand for specially formatted feed. The demand is met by standardized feed preparations with *Spirulina,* which also substitute the natural sources of feed available.

Spirulina in Poultry

Feed additives in the poultry market, with special attention to the effective colouring of yolk represents major part of this fast growing industry. A few studies on the use of very effective agent in inducing the yolk colour have been reported (Saxena *et al.,* 1982, 1983). Thus, *Spirulina* can be effectively used as an feed additive to enhance the colour of the yolk and thus the market value of the eggs.

Aquaculture

Spirulina *feed in fish nutrition*

Spirulina formulated feed increases the growth rate of many species. It improves the palatability of the feed. It has also been reported that fish fed with *Spirulina* had less abdominal fat (Kato, 1989). The feed conversion was also improved. It has been claimed that the fishes grown with feed containing *Spirulina* are of better quality in terms of better flavour, firmer flesh, and brighter skin colour than the control fish (Hirano, 1985: Suyama, 1985; Mori, 1987).

Another very important effect relates to reduction in the rate of mortality of fingerlings or post-larval stages. An addition of 0.5 to 1 percent of *Spirulina* in the feed has a very significant effect on growth (improvement of 17-25 percent) and reduction of mortality (30-50 percent) depending on species and *Spirulina* concentration. It is important to note that one of the most crucial stages in modern intensive aquaculture is the survival of the initial inoculum. Improvement in the survival rates may provide a significant improvement in the economic performance of any aquaculture farm. A report on the use of *Spirulina* in aquaculture in Japan concluded that *Spirulina* improves the cost/performance ratio of the fish feed, the largest expense in fish production.

Uses in crustaceans

A specific application of *Spirulina*, mainly as a colorant and pigmentation agent in the diet of the black tiger prawn was suggested by Liao *et al.* (1993). They reported that incorporating 3 percent *Spirulina* in the diet of the prawn for 14 to 28 days resulted in a marked increase in the carotenoid content in the carapace and suggested that zeaxanthin, a major component of *Spirulina* carotenoids, is rapidly converted into astaxanthin by the prawn. Comparatively 3 percent *Spirulina* was the most effective pigmentation agent, when compared *Spirulina* with the use of other carotenoid has been containing feed.

Intensive work was carried out to test the utilization of *Spirulina* as mixed feed for abalone, scallops and penaeid shrimp (Zhou *et al.*, 1991). It was concluded that *Spirulina* mixed feed made a good supplement in the cultivation of parent scallops because it proved useful for the normal development of the scallop gonads which achieved a higher fecundity and hatchery rate. Thus, the use of *Spirulina* mixed feed enhanced the production of the bay scallops by giving them an advantage of about one month in season. Moreover, mixed feed greatly simplified the feeding procedure, precluding the change of temperature, which would normally result from pouring large quantities of five different microalgae into the production tanks. It was calculated that *Spirulina* mixed feed was cheaper than live microalgae and some hatcheries have already adopted it in their production of bay scallop larva.

Medicinal Uses

Based on preliminary research scientists hope that the use of *Spirulina* and its extracts may reduce or prevent cancer and viral disease. It has been established that the bacterial and parasitic infection may be prevented or better treated, and also wound healing may be improved with *Spirulina.* It is also concluded that it acts in the following ways:

(i) strengthen and improve the immune system,
(ii) phycocyanins build the blood cells,
(iii) increases anti-viral activity, and
(iv) exhibits anti-cancer activity.

Spirulina consumption benefits

Studies have shown that *Spirulina* consumption for 4 weeks reduced the serum cholesterol levels in human beings by 4.5% (Henrikson, 1994) and significantly reduced the body weight by 1.4 ± 0.4 kg after four weeks (Becker, *et al.*, 1986). These reports indicated no changes in clinical parameters (blood pressure) or in biochemical variables (haematocrit, haemoglobin, blood cells, sedimentation rate) and absence of adverse effects. The reduction of cholesterol is partly due to high content of gamma-linolenic acid in cyanobacteria (Henrikson, 1994).

The β-carotene is one of the most effective substances to counteract those free radicals that alter cells causing cancer (Fedkovic, *et al.*, 1993; Schwartz, *et al.*, 1990). Studies at the Harvard University School of Dental Medicine found a reduction in mouth cancer when β-carotene extracts, obtained from *Spirulina*, are consumed. The β-carotene solution applied to oral cancer tumorus in hamsters reduced the tumor number and size and in some cases they disappeared (Schwartz and Shklar, 1987; Schwartz, *et al.*, 1988). *Spirulina* extract induces the tumor necrosis factor in macrophages, suggesting a possible tumor destruction mechanism (Shklar and Schwartz, 1988).

An extract of sulphated polysaccharides called Calcium-Spirulan, (Ca-Sp) made up of thamnose, ribose, mannose, fructose, galactose, xylose, glucose, glucuronic acid, galacturonic acid, and calcium sulfate, obtained from *Spirulina*, showed activity against HIV, Herpes simplex virus, human cytomegalovirus, influenza A virus, mumps virus and measles virus (Henrikson R, 1994; Haysashi, 1996b). Current investigation in this field is searching for extracts that inhibits the AIDS virus replication (Ayehunie, *et al.*, 1998) and allows these patients to improve their health.

Spirulina excretes variable quantities of products from its metabolism such as organic acids, vitamins and phytohormones. Cell extract of *S. maxima* has shown antimicrobial activity against *Bacillus subtilis, Streptococcus aureus, Saccharomyces cerevisiae,* and *Candida albicans*. The presence of high quantities of acrylic acid in *Spirulina* was substantiated at the end of the seventies. This substance shows anti-microbial activity, in a 2 mg/l of biomass concentration (Balloni *et al.*, 1980).

Lactobacillus population in human gastrointestinal tract is increased by *Spirulina* consumption. This means that there was improvement in food digestion and absorption, protection against intestinal bacterial infections and stimulation of immune system (Henrikson, 1994; Schiffrin, *et al.*, 1997). Immune system modulation is due to interference in the production of which end cytotoxicity (Hirahashi, *et al.*, 2002).

Spirulina contains vitamin A, important in preventing eye diseases; iron and vitamin B_{12} useful in treating hypoferric anaemia and pernicious anaemia, respectively; γ-linolenic acid, appropriate in treatment of atopic child eczema; to alleviate premenstural syndrome, and in the stimulation of immune system (Pascaud, 1993). *Spirulina* also has a positive effect on cardiac disease, Parkinsons's disease, malnutrition, sclerosis (Richmond A, 1992; Fox, 1993, Fox, 1998; Thein, 1993) and wounds cure (Richmond, 1992). Other benefits are attributed to *Spirulina*: anti-arthritic effect due to the anti-inflammatory and antioxidative properties of phycocyanin (Ramirez, *et al.*, 2002), anti-atherogeic property (Kaji, *et al.*, 2002), tumour burden inhibition (*Dasgupta, et al.,* 2001), chemo protective and radio-protective effect (Zhang, *et al.,* 2001), and antioxidant activity on lead-induced toxicity in rats (Upasani, *et al.,* 2001).

In Mexico, *Spirulina* is used to enrich candies. In Australia and New Zealand beverages of this substance are marketed. In Japan, India and Singapore, *Spirulina*-enriched appetizers are sold specially to pregnant women, children and elderly people. *Spirulina* is not only food, but also a natural colour agent in Japanese chewing gums. Countries such as Chile, France, Cuba, Germany, Switzerland, Spain, Portugal, Sweden, Holland, Belgium, Denmark, United Kingdom,

Australia and New Zealand market food complements, which include *Spirulina* as the main component. Internationally, skin care products, shampoos, dyes, masks, creams and tonics containing this microorganism are also marketed. In Sweden low caloric bread enriched with *Spirulina* is sold (Henrikson, 1994).

Spirulina has been studied as an animal cell-growth stimulant (Kerby and Rowell, 1992) and in the treatment of residual waters using alginate (Canizares, *et al.*, 1993; Patnaik, *et al.*, 2001). Phycocyanin shows activity on vegetable cell cultures with production of secondary metabolites damage in DNA and hence it may be used as a therapeutic agent (Bhat, *et al.*, 2001). *Spirulina* has been studied as a feed for broiler chicks (Balamurugan and Balakrishnan, 2007).

Spirulina is used in Japan and Taiwan as aquarium fish food, in United States to enhance the colour, the growth and sexual maturation of Canaries and exotic birds (Saxena, *et al.*, 1983). Cattle and horse breeders affirm that when adding *Spirulina* to silage, the quantity of sperms in males and the fertility in females are increased (Henrikson, 1994). *Labeo rohita* (rohu), an Indian carp, showed greater growth after being fed with *Spirulina* (Nandeesha, *et al.*, 2001). In chickens, *Spirulina* increases the mononuclear phagocyte system function thereby enhancing their disease resistance (Al-batshan, *et al.*, 2001). More benefits of *Spirulina* are still being under investigation throughout the world.

CONCLUSION

It is truly a comprehensive supplement for improving health and optimizing body functions. From the foregoing paragraphs it is obvious that *Spirulina* certainly forms a food, feed and food additive and it is a proven protein source. Its production is comparatively easier and cheaper than other protein sources. Now-a-days *Spirulina* is used as health food and food supplements. Nevertheless there are arguments, who are against the use of non-conventional foods, regaling the levels of safety of using *Spirulina* harvested from open systems, where sewage water is used, as it is capable of biomagnifying the available heavy metals and the pesticide molecules available in the contaminated water. Moreover, the nucleic acid content is directly proportional to the protein content. When the nucleic acid is consumed it leads to accumulation of uric acid that leads to gout. Hence, it becomes essential that the nucleic acid should remove when it is proposed as human food which involves heavy expenditure. However, the feed formulations with *Spirulina* do not warrant such stringent measures for health standards. Hence, they could be used readily with animal and poultry feeds.

REFERENCES

Al-Batshan, H.A., Al-Myfarrej, S.I., Al-Homaidan, A.A. and Qureshi, M.A. (2001). *Immunophar. Immunotox.l* **23:** 281-289.

Anusya Devi, M. and Venkataraman, L.V. (1983). *Nutr. Rep. Int.* **28:** 1029.

Anusya Devi, M., Rajashekaran, T., Becker, E.W. and Venkataraman, L.V. (1979). *Nutr. Rep. Int.* **19:** 785.

Ayala, F. and Vargas, T. (1987). *European J. Appl. Microbiol.* Biotechnology **15:** 198-199.

Ayehunie, S., Belay, A., Baba, A. and Ruprecht, R.M. (1998). *J. Acquire Immune defic Syndr Hum retrovirol* **18:** 7-12.

Balamurugan, S. and Balakrishnan, S. (2007). *Geobios.* **34(1):** 63-64.

Balloni, W., Tomaselli, L., Giovannetti, L. and Margheri, M.C. (1980). *In:* Biologia fondamentale del genere Spirulina. (Ed. Cantarelli) CNR, Tipografia Coppini. 49-82.

Becker, E.W. (1984). *In: Handbook of microalgal mass culture.* (Ed. Richmond) CRC Press. 339-408.

Belay, A. (1997). *In: Spirulina platensis* (Arthrospira) (Ed. Vonshak): Physiology, cell-biology and biotechnology. Taylor and Francis. London. 131-158.

Bhat, V.V. and Madystha, K.M. (2001). *Biochem. Biophys. Res Commun.* **285:** 262-265.

Borowitzka, M. (1998). *J. Biotech,* **70:** 313-321.

Canizares, R.O., Dominguez, A.R., Rivas, L., Montes, M.C., Travieso, L. and Benitez. F. (1993). *Biotech Letters* **15:** 32-326.

Chronakis, I.S. (2001). *Bioresour Technol.* **77:** 19-24.

Ciferri, O. and Tiboni, O. (1985). *Ann Rev Microbiol.* **39:** 503-526.

Cifferri, O. (1983). *Microbiol. Rev.* **47:** 551-578.

Clement and Durand Chastel, H.L. (1972). Mexico I.F.P. Report **20:** 743.

Cohen, Z. (1997). *In: Spirulina platensis (Arthorospira)* (Ed. Vonshak): *Physiology, cell-biology and biotechnology.* Taylor and Francis. London. 175-204.

Dasgupta, T., Banejee, S., Yadav, P.K. and Rao, A.R. (2001). *Moll Cell Biochem.* **226**: 27-38.

Dhillon, J.C. and Phan, P.A. (1993). **12:** 103-107.

Fedkovic, Y., Astre, C., Pinguet, F., Gerber, M., Ychou, M. and Pujol, H. (1993). In: *Spiruline* algue de vie. (Eds. Doumenge, F., Durand-Chastel, H. and Toulemont, A.) **12:** 117-120.

Fox, D. (1993). **12:** 179-185.

Fox, R. (1998). In: Algal Biotechnology (Eds. Stadler, *et al.*,) Proceedings of the 4th International Meeting of the SAA. Elsevier Applied science, London–New York. 355-364.

Gonzalez, M., Parra, O. and Cifuentes, A. (1995). *Chile* 219-250.

Hayashi, K. (1999b). *J. Nat. Prod.* **59:** 83-87.

Henrikson, R. (1994). Microalga Spirulina, superalimento del futuro. Ronore Enterprises. 2nd Ediciones Urano, Barceloan, Espana. 222.

Hirabyashi, S. and Guevarra, H.T. (1999). Dome type photobioreactor (Bio-dome) for mass cultivation of microalgae. Intl. conf. Apalgo, Italy.

Hirahashi, T., Matsumoto, M., Hazeki, K., Saeki, Y., Ul, M. and Seya, T. (2002). *Int. Immunopharmacol* **2:** 423-434.

Hirano, T. (1985). *J. Tokyo Univ. Fish.*, **72:** 21.

Hoshaw, R. and Rosowski, J. (1979). *In:* Handbook of hycological methods. Culture methods and growth measurements. (Ed. Stein, J.) Cambridge University Press. 53-60.

Jourdan, J.P. (1993). *In: Spiruline* algue de vie. (Eds. Doumenge, F., Durand-Chastel, H., Toulemont, A.) Bulletin de l'Institut Oceanographique Monaco. Musee Oceanographique. Numero speial. **12:** 191-194.

Kaji, T., Fujiwara, Y., Inomata, Y., Hamada, C., Yamamoto, C., Shimada, S., Lee, J.B. and Hayashi, T. (2002). *Life Sci.* **70:** 1841.

Kato. T. (1989). Cherry salmon fed Spirulinak, Dainippon Ink and Chemicals, Inc., July, 1989, Tokyo, Japan, unpublished paper.

Kerby, N. and Rowell, P. (1992). *In:* Photosynthetic prokaryotes. (Eds. Mann, N. and Carr, N.) Plenum Press. 93-120.

Liao, W.L., Nur-E-Borhan, S.A., Okada, S., Matsui, T. and Yamaguchi, K. (1993). *Nippon suisan gakkaishi*, **59:** 165.

Mori, T. (1987). *Bull. Jap. Soc. Fisheries Sci.*, **53:** 133.

Nandeesha, M.C., Gangadhara, B., Manissery, J.K. and Venkataraman, L.V. (2001). *Bioresou Technol.* **80:** 117-200.

Pascaud, M. (1993). *In:* Spiruline algue de vie. (Eds. Doumengue, F., Durand-Chastel, H. and Toulemont, A.) **12:** 49-57.

Pirt, S.J., Lee, Y.K., Walach, M.R., Pirt, MoW., Balyuzi, H.H. and Basin, M.J. (1982). *J. Chem. Tech. Biotech.*, **33:** 35-58.

Pugh, N., Ross, S.A., Elsohly, H.N., Elsohly, M.A and Pasco, D.S. (2001). *Planta Med.* **67:** 737-42.

Puyfoulhoux, G., Rouanet, J.M., Besancon, P., Baroux, B., Baccou, J.C. and Caporiccio, B. (2001). *J. agric food Chem.* **49:** 1625-29.

Ramirez, D., Gonzalez, R., Merino, N., Rodriguez, S. and Ancheta, O. (2002). *Mediators Inflamm.* **11:** 75-90.

Ratafia, M. and Purinton, T. (1989). *Aquaculture Magazine* **6:** 32.

Riccardi, G., Sora, S. and Cifferri, O. (1981). *J. Bacteriol.*, **147:** 1002.

Richmond, A. (1986). C.R.C Hand Book of Micro algal mass cultivation. Boca Raton Florida, USA.

Richmond, A. (1992). *In: Photosynthetic prokaryotes* (Ed. Mann). 2nd ed. Plenum Press, New York and London. 181-210.

Richmond, A. (1995). *J. Arid. Land Studies* **55:** 199-202.

Rippka, R. (1988). *Meth. Enzymol.* **67:** 3-28.

Samson. R. and Leduy, A. (1985). *Can. J. Chem. Eng.* **63:** 105-112.

Saravanamuthu, R. (2007). *In:* Biodiversity and Environmental Biotechnlogy. (Eds Dwivedi, P. Dwivedi, S.K. and Kalita, M.C.) Scientific Publishers India. pp. 417-436.

Saxena, P.N., Ahmad, M.R., Shyam, R., Srivastava, H.K., Doval, P. and Sinha, D. (1982). *Avian research.* **66:** 41.

Saxena, P.N., Ahmad, M.R., Shyan, R. and Amla, D.V. (1983). *Experientia* **39:** 1077-1083.

Schwartz, J. and Shklar, G. (1987). *J. Oral Maxillofac. Surg.* **45:** 510-515.

Schwartz, J., Shklar, G., Reid, S. and Trickler, D. (1988). *Nutr. Cancer* **11:** 127-134.

Shekharam, K., Ventakaraman, L. and Salimath, P. (1987). *Phytochem.* **26:** 2267-2269.

Shklar, G. and Schwartz, J. (1988). *Eur. J. Cancer Clin. Oncol.* **24:** 839-850.

Singh, S., Patel, R. and Data, P. (2001). *In:* Algal Biotechnology. (Ed. Trivedi, P.C.) Pointer Publ., India. 80-90.

Suyama, M. (1985). *Jap. Soc. Fisheries Sci.*, **51:** 286.

Thein, M. (1993), Production of *Spirulina* in Myanmar. **12:** 175-178.

Tomaselli, L. (1997). *In: Spirulina platensis* (*Arthrospira*): Physiology, cell biology and biotechnology. (Ed. vonshak, A.) Taylor and Francis. London. 1-16.

Toyomizu, M.S., Sato, K., Taroda, H., Kato, T. and Akiaba, Y. (2001). *Br. Poul. Sci.* **42:** 197-202.

Tredici, M.R., Carlozzi, P., Gehini Zittelli, and Materassi, R. (1991). *Bioresource Tech.*, **38:** 153-160.

Tredici, R., Margheri, M.C., De Phillippis, R., Bocci, F. and Materassi, R. (1988). *Int. J. Solar Energy*, **6:** 235.

Upasani, C.D., Khera, A. and Balaraman, R. (2001). *Indian J. Exp. Boil.* **39:** 70.

Venkataraman, L.V. (2001). *In:* Algal biotechnology (Ed. Trivedi, P.C.) pp. 268-276. Pointer Pub. India.

Venkataraman, L.V. and Becker, E.W. (1986). The Indian Experience Dept. Sci. Technol., New Delhi and CFTRI, Mysore. 1-25.

Venkataraman, L.V. Suvarna Latha, G., Krishnakumari, M.K. and Plus Joseph, (1994). *J. Fd. Sci. and Technol.* **31:** 430-432.

Venkataraman, L.V. (1989). *Phykos* **28:** 231-330.

Vincenzini, M., Sili, C., Philippis, R., Ena, A. and Materassi, R. (1990). *J. Bacteriol.* **172:** 2791-2792.

Vonshak, A. (1984). In: *Handbook of microalgal mass culture.* (Ed. Richmond, A.) CRC Press. 117-134.

Vonshak, A. (1987b). *Hydrobiologia.* **83**.

Zarouk, C. (1966). Ph.D. Thesis, University of Paris, France.

Zhou,B., Liu, W., Qu, W. and Tseng, C.K. (1991). *Bioresource Technol.*, **38:** 229.

Chapter 4

Microbial Biopesticides

S. Manian, and K. Sowndhararajan
Department of Botany, Bharathiar University, Coimbatore–641 046
E-mail: manian_sellamuthu@yahoo.com

The term 'microbial biopesticides' means any mass produced and marketed microbial agents directed against pathogens, pests and weeds. These microbial agents may be antagonists, parasites, predators, competitors or may induce resistance in plants. The biopesticides are generally named as biofungicides, bioinsecticides, bioherbicides, etc. depending on the type of pests targeted. The biopesticides hail from diverse groups of microorganisms such as viruses, bacteria and fungi. The present review highlights the potential of an array of microorganisms that are effective as biopesticides against various target organisms.

INTRODUCTION

A pesticide is any substance intended to control, destroy, repel or attract a pest. Pesticide is a broad term referring to a range of products used to control pests. Under the Food and Environment Protection Act (1985), 'a pesticide is any substance, preparation or organism prepared or used, among other uses, to protect plants or woods or plant products from harmful organism, to regulate the growth of plants for protection against harmful creatures or to render such creatures harmless'. Agriculture and public health sectors use pesticides to combat various pests and diseases that affect crop plants and man. The actual production of pesticides commenced in 1952 with the manufacture of benzene hexachloride (BHC), followed by dichlorodiphenyltri-chloroethane (DDT). Since then, production of pesticides has increase tremendously. In 1958, India produced over 5000 metric tonnes of pesticides, especially insecticides like DDT and BHC (HCH). In the mid-nineties, about 145 pesticides were registered and the production was approximately 85,000 metric tonnes. Even today, the bulk of pesticide production is insecticides (Anonymous, 2002). India is presently the second largest manufacturer of basic pesticides in Asia and ranks 12th globally. Despite such a large consumption of pesticides, it is estimated that crop losses vary between 10–30% due to pests alone. In monetary terms, these losses amount to Rs. 290,000 million per year (Agnihotri, 1999).

The worldwide consumption of pesticides is about two million tonnes per year, of which 24% is consumed in the USA alone, 45% in Europe and 25% in the rest of the world. India's share is just 3.75%. The usage of pesticides in India is only 0.5 kg/ha, while in Korea and Japan, it is 6.6 and 12.0 kg/ha, respectively. Currently, the pesticides are being used on 25% of the cultivated areas. The three commonly used pesticides, HCH (only gamma-HCH is allowed), DDT and malathion account for 70% of the total pesticides consumption. Out of the total pesticides consumed in India, 80% are insecticides, 15% are herbicides, 1.46% fungicides and less than 3% are others. In comparison, the worldwide consumption of herbicides is 47.5%, insecticides 29.5%, and fungicides 17.5% and others account for 5.5% (Gupta, 2004).

BIOPESTICIDES

Most of the chlorinated, non-degradable chemical pesticides leave residues in various living systems for prolonged periods of their lifespan and are presumably responsible for a variety of known and unknown toxic symptoms. Even when present in minute quantities, their variety, toxicity, and persistence have an adverse effect on ecological systems such as birds, fish, and trees with which human welfare is inseparably bound (Gupta, 1985, 1986). Furthermore, the repeated use of such chemicals encourages the development of resistance in the target organisms (Goldman *et al.,* 1994), and has a negative effect on beneficial organisms. In addition, some chemical pesticides may predispose plants to greater damage from root pathogens (Cook *et al.,* 1996). Pesticides may also leach into and pollute groundwater supplies. Such concerns are driving the search for more environment friendly methods to control plant disease that will contribute to the goal of sustainability in agriculture. The increasing demand for a steady food supply to the growing world population will require controlling of plant diseases that reduce crop yield substantially. The current practices for controlling plant diseases are based largely on development of resistant varieties and application of synthetic pesticides (Emmert and Handelsman 1999). It is desirable to replace chemical pesticides with materials that possess the following three criteria: high specificity against the targeted plant pathogens, easy degradability after effective usage, and low cost of mass production. Products produced biologically or the microbial cells themselves, if they fulfill these criteria, are called biological control (biocontrol) agents or biological pesticides. However, of the approximately 80 commercial biocontrol agents (BCAs) available today for use against crop diseases (Paulitz and Belanger, 2001), many of them seldom meet all the said requirements. Biological control of plant pathogens involves harnessing disease-suppressive microorganisms to improve plant health (Handelsman and Stabb, 1996). It seems to be a reliable alternative to chemical fungicides, which have raised serious concerns of food contamination and environmental pollution. Besides, biocontrol is eco-friendly, safe and may provide long-term protection to the crop.

The term biopesticide is misleading (Crump *et al.,* 1999) as the use of biological agents for pest control need not necessarily be complete pest eradication, but rather suppression, thus allowing the crop to adequately develop such that the deleterious effects of the pests are minimal and do not affect crop yield. Biopesticides mean 'any mass produced and marketed natural enemy, including predators, parasitoids, nematodes and microbial agents' (Waage, 1997). Here, biological control is achieved by regulation of a species which has reached the level of a pest by another living organism added to the environment to suppress pest densities. So it is ecologically based management of pests by naturally occurring, or genetically modified enemies. Natural enemy choice will vary greatly depending on the target species. Apart from predators, insects have many natural pathogens. These include bacteria, fungi, nematodes and viruses. These pathogens when applied artificially as microbial pesticides may effectively suppress pests. The population of microbial pathogens and/or predators increases because pests are used as a nutritional source and this, in turn, leads to the gradual decline of pest population. Biological control can be potentially permanent because the natural enemies supplied from outside will establish themselves in the pest population and are likely to exert long-term protection against the target pest species (Szewczyk *et al.*, 2006).

Biofungicides

Biological control of the fungal diseases may be achieved directly by addition of antagonists or indirectly by modifying the environment to selectively favour antagonists (Leggett, 1982). Microorganisms living in vascular tissue of plants or in rhizosphere are in an appropriate position to protect the plant from deleterious organisms. The mechanism involved in the action of these microorganisms used against plant pathogens includes parasitism, cross protection, antibiosis and competition. Competition for growth space and nutrients by the beneficial bacteria can indirectly protect the plant from harmful microorganisms (Idriss *et al.*, 2002). Bacteria can produce a wide array of antibiotics that can be effective against many different fungal and bacterial pathogens. Peptide antibiotics represent the predominant class of bacterially produced antibiotics (Handelsman and Stabb, 1996).

Bacteria as Biofungicide

Saprotrophic bacteria with the ability to antagonize phytopathogens and thus prevent plant disease represent a realistic alternative to chemical fungicides. Because of their catabolic versatility, excellent root-colonizing abilities, and capacity to produce a wide range of antifungal metabolites, fluorescent pseudomonads have received particular attention (Walsh *et al.*, 2001). In addition, some *Pseudomonas* BCAs have been shown to elicit a disease-resistance response in crop species, a phenomenon known as induced systemic resistance (ISR) reviewed by Van Loon *et al.* (1998) and Pieterse and Van Loon (1999). This dual activity of *Pseudomonas* BCAs (i.e., direct antagonism to phytopathogens and induction of disease resistance in the host plants) highlights their potential as plant protection products.

Pseudomonas fluorescens, *P. putida* and *P. cepacia* are the predominant focus of research for practical applications. *Pseudomonas* spp. produce a wide range of antibiotics, which include phenazine (Thomashow and Weller, 1988), pyrrolnitrin (Kanner *et al.*, 1978), oomycin A (Howell and Stipanovic, 1979), geldanamycin (Rothrock and Gottlieb, 1984), pyroluteorin (James and Gutterson, 1986), pyocyanin (Gutterson *et al.*, 1988) and 2,4-diacetylphloroglucinol (phl) (Vincent *et al.*, 1991). The high frequency of pseudomonads containing the Phl biosynthetic locus isolated from take-all suppressive soils relative to take-all conducive soils suggests that 2,4-diacetylphloroglucinol producing pseudomonads play a role in soils naturally suppressive to *Gaeumannomyces graminis* var. *tritici* (Raaijmakers *et al.*, 1997). Given the key role of antimicrobial metabolite production in biocontrol, strategies for increasing the biocontrol efficacy of BCAs based on genetic modification for enhanced anti-microbial metabolite production have been investigated. Maurhofer *et al.* (1994) demonstrated that increased Phl and pyoluteorin production by *P. fluorescens* CHA0, via the introduction of the recombinant cosmid pME3090 bearing a 22 kb insert of CHA0 DNA, resulted in increased protection of cucumber from *Fusarium oxysporum* f.sp. *cucumerinum* and *Phomopsis sclerotioides*. Significantly, the antimicrobial metabolite overproducing strain proved to be phytotoxic to both tobacco and sweet corn. The role of antibiotics in biocontrol of plant diseases has been reviewed extensively by many researchers (Fravel, 1988; Thomashow and Weller, 1990; Dowling and O'Gara, 1994; Handelsman and Stabb, 1996; Dwivedi and Johri, 2003).

Siderophores are extracellular, low-molecular-weight compounds, which have a high affinity for ferric ions and play a role in the biocontrol (Leong, 1986). Some *Pseudomonas* strains were confirmed to suppress the activities of plant pathogens by the production of siderophore or pseudobactin (Kloepper *et al.,* 1980, 1986). Bacterial siderophores are also known to have functions to help plant growth such as supplying iron via an iron-chelating function in PGPR (Plant Growth Promoting Rhizobacteria) (Leong, 1986; Loper, 1988; Crowley *et al.,* 1991), which eventually leads to a decrease in disease occurrence. Siderophores are produced by most Gram-positive and Gram-negative bacteria and virtually all fungi (Neilands and Leong, 1986). Probably all fluorescent pseudomonads growing on low-iron (LI) substrate produce pseudobactin-like siderophores (Neilands and Leong, 1986), some of which are implicated in inhibition of root pathogens by sequestering iron (Kloepper *et al.,* 1980). *P. fluorescens* strain 3551 produces a fluorescent siderophore that contributes to its antagonism of *Pythium ultimum* (Loper, 1988).

Understanding which environmental factors are important and how these influence disease suppression is widely recognized as a key to improving the level and reliability of biocontrol (Siddiqui and Ehteshamul-Haque, 2001). HCN is a secondary metabolite commonly produced by rhizosphere pseudomonads (Cipollone *et al*., 2007). HCN production by *P. fluorescens* accounts for part of the strain's biocontrol capacity, for example the suppression of tobacco black root rot caused by *Thielaviopsis basicola*, a black root-rot fungus (Keel *et al.,* 1989; Laville *et al.,* 1998). The biosynthesis of HCN has been demonstrated in a small number of bacterial species, such as *Pseudomonas aeruginosa*, *P. fluorescens* and *Chromobacterium violaceum* (Castric, 1975; Askeland and Morrison, 1983; Knowles and Bunch, 1986). Cyanogenesis is known to be influenced by several environmental factors including iron, phosphate and oxygen concentrations (Knowles and Bunch, 1986). Iron sufficiency has been identified as an important factor for both HCN production and disease suppression (Keel *et al.,* 1989; Voisard *et al.,* 1989; Siddiqui and Shaukat, 2002). Likewise, in batch cultures of *P. aeruginosa* and *P. fluorescens*, HCN production is optimally induced when oxygen is growth-limiting (Castric, 1983; Laville *et al.,* 1998). In *Pseudomonas* spp., HCN biosynthesis is catalyzed by the membrane-bound enzyme HCN synthase, which forms HCN and CO_2 from glycine (Castric, 1977). The enzyme is sensitive to molecular oxygen and has been purified only partially from a *Pseudomonas* species (Wissing and Andersen, 1981). In *P. fluorescens*, the structural *hcnABC* genes encoding HCN synthase are clustered and probably form an operon (Laville *et al.,* 1998).

Bacillus subtilis is a Gram-positive, spore forming saprophyte capable of growing anaerobically when nitrate is available as a terminal electron acceptor (Priest, 1993). If *B. subtilis* were able to utilise nitrate under soil conditions where oxygen was limited, then this may indicate a niche where the role of *B. subtilis* in the soil has not previously been considered or exploited. Many strains of *B. subtilis* have shown the potential to be used as biocontrol agents against fungal pathogens (Fravel, 1988). Work presented by many workers suggests that the principal mechanism of this antifungal action involves the production of antibiotics (Loeffler *et al.,* 1986; McKeen *et al.,* 1986; Fravel, 1988), especially within soil microsites. However, it is likely that several mechanisms are acting in concert to achieve control including the production of volatiles, which have a significant effect on soil microbiology (Linderman and Gilbert, 1975). Production of volatiles by *B. subtilis* strains have been shown to antagonize a range of organisms including the soil-borne plant pathogens *Rhizoctonia solani* and *P. ultimum* (Wright and Tjompson, 1985; Fiddaman and Rossal, 1993). Fiddaman and Rossal (1994) have shown, for selected strains of *B. subtilis*, that active antifungal volatiles (AFV) are produced on a range of growth media and soils and the activity of these volatiles can be enhanced by adding complex carbohydrates (e.g., d-glucose and peptones) to the media.

The use of *Bacillus* species as a biocontrol agent has been relatively rare and has been studied less intensively than that of the Gram-negative *Pseudomonas*. The antagonists studied are mainly *B. subtilis* and occasionally *B. megaterium*, *B. cereus*, *B. pumilus*, and *B. polymyxa* (Aldrich and Baker, 1970; Utkhede, 1984; Clardy and Handelsman, 1994). The population dynamics and mechanism of suppression of plant pathogens in soil by *Bacillus* spp. have not been extensively investigated. *B. subtilis* produces several kinds of antibiotics, e.g., bacillomycin (Peypoux *et al.,* 1980; Chevanet, *et al.,* 1985), iturin (Delcambe *et al.,* 1977; Peypoux *et al.,* 1978a, 1978b), mycosubtilin (Peypoux *et al.*, 1986), bacilysin (Roger *et al.*, 1965; Loeffler *et al.*, 1986), fengymycin (Roger *et al.,* 1965), mycobacillin (Majumdar and Bose, 1970) and a siderophore 2,3-dihydroxybenzoylglycine (2,3-DHBG). The strains of *B. subtilis* which showed a wide suppressive spectrum on plant pathogens by producing antibiotics, also produce 2,3-DHBG (Ito and Neilands, 1958; Leong, 1986).

Alcaligenes faecalis demonstrated the ability to suppress the growth of *F. oxysporum*, via the production of a siderophore (Martinetti and Loper, 1992). However, *A. faecalis* no. 4 which showed a significant suppressive effect on the growth of 13 kinds of plant pathogens *in vitro*, produced hydroxylamine during heterotrophic nitrification (Honda *et al.,* 1999). When culture broth, cell suspension and supernatant of no. 4 were applied to sterilized soil containing *R. solani,* the damping-off of tomato seedlings caused by *R. solani* was suppressed significantly. However, in unsterilized soil, damping-off was suppressed only by the culture broth and the cell suspension treatments, presumably because hydroxylamine in the supernatant was decomposed in soil. When transposon mutagenesis was applied to no. 4 to produce a non-hydroxylamine-producing mutant, this mutant showed no suppressive effect on the growth of *R. solani* either *in vitro* or *in vivo*. These data indicate that the hydroxylamine produced by no. 4 is involved in the suppression of the plant disease (Shoda, 2000).

Streptomyces spp. have been tried as broad-spectrum biological control agents for multiple pathogens in diverse cropping systems including potato, alfalfa, soybean, and corn. Streptomycetes are common filamentous bacteria that are effective, persistent soil saprophytes and often are associated with plant roots. They are well-known producers of antibiotics and extracellular hydrolytic enzymes. These characteristics, as well as their ability to withstand desiccation and high temperatures as spores, make them attractive biological control agents. Several strains are available as commercial products (Nemec *et al.,* 1996). A naturally-occurring potato scab suppressive soil was found to have high population densities of antibiotic-producing, non-pathogenic streptomycetes (Lorang *et al.,* 1995; Liu *et al.,* 1996). These antibiotic producing strains have strong *in vitro* inhibitory activity against a wide variety of plant-pathogenic bacteria and fungi (Liu, 1992; Jones and Samac, 1996; Xiao *et al.,* 2002). Inoculation of soil with individual strains controlled potato scab in greenhouse and field experiments (Liu *et al.,* 1995). They also controlled root rot caused by *Aphanomyces euteiches* Dresh. and *Phytophthora medicaginis* Hansen on alfalfa in growth chamber and greenhouse studies (Jones and Samac, 1996; Xiao *et al.,* 2002) and reduce population densities of the root-lesion nematode (*Pratylenchus penetrans* (Cobb) Filipjev) on alfalfa roots (Samac and Kinkel, 2001). Field experiments have shown control of *Septoria* leaf spot of hybrid poplar by foliar applications of single *Streptomyces* strains originating from the suppressive soil (Shimizu, 1994; Gyenis, 2000). *Streptomyces* spp. have the potential to contribute to management of soil-borne and foliar plant pathogens on diverse plant hosts. However, their value in a multiple crop, integrated disease management system depends not only on their abilities to control plant disease, but also on their ability to colonize plant host surfaces and their potential inhibitory effects on beneficial microbes. In particular, because of their production of broad spectrum antimicrobial compounds, *Streptomyces* spp. may have negative impacts on the growth and performance of nitrogen-fixing mutualistic bacteria (Samac *et al.*, 2003).

Fungi as Biofungicide

Trichoderma spp. have been successfully used for biocontrol of pathogens on plant surfaces of cruciferous, solanaceous and gramineous plants (Rai and Singh, 1980; Scharen and Bryan, 1981; Kumar and Singh, 1985; Trosnmo, 1986; Elad and Kirschner, 1993; Sutton and Peng, 1993; Michereff *et al.,*1995). The diversity of mechanisms available to *Trichoderma* spp. for pathogen suppression (e.g., production of a wide range of broad-spectrum antifungal metabolites, mycoparasitism, and competition with pathogen for nutrients and for occupation of the infection court and crop residue) make these fungi attractive biocontrol agents. Much of the research concerning *Trichoderma* species plant interactions centers on the concept of *Trichoderma* species as common soil inhabitants (Harman *et al.*, 2004). *Trichoderma harzianum* has been effective against soil-borne *R. solani* (Fagardo *et al.,* 1998), *Fusarium* spp. (Chet *et al.,* 1981), *Pythium* spp. (Hagedorn and Rand, 1974) and *Phytophthora cactorum,* improving plant stand and growth (Chet and Baker, 1981). Intensive research into biocontrol with *T. harzianum* has been carried out under commercial conditions, and there have been some significant achievements on greenhouse crops and in vineyards (Elad *et al.*, 1994; Elad and Shtienberg, 1995). The first biocontrol agent to be commercialized, registered and used for greenhouse crops and vineyards was isolate T39 of *T. harzianum* (TRICHODEX 20P), which effectively controlled *Botrytis* diseases in greenhouse crops and in vineyards (Elad, 1994; Elad *et al.*, 1994). It was also reported to control powdery mildew (Elad *et al.*, 1998). Increasing evidence indicates that certain *Trichoderma* species have significant potential for biocontrol of plant diseases in the plant canopy where environmental conditions, microbial communities, and plant tissues differ greatly from those encountered in the root zone (Wilson 1997). For example, *T. stromaticum*, an epiphytic mycoparasite, has already proven to be beneficial for control of *C. perniciosa*, the causal agent of witches' broom of cacao (Bastos 1996; Samuels *et al.* 2000).

Fungi such as *Coniothyrium minitans* (Campbell, 1947), *Trichoderma viride* (Huang, 1980), *Talaromyces flavus* (McLaren *et al.*, 1994), *Trichothecium roseum* (Huang and Kokko, 1993), and *Epicoccum purpurascens* (Zhou and Reeleder, 1989) are known as mycoparasites or antagonists capable of attacking sclerotia or affecting mycelial growth of *Sclerotinia sclerotiorum* under natural conditions. Previous reports indicated that soil treatment with *C. minitans* was effective in reducing incidence of sunflower (*Helianthus annuus* L.) wilt (Huang, 1980; Bogdanova *et al.*, 1986; McLaren *et al.*, 1994) and lettuce (*Lactuca sativa* L.) drop (Budge and Whipps, 1991) caused by *S. sclerotiorum*. Zhou and Reeleder (1989) found that application of *E. purpurascens* significantly reduced white mold incidence of snap bean. Other reports indicated that foliar application of *C. minitans* significantly reduced viability of sclerotia of *S. sclerotiorum* produced on diseased bean plants (Trutmann *et al.*, 1982; Fokkema *et al.*, 1992).

In addition to the commonly studied fungi of the genera *Trichoderma*, *Gliocladium*, and *Penicillium* for the biocontrol of soil borne plant pathogens (Abd El-Moity and Shatta, 1981; Legett, 1982; Abd-Elrazik *et al.*, 1985; Hassan, 1992; Kay and Stewart, 1994; Mclean and Stewart, 2000; Payghami *et al.*, 2001; Metcalf and Wilson, 2001; Amin, 2003), there is an array of microfungi with biocontrol potential which has not been extensively investigated. The reasons for this include difficulty in isolation and problems with large scale growth and formulation. With advances in various aspects of biotechnology, attention is being devoted to biocontrol studies with nonpathogenic *Fusarium* (Alabouvette *et al.*, 1993), binucleate and non-pathogenic *Rhizoctonia* (Sneh, 1996), non-pathogenic and mycoparasitic isolates of *Pythium* (Fang and Tsao, 1995) *Verticillium* (Jager and Velvis, 1985), *Coniothyrium* (Whipps and Gerlagh, 1992), *Sporidesmium* (Adams and Fravel, 1990), *Laetisaria* (Lewis and Papavizas, 1992), and *Stilbella* (Lewis and Papavizas, 1993). Raju *et al.* (1999) reported the reduction of *Fusarium moniliforme*-induced diseases of maize by application of *T. harzianum, P. fluorescens* and *Chaetomium globosum*.

The plant surface-inhabiting filamentous fungi and yeasts have the potential to suppress plant diseases (Zhou and Reeleder, 1989; Madrigal *et al.*, 1994; Petersson and Schnurer, 1998; Helbig, 2002; Calvo *et al.*, 2003; Lima *et al.*, 2003). Yeasts are particularly attractive as biological control candidates as they are widely used for various food and industrial purposes and thus considerable information is available with respect to techniques for genetic manipulation, production and storage of yeast cells (Hofstein *et al.*, 1994; Hamilton *et al.*, 2003). The mechanism of action of biocontrol yeasts is not clear. Competition with pathogens for nutrients on the plant surface is one likely mode of disease suppression but mycoparasitism and production of antifungal compounds are other possible mechanisms (Suzzi *et al.*, 1995; Cook *et al.*, 1997; Filonow, 1998; Buck, 2002).

Bioinsecticides

A number of bioinsecticides (bacteria, fungi, virus, pheromones and plant extracts) have already been in use to control various types of insects responsible for the destruction of forests and agricultural crops. Botanical and microbial insecticides are highly effective, safe, and ecologically acceptable. Plants produce a diversity of biologically active substances that affect the growth and development of other organisms and can provide protection against herbivory. These plant products discourage or prevent attack from non-adapted organisms and play an important role in the ecology and physiology of phytophagous insects (Sukumar, 1993). The application of naturally occurring microorganisms with the ability to control or suppress a pest population has attracted modest research over the past several decades. Microbial pesticides are probably the most widely used and cheaper than the other methods of pest bioregulation. Insects can be infected with many species of bacteria but the species belonging to the genus *Bacillus* are the most widely used pesticides. Of these, *Bacillus thuringiensis* is the most successful. It is a Gram-positive, spore-forming bacterium with parasporal crystals. *B. thuringiensis* has developed an array of molecular mechanisms to produce pesticidal toxins, most of them coded by several *cry* genes (Agaisse and Lereclus, 1995). An extensive number of fungi are associated with a variety of insects and other arthropods, establishing diverse interactions, including the pathogenicity. Approximately 750 species belonging to 100 genera of entomopathogenic fungi have been reported, but only about ten of them have been or are now being developed for insect control (Hajek and St. Lager, 1994). Fungi belonging to the genera *Beauveria* and *Metarhizium* have been found to be effective against many insect pests including African locust (Langewald *et al.*, 1999). Some nematodes, mostly those belonging to families Steinerrnematidae and Heterorhabditidae, are successfully used to control insects (Blackburn *et al.*, 1998). Viruses of fifteen families are also known to infect insects but only those belonging to the highly specialized family Baculoviridae have been considered as potential pesticides. The viruses are very safe to people and wildlife. Their specificity is very narrow, often limited to only one species (Szewczyk *et al.*, 2006).

Entomopathogenic Fungi as Bioinsecticide

The entomopathogenic fungi can be classified into two main groups: the biotrophic fungi and necrotrophic fungi. The former fungi require living cells of their hosts, some of them being commensals that obtain nutrients from the digestive tract of the insect. This class of fungi is widespread in many regions, but are not generally used for pest control since they are either asymptomatic in insects or the changes caused by pathogens are difficult to observe (Lacey and Kaya, 2000). Necrotrophic fungi have to kill their hosts before consuming it as they are used to live at the expense of dead cells. Fungi belonging to this group can attack insects of the orders Coleoptera, Lepidoptera, Hymenoptera, Hemiptera, Orthoptera, Homoptera, Diptera and the attack can occur in several stages of their life cycles (Goettel and Johnson, 1995). Several toxic compounds have been isolated from some species of fungi belonging to the genera *Beauveria* and *Metarhizium, Nomuraea, Aspergillus, Verticillium, Paecilomyces, Isaria, Fusarium, Cordyceps, Entomophthora*. One of the substances well studied is the beauverin, a peptide isolated from *Beauveria bassiana*, with activity against mosquito larvae (Uribe and Khachatourians, 2004).

Entomogenous fungi have been used successfully to control various agricultural and pasture pests. In Brazil, they have been sprayed in large fields to control sugarcane pests (Gillespie and Claydon, 1989), whereas in Indonesia and Malaysia, they are used to control the rhinoceros beetle, a serious pest of oil palms (Munaan and Wikardi, 1986). In Australia, entomogenous fungi have been used to control the subterranean pasture pest, *Adoryphorus couloni* (Rath, 1992;

Rath *et al.*, 1995a,) and field tests have produced promising results with other Australian pests such as the plague locust, *Phaulacridium vittatum*, and the wingless grasshopper, *Chortoicetes terminifera* (Milner *et al.*, 1994; Hooper *et al.*, 1995). Entomogenous fungi have also been found to induce high mortality in tsetse flies (Kaaya, 1989; Kaaya and Munyinyi, 1995) and evidence of horizontal transmission from infected to uninfected flies has been demonstrated (Kaaya and Okech, 1990). Mycopesticides are currently being developed for control of the African locusts and grasshoppers by the Commonwealth Agricultural Bureau International (CABI) and the International Institute of Tropical Agriculture (IITA) (Lomer *et al.*, 1997).

Baculoviruses as Bioinsecticide

Baculoviruses are the major group of arthropod viruses well known due to their potential as agents of biological control of pests in agriculture and forestry. The family Baculoviridae comprises two genera: the Nucleopolyhedrovirus (NPVs) and the Granulovirus (GVs) (Van Regenmortel *et al.*, 2000). Particular virus strains have limited host ranges and can infect one or a few species, with no effect on non-target arthropods (Black *et al.*, 1997; Inceoglu *et al.*, 2001). Wild-type baculoviruses are highly specific for insects and have been applied to protect crops in the agricultural sector since 1930s (Inceoglu *et al.*, 2001). In Brazil over two million hectares of soybean are protected by baculovirus AgMNPV (Moscardi, 1999; Moscardi and Santos, 2005). The success of Brazilian project revived the interest in baculovirus as a biopesticide and gradually many countries have begun to increase the area of fields and forests to be experimentally protected by baculovirus pesticides.

Baculoviruses infect arthropods but they do not replicate in vertebrates, plants and microorganisms. However, under special conditions, they may enter animal cells. This unexpected property made them a valuable tool in the last few years for studies of transient expression of foreign genes under vertebrate promoters introduced into baculovirus genome (Boyce and Bucher, 1996; Kost *et al.*, 2005). Historically, the first viral insecticide Elcark was introduced by Sandoz Inc. in 1975 (Ignoffo and Couch, 1981). It was a preparation of *Heliothis zea* NPV. HzSNPV infects many species belonging to genera *Helicoverpa* and *Heliothis* which include many important pests attacking a wide range of crops (Chakraborty *et al.*, 1999). *Anticarsia gemmatalis* nucleopolyhedrovirus (AgMNPV) was used to control the velvet bean caterpillar in soybean (Moscardi, 1999; Moscardi *et al.*, 2002). The genera *Autographa* and *Trichoplusia* played a major role in the development of baculovirus expression system but, on the other hand, NPVs which attack members of this genera were also extensively studied as potential biopesticides (Copping and Menn, 2000). *Autographa californica* and *Anagrapha falcifera* NPVs were registered in the USA and field-tested at a limited scale. These two NPVs have relatively broad host spectrum and can be used on a variety of crops infested with *Spodoptera exigua*, *Trichoplusia ni*, *Heliothis zea* and many other pests (Hostetter and Putler, 1991; Vail *et al.*, 1993; Cunningham, 1995). Even codling moth, *Cydia pomonella*, a common pest of orchards in temperate zone is susceptible to these two broad range baculoviruses (Lacey *et al.*, 2002) though the susceptibility is much lower than the granulovirus CpGV of *C. pomonella*, its natural host. CpGV is the active component of at least five products commonly used for protection of orchards in Europe against codling moth (Carpovirusinek in France, Madexk and Granupomk in Switzerland, Granusalk in Germany, Virin-CyAP in Russia). The importance of *C. pomonella* is recognized worldwide, hence the use of CpGV gradually increases and may reach 100,000 ha annually sprayed with this virus (Bellotti, 1999).

Bacteria as Bioinsecticide

Bacillus thuringiensis (Bt) is a Gram-positive, spore forming bacterium that has insecticidal properties (also called "entomotoxicity") affecting a selective range of insect orders, namely, Lepidoptera, Diptera and Coleoptera (Burges, 1998; Chapple *et al.,* 2000). The *B. thuringiensis* gene codes for production of a protein toxin that attacks the insect midgut, stops feeding and eventually kills several species of Lepidoptera of economic importance to agriculture (Gill *et al.,* 1992; Fischhoff, 1996). Bt have no significant activity against beneficial insects, wildlife or humans (Meadows, 1993).

Screening of Bt products based on the spore crystal complex has been conducted since the 1950s, but it was only in 1970 that the HD-1 strain of Bt subspecies *kurstaki,* which is active against a large number of moth species, was discovered (Dulmage, 1970) and produced commercially by many agrochemical and fermentation companies (Baum *et al.,* 1999; Shah and Goettel, 1999). Screening programs by means of potency bioassays (Navon, 2000) resulted in highly potent new strains. Between 1970 and 1990, strains of Bt subspecies *aizawai,* active against armyworms such as *Spodoptera* spp., and subspecies *tenebrionis* (Krieg *et al.*, 1983) and *san-diego* (Hernstadt *et al.,* 1986) active against beetles were introduced into the market, and these new products extended the means of control to more insects. It has been shown in lepidopterous insects that the spores potentiated (Moar *et al.,* 1989) and synergized (Dubois and Dean, 1995) the insecticidal activity of the crystal protein. Improved knowledge of the molecular biology of Bt (Cannon, 1996) led to the use of Bt genes for manipulation of the Bt toxin (Baum *et al.,* 1999).

The *Photorhabdus* and *Xenorhabdus* bacteria live in symbiosis with nematodes that actively seek out and physically penetrate the insect cuticle (Forst and Clarke, 2002). In the case of *Photorhabdus,* once inside the insect the bacteria are physically regurgitated from the infective juvenile nematode (Ciche and Ensign, 2003). Once free inside the insect blood system (hemocoel) the bacteria then replicate and release a range of toxins to kill the insect host (Ffrench – Constant *et al.,* 2003). The toxin complexes (Tc's) are large orally active toxins that are displayed on the outer surface of the bacterium. One such group of Tc's, the PirAB binary toxins, have oral activity against mosquitoes and some caterpillar pests (Ffrench – Constant *et al.,* 2007).

Bioherbicides

A bioherbicide is defined as a plant pathogen used as a weed-control agent through inundative and repeated applications of its inoculum. Daniel *et al.* (1973) introduced the concept of bioherbicide and demonstrated that an endemic pathogen might be rendered completely destructive to its weed host by applying a massive dose of inoculum at a particular vulnerable stage of weed growth. There is considerable current interest in the use of plant pathogenic microbes as agents for the biological control of weeds. One potential approach to control weeds is to use phytotoxins or their derivatives for direct application to the noxious plants (Boyette and Abbas, 1995; Milat and Blein, 1995; Scheepens *et al.,* 1997; Evidente *et al.*, 1998; Auld and Mc Rae, 1999; Mohan Babu *et al.,* 2003).

In the United States and many other countries, the prescriptive use of plant pathogens as weed control agents is regarded as 'pesticidal use' and therefore these pathogens must be registered or approved as biopesticides by appropriate governmental agencies. Currently, five fungi and one bacterium are registered as bioherbicides in Canada, Japan, South Africa, and the United States (Charudattan, 2000). DeVinet, composed of a Florida isolate of *Phytophthora palmivora*, is used for the control of *Morrenia odorata* (strangler vine or milkweed vine) in citrus in Florida. Collegot, based on *Colletotrichum gloeosporioides* f.sp. *aeschynomene*, is used to control *Aeschynomene virginica* (northern jointvetch), a leguminous weed in rice and soybean crops in Arkansas, Mississippi, and Louisiana. BioMalt, registered in Canada for the control of *Malva pusilla* (round-leaved mallow), containing *Colletotrichum gloeosporioides* f.sp. *malvae*, is presently unavailable for commercial use. The fourth bioherbicide, Dr. BioSedget, based on the rust fungus *Puccinia canaliculata* and registered for the control of *Cyperus esculentus* (yellow nutsedge) in the United States, is also unavailable for commercial use. An isolate of *Xanthomonas campestris* pv. *poae*, a wilt-inducing bacterium, isolated in Japan from *Poa annua* (annual bluegrass or wintergrass), is registered in Japan as the bioherbicide CAMPERICOT to control annual bluegrass in golf courses. A stump-treatment product based on the wood-infecting basidiomycete, *Cylindrobasidium laeve*, under the commercial name Stumpoutt, is registered in South Africa to control resprouting of cut trees in tree plantations and in natural areas. In addition, about five or six other unregistered bioherbicides are used on small scales on regional basis in China, the Netherlands, South Africa, and the United States (Charudattan, 2000). Presently, the following pathogens are undergoing precommercial evaluations in the United States and other countries: *Alternaria destruens* for the control of dodder (*Cuscuta* spp.) in cranberries and carrots, *Ascochyta caulina* (for the control of common lambsquarters, *Chenopodium album*), *Colletotrichum truncatum* (for hemp sesbania, *Sesbania exaltata*), *Dactylaria higginsii* (for purple nutsedge, *Cyperus rotundus*), *Exserohilum monoceras* (for barnyard grass, *Echinochloa crus-galli* and other species), *Fusarium* spp. (for witchweed *Striga hermonthica* and other species), *Fusarium graminearum* (for *Egeria* spp.), *Pseudomonas syringae* pv. *tagetis* (for various weeds in the Asteraceae and a few other families), *Phomopsis amaranthicola* (for several pigweed and amaranth species, *Amaranthus* spp.), *Ralstonia solanacearum* (for tropical soda apple, *Solanum viarum* in Florida and *Hedychium gardnerianum* in Hawaii), an unnamed fungus (for dandelion *Taraxacum ozcinale* in turf), *Cercospora* nr. *caricis*, *Dactylaria higginsii* and *Puccinia romagnoliana* (for purple nutsedge) and *Colletotrichum gloeosporioides* f.sp. *cuscutae (*for dodder species, and others) (Dinoor *et al.,* 1994a, b, 1999; Muller-Scharer and Scheepens, 1997; Nof *et al.,* 1999; Rosskopf *et al.,* 1999).

The most successful examples of classical biological control of weeds is the introduction of a rust fungus, *Puccinia chondrillina*, into Australia to control rush skeleton weed (*Chondrilla juncea*). Other successful examples of classical biocontrol programs include the use of a smut fungus, *Entyloma ageratinae*, imported from Jamaica to control Hamakua pamakani (*Ageratina riparia*, Asteraceae) in Hawaiian forests and rangelands (Trujillo, 1985) and three other rust fungi, *Puccinia carduorum*, imported from Turkey and released into northeastern United States to control musk thistle, *Carduus thoermeri* (Baudoin *et al.*, 1993), *Phragmidium violaceum* to control weedy species of *Rubus* in Chile (Oehrens, 1977) and Australia (Bruzzese, 1995), and *Uromycladium tepperianum*, to control an introduced invasive tree species, *Acacia saligna* in South Africa (Morris, 1997). The fungus *Alternaria eichhornia*. (Nag Raj and Ponnappa, 1970; Elwakil *et al.,* 1990; Shabana, 1997a, b, 2000; Mohan Babu *et al.,* 2002), which infect water hyacinth has been extensively studied. This pathogen caused severe disease on water hyacinth and suppressed the entire growth of the older plant. Another fungus *Fusarium oxysporum* is well suited as a mycoherbicide (McCain and Noviello, 1984; Boyette *et al.*, 1993; Sands *et al.*, 1997). The pathogen is host-specific and will not negatively impact other plant species. This host-specificity allows the mycoherbicide to be applied to weed infestations within a crop or within a natural area. In addition, *F. oxysporum* can survive in the soil for many years, preventing re-establishment of the weed population. *F. oxysporum* was first evaluated as a mycoherbicide for control of *Opuntia megacantha* (tree cactus) in Hawaii in the 1940s (Fullaway, 1954; Wilson, 1969). Field inoculations of this pathogen were discontinued with the introduction of insect agents. Several other formae specialis of *F. oxysporum* are being developed as mycoherbicides for the control of noxious weeds or narcotic crops, including *F. oxysporum* f. sp. *orthoceras* for control of broomrape (*Orobanche* spp.) (Bedi, 1992; Thomas *et al.*, 1998; Thomas *et al.*, 1999) and *F. oxysporum* f. sp. *erythroxyli* for control of illicit coca (Sands *et al.*, 1997; Nelson *et al.*, 1997). A formae specialis of *F. oxysporum* was also investigated for control of witchweed (*Striga hermonthica*) in Africa (Abbasher *et al.*, 1995; Kroschel *et al.*, 1996; Marley *et al.*, 1999; Ciotola *et al.*, 2000).

Parthenium weed (*Parthenium hysterophorus* L.: Asteraceae), accidentally introduced into central Queensland as a contaminant of pasture seed from the United States of America in 1958, has become a serious rangeland weed (Haseler, 1976). The weed is an aggressive colonizer of disturbed land and has rapidly spread from the point of introduction to an area of more than 170,000 km^2 (Haseler, 1976; McFadyen, 1992). The pollen of *P. hysterophorus* also causes allergic rhinitis and bronchitis in humans (Lonkar *et al.,* 1974). Chemical control of this central Queensland infestation is now not economical (Holman and Dale, 1981), while mechanical control is inefficient (Haseler, 1976). Hand pulling is not recommended due to the potential of the weed to cause contact dermatitis (Anonymous, 1993). Therefore, biological control has been considered to be a major, long-term option for parthenium weed control in this region (Haseler, 1976). Several insect biological control agents have been released into central Queensland since 1978 to help suppress this weed, but none has been totally successful. *Epiblema strenuana* Walker (Lepidoptera: Tortricidae), a stem-galling moth, has been shown to seriously damage the weed (McClay, 1987); however, the level of control is reduced due to the insect being unable to cope with lengthy dry periods, a common event in the central Queensland environment (Anonymous, 1995). Consequently, a search for non-insect biological control agents such as plant pathogens has begun in recent years. The rust fungus *Puccinia abrupta* Diet. and Holw. var. *partheniicola* (Jackson) Parmelee from the semiarid, upland regions (1400–1600 m above sea level) of Mexico is considered to be a potential biological control agent for parthenium weed (Evans, 1987). Infection by this pathogen causes considerable damage by both reducing vegetative growth of young plants and by reducing seed production of older plants (Evans, 1987). In addition, the rust is very host specific (Parker *et al.,* 1994), infecting only parthenium weed and a close relative not present in Australia, *Parthenium confertum* L. Since the release of this biological control agent in central Queensland, it has become established but has not spread widely. *Puccinia abrupta* var. *partheniicola* is not thought to be well adapted to the long, hot, dry summer periods of this region of Australia.

Some herbicides and growth regulators with specific chemistry and molecular modes of action have been shown to act synergistically with biocontrol agents to enhance weed control (Christy *et al.*, 1993). Sharon *et al.* (1992) demonstrated that the suppression of biosynthesis of a phytoalexin in *Senna* (*Cassia obtusifolia* (L.)) by the herbicide glyphosate at sublethal rates led to subsequent increase in susceptibility of this weed to the mycoherbicide *Alternaria cassiae*. Another example of synergy between biopesticides and chemicals is in the application of *Puccinia canaliculata* (Schw.) Lagerh. in combination with the herbicide paraquat to achieve 99% control of *Cyperus esculentus* L., compared to only 60% control with the rust alone (Phatak, 1984). Timing of herbicide application might also be a major factor in integrating biocontrol pathogens and other chemicals in attempts to find synergy. Tank-mixture of a biocontrol agent and agricultural chemicals can be detrimental to the living organism but a sequential application may prevent inhibition of the pathogen's growth or activity. In the case of *Colletotrichum gloeosporioides* (Penz.) Sacc. f. sp. *aeschynomene*, marketed as the microbial herbicide Collego for the control of *Aeschynomene virginica* L., the tank-mixture of the fungus with the herbicides propanil and 2,4,5-T, and the fungicide fentin hydroxide inhibited fungal colonization. Although there was no synergy in this system, infection and disease development were successful when the pathogen was applied before the chemical herbicides and the fungicide (Smith, 1986). The weed pathosystem *Senecio vulgaris* L. - *Puccinia lagenophorae* Cooke was chosen as a research model to test a system-management approach of biological weed control (Muller-Scharer and Frantzen, 1996; Muller-Scharer and Scheepens, 1997). This approach aims at the reduction of competitiveness of the annual weed *S. vulgaris* (common groundsel) by stimulating disease epidemics with a relatively small initial inoculum source (Frantzen and Hatcher, 1997; Frantzen and Muller-Scharer, 1998). A survey carried out in horticultural crops, in nurseries and ruderal sites in Switzerland indicated that despite abundant use of pesticides, especially fungicides, *P. lagenophorae* still occurred at 60% of the sites investigated (Wyss, 1997). These fungicides were applied over the whole growing season and rust incidence was observed predominantly in early autumn. The rust fungus either is not controlled by fungicides or survives locally and the disease becomes epidemic later in the year. Wyss and Muller-Scharer (1999) have demonstrated the existence of race-nonspecific quantitative resistance in *S. vulgaris* to *P. lagenophorae*. This resistance is manifested during the most sensitive transition in the infection process, namely at the stage of penetration-peg formation when the fungal development was most reduced.

CONCLUSION

A successful biocontrol agent should have the ability to survive, establish and maintain a threshold population in the introduced environment below which their efficacy is impaired or insufficient. It is an established fact that the introduced biocontrol agents may not perform well in different locations due to various factors like poor viability, sensitivity and dessication and other fluctuating environmental conditions (Powell and Jutsum, 1993). However, the general belief that is proved to some extent is that the native strains do well in controlling pests and improving plant growth than the introduced strains. As such, use of microbial consortia is gaining momentum in recent days as in the combination treatment, even if one organism failed, the other organism(s) possibly will take care of the biological control of the target pest. At the same time, enhanced pest suppression could be achieved if both or all the organisms in the combination are active against the pest. The efficacy of biocontrol can further be enhanced by preparing the microbial inoculants with the addition of selected ingredients and preparing the final product in a form (formulation) that is specific for pathogen–crop complex.

REFERENCES

Abbasher, A.A., Kroschel, J. and Sauerborn, J. (1995). *Biocontrol Sci. and Technol.* **5:** 157-162.

Abd El-Moity, T.H. and Shatla, M.N. (1981). *Phytopathol.* **100:** 29-35.

Abd El-Razik, A.A., El-Shabrawy, A.M., Sallam, M.A. and Abd-El-Rahim M.H.A. (1985). *Egyptian J. Phytopathol.* **17:** 95-105.

Adams, P.B. and Fravel, D.R. (1990). *Phytopathol.* **80:** 1120-1124.

Agaisse, H. and Lereclus D. (1995). *J. Bacteriol.* **177:** 6027-6032.

Agnihotri, N.P. (1999). All India Coordinated Research Project on Pesticides Residues, Indian Council of Agricultural Research, New Delhi.

Alabouvette, C., Lemanceau, P. and Steinberg, C. (1993). *Pestic. Sci.* **37:** 365-373.

Aldrich, J. and Baker, R. (1970). *Plant Dis, Reptr.* **54:** 446-448.

Amin, M.M.M. (2003). Ph.D. Thesis, Ain Shams University, Egypt. pp. 105.

Anonymous (1993). "Pestfact: Parthenium Weed (*Parthenium hysterophorus*)." Department of Land, Brisbane, Australia.

Anonymous. (1995). Hopes hinge on new "improved" beetle. *Queensland Graingrower.*

Anonymous. (2002). *Pestic. Inf.* **28(3):** 52-55.

Askeland, R.A. and Morrison, S.M. (1983). *Appl. and Environ. Microbiol.* **45:** 1802-1807.

Auld, B.A. and McRae, C. (1999). *In:* Proceedings of New Zealand Plant Protection Society (NZPPS), New Zealand, pp. 1-4.

Bastos, C.N. (1996). *Fitopatol. Bras.* **21:** 509-512.

Baum, J.A., Timothy, B.J. and Carlton, B.C. (1999). *In:* Biopesticide Use and Delivery. (Eds. Hall, F.R. and Menn, J.J.), Humana Press, NJ. USA, pp. 189-209.

Baudoin, A.B.A.M., Abad, R.G., Kok, L.T. and Bruckart, W.L. (1993). *Biol. Control* **3:** 53-60.

Bedi, J. (1992). *Haustorium* **27:** 1-2.

Bellotti, A.C. (1999). *Annu. Rev. Entomol.* **44:** 343-70.

Black, B.C., Brennan, L.A, Dierks, P.M. and Gard, I.E. (1997). *The baculoviruses. New York7 Plenum;* pp. 341-87.

Blackburn, M., Golubeva, E., Bowen, D. and Ffrench-Constant, R.H. (1998). *Appl. Environ. Microbiol.* **64:** 3036-41.

Bogdanova, V.N., Karadzhova, L.V. and Klimenko, T.F. (1986). *Sel'skokhozyaistvennaya Biologiya (Agric. Biol.)* **5:** 80-84.

Boyce, F.M. and Bucher, N.L.R. (1996). *Proc Natl Acad Sci. USA.* **93:** 2348-52.

Boyette, C.D., Abbas, H.K. and Connick, W.J. (1993). *Weed Science* **41:** 678-681.

Boyette, C.D. and Abbas, H.K. (1995). *In:* Allelopathy: Organisms, Processes, and Applications (Eds. Inderjit, K.M., Dakshini, M. and Einhellig, F.A.), pp. 280-299.

Bruzzese, E. (1995). *In:* Proceedings of the Eighth International Symposium on Biological Control of Weeds. (Eds Delfosse, E.S. and Scott, R.R.), *DSIR/CSIRO. Melbourne, Australia*, pp. 297-299.

Buck, J.W. (2002). *Can. J. Botany* **80:** 885-891.

Budge, S.P. and Whipps, J.M. (1991). *Plant Pathol.* **40:** 59-66.

Burges H.D. (1998). *Formulation of microbial biopesticides: beneficial organisms, nematodes and seed treatments.* Dordrecht, The Netherlands: Kluwer Academic Publishers.

Calvo, J., Calvente, V., de Orellano, M.E., Benuzzi, D. and de Tosetti, M.I.S. (2003). Improvement in the biocontrol of postharvest diseases of apples with the use of yeast mixtures. *Bio Control* **48:** 579-593.

Campbell, W.A. (1947). *Mycologia* **39:** 190-195.

Cannon, R.J.C. (1996). *Biol. Rev.* **71:** 561-636.

Castric, P.A. (1975). *Can. J. Microbiol.* **21:** 613-618.

Castric, P.A. (1977). *J. Bacteriol.* **130**: 826-831.

Castric, P.A. (1983). *Can. J. Microbiol.* **29:** 1344-1349.
Chakraborty, S., Monsour, C., Teakle R. and Reid S. (1999). *J. Invertebr Pathol.* **73:** 199-205.
Chapple, A.C., Downer, R.A. and Bateman, R.P. (2000). *In:* Lacey LA, Kaya HA, editors. Field manual of techniques in invertebrate pathology. Dordrecht: Kluwer. pp. 5-37.
Charudattan, R. (2000). *In:* Emerging Technologies for Integrated Pest Management: Concepts, Research, and Implementation. (Eds. Kennedy, C.G. and Sutton, T.B.) APS Press, St. Paul, MN, pp. 269-288.
Chen, J., Gao, H.M., Lin, R.M., Ji, M.S. and Gao, Z.G. (1999). Res. Prog. *Plant Protect. Plant Nutr.* 78-84.
Chet, I. and Baker, R. (1981). *Phytopathol.* **71:** 286-290.
Chet, I., Harman, G.E. and Backer, R. (1981). *Microbiol. Ecol.* **7:** 29-38.
Chevanet, C., Besson, F. and Michel, G. (1985). *Can. J. Microbial.* **32:** 255-258.
Christy, A.L., Herbst, K.A., Kostka, S.J., Mullen, J.P. and Carlson, P.S. (1993). *In:* "Pest Control with Enhanced Environmental Safety" (Eds. Duke, S.O., Menn, J.J. and Plimmer, J.R.), *ACS Symp. Ser.* **524:** 87-100.
Ciche, T.A. and Ensign, J.C. (2003). *Appl. Environ. Microbiol.* **69:** 1890-1897.
Ciotola, M., Ditommaso, A. and Watson, A.K. (2000). *Biocontl. Sci. and Technol.* **10**: 12-145.
Cipollone, R., Frangipani, E., Tiburzi, F., Ascenzi, P. and Visca, P. (2007). *Appl. Environ. Microbiol.* **739(2):** 390-398.
Clardy, J. and Handelsman, J. (1994). *Appl. Environ. Microbial.* **60:** 2023-2030.
Cook, D.W.M., Long, P.G., Ganesh, S. and Cheah, L.H. (1997). *Ann. Appl. Biol.* **131:** 503-518.
Cook, J.R., Bruckart, W., Coulson, J.R., Goettel, S., Humber, A., Lumsden, R.D., Maddox, J.V., McManus, M.L., Moore, L., Meyer, S.F., Quimby, P.C., Stack, J.P. and Vaughn, J.L. (1996). *Biol. Control* **7:** 333-351.
Copping, L.G. and Menn, J.J. (2000). *Pest Manag Sci.* **56:** 651-76.
Crowley, D.E., Wang, Y.C., Reid, C.P.P. and Szaniszlo, P.J. (1991). *Plant Soil,* **130:** 179-198.
Crump, N.S., Cother, E.J. and Ash, G.J. (1999). *Biocontl. Sci. and Technol.* **9:** 89-97.
Cunningham, J.C. (1995). *In:* Reuveni R, *editor. Novel approaches to integrated pest management.* Boca Raton, FL7 Lewis, pp. 261-92.
Daniel, J.T., Templeton, G.E., Smith Jr., R.J. and Fox, W.T. (1973). *Weed Sci.* **21:** 303-307.
Delcambe, L., Peypoux, F., Besson, F., Guinand, M. and Michel, G. (1977). *Biochem. Society Transac.,* **5:** 1122-1124.
Dinoor, A., Boyle, C., Aust, A.H. and Eshed, N. (1994a). *Phytoparasitica* **22:** 83-84.
Dinoor, A., Ronen, Y., Eshed, N., Kleifeld, Y. and Zilberstaine, M. (1994b). *Phytoparasitica* **22:** 84.
Dowling, D.N. and O'Gara, F. (1994). *Trends Biotechnol.* **12:** 133-141.
Dubois, N.R. and Dean, D.H. (1995). *Environ. Entomol.* **24:** 1741-1747.
Dulmage, H.D. (1970). *J. Invertebr. Pathol.* **15:** 232-239.
Dwivedi, D. and Johri, B.N. (2003). *Curr. Sci.* **12:** 1693-1703.
Elad, Y. and Shtienberg, D. (1995). *Integrated Pest Management Reviews* **1:** 15-29.
Elad, Y. (1994). *Crop Protect.* **13:** 35-38.
Elad, Y., Kirshner, B., Nitzani, Y. and Sztejnberg, A. (1998). *Biocontl.* **43:** 241-251.
Elad, Y., Shtienberg, D. and Niv, A. (1994). *British Crop Protection Conference Pests and Diseases,* BCPC, Farnham, pp. 1109-1113.
Elad,Y. and Kirshner, B. (1993). *Phytoparasitica* **21:** 303-313.
Elwakil, M.A., Sadik, E.A., Fayzalla, E.A. and Shabana, Y.M. (1990). *In:* Proceedings of the VIII International Symposium on Biological Control of Weeds. Lincoln University, Canterbury, New Zealand, pp. 483-497.
Emmert, E.A.B. and Handelsman, J. (1999). *FEMS Microbiol. Lett.* **171:** 1-9.
Evans, H.C. (1987). *Trans. Br. Mycol. Soc.* **88:** 105-111.
Evidente, A., Capasso, R., Cutignano, A., Taglialatela-Scafati, O., Vurro, M., Chiara Zonno, M. and Motta, A. (1998). *Phytochemistry* **48:** 1131-1137.
Fagardo, J.E., Backman, P.A. and Wells, I.W. (1998). *Biol. Cultural Tests,* 147.
Fang, J.H. and Tsao, P.H. (1995). *Phytopathol.* **85:** 29-36.
Ffrench-Constant, R., Waterfield, N., Daborn, P., Joyce, S., Bennett, H., Au, C., Dowling, A., Boundy, S., Reynolds, S. and Clarke, D. (2003). *FEMS Microbiol. Rev.* **26:** 433-456.
Ffrench-Constanta, R.H., Andrea Dowlingb, A. and Waterfield, N.R. (2007). *Toxicon* **49:** 436-451.
Fiddaman, P.J. and Rossal, S. (1993). *J. Bacteriol.* **74:** 119-126.
Fiddaman, P.J. and Rossal, S. (1994). *J. Appl. Bacteriol.* **76:** 395-405.
Filonow, A.B. (1998). *Biocontrol Sci. Technol.* **8:** 243-256.
Fischhoff, D.A. (1996). In: *Biotechnology and Integrated Pest Management,* (Ed. Persley, G.J.) pp. 214-227. CAB International, Wallingford, UK.
Fokkema, N.J., Gerlagh, M. and Köhl, J. (1992). *In:* "Biological Control of Plant Diseases" (Eds. Tjamos, E. *et al.,*) pp. 267-271. Plenum, New York.

Forst, S. and Clarke, D. (2002). *In:* (Ed.Gaugler, R.), *Entomopathogenic Nematology.* CAB International, London, pp. 57-77.

Frantzen, J. and Hatcher, P.E. (1997). *Integr. Pest Manage. Rev.* **2:** 77-85.

Frantzen, J. and Muller-Scharer, H. (1998). *Phytopathol.* **88:** 180-184.

Fravel, D.R. (1988). *Ann. Rev. Phytopathol.* **26:** 75-91.

Fullaway, D.T. (1954). *J. Eco. Entomol.* **47:** 696-700.

Gill, S.S., Cowles, E.A. and Pietrantonio, P.V. (1992), *Annu. Rev. Entomol.* **37:** 615-636.

Gillespie, A.T. and Claydon, N. (1989). *Pesticide Sci.* **27:** 203-215.

Goettel, M.S., Johnson, D.L. and Inglis, G.D. (1995). *Can. J. Bot.* **73(Suppl. 1):** S1-S75.

Goldman, G.H., Hayes, C. and Harman, G.E. (1994). *Trends in Biotech* **12:** 478-482.

Gupta, P.K. (2004). Pesticide exposure - Indian Scene. *Toxicol.* **198:** 83-90.

Gupta, P.K. (1985). In: *Modern Toxicology.* (Eds. Gupta, P.K. and Salunkhe, D.K.) Metropolitan Book Company, New Delh*i,* pp. 1-60.

Gupta, P.K. (1986). Pesticides in the India environment. *Interprint, New Delhi,* pp. 1-206.

Gutterson, N., Ziegle, J.S., Warren, G.J. and Layton, T.J. (1988). *J. Bacteriol.* **170:** 380-385.

Gyenis, L. (2000). Ph.D. Thesis, University of Minnesota, St. Paul, MN, 91 pp.

Hagedorn, D.J. and Rand, R.E. (1974). *Ann. Rep. Bean Improve. Coop.* **17:** 28-31.

Hajek, A.E. and St. Lager, R.J. (1994). *Annu. Rev. Entomol.* **39:** 293-322.

Hamilton, S.R., Bobrowicz, P., Bobrowicz, B., Davidson, R.C., Li, H., Mitchell, T., Nett, J.H., Rausch, S., Stadheim, T.A., Wischnewski, H., Wildt, S. and Gerngross, T.U. (2003). *Science* **301:** 1244-1246.

Handelsman, J. and Stabb, E.V. (1996). *The Plant Cell* **8:** 1855-1869.

Harman, G.E., Howell, C.R., Viterbo, A. and Chet, I. (2004). *Nat. Res.* **2:** 43-56.

Haseler, W.H. (1976). *PANS* **24:** 515-517.

Hassan, M.H.A. (1992). Ph.D. Thesis, Assuit University. p. 167. Available from Fac. Agric, Assiut University, Egypt.

Helbig, J. (2002). *Bio Control* **47:** 85-99.

Hernstadt, C., Soares, G.G., Wilcox, E.R. and Edwards, D.L. (1986). *Biotechnol.* **4:** 305-308.

Hofstein, R., Fridlender, B., Chalutz, E. and Droby, S. (1994). In: (Eds. Wilson, C.L. and Wisniewski, M.E), *Biological Control of Postharvest Diseases.* CRC Press Inc, Boca Raton, Florida. pp. 89-100.

Holman, D.J. and Dale, I.J. (1981). *Queensland Agric. J.* **107:** 57-60.

Honda, N., Hirai, M., Ano, T. and Shoda, M. (1999). *Ann. Phytopathol. Sot. Jpn.* **65:** 153-162.

Hooper, G.H.S., Milner, R.J., Spurgin, P.A. and Prior, C. (1995). *J. Aust. Ent. Soc.* **34:** 83-84.

Hostetter, D.L. and Putler, P. (1991). *Environ. Entomol.* **20:** 1480-1488.

Howell, C.R. and Stipanovic, R.D. (1979). *Phytopathol.* **69:** 480-482.

Huang, H.C. (1980). *Can. J. Plant Pathol.* **2:** 26-32.

Huang, H.C. and Kokko, E.G. (1993). *Can. J. Bot.* **71:** 1631-1638.

Idriss, E.E., Makarewicz, O., Farouk, A., Rosner, K., Greiner, R., Bochow, H., Richter, T. and Borriss, R. (2002) *Microbiol.* **148:** 2097-2109.

Inceoglu, A.B., Kamita, S.G., Hinton, A.C., Huang, Q., Severson, T.F. and Kang K. (2001). *Pest Manag. Sci.* **57:** 981-987.

Ito, T. and Neilands, J.B. (1958). *J. Am. Chem. Sot.* **80:** 4645-4647.

Jager, G. and Velvis, H. (1985). *Neth. J. Plant Pathol.* **91:** 49-63.

James, D.W. and Gutterson, N.I. (1986). *Appl. and Environ. Microbiol.* **52:** 1183-1189.

Jones, C.R. and Samac, D.A. (1996). *Biol. Control* **7:** 196-204.

Kaaya, G.P. (1989). *Acta Trop.* **46:** 107-114.

Kaaya, G.P. and Munyinyil, D. (1995). *J. Invert. Pathol.* **66:** 237-241.

Kaaya, G.P. and Okech, M.A. (1990). *Entomophaga* **35:** 46-57.

Kanner, D., Gerber, N.N. and Bartha, R. (1978). *J. Bacteriol.* **134:** 690-692.

Kay, J. and Stewart, A. (1994). *Plant Pathol.* **43:** 371-377.

Keel, C., Voisard, C., Berling, C.H., Kahr, G. and Defago, G. (1989). *Phytopathol.* **79:** 584-589.

Kloepper, J.M., Laverry, R. and Lozano, J.C. (1986). *J. Phytopathol.* **117:** 17-25.

Kloepper, J.W., Leong, J., Teintze, M. and Scroth, M.N. (1980). *Nature* **286:** 885-886.

Knowles, C.J. and Bunch, A.W. (1986). *Ad. Microbial Physiol.* **27:** 73-111.

Kommendahl, T. and Chang, I.P. (1975). *Phytopathol.* **65:** 296-300.

Kost, T.A., Condreay, J.P. and Jarvis, D.L. (2005). *Nat. Biotechnol.* **23:** 567-75.

Krieg, A., Huger, A.M., Langenbruch, G.A. and Schnetter, W. (1983). *Bacillus thuringiensis* var. *tenbrionis*: Ein neuer, gegenuK ber Glarven von Coleoptaran Wirksamer Pathotype. *Z. angew. Entomol.* **96:** 500-508.

Kroschel, J., Hundt, A., Abbasher, A.A. and Sauerborn, J. (1996). *Weed Research* **36:** 515-520.

Kumar, K. and Singh, H. (1985). *Indian J. Mycol. Plant Pathol.* **13:** 24-27.
Lacey, L.A. and Kaya, H.K. (2000). Application and evaluation of pathogens for control of insects and other invertebrate pests. Dordrecht7 Kluwer
Lacey, L.A, Vail P.V. and Hoffmann, D.F. (2002). *J. Invertebr. Pathol.* **80:** 64-68.
Langewald, J., Ouambama, Z., Mamadou, A., Peveling, R., Stolz, I. and Bateman, R. (1999). *Biocont. Sci. Technol.* **9:** 199-214.
Laville, J., Blumer, C., Von Schroetter, C., Gaia, V., De´fago, G., Keel, C. and Haas, D. (1998). *J. Bacteriol.* **180:** 3187-3196.
Leggett, M. (1982). Proceedings of the 29th annual meeting, Canadian Pest Management Society; 12-15, July, 1982. Vancouver, British Columbia, (C.F. CAB abstracts 1984-1986).
Leong, J. (1986). *Ann. Rev. Phytopathol.* **24:** 187-209.
Lewis, J.A. and Papavizas, G.C. (1992). *Soil Biol. Biochem.* **27:** 1075-1079.
Lewis, J.A. and Papavizas, G.C. (1993). *Biocontrol Sci. Technol.* **3:** 3-11.
Lima, G.F., Decurtis, R. and Castoria, V. Decicco. (2003). *European J. Plant Pathol.* **109:** 341-349.
Linderman, R.G. and Gilbert, R.G. (1975). In: (ed. Bruehl, G.W.), *Biology and Control of Soil-Borne Plant Pathogens.* American Phytopathological Society, St Paul, MN, pp. 90-99.
Liu, D. (1992). Ph.D. Thesis, University of Minnesota, St. Paul, MN, pp 138.
Liu, D., Anderson, N.A. and Kinkel, L.L. (1995). *Phytopathol.* **85:** 827-831.
Liu, D., Anderson, N.A. and Kinkel, L.L. (1996). *Can. J. Microbiol.* **42:** 487-502.
Loeffler, W., Tschen, J.S.M., Vanittanakcom, N., Kugler, M., Knorpp, E., Hsieh, T.F. and Wu, T.G. (1986). *J. Phytopathol.* **115:** 204-213.
Lomer, C.J., Prior, C. and Kooyman, C. (1997). *Mem. Entomol. Soc. Canada* **171:** 265-286.
Lonkar, A., Mitchell, J.C. and Calnan, C.D. (1974). *Trans. St. John's Hospital Dermatol. Soc.* **60:** 43-53.
Loper, J.E. (1988). *Phytopathol.* **78:** 166-172.
Lorang, J.M., Liu, D., Anderson, N.A. and Schottel, J.L. (1995). *Phytopathol.* **85:** 261-268.
Madrigal, C., Pascual, S. and Melgarejo, P. (1994). *Plant Pathol.* **43:** 554-561.
Majumandar, S.K. and Bose, S.K. (1970). *Arch. Biochem. Biophy.* **90:** 154-158.
Marley, P.S., Ahmed, S.M., Shebayan, J.A.Y. and Lagoke, S.T.O. (1999). *Biocontrol Science and Technology* **9:** 159-163.
Martinetti, G. and Loper, J.E. (1992). *Can. J. Microbial.*, **38:** 241-247.
Maurhofer, M., Keel, C., Haas, D. and Défago, G. (1994). *Eur. J. Plant. Pathol.* **100:** 221-232.
McCain, A.H. and Noviello, C. (1984). *In:* Proceedings of the VI International Symposium on Biological Control of Weeds, Vancouver, Canada (Ed. Delfosse, E.S.). Agriculture Canada, pp. 635-642.
McClay, A.S. (1987). *Entomophaga* **32:** 23-34.
McFadyen, R.E. (1992). *Crop Protect.* **11:** 400-407.
McKeen, C.D., Reilly, C.C. and Pusey, P.L. (1986). *Phytopathol.* **76:** 136-139.
McLaren, D.L., Huang, H.C., Kozub, G.C. and Rimmer, S.R. (1994). *Plant Dis.* **78:** 231-235.
Mclean, K.L. and Stewart, A. (2000). *J. Crop Rot Horticulture Sci.* **28:** 115-122.
Meadows, M.P. (1993). *In:* Bacillus thuringiensis, An Environmental Biopesticide: Theory and Practice, (Ed. Entwistle, P.F., Cory, J.S. Bailey, M.J. and Higgs, S.) pp.193-220. John Wiley, New York.
Metcalf, D.A. and Wilson, C.R. (2001). *Plant Pathol.* **50:** 249-257.
Michereff, N., Da Siveira, S., Reis, A. and Mariano, R. (1995). *Mycopathology* **130:** 103-108.
Milat, M.L. and Blein, J.P. (1995). *J. Chromatogr.* **699:** 277-283.
Milner, R.J., Hartley, T.R., Lutton, G.G. and Prior, C. (1994). *J. Austr. Entomol. Soc.* **33:** 165-167.
Moar, W.J., Trumble, J.T. and Federici, B.A. (1989). *J. Econ. Entomol.* **82:** 1593-1603.
Mohan Babu, R., Sajeena, A., Seetharaman, K., Vidhyasekeran, P., Rangasamy, P., Som Prakash, M., Senthil Raja, A. and Biji, K.R. (2002). *Crop Prot.* **21:** 1083-1085.
Mohan Babu, R., Sajeena, A., Vidhyasekaran, P., Seetharaman, K. and Reddy, M.S. (2003). *Phytoparasitica,* **31(3):** 265-274.
Morris, M.J. (1997). *Biol. Control.* **10:** 75-82.
Moscardi, F., Morales, L. and Santos, B. (2002). *Proceedings of the VIII international colloquium on invertebrate pathology and microbial control and XXXV annual meeting of the Society for Invertebrate Pathology. Brazil7 Foz do Iguassu.* pp. 86-91.
Moscardi, F. and Santos, B. (2005). *Proceedings of the IX Simposio de Controle Biologico. Brazil7 Recife* pp. 42.
Moscardi, F. (1999). *Annu. Rev. Entomol.* **44:** 257-89.
Muller-Scharer, H. and Scheepens, P.C. (1997). *Int. Pest Manage.* **2:** 45-50.
Muller-Scharer, H. and Frantzen, J. (1996). *Weed Res.* **36:** 483-491.

Munaan, A. and Wikardi, E.A. (1986). *In:* Biological Control in the Tropics, M.Y. Hussein, and A.G. Ibrahim (Eds), pp. 149-157, Penerbit Universiti Partanian, Malaysia, Serdang.

Nag Raj, T.R. and Ponnappa, K.M. (1970). *Trans. Br. Mycol. Soc.* **55(1):** 123-130.

Neilands, J.B. and Leong, S.A. (1986). *Ann. Rev. Plant Physiol.* **37:** 187-208.

Nelson, A.J., Elias, K.S., Arevalo, E., Darlington, L.C. and Bailey, B.A. (1997). *Phytopathol.* **87:** 1220-1225.

Nemec, S., Datnoff, L.E. and Strandberg, J. (1996). *Crop Protect.* **15:** 735-742.

Nof, E., Rubin, B. and Dinoor, A. (1999). *Phytoparasitica* **27:** 159-60.

Oehrens, E. (1977). *FAO Plant Protect. Bull.* **25:** 26-28.

Parker, A., Holden, A.N.G. and Tomley, A.J. (1994). *Plant Pathol.* **43:** 1-16.

Paulitz, T.C. and Belanger, R.R. (2001). *Annu. Rev. Phytopathol.* **39:**103-133.

Payghami, E., Massiham, S., Ahary, B.V. and Motallebi, A. (2001). *Acta Agron Hung* **49:** 393-395.

Petersson, S. and Schnurer, J. (1998). *Can. J. Microbiol.* **44:** 471-476.

Peypoux, F., Besson, F. and Michel, G. (1980). *J. Antibiot.* **33:** 1146-1149.

Peypoux, F., Besson, F., Michel, G., Delcambe, L. and Das, B.C. (1978b). *BICHA Biochem.*, **17:** 3992-3996.

Peypoux, F., Guinand, M., Michel, G., Delcambe. L., Das, B.C. and Lederer, E. (1978a). Biochem. **17:** 3992-3996.

Peypoux, F., Pommier, M.T., Marion, D., Ptak, M. and Michel, G. (1986). *J. Antibiot.* **39:** 636-641.

Phatak, S.C. (1984). *Am. Veg. Grower* **32:** 44-46.

Pieterse, C.M.J. and van Loon, L.C. (1999). *Trends Plant Sci.* **4:** 52-58.

Powell, K.A. and Justum, A.R. (1993). *Pesticide Sci.* **37:** 315-321.

Priest, F.G. (1993). *In:* Sonenshine, *Bacillus Subtilis* and other Gram-positive Bacteria; Biochemistry, Physiology and Molecular Genetics. (Eds. Hoch, A.L. and Losick, R.) *American Society for Microbiology, Washington, DC,* pp. 3-16.

Raaijmakers, J.M., Weller, D.M. and Thomashow, L.S. (1997). *Appl. Environ. Microbiol.* **63:** 881-887.

Rai, B. and Singh, D. (1980). *Trans. Br. Mycol. Soc.* **75:** 363-369.

Raju, N.S., Niranjana, S.R., Janardhana, G.R., Prakash, H.S., Shetty, H.S. and Mathur, S.B. (1999). *J. Sci. Food Agric.* **79:** 206-212.

Rath, A.C., Worledge, D., Koen, T.B. and Rowe, B.A. (1995a). *Biocontr. Sci. Tech.* **5:** 439-451.

Rath, A.C. (1992). *In:* The use of pathogens in Scarab Pest Management, (Eds. Jackson, T.A. and Glare, T.R.), pp. 217-227, Andover.

Roger, H.J., Lomalina, N. and Abraham, E.P. (1965). *Biochem. J.* **97:** 579-586.

Rosskopf, E.N., Charudattan, R. and Kadir, J.B. (1999). *In:* Handbook of Biological Control (Eds. Fisher, T.W., Bellows, T.S., Caltagirone, L.E., Dahlsten, D.L., Hu!aker, C. and Gordh, G.) *Academic Press, San Diego, CA,* pp. 891-918.

Rothrock, C.S. and Gottlieb, D. (1984). *Can. J. Microbiol.* **30:** 1440-1447.

Samac, D.A., Willert, A.M., McBride, M.J. and Kinkel, L.L. (2003). *Appld. Soil Ecol.* **22:** 55-66.

Samac, D.A. and Kinkel, L.L. (2001). *Plant Soil* **235:** 35-44.

Samuels, G.J., Pardo-Schultheiss, R.A., Hebbar. K.P., Lumsden, R.D., Bastos, C.N., Costa, J.C. and Bezerra, J.L. (2000). *Mycol. Res.* **104:** 760-764.

Sands, D.C., Ford, E.J., Miller, R.V., Sally, B.K., McCarthy, M.K., Anderson, T.W., Weaver, M.B., Morgan, C.T., Pilgeram, A.L. and Darlington, L.C. (1997). *Plant Disease* **81:** 501-504.

Scharen, A. and Bryan, D. (1981). *Phytopathol.* **71:** 902-903.

Scheepens, P.C., Kempenaar, C., Andreasen, C., Eggers, T., Netland, J. and Vurro, M. (1997). *Integrated Pest Manage. Rev.* **2:** 71-76.

Shabana, Y.M. (1997a). *Z. Pflanzenkr. Pflanzenschutz* **104(3):** 231-238.

Shabana, Y.M. (1997b). *Z. Pflanzenkr. Pflanzenschutz* **104(3):** 239-245.

Shabana, Y.M., Elwakil, M.A. and Charudattan, R. (2000). *Z. Pflanzenkr. Pflanzenschutz* **107:** 617-626.

Shah, P.A. and Goettel, M.S. (Eds.), (1999). *Directory of microbial control products and services.* (Eds. Shah, P.A. and Goettel, M.S.) Society of Invertebrate Pathology, Gainesville, *FL 32614-7050, USA,* pp. 31.

Sharon, A., Amsellem, Z. and Gressel, J. (1992). *Plant Physiol.* **98:** 654-659.

Shimizu, K. (1994). Biological control of *Septoria* leaf spot onhybrid poplars using *Streptomyces* spp. M.S. Thesis, University of Minnesota, St. Paul, MN, 48 pp.

Shoda, M. (2000). *J. Biosci. and Bioengi.* **89(6):** 515-521.

Siddiqui, I.A. and Ehteshamul-Haque, S. (2001). *Plant Soil* **237:** 81-89.

Siddiqui, I.A. and Shaukat, S.S. (2002). *Biology and Fertility of Soils* **36:** 260-268.

Smith, R.J. (1982). *In:* "Biological Control of Weeds with Plant Pathogens" (eds. Charudattan, R. and Walker, H.L.), pp. 189-203. Wiley, New York.

Sneh, B. (1996). *In:* Rhizoctonia species: Taxonomy, Molecular Biology, Ecology, Pathology and Disease Control (Eds.), Sneh, B. *et al.,* pp. 473-483. Kluwer Academic, Dordrecht.
Sukumar, K. (1993). *In:* Chemical Ecology of Phytophagous Insects. (eds. Ananthakrishnan, T.N. and Raman, A) Oxford and IBH, New Delhi, pp. 82-89.
Sutton, J. and Peng, G. (1993). *Ann. Rev. Phytopathol.* **31:** 473-479.
Suzzi, G., Romano, P., Ponti I. and Montuschi, C. (1995). *J. Appl. Bacteriol.* 78, no. **3:** 304-308.
Szewczyk, B., Hoyos-Carvajal, L., Paluzek, M., Skrzecz, I. and de Souza, M.L. (2006). *Biotechnol. Advan.* **24:** 143-160.
Thomas, H., Heller, A., Sauerborn, J. and Muller-Stover, D. (1999). *Annals of Botany* **83:** 453-458.
Thomas, H., Sauerborn, J., Muller-Stover, D., Ziegler, A., Bedi, J.S. and Kroschel, J. (1998). *Biological Control* **13:** 41-48.
Thomashow, L.S. and Weller, D.M. (1988). *J. Bacteriol,* **170:** 3499-3508.
Thomashow, L.S. and Weller, D.M. (1990). *Plant Soil* **129:** 93-99.
Trosnmo, A. (1986). *In: Microbiology of the Phyllosphere.* (Eds. Fokkema, N. and Der Heuvel, J.) Cambridge University Press, Cambridge.
Trujillo, E.E. (1985). *In: Proceedings of the VI International Symposium on Biological Control of Weeds.* (Ed. Delfosse, E.S.). Agriculture Canada, Ottawa, pp. 661-671.
Trutmann, P., Keane, P.J. and Merriman, P.R. (1982). *Trans. Br. Mycol. Soc.* **78:** 521-529.
Uribe, D. and Khachatourians, G.G. (2004). *Mycol. Res.* **108:** 1070-1078.
Utkhede, R.S. (1984). *Can. Bot.* **62:** 1032-1035.
Vail, P.V., Hoffmann, D.F., Street, D.A., Manning, J.S. and Tebbets, J.S. (1993). *Environ. Entomol.* **22:** 110-45.
Van Loon, L.C., Bakker, P.A.H.M. and Pieterse, C.M.J. (1998). *Annu. Rev. Phytopathol.* **36:** 453-483.
Van Regenmortel, M.H.V., Fauquet, C.M., Bishop, D.H.L., Cartens, E.B., Estes, M.K. and Lemon, S.M. (2000). Virus taxonomy. Seventh report of the international committee on taxonomy of viruses. New York 7 Academic.
Vincent, M.N., Harrison, L.A., Brackin, J.M., Kovacevich, P.A., Mukerji, P., Weller, D.M. and Pierson, E.A. (1991). *Appl. and Environ. Microbiol.* **57:** 2928-2934.
Voisard, C., Keel, C., Haas, D. and De´fago, G. (1989). *The EMBO J.* **8:** 351-358.
Waage, J.K. (1997). *In: BCPC Symposium Proceedings Ni.* 68: Microbial Insecticides: Novelty or necessity? pp.11-19.
Walsh, U.F., Morrissey, J.P. and O'Gara, F. (2001). *Curr. Op. Biotechnol.* **12:** 289-295.
Whipps, J.M. and Gerlagh, M. (1992). *Mycol. Res.* **96:** 897-907.
Wilson, M. (1997). *J. Ind. Microbiol. Biotechnol.* **19:** 188-191.
Wilson, C.L. (1969). *Ann. Rev. Phytopathol* **7:** 411-434.
Wissing, F. and Andersen, K.A. (1981). *In:* Cyanide in biology. (Eds. Vennesland, B., Conn, E.E., Knowles, C.J., Westley, J. and Wissing, F.), Academic Press Ltd., London, England, pp. 275-288.
Wright, S.J.L. and Tjompson, R.J. (1985). *FEMS Microbiol. Lett.* **30:** 263-267.
Wyss, G.S. (1997). Ph.D. thesis, ETH Zü rich, Switzerland.
Wyss, G.S. and Muller-Scharer, H. (1999). *Can. J. Bot.* **77:** 361-369.
Xiao, K., Samac, D.A. and Kinkel, L.L. (2002). *Biol. Control.* **23:** 285-295.
Zhou, T. and Reeleder, R.D. (1989). *Plant Dis.* **73:** 639-642.

Chapter **5**

Biotechnological Potential of Cyanobacteria and their Industrial Applications

N. THAJUDDIN

Department of Microbiology, Bharathidasan University, Tiruchirappalli–620 024
E-mail: nthaju2002@yahoo.com

The first use of cyanobacteria by humans dates back 2000 years to the Chinese, who used *Nostoc*, *Spirulina* and *Aphanizomenon* species to survive during famine. However, the large-scale cultivation of useful cyanobacteria is only a few decades old. Indeed, among the 10,000 species of micro algae including cyanobacteria that are believed to exist, only a few thousand strains are kept in culture collections, a few hundred are investigated for chemical content and just a handful are cultivated in industrial qualities. There has been a renewed interest in these organisms because of their possible use in mariculture, food, feed, fuel, fertilizer, colourant, production of various secondary metabolites including vitamins, toxins, enzymes, pharmaceuticals, pharmacological probes and pollution abatement. With the decreasing agriculture water supplies, land resources and other environmental concerns exacerbate the challenges we face to meet the nutritional requirements for the explosive population, we have to turn our attention towards the possibility of not only identifying the potential new cyanobacterial strains from the environment but also to improve the useful strains through genetic manipulations for attaining sustained maximal productivity. Until recently, improvement in yield and other capabilities of cyanobacteria have been achieved by selection of strains and improvements in culturing and processing procedures. Very little work has been done so far by adopting a genetic-engineering approach to improve cyanobacterial production. It is important to transform the cyanobacteria genetically in a stable, facile and predictable manner, in order to explore the full biotechnological potential of these organisms.

INTRODUCTION

Cyanobacteria also referred as blue–green algae, are the pioneer oxygenic phototrophs on earth whose distribution around the world is revealed only by bacteria (Adams, 2000). Fossil evidence points to their presence in geographically diverse regions during the Precambrian (2 to more than 3.5 billion years ago). They are a large and morphologically diverse group of phototrophic prokaryotes with such wide ecological tolerance that they occur in almost every habitat on earth. This versatility may explain the remarkable lack of morphological (and presumably physiological) change to be seen when comparing 3.5 billion years old fossilized cyanobacteria and their modern day counterparts. Their long evolutionary history has been marked by key geochemical and biotic transitions, including the creation of an oxygenic photosynthesis (Holland, 1977), a prerequisite for the development and proliferation of metabolically complex microbial and higher eukaryotic life forms. Indeed, the plastids of higher plants and other photosynthetic eukaryotes are thought possibly to have arisen from a single common ancestor, itself the result of an endosymbiosis between a phagotrophic host and a cyanobacterium (Loffelhandt and Bohnert, 1977).

Several names are commonly used for this organism, including Cyanobacteria, Cyanophyceae, Cyanophytes, Myxophyceae, Cyanoprokaryotes and Blue-green algae. They were originally considered algae because of their microscopic morphology, pigmentation and oxygen evolving photosynthesis in which two photosystems (PS II and PS I) are connected in series. The genome size of cyanobacteria, representative of all major taxonomic groups, lie in the range of 1.6×10^9 to 8.6×10^9 daltons. The majority of unicellular cyanobacteria contain genomes of 1.6×10^9 daltons, comparable in size of those of other bacteria (1.0 to 3.6×10^9 daltons) (Herdman *et al.*, 1979). With the advent of electron microscopy and several other sophisticated instruments and techniques, it was realized that the so called 'blue-green algae' are really Gram-negative bacteria in their cell structure and chemistry but form a connecting link with algae by photosynthetic mechanism. The discovery about the prokaryotic nature of cyanobacteria coupled with the discovery that a number of these organisms are unique in not only having a tropic independence for carbon but also nitrogen, evoked considerable interest on these organisms around the world. Cyanobacteria are known to occur in oxic as well as anoxic environments. Several species can switch to the typical bacterial anoxygenic photosynthesis using sulphide as an electron donor. Other species assimilate sugars and other organic compounds in light (photoheterotrophy) (Rippka, 1972). In dark, cyanobacteria gain energy by respiring endogenous carbohydrates which are accumulated in the light. However, under anoxic conditions some species are able to cover maintenance requirement by fermentation. In few cases chemoorganotrophy has been observed in cyanobacteria. Another important capacity of cyanobacteria is to grow under a very low water potential. Such species can either resist desiccation very well and grow in arid environments such as deserts, or tolerate high salinity and grow in hypersaline ponds (Thajuddin and Subramanian, 1992; Thajuddin *et al.,* 2002).

The cells of cyanobacteria contain only unstacked lamellae (thylakoids) where the electron transport of photosynthesis and respiration occur (Stewart, 1980; Cohen–Bazire and Bryant, 1982). The photosynthetic pigment located in these membranes includes chlorophyll a (a higher plant pigment) and the water-soluble phycobiliprotein pigments, phycocyanin, phycoerythrin and allophycocyanin. These are localized in the cell centre (centroplasm) which contains nucleic acids, ribosomes, cyanophycin (protein) granules and lipid droplets are distributed throughout the cytoplasm. Photosynthetic phosphorylation (producing ATP) occurs via electron transport through photosytem II and photosystem I as in higher plants and chloroplasts (Paden and Cohen, 1982). While water is the main electron donor for cyanobacterial photosynthesis, several species are able to shift between H_2O and H_2S as electron donors. The final electron acceptor in photosynthetic phosphorylation is $NADP^+$ and the NADPH synthesized by this process is used to reduce CO_2 to form carbohydrates by the action of Ribulose-1, 5–bis-phosphate carboxylase and the other Calvin-Benson (C3) cycle enzymes (Smith, 1982). Recent evidence indicates that the phosphoenol pyruvate (PEP) carboxylase may participate in CO_2 fixation in the filamentous N_2 fixing cyanobacterium *Anabaena variabilis* (Lem *et al.*, 1985). The importance of this lies in the observation that this enzyme has kinetic characteristics that are similar to the PEP carboxylase enzyme from C4 higher plants (*Zea mays*), which fixes CO_2 very efficiently and would be important in environments with low CO_2 and HCO_3^- concentrations.

In many environments, cyanobacteria are the primary producers at the base of the food web of ecosystems. This is the case in many marine waters; hyper saline; brackish waters; soda lakes; freshwater; in paddy fields, in soils; in deserts; on cave walls; in hot springs; in polar regions and other extreme environments. Cyanobacteria are also known as symbionts in a variety of other organisms (example – the marine diatom *Rhizosolenia*, leaves of *Azolla* and the roots of *Cycas)* (Rai, 1990). In addition to their widespread geographic distribution, cyanobacteria have probably displayed a major role throughout the biological history of the earth.

POTENTIAL APPLICATIONS IN BIOTECHNOLOGY

Cyanobacteria are considered to be one of the potential organisms which can be useful to mankind in various ways. Though they have a long history of existence, and were till recently in the oblivion uncared and unrecognized, have shot into fame and popularity owing to a host of their innate properties that make them ideal organisms for use in a variety of ways to meet our needs and to promise us a bright future. Cyanobacteria constitute a vast potential resource in varied areas such as mariculture, food, feed, fuel, fertilizer, medicine, industry and combating pollution, since the usefulness of cyanobacteria for these purposes has been established (Emodi, 1978; Mitsui *et al.*, 1981; Venkataraman, 1983; Gustafson *et al.*, 1989; Moore *et al.*, 1984; Carmichael, 1992; Bender, 1994; Patterson, 1996; Prabaharan and Subramanian, 1995; Sundararaman *et al.*, 1996; Kulik, 1995; Subramanian and Uma, 1996; Jaki, 2001).

Food and Feed

Due to the demand for animal protein as food for humans, algal protein either as a supplement or alternative source has received worldwide attention. Cyanobacteria have been used as a food source in various parts of the world since ancient time (Fox, 1985). The cyanobacteria are harvested from alkaline ponds where they grow in abundance and then dried and consumed as a source of proteins and vitamins. In fact, dried cyanobacteria and cyanobacterial tablets are now sold in health food stores in Japan, North America and Europe, because they are recognized for their nutritional value. Cyanobacterial single cell proteins (SCP) have also been considered for use as supplement to livestock feed (Litchfield, 1983). Some strains of *Anabaena* and *Nostoc* are consumed as human food in Chile, Mexico, Peru and Philippines. *Nostoc commune* contains a high amount of fiber and moderate amounts of protein. This suggested its potential use as a new dietary fiber source, which plays an important physiological and nutritional role in human diets (Jeraci and Vansoest, 1986). *Spirulina* is used as a food supplement because of its excellent nutrient composition and better digestibility due to thin cell wall. It has the highest protein content

(60–70%), 20% carbohydrate, 5% lipids, 7% minerals and 6% moisture. It is also a rich source of β carotene, thiamine, riboflavin and is the richest source of vitamin B_{12}. It is commercially available in the market in the form of powder, granules or flakes and as tablets and capsules. The protein of *Spirulina maxima* and *Anabaena cylindrica* is easily digestible and approximately 65% of the protein is assimilatible. The protein amino acid content compares favorably with egg protein, soybean, and a standard reference protein (Volesky *et al.,* 1970; Pirt, 1983; Fox, 1985) although *Spirulina* is deficient in arginine and histidine. Trace elements and iodine found in cyanobacteria are also important when considering the nutritional quality. Despite of high nutritional quality of cyanobacteria, there were data which indicate that some species of cyanobacteria contain very fast acting and deadly toxins. Schantz (1970) have described the biological and chemical properties of toxins produced by *Microcystis aeruginosa, Anabaena flos-aquae, Aphanizomenon flos-aquae*, and *Schizothrix calcicola.* Among the many cyanobacterial secondary metabolites, a number of potent toxins are responsible for a wide array of human illness, mammal and bird morbidity and mortality and extensive fish kills. Therefore, extensive testing of new cyanobacterial species for toxins prior to their use as food is absolutely required. Table 1 presents a comparison of the general compositions of human food sources with that of different micro algae.

Table 1: General composition of different human food sources and algae (percentage of dry matter) (Becker, 2004).

Commodity	Protein	Carbohydrate	Lipid
Baker's yeast	39	38	1
Meat	43	1	34
Milk	26	38	28
Rice	8	77	2
Soyabean	37	30	20
Anabaena cylindrica	43-56	25-30	20
Chlamydomonas rheinhardii	48	17	21
Chlorella vulgaris	51-58	12-17	14-22
Dunaliella salina	57	32	6
Porphyridium cruentum	28-39	40-57	9-14
Scenedesmus obliquus	50-56	10-17	12-14
Spirulina maxima	60-71	13-16	6-7
Synechococcus sp.	63	15	11

In addition to its use in human nutrition, cyanobacteria can be incorporated into the feed for wide variety of animals ranging from fish (aquaculture) to pets and farm animals. In fact, 30% of the current world production of *Spirulina/Arthrospira* is used as feed supplement (Yamaguchi, 1997). The importance of algae in this domain is not surprising as they are the natural food sources of these animals. To provide better balanced nutrition and improve animal growth, several reports advice mixing species which contains more amount of proteins, carbohydrates and lipids etc. This gives better result than a diet composed of only one algal species.

Many nutritional and toxicological evaluations have proved the suitability of algal biomass as feed supplement. *Arthrospira (Spirulina)* is largely used in this domain and concerns many types of animal: cats, dogs, aquarium fish, ornamental birds, horses, cows and breeding bulls. Microalgae positively affect the physiology (by providing a large profile of natural vitamins, minerals and essential fatty acids; improved immune response and fertility; and better weight control) and their external appearance (resulting in health skin and a lustrous coat) of animals (Certik and Shimizu, 1999).

A large number of marine nitrogen fixing cyanobacteria have been tested for their nutritional value with the hybrid *Tilapia* fish fry and majority were acceptable as single ingredient feeds. Very high growth rates of *Tilapia* fish using marine cyanobacteria with indoor and outdoor cultures were also reported (Mitsui *et al.*, 1983). In studies conducted in our laboratory, the marine cyanobacterium *Phormidium valderianum* BDU 30501 was chosen as a complete aquaculture feed source based on the nutritional qualities and non-toxic nature with animal model experiments. A technology for mass cultivation of this strain and production of pellet feed was also developed and transferred to M/S ABL Biotechnologies, Chennai, who would commercially produce and market the marine cyanobacterial feed.

Fine Chemicals

A variety of fine chemicals such as pigments, vitamins and enzymes with varied applications can be obtained on a commercially viable scale from cyanobacteria. A number of cyanobacteria are known to be rich sources of vitamins and many are known even to excrete them into the milliew (Borowitzka, 1988). Some of the marine cyanobacteria appear to be potential source for large scale production of vitamins of commercial interest such as vitamins of the B-complex group and vitamin E (Borowitzka, 1988).

Some cyanobacterial species are established in the skin care market, the main ones being *Arthrospira* (Stolz and Obermayer, 2005). Cyanobacterial extracts can be mainly found in face and skin care products (e.g., anti-aging cream, refreshing or regenerant care products, emollient and as an anti-irritant in peelers). Micro algae are also represented in sun protection and hair care products. One example of commercially available product and its properties claimed by the company, a protein-rich extract from *Arthrospira*, repairs the signs of early skin aging, exerts a tightening effect and prevents stria formation (Protulines, Exsymol S. A. M., Monacco). Some cyanobacteria associated with corals were found to produce UV absorbing compounds such as mycosporin-like amino acids. *Oscillatoria* sp. NKBG 091600 grew well in the presence of UV irradiation, and were found to produce high levels of UV-A absorbing compound, biopterin glucoside (BG) in response to UV-A irradiation (Matsunaga *et al.*, 1993; Wachi *et al.*, 1995) and found that intracellular levels of BG increased with increasing UV-A light intensities. BG is a water-soluble compound and is used in skin formulations.

The carotenoids and phycobiliproteins characteristic of cyanobacteria have high commercial value. They are used as natural food colourants, as food additives to enhance the colour of the flesh of Salmonid fish and to improve the health and fertility of cattle and in the cosmetic industries (Emodi, 1978). In addition, phycobiliproteins are widely used in industry and clinical or research immunology laboratories. Indeed, their properties (high molar absorbance coefficients, high fluoresecence quantum yield, large stokes shift, high oligomer stability and high photostability) make them very powerful and highly sensitive fluorescent reagents. They can serve as labels for antibodies, receptors and other biological molecules in a fluorescence-activated cell sorter and they are used in immunolabelling experiments and fluorescence microscopy or diagnostics (Spolaore *et al.*, 2006). Many marine cyanobacteria, especially the feed grade *Phormidium valderianum* has been found to be an excellent source of phycocyanin, a blue natural colorant useful as a phycofluor in diagnostics, a technology for the inexpensive production of which has been developed and transferred to M/S ABL Biotechnologies, Chennai.

Cyanobacteria being photoautotrophs, have the ability to photosynthetically transform simple, labeled compounds such as $^{14}CO_2$, $^{13}CO_2$, $^33H_20$, $^{15}NO_3$ into complex organic compounds. Isotopically labeled cyanobacterial metabolites such as sugars, lipids, amino acids are commercially available (Patterson, 1996).

A number of important enzymes are known to be produced in sufficient amounts by cyanobacteria to be exploited as useful commercial ventures. In our laboratory, marine cyanobacteria useful for the large-scale production of enzymes such as β-lactamase, protease and lipase have been identified and characterized (Prabaharan *et al.*, 1994) and the technology developed for beta lactamase already transferred to M/S ABL Biotechnologies, Chennai. Several common as well as unique sequence specific endonucleases are known from different cyanobacteria namely *Anabaena cylindrica* (Acy I), *Anabaena flos-aquae* (Afl I and Afl III), *Anabaena variabilis* (Ava I and Ava II), *Anabaena variabilis UW* (Avr II), *Nostoc sp. PCC* 7524 (Nsp C I), *Microcoleus sp. UFEX* 2220 (Mst II), which can be made available in the market at a lesser cost since relative biomass production of cyanobacteria is much less expensive than bacteria (Elhai, and Wolk, 1988). The extracellular phosphatase activities of cyanobacterial isolates have also been reported, which has the capacity of mineralizing organic phosphorous per unit biomass due to alkaline phosphatase activity (Giraudet *et al.*, 1997). A number of studies have been made for the detection of enzymes including chitinase, L-asparaginase, L-glutaminase, amylase, protease, lipase, cellulase, urease and superoxide dismutase (Wikstrom *et al.*, 1997). Photoproduction of ammonia and amino acids in large amounts by cyanobacteria, which can be harvested continuously over longer periods, has also been described (Hall and Grassi, 1989).

Analysis of extracellular growth promoting substances liberated by *Nostoc muscorum* and *Hapalosiphon fontinalis* in the external medium was found to be rich in several amino acids like serine, arginine, glycine, aspartic acid, threonine, glutamic acid, cystine, proline, valine, ornithine, lysine, histidine and isoleucine (Misra and Kaushik, 1989). In addition, cyanobacteria can be rich sources of several polyols, polysaccharides, lipids, fatty acids, halogenated compounds, etc. with varied properties employable as flocculants, surfactants, and so on (Becker, 1994).

Pharmaceuticals

The fact that cyanobacteria, in general, and marine forms, in particular, are the richest sources of known and novel bioactive compounds including toxins with wide pharmaceutical application is unquestionable (Thajuddin and Subramanian, 2004; Jha and Zi-rong, 2004). Some of the marine cyanobacteria appear to be potential sources for large scale production of vitamins of commercial interest such as vitamin B complex group and vitamin E. Anti-HIV activity has been observed with the compound extracted from *Lyngbya lagerheimii* and *Phormidium tenue* (Gustafson *et al.,* 1989). More than 50% of the 100 isolates from marine sources are potentially exploitable bioactive substances. These substances tested were either the ones that killed cancer cells by inducing apoptotic death, or those that affected cell signaling through activation of the members of protein kinase–c family of signaling enzymes. Rickards *et al.*, (1999) reported that the cell extracts of *Calothrix* isolates inhibited the growth *in vitro* of a chloroquine-resistant strain of the malaria parasite, *Plasmodium falciparum* and of human HeLa cancer cells in a dose-dependent manner. Bioassay directed fractions of the extracts let to their isolation and structural characterization of Calothrixin A (I) and B (II), pentacyclic metabolites with an indole [3, 2 – j] phenanthridine alkaloids which exerts their growth inhibitory effects at nanomolar concentrations.

A massive program of screening the extracts from the large culture collection of marine cyanobacteria in our research facility for antiviral, antibacterial, antifungal and immuno-modulatory activities has been going on with great success. An immunosuppressive linear peptide microcolin-A, which at nanomolar concentrations suppresses the two-way murine mixed lymphocyte reaction, has been isolated from *Lyngbya majuscula* (Koehn *et al*., 1992). Another compound, Curacin-A isolated from the organic extracts of Curacao collections of *Lyngbya majuscula* has been found to be an exceptionally potent antiproliferative agent as it inhibits the polymerization of the tubulin which shows some selectivity for colon, renal and breast cancer-derived cell lines (Carte, 1996). A compound has been purified from marine *Oscillatoria laete-virians BDU* 20801 showing anti-candida activity (Deth, 1999). An immunopotentiating property compound with male anti-fertility (Raghavan *et al*., 2002), without being toxic to other systems in a mice model was exhibited by the crude and purified extracts of *Oscillatoria willei BDU* 130511 (Deth, 1999). Medically important gamma linolenic acid (GLA) is relatively rich in cyanobacteria, namely *Spirulina platensis* and *Arthrospira* sp. which is easily converted into arachidionic acid in the human body and arachidionic acid into prostaglandin E2 (Euler and Eliassen, 1967). Prostoglandin E2 has lowering action of blood pressure and the contracting function of smooth muscle and plays a very important role in lipid metabolism.

Research activities concerning the investigation of products of metabolism from plants and other groups of microbes were undertaken not only for better understanding of nature but also to discover metabolites of possible use for human in different fields of interest. In such research activities, cyanobacteria are found to be a rich source for various products of commercial, pharmaceutical or toxicological interest: Primary metabolites, such as proteins, fatty acids, vitamins or pigments (Borowitzka, 1988 a, b; 1995), various secondary metabolites with different bioactivities (antifungal, antiviral, antibiotic and other) (Patterson *et al*., 1994; Falch, 1996) cyanotoxin like hepatotoxin microcystins and saxitoxins (Carmichael, 1992) were isolated from cyanobacteria.

Most of the cyanobacterial metabolites are accumulated in the cyanobacterial biomass. Moreover, cyanobacterial excrete various organic compounds in their environment as exometabolites, e.g., some antibacterial titerpenoides in *Nostoc commune* (Jakai *et al.,* 1999; 2000) or antifungal peptides in *Tolipothrix byssoidea* (Jakai *et al.,* 2001). Volk and Furkert (2006) identified the antimicrobial activities (antialgal, antibacterial and antifungal) of two exometabolites in different concentration, namely non-harmane (9H-pyrido(3,4-b)indole) and 4,4'-dihydroxybiphenyl from two cyanobacterial isolates such as *Nodularia harveyana* and *Nostoc insulara.* Kreitlow *et al.* (1999) tested the hydrophilic and lipophilic extract of twelve cyanobacterial strains isolated from fresh and brackish water and two water blooms collected during summer from Baltic Sea and found that most of the extract inhibited the growth of *Micrococus flavus, Staphylococcus aureus* and *Bacillus subtilis*.

There are three cyanobacterial orders that produce the vast majority of toxic compounds, these being Chroococcales, Oscillatoriales and Nostocales. Table 2 demonstrates quite succinctly that difference species coming under Nostocales produce over 120 natural products, these almost entirely being secondary metabolites, which can be classified as biotoxins (Burja *et al.,* 2001). The following table shows different compounds isolated from marine cyanobacteria.

Table 2. Production of various compounds by members of Nostocales.

Source	Compound	Activity	Reference
Order Chroococcales			
Microcystis aeruginosa	Microviridin Toxin BE-4, Siatoxin	Antibiotic, anticancer	Armente and Carmichael, 1996; Domingos *et al.* 1999; Shi *et al.*, 1999
Synechococcus sp.	Lipopeptide	Antibiotic	Ohta *et al.*, 1994
Synechocystis trididemni	Didemnin	Anticancer, antiviral	Rinehart *et al.*, 1981; Chun *et al.*, 1986
Order Pleurocapsales			
Hyella caespitosa	Carazostatin	Antifungal	Cardellina *et al.*, 1979
Order Oscillatoriales			
Lyngbya lagerheimii	Sulfolipid	Anti HIV activity	Gustifson *et al.*, 1989
Lyngbya majuscula	Sulfolipid amide (bromo, chlor and pyrrole) fatty acid (chloro sulfo thiazoline) lipopeptides	Anti HIV, anticancer, Antifungal, antimicrobial	Gerwick *et al.*, 1994; Luesch *et al.* 2000; Mynderse *et al.*, 1988; Mitchell *et al.*, 2000; Milligan *et al.*, 2000
Oscillatoria acutissima	Acutiphycin	Anticancer	Barchi *et al.*, 1984
Oscillatoria nigroviridis	Oscillatoxin	Anticancer	Moore *et al.*, 1984
Phromidium ectocarpi	Hierridin, 2, 4-dimethoxy-6-heptadecyl-phenol	Antiplasmodial, antibiotic	Murakami *et al.*, 1991
Phormidium tenue	Sulfolipid	Anti-HIV activity	Gustafson *et al.*, 1989; Murakami *et al.*, 1991
Schizothrix calcicola	E-1-chlorotridec-1 ene-6, 8 diol	Antibiotic, anticancer	Harrigan *et al.*, 1999; Mynderse and Moore, 1978
Spirulina platensis	Calcium spirulan	Anticancer, anti-HIV activity	Yang *et al.*, 1987; Campbell *et al.*, 1982, Bhat and Madyastha 2000
Order Nostacales			
Anabaena flos-aquae	Lipopeptide, alkaloid	Antibiotic, anticancer	Matsunaga *et al.*, 1989; Ma and Led 2000
Anabaena variabilis	Lipopeptide	Antibiotic	Carmichel *et al.*, 1975
Aphanizomenon flos-aquae	Aphanorphine, Siatoxin	Antibiotic, anticancer	Moore, 1978
Aulosira fertilisima	Aulosirazole	Anticancer	Stratmann, 1994
Calothrix sp	Calothrixin	Antimalarial, anticancer	Issa, 1999
Cylindrospermum licheniforme	Cylindrocyclophane	Anticancer	Moore *et al.*, 1990
Hormothomnion enteromorphoides	Hormothomnin	Cytotoxic, antibiotic	Gerwick, 1990
Nostoc sp.	Cryptophycins, Nostophycin	Anticancer, Cytotoxic, Antifungal, Antibiotic	Terimurtulu *et al.*, 1994; Yang and Shimizu, 1993
Nostoc commune	Nostodione, Microsporin	Antifungal, Antibiotic	Bohm *et al.*, 1995; Jaki *et al.*, 1999, 2000
Nostoc elipsosporum	Cyanovirin	Anti-HIV, Antiviral	Dey *et al.*, 2000; Esser *et al.*, 1999
Nostoc sphaericum	Staurosporine	Antiviral	Wipf and Venkataraman, 1996
Nostoc spongiaeforme	Nostocine A	Antibiotic, Antialgal	Hirata *et al.*, 1996, 1996a
Scytonema pseudohofmanni	Scytophycine	Antifungal, Antiviral	Moore *et al.*, 1986a, 1986b; 1989
Tolypothrix nodosa	Tolyporphin	Antibiotic	Prinsep, 1992
Tolypothrix tenuis	Toyocamycin	Antifungal	Moore, 1982
Order Stigonematales			
Fischerella muscicola	Fischerellin	Antifungal, Herbicidal	Srivastava *et al.*, 1998, 1999.
Haplosiphon fontinalis	Alkaloids, Indole	Antibiotic, Antifungal	Moore *et al.*, 1987
Haplosiphon Welwitschii	Hapalosin, Welwisthtin	Anticancer, Antibiotic	Zhang and Smith, 1996; Smith *et al.*, 1995
Stigonema dendroideum	Dendroamide	Antibiotic	Ogino *et al.*, 1996

The most significant results have been the discoveries of borophycin, cryptophycin 1 and 8 and cyanovirin. Borophycin is a boron-containing metabolite isolated from marine strains of the cyanobacteria *Nostoc linckia* and *Nostoc spongiaeforme* var. *tenue* (Banker and Carmeli, 1998). Borophycin, which is related both to the boron containing boromycins isolated from a terrestrial strain of *Streptomyces antibioticus* and to the aplasmomycins isolated from a marine strain of *Streptomyces griseus* (actinomycetes), exhibits potent cytotoxicity against human epidermoid carcinoma (LoVo) and human colorectal adnocarcinoma (KB) cell lines (Davidson, 1995), and has recently been found to exhibit anitimicrobial activity.

Cryptiphycin 1 was first isolated from *Nostoc* sp. ATCC 53789 by researchers at Merck, and found to be a potent fungicide. It was, however, found to be too toxic, and disregarded as a natural product lead. Subsequently, Moore (1996) isolated same compound from *Nostoc* sp. GSV 224, which exhibited potent cytotoxicity against human tumor cell lines and good activity against a broad spectrum of drug sensitive and drug-resistant murine and human solid tumors. Nevertheless, cryptophycin 1 again appeared to be too toxic to become a clinical candidate. This led to a detailed structure-function study by Carmichael (1992), and resulted in the isolation of cryptophycin 8, a semisynthetic analogue which proved to have a greater therapeutic efficiency and lower toxicity than 14 *in vivo*. Although neither cryptophycin, nor any of its analogs have entered clinical trials to date, interest in these compounds continues. Recently, Eli Lilly and Corporation (Indianapolis, In, USA) (US Patent No. 6,020,512) have shown interest in developing a novel method for the production of cryptophycin intermediates, presumably for further study. In any case, the work of Moore and Carmichael (US Patent Nos. 6,013,626, 5,955,423 and 5,952,298) shows quite eloquently that secondary metabolites in their natural form will not necessarily lead directly to commercialization, but the activity induced by these natural compounds may be exploited in structure-function studies to develop analogues with greater and more focused activities.

Cyanovirin (CV-N, cyanovirin-N), a 101 amino acid protein, has recently been placed on an accelerated track for clinical development for use as a virucidal agent. The viron-receptor binding activity of cyanovirin was first discovered in 1997 from an extract of *Nostoc ellipsosporum* (F90783, Q68D170) isolated in the early 1980s. The USA's National Cancer Institute (NCI) discovered that cyanovirin is a fusion inhibitor of HIV, somehow preventing the surface receptor of HIV from undergoing the conformational changes that enable the viral and cell membranes to fuse, thus allowing the transfer of the viral genetic material into the host. CV-N was found to have potent activity against all immunodeficiency viruses (HIV-1, M and T-tropic strains of HIV-1, HIV-2, SIV (simian) and FIV (feline)) (Blakeslee, 1998). Recently, several patents have been lodged (US Patent Nos. 6,015,876, 5,998,597, 5,962,653, 5,843,882 and 5,821,081) to protect this new method of HIV prevention. Initial observations suggest that CV-N proteins are not present in other species of cyanobacteria; there are no known open reading frame homologs and that the normal biological function is unknown. A systematic search for such analogs has not, however, been conducted, and it would seem highly improbable that such a protein would have evolved in only one of the 100 or more species of the genus *Nostoc*, or indeed within related species. The use of CV-N and any analogs, if discovered, in structure–function experiments could lead to an entirely new class of anti-HIV drug being developed.

Antifungal activity and algicidal is common among cyanobacteria, leading some researchers to hypothesize that the cyanobacteria may find use as biocontrol agents. While most reports of antifungal effects of cyanobacteria have been demonstrated using fungi of biomedical, rather than agricultural importance, at least one research group has discovered and characterized the effects of *Nostoc muscorum* extracts on the growth of plant pathogenic fungi particularly *Sclerotina sclerotiorum* and *Rhizoctonia solani* (Cano *et al.*, 1990). The potential of cyanobacteria as biological control agents concludes that cyanobacteria, since they produce antibacterial and antifungal materials, do not pose a threat to the environment, and can be grown in mass culture, are suitable candidates for exploitation as biocontrol agents for plant pathogenic bacteria and fungi (Kulik, 1995).

Biofertilizer

De (1939) attributed the inherent fertility of tropical rice field soils to the activity of N_2 fixing cyanobacteria. A large variety of cyanobacterial strains colonize the rice field soils and some species are capable to fix atmospheric nitrogen using their specialized cells called heterocysts. Several non-heterocystous cyanobacteria are also capable of fixing atmospheric nitrogen under micro aerophilic conditions. *In situ* estimations using acetylene reduction technique have shown an addition of 18–15 kg N ha^{-1} yr^{-1} due to the activity of diazotrophic cyanobacteria (Watanabe and Cholitkul, 1979). The role of N_2 fixing cyanobacteria in the maintenance of the fertility of rice fields has been well substantiated and documented all over the world. In India alone, the beneficial effects of cyanobacteria on yield of many rice varieties have been demonstrated in a number of localities (Venkataraman, 1981). Beneficial effects of cyanobacterial inoculation have also been reported on a number of other crops such as barley, oats, tomato, radish, cotton, sugarcane and maize, chilli and lettuce (Dadhich *et al.*, 1969; Kaushik and Venkataraman, 1979). The importance of *Azolla* as an organic fertilizer in rice cultivation is well appreciated and practiced in several countries. The cyanobacterial symbiont *Anabaena-azollae* found within a center cavity, fixes atmospheric nitrogen estimated between 120–312 kg N_2 per hectare. *Azolla* supplies substantial amounts of green manure between 150–300 tons per hectare per year, which supports the growth of soil microorganisms including heterotrophic N_2 fixers (Kannaiyan, 1985).

The major interest in the use of cyanobacteria as a bacterial fertilizer is in the Third World where chemical fertilizers may be prohibitively expensive. For example, there are two main lines of field research being conducted in India with free-living cyanobacteria (Stewart, 1982). These include (i) the stimulation of cyanobacterial nitrogen-fixation by addition of trace amounts of minerals (Mo, P, K, Ca) and (ii) selection and production of cyanobacterial strains and inoculants. Other research is directed towards the development of cyanobacterial strains that are resistant to pesticides (used to prevent animal grazing on cyanobacteria) or herbicides (Vaisampayan, 1984), have high nitrogenase activity and, to a lesser extent, specific strains for use as inoculants in different countries with different environmental conditions. Ramos *et al.*, (1984) have recently described conditions which allow for sustained photoproduction of ammonia for fertilizer by the cyanobacterium *Anabaena* ATCC 33047. A system of this sort has the potential of providing biologically fixed nitrogen which could be used instead of a chemical nitrogen source. The water fern, *Azolla* is used as 'green manure' in China, Vietnam, Indonesia, the Philliphines and United States. *Azolla*, which forms a symbiotic relationship with the nitrogen-fixing cyanobacterium *Anabaena*, can add 60-80 kg N per hectare of crop and is often grown in rotation with rice crops or in between rows of rice (Stewart, 1980). The fixed nitrogen is released only when *Azolla* is incorporated into soil and decomposes (Fox, 1985).

The use of algae and cyanobacteria in waste treatment could prove beneficial in different ways since they will bring about oxygenation and mineralization, in addition to serving as food source for aquatic species. Microbial mats dominated by cyanobacteria are effective in removing metals (Cd, Cu, Zn, Cr, and Mn) from water by a mechanism involving exopolysaccharide (Bender *et al.*, 1989; 1994). Naturally occurring aromatic hydrocarbons (Cerniglia *et al.*, 1979; Narro *et al.*, 1992) and xenobiotics can be degraded by cyanobacteria, and cyanobacteria can be genetically engineered to enhance their biodegradative ability (Kuritz and Wolk 1995). Microbial mats rich in cyanobacteria facilitate the remediation of oil-polluted waters and desert in the region of the Arabian gulf by utilizing crude oil and individual *n*-alkanes as sources of carbon and energy (Sorkhoh *et al.*, 1995). Cyanobacteria can also perform syntheses that are of biotechnological significance. Like many genera of eubacteria, they synthesize polyhydroxyalkanoates (PHAs), a thermoplastic class of biodegradable polyesters that includes polyhydroxybutyrate (PHB). PHAs are carbon- and energy-storage compounds that are deposited in the cytoplasm as inclusions. Miyake and colleagues (Miyake *et al.*, 2000) isolated a Tn*5* insertion mutant of *Synechococcus* sp. MA19 with enhanced accumulation of PHB. By using the genomic sequence of *Synechocystis*, Hein *et al.* (1998) identified and characterized a gene encoding PHB synthase. Two related genes, a PHA specific β-ketothiolase and an acetoacetyl-CoA reductase, have been identified and characterized (Taroncher-Oldenburg *et al.*, 2000). C-phycocyanin and allophycocyanin are phycobiliproteins, pigmented components of the photosystem-II antenna structure, the phycobilisome. Phycobiliproteins, coupled to monoclonal and polyclonal antibodies to form fluorescent antibody reagents, are valuable as fluorescent tags in cell sorting, studies of cell surface antigens, and screening of high-density arrays. *Spirulina* is a convenient and inexpensive source of allophycocyanin and C-phycocyanin. As an alternative approach, genetic engineering of *Anabaena* 7120 has permitted the *in vivo* production of stable phycobiliprotein constructs bearing affinity purification tags, usable as fluorescent labels without further chemical manipulation (Cai *et al.* 2001). Cyanobacteria, since they inhabit the same ecological niches as mosquito larvae and are eaten by them (Thiery *et al.*, 1991; Avissar *et al.*, 1994), are attractive candidates for mosquito control. Transgenic mosquitocidal cyanobacteria, filamentous as well as unicellular, have been engineered (Tandeau de Marsac *et al.*, 1987 Angsuthanasombat and Panyim 1989; Murphy and Stevens 1992; Xu *et al.*, 2000; Wu *et al.*, 1997). A high level of toxicity was observed with recombinant clones of *Anabaena* 7120 bearing two δ-endotoxin genes (*cryIVA* and *cryIVD*) and gene *p20* of *Bacillus thuringiensis* subsp. *israelensis* (Wu *et al.*, 1997). Outdoor tests indicated that genetically altered *Anabaena* 7120 could keep containers with natural water from being inhabited by *Culex* larvae for over 2 months (Xu *et al.*, 2000). Although one might expect laboratory strains to have low competitive ability compared with indigenous species that same characteristic may help to prevent unwanted spread of a genetically modified microorganism. Notably, a lyophilized (and presumably nonproliferative) preparation of the recombinant cells retained the same high mosquitocidal activity as the original culture.

Liquid suspension cultures or immobilized cells of cyanobacteria offer opportunities for photoproduction of molecular hydrogen (Gisby *et al.*, 1987; Lindblad 1999). Cyanobacteria possess several enzymes directly involved in hydrogen metabolism: (1) nitrogenase(s), catalyzing the production of H_2 as a side product of reduction of N_2 to NH_3; (2) an uptake hydrogenase, catalyzing the consumption of H_2 produced by the nitrogenase; and (3) a bidirectional hydrogenase, which has the capacity both to take up and to produce H_2 (Papen *et al.*, 1986; Schmitz *et al.*, 1995; Tamagnini *et al.*, 2000). To maximize H_2 production, mutants of *A. variabilis* strain ATCC 29413 defective in H_2-utilization were isolated by Polukhina-Mikheeva and Koksharova (Mikheeva *et al.*, 1994). Two mutants altered in hydrogen metabolism were characterized and one of them, PK 84, has been used for hydrogen production in an automated helical tubular photobioreactor (Borodin *et al.*, 2000). H_2-production aside, no biotechnological use has yet been made of the capacity of heterocyst-forming cyanobacteria while growing in air to support reactions that require microoxic conditions.

Using the marine cyanobacteria *Oscillatoria* sp. BDU 10742 and *Aphanocapsa sp.* BDU 16 and a halophilic bacterium *Halobacterium* US 101, Uma and Subramanian (1990) could treat ossein factory effluent and reduce calcium and chloride level significantly at field levels to enable 100% survival and multiplication of *Tilopia* fish with only the cyanobacteria as the feed source. Shashireka *et al.* (1997) found that another marine cyanobacterium *Phormidium valderianum* BDU 30501 was able to tolerate and grow at a phenol concentration of 50 mg/l and remove 38 mg/l with in a retention period of 7 days. This result opens the possibility of treating a variety of phenol containing effluents. The same organism was used to study the condition and reagents optimal sorption/desorption of heavy metal ions (Cd^{2-}, Co^{2-}) (Karna *et al.*, 1999). Another marine cyanobacterium *Oscillatoria boryana* BDU 92181 was found to effectively degrade and metabolize melanoidin pigment which is abundant in distillery effluents (Kalavathi *et al.*, 2001). Studies at the National Facility for Marine Cyanobacteria (NFMC) have identified suitable cyanobacteria for treating a number of noxious effluents containing organophosphorus pesticides, detergents, antibiotics, etc. and even degradation of solid wastes like coir pith by the lignolytic action of certain cyanobacteria (Malliga *et al.*, 1996).

Cyanobacteria have long been considered either as organisms of academic curiosity or as organisms of nuisance value. Pioneering work in India and subsequently other countries have raised the status of these microbes to the level of biotechnologically most useful ones. Therefore, in tropical countries like India, it is essential not only to understand and preserve the biodiversity of cyanobacteria found in all habitats but also to gainfully exploit them for different biotechnological applications including pollution abatement.

REFERENCES

Adams, D.G. (2000). In: *Prokaryotic Development* (Eds. Brun, Y.V. and Shimkets, L.J.), ASM Press, Washington DC, pp. 51-81.

Angsuthanasombat, C. and Panyim, S. (1989). *Appl. Environ. Microbiol.* **55:** 2428-2430.

Arment, A.R. and Carmichael, W.W. (1996). *J. Phycol.,* **32:** 591-597.

Avissar, Y.J., Margalit, J. and Spielman, A. (1994). *J. Am. Mosquito. Control Assoc.* **10:** 45-50.

Banker, R. and Carmeli, S.J. (1998). *Nat. Prod.* **61:** 1248-1251.

Barchi, J.J., Moore, R.E. and Patterson, G.M.L. (1984). *J. Am. Chem. Soc.,* **106:** 8193-8197.

Becker, E.W. (1994). Microalgae – Biotechnology and Microbiology, Cambridge Univ. Press, UK,

Becker, W. (2004). Microalgae in human and animal nutrition, in (Ed. Richmond, A.) Handbook of microalgal culture, Blackwell, Oxford, 312-351.

Bender, J.E.R. Archibold, V. Iveanusi, and J.P. Gould. (1989). *Water Sci. Technol.* **21:** 1661-1665.

Bender, J., Rodriguez-eaton, S., Ekanemesang, U.N. and Philippes, P. (1994). *Appl. Environ. Microbiol.*, **7:** 2311-2315.
Bhat, V.B. and Madyastha, K.M. (2000). *Biochem. Biophys. Res. Commun.,* **275:** 20-25.
Blakeslee, D.J. (1998). *Am. Med. Assoc.* (Online Suppl.).
Bohm, G.A., Pfleiderer, W., Boger, P. and Scherer, S.J. (1995). *J. Biol. Chem.,* **270:** 8536-8539.
Borodin, V.B., Tsygankov, A.A., Rao, K.K. and Hall, D.O. (2000). *Biotechnol Bioeng* **69:** 478-485.
Borowitzka, M.A. (1988). *In:* Microalgal Biotechnology (Eds Borowitzka, M.A. and Borowitzka, L.J.), Cambridge University Press, pp. 153-196.
Borowitzka, M.A. (1988a). *In:* Microalgal Biotechnology. (Eds. Borowitzka, M.A. and Borowitzka, L.J.) Cambridge University Press, Cambridge, pp. 211-217.
Borowitzka, M.A. (1988b). *In:* Microalgal Biotechnology. (Eds. Borowitzka, M.A. and Borowitzka, L.J.) Cambridge University Press, Cambridge, pp. 257-287.
Borowitzka, M.A. (1995). *J. Appl. Phycol.* **7:** 3-15.
Burja, A.M., Banaigs, B., Mansour, E.A., Burgess, J.G. and Wright, P.C. (2001). *Tetrahedron,* **57:** 9347-9377.
Cai Y.A., Murphy, J.T., Wedemayer, G.J. and Glazer, A.N. (2001). *Anal Biochem* **290:** 186-204.
Campbell II, J., Stevens, Jr. S. and Blakwill, D.L. (1982). *J. Bacteriol.,* **149:** 361-363.
Cano, M.S., de, Mule, M.C.Z., de, Caire, G.Z., de and Helperin, D.R., de, (1990). *J. Appl. Phycol,* **2:** 79-81.
Cardellina II, J.H., Kirkup, J.H., M.P., Moore, R.E., Mynderse, J.S., Seff, L. and Simmons, C.J. (1979). *Tetrahedron Lett.,* 4915-1916.
Carmichael, W.W. (1992). *J. Appl. Bacteriol.* **72:** 445-459.
Carmichael, W.W., Biggs, D.F., Gorham, P.R. (1975). *Science,* 542-544.
Carte, B.K. (1996). *Bioscience.* **46:** 271-286.
Cerniglia, C.E., Gibson, D.T. and Van Baalen, C. (1979). *Biochem. Biophys. Res. Commun.* **88:** 50-58.
Certik, M. and Shimizu, S. (1999) *J. Biosci. Bioeng.* **87:** 1-14.
Chun, H.G., Davies, B., Hoth, D., Suffness, M., Plowman, J., Flora, K., Groesjaber, C. and Jones, B. (1986). *Invest. New Drugs,* **4:** 279-284.
Cohen–Bazire, and Bryant, D.A. (1982). *In:* The biology of cyanobacteria, (Eds. Carr, N.G. and Whitton, B.A.) University of California Press, 143-190.
Dadhich, K.S., Verma, A.K. and Venkataraman, G.S. (1969). *Plant Soil* **31:** 377-379.
Davidson, B.S. (1995). *Curr. Opin. Biotechnol.,* **6:** 284-291.
De, P.K. (1939). *Proc. R. Soc. B.,* **127:** 121-139.
Deth, S. K. (1999). Ph.D thesis, Bharathidasan University, Thiruchirappalli, India.
Dey, B., Lerner, D.L., Lusso, P., Boyd, M.R., Elder, J.H. and Berger, E.A. (2000). *J. Virol.* **74:** 4562-4569.
Domingos, P., Rubim, T.K., Molica, R.R., Azevedo, S.M.F.O. and Carmichael, W.W. (1999). *Environ. Toxico.,* **14:** 31-35.
Elhai, J. and Wolk, C.P. (1988). *Methods Enzymol.,* **167:** 747.
Emodi, A. (1978). *Food Technol.,* **32:** 38-42.
Esser, M.T., Mori, T., Mondor, I., Sattentau, Q.J., Dey, B., Berger, E.A., Boyd, M.R. and Lifson, J.D. (1999). *J. Virol.* **73:** 4360-4371.
Euler, U.S. and Eliassen, R. (1996). Prostaglandins, Academic Press, New York, 1967.
Falch, B. (1996). *Pharm. i. u. Zeit.* **25:** 311-319.
Fox, R.D. (1985). *The Futurist,* **19:** 30-35.
Gerwick, W.H. (1990). US Patent number 4,935,529, Oregon State University, United States of America.
Giraudet, H., Berthon, J.L. and Buisson, B. (1997). *Vie. Life Sci.,* **320:** 451-458.
Gisby, P.F., Rao, K. and Hall, D.O. (1987). *Methods Enzymol.* **135:** 440-454.
Gustafson, K.R., Cardellina, J.H., Fuller, R.W., Weislon, O.S., Kiser, R.F. and Snader, K.M. (1989). *J. Nat. Caner Inst.,* **81:** 1254.
Gustafson, K.R., Cardellina, J.H., Fuller, R.W., Ewislow, O.S., Kiser, R.F., Snader, K.M., Patterson, G.M.L. and Boyd, M.R. (1989). *J. Natl. Cancer Inst.,* **81:** 1254-1258.
Hall, D.O. and Grassi, G. (1989). Photoconvertion Processes for Energy and Chemicals (Eds. Hall, D.O. and Grassi, G.), Elsevier, Amsterdam, pp. 28-45.
Harrigan, G.G., Luesch, H., Yoshida, W.Y., Moore, R.E., Nagle, D.G., Biggs, J., Park, P.U. and Paul, V.J. (1999). *J. Nat. Prod.,* **62:** 464-467.
Hein, S., Tran, H., Steinbüchel, A. (1998). *Arch. Microbiol.* **170:** 162-170.
Herdman, M., Janvier, M., Rippka, R. and Stanier, R.Y. (1979). *J. Gen. Microbiol.,* **111:** 73-85.
Hirata, K., Nakagami, H., Takashina, J., Mahmud, T., Kobaysahi, M. In, Y: Ishida, T. and Miyamota, K. (1996). *Heterocycles* **43:** 1513-1519.

Hirata, K., Takashina, J., Nakagami, H., Ueyama, S., Murakami, K., Kanamori, T. and Miyamota, K. (1996). *Biosci. Biotechnol. Biochem.* **60:** 1905-1906.

Holland, J. (1977). *The Chemistry of the Atmosphere and Oceans*, John Wiley & Sons, New York, vol. I.

Issa, A.A. (1999). *Envrion. Toxicol. Pharmacol.*, **8:** 33-37.

Jaki, B., Orjala, J., Heilmann, J., Linden, A., Vogler, B. and Sticher, O. (2000). *J. Nat. Prod.* **63:** 339-343.

Jaki, B., Orjala, J. and Sticher, O. (1999). *J. Nat. Prod.* **62:** 502-503.

Jaki, B., Zerbe, O., Heilmann, J. and Sticher, O. (2001). *J. Nat. Prod.*, **64:** 154-158.

Jeraci, J. and Vansoest, P. (1986). *In:* Handbook of Dietary Fiber in Human Nutrition (Ed. Spiller, G.) CRC Press, Boca Raton, pp. 299-303.

Jha, R.K. and Zi-rong, X. (2004) *Mar. Drugs.* **2:** 123-146.

Kalavathi, F.D., Uma, L. and Subramanian, G. (2001). *Enz. Microb. Technol.*, **29:** 249-251.

Kannaiyan, S., (1985). Algal Biofertilizers for Low Land Rice, TNAU, Coimbatore, pp. 14.

Karna, R.R., Uma, L., Subramanian, G. and Mohan, P.M. (1999). *World J. Microbiol. Biotechnol.*, **15:** 729-732.

Kaushik, B.D. and Venkataraman, G.S. (1979). *Plant Soil*, **52:** 135-137.

Koehn, F.E., Longley, R.E. and Reed, J.K. (1992). *J. Nat. Prod.* **55:** 613-619.

Kreitlow, S., Mundt, S. and Lindequist, U. (1999). *J. Biotechnol.*, **70:** 61-63.

Kulik, M.M. (1995). *European J. Plant Pathol.* **101:** 585-599.

Kuritz, T. and Wolk, C.P. (1995). *Appl. Environ. Microbiol.* **61:** 234-238.

Lem, N.W., Penrose, D.M. and Glick, B.R. (1985). *Plant Physiol (Suppl.)*, **77:** 47.

Lindblad, P. (1999). *Biotechnol. Appl.* **16:**141-144.

Litchfield, J.H. (1983). *Science*, **219:** 740-746.

Loffelhandt, W. and Bohnert, H.J. (1977). In: *The Molecular Biology of Cyanobacteria* (Ed. Bryant, D.A.), Kluwer Academic Publishers, Dordrecht, The Netherlands, pp. 65-89.

Luesch, H., Yoshida, W.Y., Moore, R.E., Paul, V.J. and Mooberry, S.L. (2000). *J. Nat. Prod.*, **63:** 1106-1112.

Ma, L.X. and Led, J.J. (2000). *J. Am. Chem. Soc.*, **122:** 7823-7824.

Malliga, P., Uma, L. and Subramanian, G. (1996). *Microbios*, **86:** 175-183.

Matsunaga, S., Moore, R.E., Niemczura, W.P. and Carmichael, W.W. (1989). *J. Am. Chem. Soc.*, **111:** 8021-8023.

Matsunaga, T., Burgess, J.G., Yamada, N., Komatsu, K., Yoshida, S. and Wachi, Y. (1993). *Appl. Microbiol. Biotechnol.* **39:** 250-253.

Mikheeva, L.E., Koksharova, O.A. and Shestakov S.V. (1994). *Vest Mosk Univ Ser XVI Biol No* **2:** 54-57.

Milligan, K.E., Marquez, B., Willamson, R.T., Davies-Coleman, M. and Gerwick, W.H. (2000). *J. Nat. Prod.*, **63:** 965-968.

Misra, S. and Kaushik, B.D. (1989). *Proc. Indian Natl. Sci. Acad.*, **55:** 295-300.

Mitchell, S.S., Faulkner, D.J., Rubins, K. and Bushman, F.D. (2000). *J. Nat. Prod.*, **63:** 279-282.

Mitsui, A., Enternmann, B. and Gill, K. (1983). *In:* Proceedings of the 2nd North Pacific Aquaculture System, Tokyo University, Japan, pp. 323-340.

Mitsui, A., Murray, R., Entenmann, B., Miyazawa, K. and Polk, E. (1981). *In:* Biosaline Research. A Look to the Future (Ed. San Pietro, A.), Plenum Press, New York, pp. 215-225.

Miyake, M., Takase, K., Narato, M., Khatipov, E., Schnackenberg, J., Shirai, M., Kurane, R. and Asada, Y. (2000). *Appl. Biochem.* Biotechnol. **84:** 991-1002.

Moore, B.S., Chen, J.L., Patterson, G.M.L., Moore, R.E., Brinen, L.S., Kato, Y. and Clardy, J. (1990). *J. Am. Chem. Soc.*, **112:** 4061-4063.

Moore, R.E. (1978). *Persp. Mar. Nat. Prod. Chem. Biol.* **1:** 52.

Moore, R.E. (1982). *Pure Appl. Chem.*, **54:** 1919-1934.

Moore, R.E. (1996). *J. Ind. Microbiol.* **16:** 134-143.

Moore, R.E., Banarjee, S., Bornemann, V., Caplan, F.R., Chen, J.L., Corley, D.G., Larsen, L.K., Moore, B.S., Patterson, G.M.L., Paul, V.J., Stewart, J.B. and Williams, D.E. (1986). *Pure Appl. Chem.* **61:** 521-524.

Moore, R.E., Blackman, A.J., Cheuk, C.E., Mynderse, J.S., Matsumoto, G.K, Clardy, J., Woodard, R.W. and Craig, J.C. (1984). *J. Org. Chem*, **49:** 2484-2489.

Moore, R.E., Corbett, T.H., Patterson, G.M.L. and Valeriote, F.A. (1996). Curr. Pharm. Es. **2:** 317-330.

Moore, R.E., Patterson, G.M.L., Mynderse, J.S., Barchi, J., Norton, T.R., Furusawa, E. and Rurusawa, S. (1986). *Pure Appl. Chem.* **58:** 263-271.

Moore, R.E., Yang, X.Q.G. and Patterson, G.M.L. (1987). *J. Org. Chem.* **52:** 3773-3777.

Murakami, M., Matsuda, H., Makabe, K. and Mamaguchi, K. (1991). *Tetrahdron Lett.*, **32:** 2391-2394.

Murakami, N., Morimoto, T., Imamura, H., Ueda, T., Nagai, S., Sakkibara, J. and Yamada, N. (1991). *Chem. Pharm. Bull.*, **39:** 2277-2281.

Murphy, R.C. and Stevens, S.E. (1992). *Appl. Environ. Microbiol.* **58:** 1650-1655.

Mynderse, J.S., Hunt, A.H. and Moore, R.E. (1988). *J. Nat. Prod.,* **51:** 1299-1301.

Mynderse, J.S. and Moore, R.E. (1978). *J. Org. Chem.,* **43:** 2301-2303.

Narro, M.L., Cerniglia, C.E., Van Baalen, C. and Gibson, D.T. (1992). *Appl Environ Microbiol* **58:** 1351-1359.

Ogino, J., Moore, R.E., Patterson, G.M.L. and Smith, C.D. (1996). *J. Nat. Prod.* **59:** 581-586.

Ohta, S., Chang, T., Kawashima, A. Nagate, T., Murase, M., Nakanishi, H., Miyata, H. and Kondo, M. (1994). *Bull. Environ. Contam. Toxico.,* **52:** 4245-4248.

Paden, E. and Cohen, Y. (1982). In: The biology of cyanobacteria, (eds. Carr, N.G. and Whitton, B.A.), University of California Press, pp. 215-237.

Papen, H., Kentemich, T., Schmülling, T. and Bothe, H. (1986). *Biochimie* **68:** 121-132.

Patterson, G.M.L. (1996). *J. Sci. Ind. Res.,* **55:** 669-684.

Patterson, G.M.L., Larsen, L.K. and Moore, R.E. (1994). *J. Appl. Phycol.,* **6:** 151-157.

Pirt, S.J. (1983). *Chem. and Ind.* 749-755.

Prabhakaran, D. and Subramanian, G. (1995). *Physiol. Mol. Biol. Plants,* **1:** 45-57.

Prabhakaran, D., Sumathi, M. and Subramanian, G. (1994). *Curr. Microbiol.,* **28:** 315-320.

Prinsep, M.R., Moore, R.E., Levine, I.A. and Patterson, G.M.L. (1992). *J. Nat. Prod.* **55:** 140-142.

Raghavan, C., Kadalmani, B., Thirunalasundari, T., Subramanian, G. and Akbarsha, M.A. (2002). In: Proceedings of the XX symposium. Report. Biol. Comp. Endocrinol., Bharathidasan University, Tiruchirapalli, pp. 137-138.

Rai, A.N. (1990). Handbook of Symbiotic Cyanobacteria (Ed. Rai, A.N.), CRC Press, Boca Raton,

Ramos, J.L., Guerrero, N.G. and Losada, M. (1984). *Appl. Environ. Microbial.* **48:** 114-118.

Rickards, R.W., Rothschild, J.M., Willis, A.C., de Chazal, N.M., Kirk, J., Kirk, K. Saliba, K.J. and Smith, G.D. (1999). *Tetrahedron,* **55:** 13513-13520.

Rinehart Jr. K.L., Gloer, J.B., Hughes, Jr. R.G., Renis, H.E., McGovren, J.P., Swynenberg, E.B., Stringfellow, D.A., Kuentzel, S.L. and Li, L.H. (1981). *Science,* **20:** 1-27.

Rippka, R. (1972). *Arch. Microbiol.,* **87:** 93-98.

Schantz, E.J. (1970). (Ed. Zajick, J.E.) Properties and products of algae, Plenum Press, 83-96.

Schmitz, O., Boison, G., Hilscher, R., Hundeshagen, B., Zimmer, W., Lottspeich, F. and Bothe, H. (1995). *Eur. J. Biochem.* **233:** 266-276.

Shashirekha, S., Uma, L. and Subramanian, G. (1997). *J. Ind. Microbiol. Biotechnol.,* **19:** 130-133.

Shi, L., Carmichael, W.W. and Kennelly, P.J. (1999). *J. Boil. Chem.,* **274:** 10039-10046.

Smith, A.J. (1982). In: The biology of cyanobacteria, (Eds. Carr. N.G. and Whitton, B.A.) University of California Press, 47-85.

Smith, C.D., Zilfou, J.T., Stratmann, K., Patterson, G.M.L. and Moore, R.E. (1995). *Mol. Phamacol.* **47:** 241-247.

Sorkhoh, N.A., Al-Hasan, R.H., Khanafer, M. and Radwan, S.S. (1995). *J. Appl. Bacteriol.* **78:** 194-199.

Spolaore, P., Joannis-Cassan, C., Duran, E. and Isambert, A. (2006). *J. Biosci. Bioeng.* **101:** 87-96.

Srivastava, A., Juttner, F. and Strasser, R.J. (1998). *Biochem. Biophys. Acta* **1364:** 326-336.

Srivastava, V.C., Manderson, G.J. and Bhamidimari, R. (1999). **143:** 309-317.

Stewart, W.D.P. (1980). **34:** 497-536.

Stewart, W.D.P. (1980). *Ann. Rev. Microbial.* **34:** 497-536.

Stewart, W.D.P. (1982). **31:** 5-44.

Stolz, P. and Obermayer, B. (2005). **120:** 99-106.

Stratmann, K., Belli, J., Jensen, C.M., Moore, R.E. and Patterson, G.M.L. (1994). *J. Org. Chem.,* **59:** 6279-6281.

Subramaninan, G. and Uma, L. (1996). *J. Sci. Ind. Res.,* **55:** 685-692.

Sundararaman, M., Subramanian, G., Averal, H.I. and Akbharsha, M.A. (1996). *Phytotherapy Res.,* **10:** 9-12.

Tamagnini, P., Costa, J-L., Almeida, L., Olivera, M-J., Salema, R. and Lindblad, P. (2000). *Curr. Microbiol.* **40:** 356-361.

Tandeau de Marsac, N., de la Torre, F. and Szulmajster, J. (1987). *Mol. Gen. Genet.* **209:** 396-398.

Taroncher-Oldenburg, G. and Stephanopoulos, G. (2000). *Appl. Microbiol. Biotechnol.* **54:** 677-680

Thajuddin, N. and Subramanian, G. (2004). *Current Science*, **89:** 47-57.

Thajuddin, N. and Subramanian, G. (1992). *Bot. Mar.,* **35:** 305-311.

Thajuddin, N., Nagasathya, A., Chelladevi, R. and Saravanan, P. (2002). *Seaweed Res. Utiln.*, **24:** 1-11.

Thiery, I., Nicolas, L., Rippka, R., Tandeau, and de Marsac, N. (1991). **57:** 1354-1359.

Trimurtulu, G., Ohtani, I., Patterson, G.M.L., Moore, R.E., Corbett, T.H., Valeriote, F.A. and Demchik, L. (1994). *J. Am. Chem. Sco.,* **116:** 2729-2737.

Uma, L. and Subramanian, G. (1990). *In:* Proceedings of the National Symposium on Cyanobacterial N_2 Fixation, IARI, New Delhi, pp. 437-444.

Vaisampayan, A. (1984). *New Phytol.* **96:** 7-11.

Venkataraman, G.S. (1981). Blue–green algae for rice production. FAO Soil Bull., **16:** 33-42.

Venkataraman, L.V. (1983). *In*: A Monograph on *Spirulina platensis*–Biotechnology and Application, DST, New Delhi.
Volesky, B., Zajick, J.E. and Knettig, E. (1970). Algal products in (Ed. Zajick, J.E.). *Plenum Press,* 49-82.
Volk, R.B. and Furkert, F.H. (2006). *Microb. Res.,* **161:** 180-186.
Wachi, Y., Burgess J.G., Iwamoto, K., Yamada, N., Nakamura, N. and Matsunaga, T. (1995). *Biochim. Biophy. Acta,* **1244:** 165-168.
Watanabe, I. and Cholitkul, W. (1979). *In:* Nitrogen and Rice, IRRI, Los Banos, Philippines, pp. 223.
Wikstrom, P., Szwajeer, E., Brodelius, P., Nilsson, K.N. and Mosbach, K. (1997). *Biotech. Lett.,* **4:** 153.
Wipf, P. and Venkataraman, S.J. (1996). *Org. Chem.* **61:** 6517-6522.
Wu, X., Vennison, S.J., Liu, H., Ben-Dov, E., Zaritskym A. and Boussiba, S. (1997). *Appl. Environ. Microbiol.* **63:** 4971-4975.
Xu, X., Yan, G., Kong, R., Liu, X. and Yu, L. (2000). *Curr. Microbiol.* **41:** 352-356.
Yamaguchi, K. (1997). *J. Appl. Phycol.,* **8:** 487-502.
Yang, H.N., Lee, E.H. and Kim, H.M. (1997). *Life Sci.,* **61:** 1237-1244.
Yang, X. and Shimzu, Y. (1993). *Tetrahedron Lett.,* **34:** 761-764.
Zhang, X. and Smith, C.D. (1996). *Mol. Pharmacol.* **49:** 288-294.

Chapter **6**

Biotechnological Potentialities of Higher Fungi

A. PANNEERSELVAM
Department of Botany and Microbiology, A.V.V.M. Sri Pushpam College (Aut.)
Poondi–613 503, Thanjavur, Tamil Nadu
E-mail: panneer_1959@yahoo.com

Fungi have been extensively used in the classical biotechnology. Yeast is the main organism used in the breweries and bakeries. Similarly, mushrooms are being directly consumed as protein source since time immemorial. The discovery of penicillin is new impetus to the pharmaceutical industry. Thus, the fungi, especially the higher fungi, are the source of food, food supplements, and feed source of industrial microorganisms which could potentially be used in the industries that produce antibiotics, organic acids, vitamins, enzymes, proteins, etc. The fungi are also used as tools of genetic engineering, as the yeast plasmids are used in transformation. The fungi can also potentially be used to solve environmental problems such as recycling of lignocellulosic wastes. The crop residues can be recycled and used to produce mycoprotein through mushroom cultivation. There are many mushrooms recommended for the cultivation as a source of mycoprotein. Among them *Calocybe indica* is the most suitable organism, that can be cultivated in the tropical environment. The present article deals with the biotechnological potentialities of higher fungi with particular reference to the cultivation of *C. indica.*

INTRODUCTION

Fungi, one of the important groups of biodiversity, are the source of vitamins, pharmaceuticals, organic acids, food, etc. There are many fungi yet to be explored and identified from different habitats. The unicellular fungus, yeast is used in industries and also as single cell protein. Similarly, there are many higher fungi which are also the source of mycoprotein. There are mushrooms which are edible and directly consumed as fungi.

Since ancient times, mushrooms have been treated as a special kind of food. Greeks believed from the very beginning that mushrooms provide strength to warriors in battle. Some praised mushrooms as a delicacy and the Romans considered it as food of the Gods, and it is only because of the taste and flavour of mushrooms. Only after thorough analysis, it is known that mushrooms are a good source of delicious food with high nutritional attributes. In this article, the biotechnological potentialities of higher fungi with particular reference to edible mushrooms are dealt with.

The mushroom cultivation technology has many advantages, which can be summarized as follows:

Mushrooms grow independent of sunlight and without fertile land, and hence, they do not compete with field or fruit crops, but provide an additional avenue for increasing food supply. The technology of mushroom cultivation can be adopted as small-scale industry and the produce has huge export potential. There is a world market for 14 lakh tonnes of mushrooms per annum against which India is exporting only about 2,400 tonnes. The world demand is likely to go up more than 30 lakh tonnes by 2010 and this technology could help to earn foreign exchange.

The technology of mushroom growing offers vast potential of rural employment. It does not involve sophisticated equipment, labour and cost. The required raw materials are crop residues and agricultural wastes. Hence, it is one of the low cost technologies most suitable for rural projects, and can provide employment in both the semi-urban and rural areas.

While the availability of land is the limiting factor for most of the crop production operations, it is not so for mushroom production. Mushrooms require little space because they can be grown under cover, inside mushroom houses, in basements, garbage or other suitable areas within the house. For outdoor cultivation, mushrooms have been grown on beds constructed from layers of various types of substrate or in tiers on wooden frames. In addition to floor, air space can also be utilized for higher production. They may be located on any unused land or available space between trees or beside buildings, or houses.

Mushroom technology is a women-friendly operation. Women in farming systems contribute a high percentage of labour in the field, especially in the developing countries. There is a need to diversify the income opportunities available to women. Technologies that can be followed on the homestead can provide them with additional food and income and thus enhance family welfare. Thus, mushroom growing is one of the agricultural activities in which women can play a vital role without sacrificing their household responsibilities, and thus it is one of the avenues for women entrepreneurship programmes and women empowerment.

BIOTECHNOLOGICAL POTENTIALITIES OF HIGHER FUNGI

Medicinal Value

Pharmaceuticals worth about 700 million are produced annually in Japan from *Lentinus corious*, *Schizophyllym* and *Ganoderma.* Mushroom extracts have a high amount of retene that has an antagonistic effect on some forms of tumours. Some mushroom extracts induce formation of interferon, a defence mechanism against viral infections, and have hypocholesteroeic activity (lowering cholesterol levels). Further, compounds extracted from mushrooms have antifungal and antibacterial properties.

Mushroom fungi are of considerable interest due to their use in medicine, including the synthesis of vitamins, antibiotics, production of plastic materials and destruction of organic wastes (Gray 1959). The main medical ingredient of the mushrooms is polysaccharides which in turn consist of large number of monosaccharides. Yet all are immunostimulants and all of them have the potentiality of strengthening the health and increasing the immunity. Polysaccharides produced by various fungi, algae, bacteria show specialized properties, which are for many industrial purposes. Filamentous fungi due to their ability to secrete large amount of protein are used extensively for the production of industrial enzymes. There is tremendous scope of microbial lipases in the chemoenzymatic synthesis of novel drugs such as antimicrobial, antitumour, non-steroidal, anti-inflammatory and cardiovascular drugs, which has been elaborately discussed (Raman *et al.,* 2000).

Table 1 summaries the medicinal values of mushrooms, their ability to cure diseases and their use in drug designing and application in pharmaceutical industries.

Table 1: Medicinal value of mushrooms.

Mushroom	Medicinal Value
Agaricus bisporus	Stimulating digestion, curing hypertension.
Auricularia polytricha	Strengthening health, helping blood circulation, curing stomach ailment.
Boletus edulis	Cause muscles and joints to relax.
Coriolus versicolor	Curing chronic diseases.
Ganodema lucidum	(Mushroom of immortality–Japan. Longevity mushroom–Korea) Rejuvenating effect, relieve stress, increase working efficiency, curing hypertension, antiviral effects, anti-HIV, anti-Hepatitis B, Epstein Barr virus.
Grifola frondosa	Strengthening health and resisting disease, anti-tumour activity.
Hericium erinaceus	Beneficial to heart, kidneys liver, lungs, spleen, gastric ulcers.
Lentinus edodes	Strengthening health and resisting disease, blood pressure, hyperacidity, anti-tumour effect, anticancerous and anti-virus.
Pleurotus ostreatus	Cause muscles and joints to relax.
Tremella fuciformis	Strengthening health and resisting diseases.
Tricholoma mongolicum	Beneficial to stomach and intestines.
Volvariella volvaceae	Cures blood pressure.

Antioxidant Activity

Mushrooms have the potentiality of antioxidant activity. It was found that the peroxide value in the sunflower oil before the addition of the mushroom extracts and BHT was 2.3, meg kg^{-1}. The addition of BHT and mushroom (*Pleurotus florida* and *Pleurotus eous*) extracts reduced the peroxide value. It was found that the peroxide value of *Pleurotus florida* extract added sun flower oil was 32.2 meq kg^{-1},while it was 40.6 meq^{-1} in *Pleurotus eous* extract added sunflower oil and 51.3 meq^{-1} in BHT added sunflower oil, whereas it was 86.3 meq kg^{-1} in control. The peroxide value of the oil incorporated with *Pleurotus florida* extract was increased in the subsequent weekly analyses upto 894 on 28th day. It was also found that the extract of *Pleurotus eous* increased the peroxide value to 89.6 and BHT influenced it to 1110.3 meq kg^{-1} day. However, it was 1341.5 meq kg^{-1} on 28th day in control.

Though there was an increase in both control and oil added with BHT and mushroom extracts showed the reduced peroxide values, it was in the following order: BHT > *Pleurotus eous* > *Pleurotus florida*. Hence, it can be concluded that mushrooms such as *Pleurotus* have antioxidant potentialities.

Environment

Higher fungi are capable of recycling the organic waste in the environment. Mushrooms are capable of degrading agro-waste. Mushrooms are grown on organic substrates, either raw or composted. These substrates are mostly waste materials from farms, plantations or factories. These are otherwise useless by-products which can be recycled to produce additional food in the form of mushrooms for human consumption. In the process, the agricultural and industrial wastes, by-products, etc. are recycled and thus the problem of environmental pollution is minimized to a greater extent. Examples of such materials are straw, corn cobs, sawdust, bagasse, poultry wastes, coconut husks, tree bark and leaves. The compost used for mushroom growing can also be recycled and used as animal feeds, soil conditioners and fertilizer.

Pleurotus spp. have been used to treat the pollutants including polycyclic aromatic hydrocarbons, polychlorinated hydrocarbons and biphenols dioxanes, etc. in soil (Huttermann, *et al*., 1987). Manganese peroxidase is produced by *Pleurotus*, which oxidizes bivalent manganese in the presence of H_2O_2 and decompose lignin model compound (Oji Paper, 1996). *P. orygit* is of biotechnological interest due to its ability to remove lignin selectively from wheat and produces high molecular weight manganese peroxidase enzymes (Maltinez, *et al.*, 1996). Several basidiomycetes and white rot fungi are known to degrade cellulose, hemicellulose and lignin (Ramaswami, *et al.*, 1985, Ramachandran and Sinha, 1993) *Pleurotus ostreatus* has been used to transform penicillin V and G into penicilloic acid (Stoppoke *et al.*, 1981). Labropoulosi (2000) reported that *Pleurotus ostreatus* had diphenolization and detoxification activities.

Industrial Effluent Treatment

The quantity and quality of many kinds of effluents that are let out of the industries pose serious environmental problems both in water and land. The effects of the textile effluent are severe on land. The high suspended solids and colloidal substances may clog the pores of the soil, thus reducing the permeability levels of the soil (Mohammad and Khan, Ansanullah, 1989). The high level of alkalinity enhances the sodium content of the effluent, which displaces the divalent cations, such as Ca and Mg. It also hardens the texture of the soil and also acts as deflocculatting agents and deprives the soil of its water-holding capacity. Colours of the effluent are bacteriostatic in nature and many of them are toxic to the microorganism.

Metals found in the effluents also cause environmental problems. The challenge does not become limited only to their removal from the environment but also extends to finding efficient and economic ways of using them in textile factory. Hence, it becomes essential to find out ways and means to contain or reduce the pollutants to the safety level and also the cheapest means of achieving this goal (Govindan, 1982).

Treatment of industrial effluents is broadly classified into primary, secondary and tertiary. Primary treatment is employed to remove suspended solids, to certain extent odour and to bring the pH to the neutral range. It includes screening, equalization, neutralization, coagulation and sedimentation. Specific toxic substances such as phenols are given special treatment. Secondary treatment is a biological process applied to reduce BOD and COD. Trickling filtration, activated sludge process, oxidation ditch and oxidation ponds are some of the common secondary treatments. Tertiary treatment aims at further purification of waste water (i.e.) final polishing, reduction of dissolved solids content and disinfection of effluent (Manivasakam, 1987).

The treatment of effluent may be further broadly classified into physical, chemical and biological methods. By physical method, it is possible to remove about 80-90 percent suspended solids and 10-15 percent of BOD from the waste water (Corning 1976). Chemical methods such as chemical precipitation (Kakabodse, 1979), co-precipitation, coagulation method, evaporation recovery process, conventional method, electrodialysis process, ion exchange process and adsorption process are also adopted.

Nema *et al.*, (1984) made laboratory studies on chemical treatment of synthetic textile mill waste water by precipitating some of the constituents using alum and ferrous sulphate and reported that the reduction of COD could be achieved in the range of 36-64 percent in the case of alum and 10-64 percent in the case of ferrous sulphate. Their studies on biological treatment of waste water by activated sludge and aerated lagoon have shown that the removal of 85 and 80-90 percent of BOD was achieved in the case of activated sludge and aerated lagoon system, respectively.

Biological treatments such as anaerobic digestion, trickling filters, high rate biological filters and activated sludge process are available. These are advantageous because fungi, bacteria, yeast and algae are capable of absorbing metals in high amount due to their peculiar surface properties (Manivasakam, 1987). Biohydrometallurgy is one such recent technology which deals with microbial methods like bioleaching, bioprecipitation, bioabsorption, etc. (Brierley *et al.*, 1996). Some microorganisms are known to accumulate metals intercellularly by a metabolism-dependent process known as bio-accumulation.

Membrane filtration, electrokinetic coagulation, ion exchange, adsorption and surfactant application method are being used for decolorization of textile effluents, and each technique has its own limitations. Recently, the method of ozonation has received attention. Colour removal of all classes of dyes used in the textile industry has been achieved to some extent with ozonation (Thakur *et al.*, 1998). Commercial azo., triaryl methane, antra quinonic and indigoid textile dyes are efficiently decolorized with enzyme preparations from *Pleurotus ostreatus, Schizophyllum commune, Neurospora crassa, Polyporus sp, Sclerotium rolfsli, Trametes villosa* and *Myceliophtora thermophila.*

Silva and Paulo (2000) reported that the nature of substituents on the dyes with benzene rings influences the enzyme activity, and hydroxyl and amino groups enhance decolorization. They have also found that the presence of lignin peroxidase and manganese peroxidase in addition to laccase increases decolorization upto 25%. The effect of textile dyeing auxiliaries depends on the individual enzymes. The enzymes produced by *Polyporus spp.* and *T. villosa* are inhibited upto 20 percent by copper and iron chelating agents and anionic detergents, while the *S. commune* enzymes lose upto 70 percent of their activity in the presence of each compound. It has been reported that the lignolytic fungus *Pleurotus ostreatus* could be used for the removal of cotton blue, malachite green, aniline blue, fast green, methyl green, methyl violet and were observed to be a good substrates of manganese peroxidase, (manganese independent peroxidase and phenol oxidase).

Mycoprotein

Mycoprotein ("myco" in Greek means "fungi") is a rich source of protein derived from fungi such as mushroom, truffles, and morels. The fungus used in all "Quorn" products is *Fusarium venenatum*. Mycoprotein is considered as rich and high quality protein, and received attention in the recent years. It is the fungal source of protein which contains high-quality protein, and becomes alternative source of vegetable protein. It contains all the nine essential amino acids and hence it is a high quality protein. It has naturally low fat content, zero cholesterol, low calorific value and essential dietary fibre.

Mycoprotein is actually a fungus that grows in huge fermentation vats. It is close to mushrooms and jelly fish. Mycoprotein products have almost as much protein as an egg, more fibre than a cooked potato, two-thirds of the fat of skinless chicken breast and no cholesterol. Quorn brand of meat-free foods and mycoscent brand of low-sodium flavouring are made from mycoprotein. Britains Marlow Foods has already sold nearly one billion packages of Quorn products throughout Europe over the past seventeen years. The fungus variety is collectively called "mycoprotein". Generally, the term "mushroom" has often been used in a user-friendly way of helping people to understand as to what they are eating. Nowadays creating awareness about the food products has become essential to popularize any new kind of food. In the case of fungi they are often being referred as contaminants of stored grains, other food products and associated with undesirable conditions such as diseases. Hence, making the people to accept convincingly such products became different at the beginning. So fungal source of mycoprotein has often been sold in the market as to Quorn products. The 'Quorn', after gracing the dinner tables in Europe for the last 17 years, has reached the United States and is now also available as frozen products. Potential buyers of this mycoprotein or mold may be surprised by the nutritional data on food labels. The Fettuccine Alfredo with Chicken-style Tenders and Broccoli have only 360 Calories with a 4 grams of dietary fibre and 17 grams of protein, but contains 16 grams of fat, 9 saturated, 45 milligrams of cholesterol, and a whopping 920 milligrams of sodium (38% of the daily allowance).

Mushroom Proteins

A sufficient calorie intake does not guarantee a good standard of nutrition. Food containing minerals, vitamins and enough of the right kind of protein is necessary in addition to that furnishing energy. Protein deficiency is not only a future problem but also an existing reality. Generally, protein is synthesized by green plants. But the presently available protein is not sufficient for human beings. Mushrooms form alternative source of nutrients in the form of protein, carbohydrate, valuable salts and vitamins. The nutritional value of mushroom lies between meat and vegetables. The mushroom protein is similar to that of muscle protein. Studies suggest that mushroom is well suited to supplement diets, which lack protein and in the sense they have rightly been called "vegetable meat".

Vast literature is available particularly on the protein content of mushroom. Generally, mushrooms can be compared more favourably with other crops in terms of yield per unit area. Cereals can give an annual yield of 3000 to 6000 kg/hectare whereas button mushrooms may give upto 2 million kg/hec. The produce of an acre of land can be transformed into ten tonnes of fungus protein. The following data show yields of dry protein per unit area utilized for forming beef and *Agaricus bisporus* (Cook, 1977).

Thus, mushrooms can produce highest quantity of protein per unit area from agro- waste. Mushrooms are not only very good source of vegetable protein, but they also possess all other nutritional requirements:

(i) They are good source of high quality proteins and are rich in vitamins and minerals.

(ii) Mushrooms contain 20-35% protein (dry weight basis), which is higher than that of the quantity of proteins found in vegetables and fruits, and is superior in quality.

(iii) Mushrooms are rich in lysine and tryptophan, the two essential amino acids that are deficient in cereals; they also contain fairly good amount of vitamin C and B complex (thiamine, riboflavin and niacin), and minerals such as potassium, phosphorus and sodium. Since potassium:sodium ratio is very high in mushrooms, it is ideal for patients of hypertension.

(iv) They are the low calorie food with very little fat and sugars and also without starch and cholesterol (Table 2).

Approximate annual yield of dry protein (kg/1 hec.)

Beef, cattle, conventional agriculture	-	78
Fish, intensive pond reading	-	675
Agaricus bisporus	-	65,000

Table 2. Proximate analysis of edible mushrooms (Percent on fresh weight basis) (Zakia, 1976).

Name	Moisture	Ash	Protein	Fat	Crude fibres
Agaricus bisporus	89.5	1.25	3.94	0.19	1.09
Lepiota sp	91.0	1.09	3.3	0.1	0.86
P. ostreatus	92.5	—	2.15	—	—
Termitomyces sp.	91.3	0.81	4.1	0.22	1.13
Volvariella diplasia	90.4	1.10	3.90	0.25	1.57
V. volvcea	88.4	1.46	4.98	0.74	1.38
Pleurotus sp.	90.0	0.97	2.78	0.65	1.08

PREDICTIVE ANALYSIS OF THE PROTEIN QUALITY OF EDIBLE MUSHROOMS

Since mushrooms form the important source of vegetable protein, the determination of the quality of such protein becomes essential. Mushroom, like other proteinaceous foods, bears a non-essential fraction of protein which does not come into use in biological system. *Pleurotus citrinopileatus* was analyzed for amino acid composition to determine its protein quality as compared to whole egg and FAO reference protein. An assessment of protein quality standard (EAA-Index, Amino Acid score, Biological value and nutritional index) indicates the good nutritional significance of the test mushroom, which can also be compared with protein standard quality of different edible mushrooms and the superiority of a particular mushroom in terms of quality of protein.

In the absence of properly designed biological method of feeding trial, there has been considerable difficulty to evaluate the standard of a mushroom protein. As a legible alternative, several biochemical methods have been advanced to make a predictive analysis based on the essential amino acid composition. It is well documented that the standard of a proteinaceous food depends on its amino acid composition in relation to the protein content and digestibility. Amino acid composition may also serve as a good relative measure to compare mushroom with other foodstuffs of established nutritive value. Based on this principle, Oser (1951, 1959) proposed essential EAA Index to rate the dietary protein in terms of highly nutritive reference protein. Bano and co-workers calculated the protein quality and nutritive value of several species of *Pleurotus* (Bano, 1981). A comprehensive study on the nutritive value of Saudi Arabian truffles was conducted (Sawaya, 1985) and their digestibility was found to run between 82.8-86.7 percent.

Fruiting bodies of the mushroom are usually collected immediately before spore maturity, dried (105°C) to constant weight and ground through mesh sieve before employing for chemical analysis. Amino acid content of the mushroom is analyzed using Techicon sequential multipurpose amino acid analyzer, which has been modified and described by Gopinathan (1986), and the Essential Amino Acid Index (EAAI) is determined.

Essential Amino Acid Index is the ratio of essential amino acids contained in a food to the essential amino acid content in reference protein. Amino acid score, also called chemical score, is considered as second alternative to the animal feeding studies for the determination of nutritional value. It is based on the amount of limiting amino acid present in the test protein in relation to its presence in reference protein. Amino acid score is calculated using the following formula:

$$\text{Amino acid score (AAS)} = \frac{\text{mg amino acid/gm test protein}}{\text{mg amino acid/gm reference protein.}}$$

Biological value is the calculated measure, which helps to predict the nitrogen retention by the feeding organisms and thereby indicates the utilizable fraction of a protein food. Based on the values of EAA-index, Biological values are calculated following the regression equation used by Oser, (1951).

Biological value (BV) = 1.09 (EAA-Index) - 11.7.

Qualitative as well as quantitative variations of protein values in different species of mushroom make it difficult to compare their nutritional status. To overcome this problem, Crisan and Sand advanced an equation considering all the factors with equal importance.

$$\text{Nutritional Index (N1) EAA} = \frac{\text{Index x protein\%}}{100}$$

Quantitative data of amino acid profiles of the mushrooms in crude protein nitrogen show variations in relation to the species. It was found that ***P. sajor-caju*** marginally exceeded over *P.citrinopileatus* except that of phenylalanine and threonine content. The phenylalanine content (460 mg/g) in the latter species was remarkably higher than the former (363 mg/g), whereas threonine contents were almost equal.

Studies on the amino acid score and sequences of limiting amino acids indicated that sulphur containing amino acids were the most limiting amino acids in both the fungi when whole egg was used as reference protein. In a study with *P.citrinopileatus* and *P. sajor-caju*, it was found that the amino acid score was 46 and 53, respectively with the sequence of limiting amino acids as sulphur containing amino acids (SC), Isoleucine (Ileu) and Valine (Val) in the former species, whereas, Ileu and Leu (Leucine) were the sequence found in the latter species. But, based on AO reference pattern, Ileu and Leu become the limiting amino acids in two species with amino acid score of 74 and 79, respectively. These sequence of limiting amino acid are changed to Ileu, sc and Val in *P. citrinopileatus* Leu, Ileu and sc in *P. sajor-caju.*

EAA-Indices of *P. citrinopileatus* based on whole egg and EAA reference pattern were 68.41 and 92.55 respectively when compared to the slightly higher values obtained in *P. sajor-caju* were 72.04 and 97.50, respectively. Thus, it has been concluded from the calculated values that similar dietary standard of protein is found in both the species of mushrooms. Biological value (BV) and Nutritional Index (N_1) are used along with the values of EAA-Index and Amino acid score to evaluate and compare the nutritive value of the mushroom species. Predicted values of all the parameters in the two species of mushroom varied, indicating a good correlation from the nutritional point of view.

Thus, it is obvious that protein is the most critical component contributing to the nutritional value food items such as mushroom. The determination of its true nutritive value in relation to protein is a complex task because it contains a significant amount of non-protein nitrogen in the chitinous cell well that interferes during the estimation of protein. In the absence of sufficiently critical mechanical device or biological feeding studies, most reliable measure presently available for determining the nutritive value of mushroom protein appears to be their amino acid content, especially the essential amino acid composition. Analytical data show that, all the essential amino acids required for human nutrition are present in adequate quantities in *P. citrinopileatus* and thus ensure its standard of higher protein equality.

Calocybe indica (ib)

In this section, the attention of the readers is drawn towards *Calocybe indica* , which is one of the most suitable mushrooms that could be cultivated in the tropical parts of the country. *Calocybe indica* is an another important edible mushroom that grows well over the paddy straw substrates and coconut coir pith with very good bioefficiency. It is popularly known as milky mushroom, which is more sturdy and has longer shelf-life, under normal conditions, than *Pleurotus*. It is indeed one of the most suitable mushrooms that can be recommended for cultivation in the tropical climate, because the optimum temperature for the crop yield is around 35°C. It is not only a food grade or edible mushroom but also one of the most efficient candidates, which could be exploited for the production of organic manure through the process of composting. Hence, the battery of enzymes produced by this mushroom can be exploited for the recycling of crop residues and agricultural wastes.

The nutritional value of the mushroom has been estimated and protein content has been found to be in the range of 17-24 mg per gram. The protein content of this mushroom varied depending on the substrate used. For example, it is 17.6 mg, 19.2 mg, 20.4 mg and 23.5 mg per gram of the fresh wight of the mushroom when grown on wood shavings, coir pith, paddy straw and banana trash respectively. Similarly, there is also variations in the total free amino acid content of the mushroom. Nevertheless, the amino acid content is maximum when it is grown on substrates such as banana trash, followed by coir pith, wood shavings and paddy straw.

The carbohydrate content of *C. indica* is in the range of 25-35 mg per gram fresh weight of the mushroom. It also varies depending on the substrate used for cultivation. In our study it was found that the maximum carbohydrate content was recorded with the milky mushroom cultivated on paddy straw substrates follwed by coconut coir pith, banana trash and wood shavings.

C. indica contains essential vitamins such as niacin, ascorbic acid, riboflavin and thiamine. Comparatively niacin is found proportionately higher than other amino acids.

Fatty Acid Profile

Fatty acids are also one of the important constituents of diet. There are three essential fatty acids, namely linoleic, linolenic and arachidonic required in the human diet for proper growth and development. Unsaturated fatty acids in the diet show decreased incidence of atherosclerosis, heart disease, cancer and other significant health problems.

Lipids are used by many organisms, including human beings for energy storage. Fatty acids are stored in organisms in the form of triglycerols or fats.

Fatty acids on oxidation provide more metabolic water than other metabolic fuels. This is helpful to mammals in overcoming dry environment.

There are many health benefits from dietary and supplemented use of the omega-3 acids, which include:

(i) prevention of cardiovascular disease, lessening of angina (chest pain) and lowering of blood pressure. The benefits are thought to be due to inhibition of platelet formation which has the effect of thinning the blood and thus reducing the risk of blood clotting;

(ii) prevention of some cancers including colon, breast, prostate and others; and

(iii) treatment of joint stiffness and swelling caused by rheumatoid arthritis.

Fatty acid analysis revealed the presence of altogether eighteen fatty acids. There is qualitative and quantitative variation in the fatty acid content of *C. indica,* when grown on different substrates. *C. indica* exhibited, 13. 12 and 8 different kinds of fatty acids when grown on wood shavings, coir pith and banana trash, respectively. Lauric acid, pentadecanonic acid, palmitic acid and heptadecanoic acid were commonly found in *C. indica* grown on all the four substrates. The presence of individual fatty acids in different mushroom samples also showed variations. In the mushroom grown on paddy straw, undeconoic acid (12,54 mg/g), non-deconic acid (6.92 mg/g), elaidic acid (6.47 mg/g), nanonoic acid (4.02 mg/g) and palmitic acid (3.56 mg/g) were commonly found. Similarly, the mushrooms from coir pith showed nanonoic acid (11.17 mg/g), elaidic acid (3.96 mg/g), non-deconoic acid (3.50 mg/g) and palmitic acid (2.24 mg/g); from wood shavings had non-decanoic acid (4.98 mg/g), palmitic acid (1.75 mg/g) and elaidic acid (1.61 mg/g) and in banana trash, nanonoic acid (6.74 mg/g) and lauric acid (4.47 mg/g). Other substrates of nutritive value found in *C. indica* are provided in Table 3 given below.

Table 3. Growth and yield performance of *Calocybe indica* on different substrates (n = 5).

Substrates (250 g dry wt)	DSR	DFPH	DFFB	DFH	TD	Yield (fresh wt. in g)	Bioefficiency (%)
Paddy straw	8.2	11.3	10.0	5.3	34.8	315.8	126.2
Coir pith	10.1	13.0	6.1	7.2	36.4	276.2	110.4
Wood shavings	12.2	15.1	12.1	6.1	45.5	117.5	47
Banana trash	9.3	10.2	9.0	4.0	32.5	365	142

Table 4. Moisture, protein, total free amino and carbohydrate contents of *C. indica* cultivated on different substrates (n = 5).

Substrate	Moisture (%)	Protein (mg/g)	Total free amino acid (mg/g)	Carbohydrates (mg/g)
Paddy straw	87.9	20.4	3.00	32
Coir pith	89.1	19.2	4.08	30
Wood shavings	85.4	17.6	4.92	26
Banana trash	86.1	23.5	5.10	29

Table 5. Crude fat, crude fibre, nitrogen and ash contents of *C. indica* cultivated on different substrates (n=5).

Substrate	Crude fat (%)	Crude fibre (%)	Nitrogen	Ash
Paddy straw	3.25	26	4.75	6
Coir pith	2.1	31.7	3.74	7
Wood shavings	3.78	33	3.32	8.1
Banana trash	4.34	27	5.49	5.4

Table 6. Mineral content of *C. indica* cultivated on different substrates (mg/100 g fruit bodies) (n=5).

Substrate	Iron (mg/100 g)	Manganese (mg/100 g)	Zinc (mg/100 g)	Copper (mg/100 g)
Paddy straw	1.8	0.25	2.23	1.07
Coir pith	1.67	0.21	1.31	0.53
Wood shavings	3.08	0.35	1.28	0.32
Banana trash	1.[illegible]0	0.33	0.90	0.20

Table 7. Vitamins content of (mg/100 g fresh weight) *C. indica* cultivated on different substrates (n=5)

Substrate	Thiamine (Vit.B_1)(mg/100 g)	Riboflavin (Vit.B_2)(mg/100 g)	Niacin (Vit.B_3)(mg/100 g	Ascorbic acid (Vit.C)(mg/100 g)
Paddy straw	0.04	0.26	2.8	0.61
Coir pith	0.11	0.24	1.86	0.32
Wood shavings	0.02	0.12	0.92	0.13
Banana trash	0.08	0.38	1.87	0.22

Table 8. Enzyme content (fresh wt.) in the fruit bodies of *C. indica* cultivated on different substrates (n=5)

Substrate	Cellulase (Cx)[1]	Pectinase[2]	Polyphenol oxidase[3]
Paddy straw	13	12	4
Coir pith	15	16	2
Wood shavings	9	10	5
Banana trash	21	18.2	2.4

Table 9. Cellulose, hemicellulose, lignin contents and energy value of different substrates used for the cultivation of *C. indica*

Substrates (250 g dry wt)	Cellulose(%) (A)		Hemicellulose (%) (B)		Lignin (%) (C)		Energy value of substrate (kcal/100 g substrate) (A+B+C)
	Cellulose content	Energy value	Hemicellulose content	Energy value	Lignin content	Energy value	
Paddy straw	38.8	162.96	22.59	94.88	18.73	133	391
Coir pith	34.16	143.47	27.5	115.5	29.12	207	466
Wood shavings	36.79	154.52	15.63	65.65	22.51	150.8	380
Banana trash	50.89	213.74	19.94	83.75	19.45	138	435

Table 10. Energy recovery of various agro residues in fruit bodies of *C.indica*

Substrate	Energy value of substrate (kcal/100 g substrate)	Energy value of mushroom (kcal/100 g mushroom)	Yield of mushroom (g/100 g substrate)	Energy value of mushroom (kcal/100 g substrate) (2 x 3/100)	Energy recovery of substrate in mushroom (%) 4x100/1
	(1)	(2)	(3)	(4)	
Paddy straw	391	239.4	15.27	36.56	9.35
Coir pith	466	204.9	112.03	24.65	5.29
Wood shavings	380	182.6	6.86	12.53	3.30
Banana trash	435	243.8	19.74	48.113	11.06

Table 11. Carbohydrate content (%) in different substrates and spent substrates

Substrates	Carbohydrate content		% loss
	Substrate	Spent substrate	
Paddy straw	16	10.5	34.4
Coir pith	11.4	8.2	28.1
Wood shavings	8.6	7.1	17.5
Banana trash	20.9	12.6	39.7

Table 12. Micronutrients in different substrates and spent substrates

Substrate	Calcium (%)			Magnesium (%)			Iron (%)			Copper (%)		
	S	SS	% gain	S	SS	% gain	S	SS	% gain	S	SS	% loss
Paddy straw	0.41	0.59	43.9	0.47	0.59	25.5	0.07	0.10	42.8	0.03	0.026	13.4
Coir pith	0.82	0.99	20.7	0.27	0.32	18.5	0.41	0.49	19.5	0.018	0.016	11.2
Wood shavings	0.29	0.33	13.8	0.65	0.71	9.2	0.07	0.09	28.5	0.04	0.03	25
Banana trash	0.84	1.19	41.6	0.81	0.95	17.2	0.12	0.19	58.3	0.03	0.025	16.7

Abbr.: S-Substrate; SS-Spent substrate

Table 13. Percentage loss of cellulose, hemicellulose and lignin contents in four different substrates caused by *C. indica*

Substrate	Cellulose (%)			Hemicellulose (%)			Lignin (%)		
	S	SS	% loss	S	SS	% loss	S	SS	% loss
Paddy straw	38.8	27	30.4	22.59	17.75	21.4	18.73	11.32	39.6
Coir pith	34.16	25.6	25.1	27.5	18.72	32	29.12	22.24	23.6
Wood shavings	36.79	31.49	14.4	15.63	12.72	18.6	22.51	18.20	19.2
Banana trash	50.89	32.04	37	19.94	15.26	23.5	19.45	13.23	32

Besides being an alternative source of food, *C. indica* is also an important source of some important extracellular enzymes such as cellulase, pectinase, polyphenol oxidase.

ENZYMES

Cellulase

Analysis of cellulase, endo 1-4 glucanase production by content of *C. indica* showed greater variations as influenced by different substrates. Mushroom cultivated on banana trash recorded the maximum concentration glucose/min/followed by coir pith and paddy straw. The lower concentration of protein was recorded in *C. indica* cultivated on wood shavings.

Pectase

Pectase content showed similar trend as that of cellulase, which was maximum with *C. indica* grown on banana trash followed by coir pith paddy straw and wood shavings.

Polyphenol Oxidase

It is interesting to note that the polyphenol oxidase produced by *C. indica* was maximum on the substrate, wood shavings.

Among the three enzymes viz., cellulase, polyphenol oxidase and pectase, polyphenol oxidase was much less, while the other two enzymes were produced relatively in high amount with less variation. These enzymes are involved in the biotransformation of lignocellulosic organic substrates.

Degradation of Lignocellulosic Substrates by Fungi

Enzymic catalaysis is generally considered to offer several potential advantages over chemical processing. There is greater substrate and reaction specificity, lower energy requirements, lower pollution generation, higher yield of desired products and opportunities for transformations which are not feasible with chemical reagents.

Studies conducted using several species of fungi isolated from the retting yards, coir industrial environment, coir decomposing heaps and known polymer reducing microorganisms showed that few of the laccase positive fungi have detoxified the phenolic substances. Besides laccase, a variety of biopolymer degrading enzymes were released by *Pleurotus sp.* during its growth and multiplication on coir pith. Species of pleurotus is reported to produce laccase enzyme, three peroxidase positive proteins and two glucose oxidases (Oji Paper, 1996). Cellulose and hemicellulose present in the coir pith supported the initial growth of the fungus and these two were the co-substrates for lignin degradation (Martinez *et al.*, 1996 and Ouseph *et al.*, 2001).

Fast substrate degradation and conservation of nitrogen besides the micronutrients are the criteria for selecting an organism for improving the decomposition rate of coir pith. Bisaria *et al.* (1983) showed ***P. sajor-caju*** as one of the most efficient lignin degrading organisms or lignocellulosic waste like coir pith. Theradi Mani and Marimuthu (1992) studied the capacity of several *Pleurotus spp.* and other edible fungi for their production of cellulases and laccases in order to select the most efficient strains.

Compost

Production of enzymes and lignocellulosic degradation of paper mill sludge have been studied and established with *P. sajor-caju* during its growth phase (Kannan *et al.,* 1990).

This fungus has also been found as a good decomposer and used in the degradation of rice husk. The recovery of lignocellulose-degrading enzymes from spent mushroom compost was assessed using a variety of physical and chemical treatments. Maximum recovery of active xylanase and its activity were found in extracts from spent compost which had been physically treated by blending. A range of lignocellulose-degrading activities has been studied in blended compost extracts. Peroxidases, the xylan debranching enzymes acetyl esterase, arabiofuranosidase, and the cellulose degrading activities due to the production of endoglucanase, cellobiohydrase- and β-glucosidase have been reported from the extracts of the composition (Ball and Jackson 1995, Singh *et al.*, 2001).

It is obvious from the foregoing paragraphs that the edible mushrooms are very good source of nutrients and extracellular enzymes, which are involved in the disintegration of lignocellulosic composts. This potentiality could be successfully exploited in the biotransformation of lignocellulosic organic wastes into composts. These composts can be used as organic manures in the organic farming system.

Thus, the higher fungi are the good source of enzymes, whose production could be optimized and industrially exploited. The organisms can be directly used for the bioconversion of lignocellulosic organic wastes into simple sugars through the process of saccharification, which could be used for ethanol production. Hence, such organisms could be considered to reduce, recycle and reuse the organic wastes.

Marimuthu *et al.* (1999) studied the importance of coir pith decomposition and transformation into compost by *Pleurotus* for preparing biomanure. They have clearly established that the composed coir pith has:

(i) improved the physical properties of soil;
(ii) increased the quality of the field and the yield of the horticultural crops;

(iii) solved the problems of degradation of soil; and
(iv) the advantage of being used as organic amendment thus enhancing the soil fertility.

Composting Method

A simple and easy method of composting of the coconut coir pith has been standardized. One tonne of raw coconut coir pith with 5 kg of urea and five bottles of ***Pleurotus spawn*** (1.5 kg of Sorghum grain spawn) are required (Nagarajan, *et al.,* 1988) to make compost within a short period of time. The period for the preparation of coir pith compost is as follows:

- About 100 kg of raw coir pith previously moistened (200%) and spread uniformly in a shaded place on the floor in an area of 5 × 3 m.
- Remove the spawn from one bottle and spread uniformly over the surface of the moistened coir pith.
- Spread the second layer of 100 kg moistened coir pith over the first layer and sprinkle 1kg of urea uniformly over the second layer.
- Repeat this process of sand witching of coir pith and mushroom spawn followed by another layer of coir pith and urea upto one metre height.
- Sprinkle water now and then to maintain 200 percent moisture (by touching the coir pith one should feel the wetness).
- Allow the heap to decompose for a month; the well decomposted coir pith will be black in colour, which is used as organic manure.

Spawn is nothing but the mushroom fungal growth maintained on grain based medium. Spawn is prepared as per the method of Marimuthu *et al.* (1989) and can be used for the cultivation of edible mushroom as well as for coir pith composting.

Production of *C. indica* as Edible Mushroom

This fungus can be grown on a wide range of lignolytic and cellulolytic wastes. The highest biological efficiency of 120% has been reported on radish waste (after seed extraction). However, due to the limited availability of radish waste, the same cannot be recommended as a substrate for the commercial production. Hence, paddy straw, which is available in plenty in Tamil Nadu, is recommended as the substrate for its cultivation.

Substrate preparation: Paddy straw is chopped up in small pieces and soaked in clean fresh water for 6-8 hours. This process helps to make the paddy straw softened.

Pasteurization of the substrate: The process helps to eliminate the harmful microbes. It is done in two ways, which include hot water treatment and steam pasteurization.

Hot water treatment: Water is boiled in large tank and the paddy straw pieces filled in gunny bags are submerged in this hot water. The temperature of the water is maintained at 80-90°C for 40-60 minutes.

Steam pasteurization: Wet straw is filled inside an insulated room either in perforated shelves or in wooden trays. Steam is released under pressure from a boiler and the temperature within the room is maintained at 65-70°C for 5-6 hours. Air inside the room is to be circulated in order to ensure the uniform temperature in the room and the substrate. After pasteurization the substrate is shifted to spawning room for bag filling and spawning.

Bag filling and spawning: After pasteurization the substrate is cooled on a wire mesh at room temperature. Once cooled, the straw is filled in the bags and spawned layer wise simultaneously. Each bag (35 × 45cm) can hold 2.5 to 3.0 kg of wet substrate. Spawning is done on 4% wet substrate. It involves the broadcasting of spawn on substrate. After spawning the bags are packed with gentle pressure; care should be taken to see that the bags are not packed tightly. Then they are plugged with non-absorbent cotton by making a neck of PVC pipe (diameter 3.5 × 1.5 length) and the bags are then shifted to spawn running room.

Surface spawning: Pasteurized substrate is filled in bags and spawn is broadcast on the top and ruffled. This process is followed when bags with substrate are sterilized in autoclave/ steam sterilizer.

Layer spawning: This method is adopted when hot water or steam pasteurization is done. In this method the substrate is spread in layers (about 8-10 cm) and each layer is followed by a layer of spawn. This process is continued till the whole bag is filled up in 3-4 layers. The polythene bag (35 × 45 cm) holds about 3 kg of wet substrate.

Thorough spawning: In this method spawn is thoroughly mixed with the substrate and filled in bags.

Spawn running: This is the phase of growth and establishment of mushroom mycelium over the substrate. Fresh air requirement at this stage is negligible. High carbon dioxide favours growth and the room should be kept dark. A temperature in the range of 28-35°C should be maintained. Constant inspection must be carried out and any contaminated bag must be discarded. Spawn running is completed when the entire substrate is covered with the mushroom mycelium. This may take 20-25 days.

Casing

After the completion of spawn running the bags are cased. Casing is the process of putting a layer of pasteurized soil on the substrate to induce fructification in addition to giving physical support to the fruiting bodies.

Pasteurization of Casing Soil

Casing material is chemically pasteurized 2 weeks before casing by drenching it with formaldehyde solution (5%), so that at the time of casing it is free from formalin. However, steam pasteurization (65°C for 2 hours) is most effective. Casing material (moist) is filled in trays with 2-3 inches layers and stacked inside pasteurization chamber one above the other in rows and 50 cm above the ground. There should be gap between trays and wall (about 30 cm) to facilitate steam movement. Steam is released at ground level and inside air should be circulated with the help of blower connected to duct opening at ceiling level and 20-30 cm from ground level to maintain uniform temperature. Casing material's temperature should be maintained at 65°C for 2 hours. Temperature should be monitored from outside with the help of dial/digital thermometer with probes/sensors inserted in the casing material at top, middle and bottom level trays. Pasteurization can also be done in autoclave at 15 lb pressure for 30 minutes. When cultivation is done on a small scale, soil can be pasteurized in pressure cooker too. However, for large-scale production a separate pasteurization room is needed. The fully colonized bags are shifted to the cropping room and opened. The plastic is rolled downwards and the substrate is leveled. A layer of cool pasteurized soil (2-3 cm depth) is spread uniformly on the top of the substrate. Immediately after spreading the soil, the bags must be sprayed with a solution of bavistin (0.1%) as an additional precautionary measure. Watering must be carried out from next day onwards.

Cropping

It takes 15-20 days after casing for the induction of pin head initials. During this phase a temperature of 28-35°C (optimum 30-32°C), R.H. of 80-90% and good cross-ventilation must be maintained. Watering must be carried out with a sprayer and care must be taken to avoid overwatering. Light (diffused or fluorescent tubes) can be used on 12 hour cycle. If the stalks tend to become long with a small pileus, fresh air must be introduced. It takes 8-10 days for the maturation of the fruit initials. The mushrooms are harvested when the pileus diameter is 6-10 cm. The stalk is gently cut at the base to remove the adhering soil and other debris. Mushrooms can be marketed fresh or can be stored in PP bags with holes in refrigerator for 20-25 days.

CROP MANAGEMENT

During substrate preparation: The substrate is the major source of contamination. Therefore, the storage and chopping of straw should be done away from the growing rooms. Workers involved in chopping and handling of straw should not do the bag filling or spawning.

During pasteurization: The substrate should be properly pasteurized and the temperature of the water must be maintained at 80-90°C for 40-60 minutes. The pasteurization chamber must be near the bag filling room so that the pasteurized straw is minimum exposed to the air to avoid air contamination.

During bag filling, spawning and spawn running: Bag filling must be carried out after draining excess of water and cooling the straw to room temperature. The bag filling room should be a closed room with highly restricted entry. It must be sprayed with 1-2% formaldehyde, 0.1% bavistin and 0.1% nuvan (individually) twice a week. The workers doing bag filling should not have entered any contaminated area and their hands and instruments used for loosening the spawn must be cleaned with alcohol. The spawned bags must be incubated at the right temperature in completely closed spawn running room. Constant inspection of the bags must be carried out and any contaminated bag must be either chemically treated (0.1% bavistin) or must be discarded from the room, as such bags serve as the source of secondary infection.

During cropping: Proper temperature, humidity and light must be maintained in the cropping rooms. Proper cross-ventilation should also be provided to prevent the built up of carbon-dioxide which adversely affects the fruiting bodies. Care should be taken neither too underwater or to over water the crop. A fine wire mesh must be provided at the doors and windows to prevent the entry of flies.

Stage of Mother Spawn Preparation using Sorghum Grains

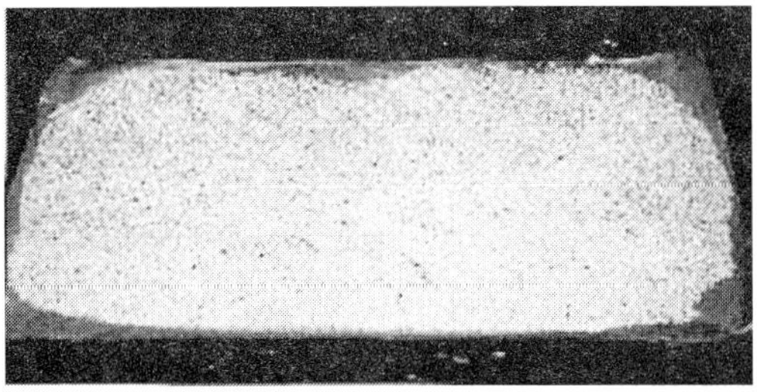

Half Cooked Sorghum Grains

Initial Stage (just inoculation)

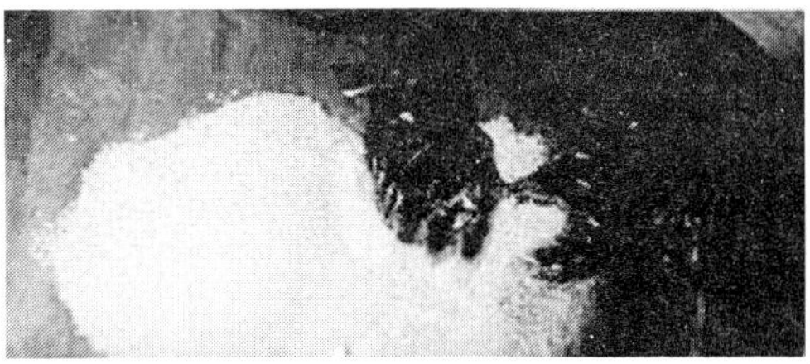

Mixed with Calcium Carbonate

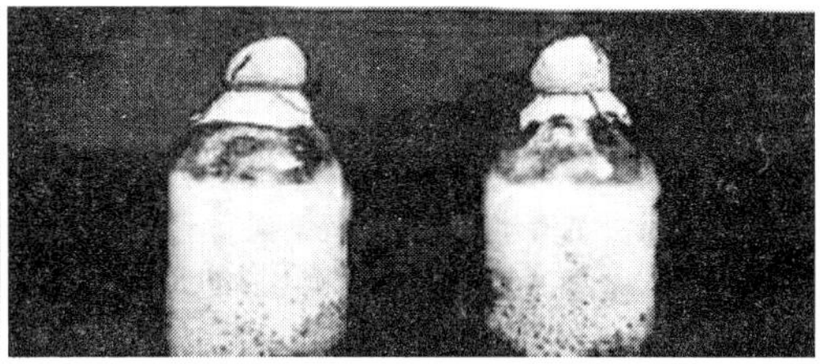

After 10 days of inoculation

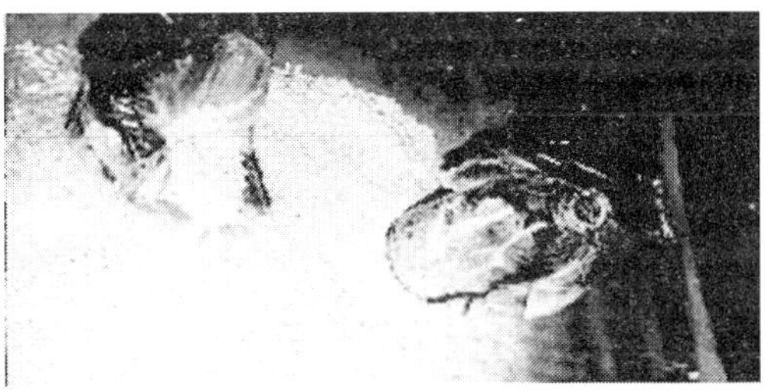

Filled in a Saline Bottle

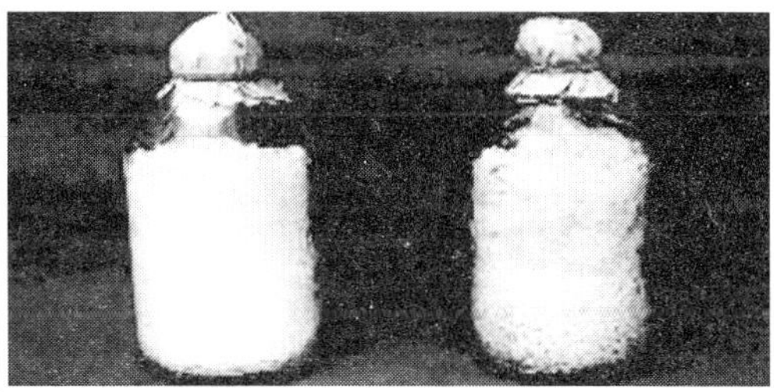

After 15 days of inoculation

Oyster Mushrooms

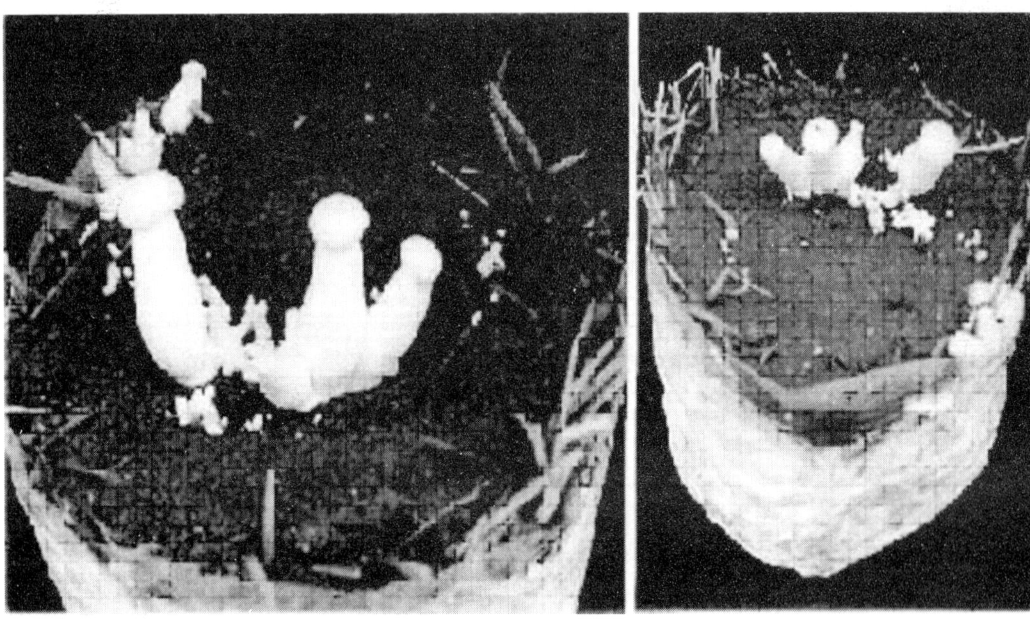

Milky Mushrooms

P. djamor

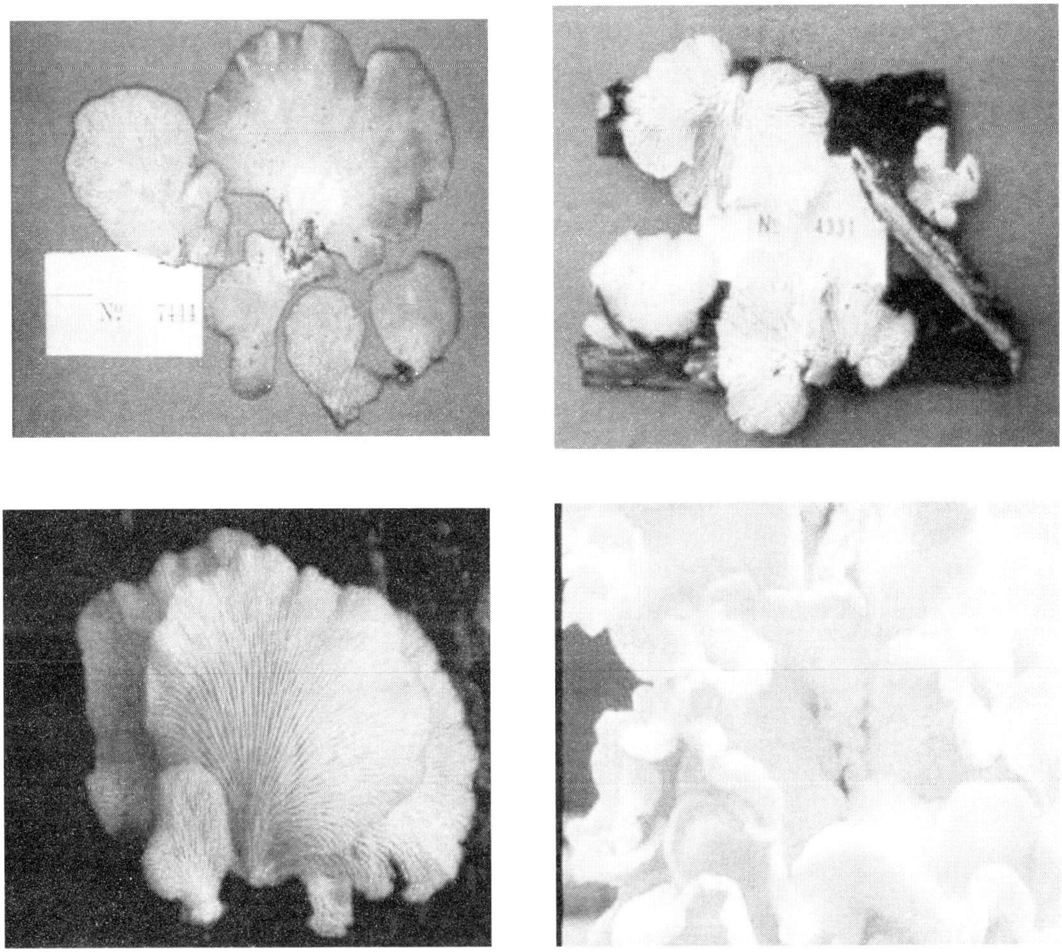

Post-cropping period: The substrate left over after complete cropping is referred to as spent substrate. The spent substrate must be sprayed with formaldehyde (4%), bavistin (0.1%) and nuvan (0.1%) individually, 24-48 hours prior to discarding. This helps in the reduction of pathogen/weed moulds/flies population in and around the farm. After discarding the substrate the cropping rooms must be washed, dried and sprayed again with the earlier mentioned chemicals. New crop must be put in the cropping rooms only after a break of 4-5 days.

CONCLUSION

Thus, it is obvious from the foregoing paragraphs that the higher fungi have the scope of application in various fields. There is no doubt that they are the industrial process organisms in the brewing, backing, pharmaceutical and many other industries. Yeasts have been used in the alcohol industries even before their involvement in the alcohol production was established by Louis Pasteur, and formed the pioneer organism in the classical biotechnology. Similarly, there are many organisms utilized in various industries, and many more organisms remain unexplored and underutilized. There are potentialities for manipulating the existing industrial organisms to increase their yield. So it is high time to explore the microorganisms, especially fungi, from diverse environments in order to properly exploit them for the welfare of human beings.

REFERENCES

Ball, and Jackson. (1995). *Bioresources Technology* **54:** 311-314.
Bisaria, R., Madan, M. and Mukhopadyaya, S.N. (1983). *Biotech. Lr.,* **5:** 811-12.
Brieley, J.A., Goyak, G.M. and Brieley, C.L. (1996). *In:* Immobilisation of ions by bisorption (Eds. Eccles, H. and Hunt, S.), Ellis, pp. 105-107.
Corning, S.T. and Milles, P.J. (1989). Edible Mushrooms and their Cultivation, Boca-Raton: CRC Press, pp. 345.
Corning, D.R. (1976). *J. Soc. Leather Tech. Chemists,* **62(3)**.
Ghosh, N. and Chakravarty, D.K. (1990). *J. Fd. Sci. Technol.* **27(4):** 236-238.
Govindan, V.S. (1982). *J.* Singapore. 22-37.
Gray, W.D. (1959). The Relation of Fungi to Human Affairs, Henry Hold, New York.
Hutterman, A. and Trojanowski, J. (1987). Erich-Schmidtverlag, Berlin, 205-218.
Kakabodse, G. (1979). Applied Science Publishers Ltd., London.
Kannan, K.G., Oblisami, M. and Ad Loganathan, B.G. (1990). *Biological wastes,* **33:** 1-8.
Labropoulosi, D.A. (2000). Dept. Techn. Educ. Inst. Athen, IFT Annual Meeting.
Mohamud, B.H. and El-Kattan, M.N. (1989). *Mush. J. Tropics.,* **9:** 37-42.
Manivasakam, N. (1987). Sakthi Publications, Kovaiputhur-Coimbatore.
Marimuthu, T., Nallathambi, P. and Nagarajan, R. (1999). *Agriculture,* 30-40.
Martinez, M.J., Bockle, S., Camareso, Gvillen, F. and Martinez, A.T. (1996). *American Chem. Soc.* **211:** meet USA.
Nema, P., Mehta, C.G. and Rajagopalan, S. (1984). *Ind. J. Environ. Health.,* **26(1):** 232-334.
Oser, B.L. (1951), *J. Am diet Ass.* **1:** 27, 396.
Quseph, A., Geetha, D. and Sudharban, M. (2001). *Mushroom Research.* **10(1):** 37-40.
Paper, O. (1996). *Biotechnol. Abst., 121 P (Abst.).*
Ramachandran, S. and Sinha, S. (1993). *J. Microbiology,* **33(1):** 25-34.
Raman, P., Agarwal, P. and Upadhyay, S. (2000). *J. CSIR,* **59:** 977-87.
Ramaswami, P.P., Sree Ramalu, U.S. and Kothandaram, G.V. (1985). Workshop on Coir Research, Coir House, Cochin, Kerala, India, pp. 55-63.
Sharma, S.R. (1992). *Mushroom Res.,* **1:** 119-121.
Silva, and Paulo, (2000). *Textile Research J.* **70(5):** 409-414.
Stoppoke, E., Wagner, F. and Zadrial, I. (1981). *European J. Appl. Microbial. Biotech.,* **13:** 60-61.
Sujatha Lakshmi, (2002). National Seminar and 12th Southern Regional Conference on Microbial inoculants, Annamalai University, India.
Suri, V.K. and Paramanik, P. (1995). *Indian Textile J.* **35(2):** 109-113.
Thakur, M., Shelke, V. and Swatipatil. (1998). *Indian Textile J.* **9(2):** 60-65.
Theradimani, M. and Marimuthu, T. (1992). *Mush. Res.,* **1:** 49-51.

Chapter 7

Cordyceps sinensis (*Yarsa gamboo*): A High Value Medicinal Mushroom

R.C. Gupta[1] and P.S. Negi[2]

[1]*Department of Botany, Kumaun University, SSJ Campus, Almora*

[2]*Defence Agricultural Research Laboratory, Pithoragarh (UK)*

E-mail: rcgupta_alm@rediffmail.com

Tibetans know the medicinal mushroom commonly as 'Yarsa gamboo', "summer grass-winter worm". Botanically the fungus is *Cordyceps* and is an entomophagous (feeding on insect), belonging to Pyrenomycetes, Ascomycotina in the family Clavicipitaceae. *Cordyceps sinensis* is a fungus parasitizing the larvae of a moth of the genus *Thitarodes (Hepialus),* which lives in alpine grasslands. After the insect is completely mummified and emptied of nutrients, leaving behind the larval exoskeleton filled with the *Cordyceps* mycelium, the fungus will develop a fruiting body out of the head above the eyes, where the larvae has a horn-like protuberance in early spring. The 5-10 cm long, brown, club-shaped fruiting body grows above ground to have its propagules dispersed by the wind in order to find a new host insect. The stroma is nearly twice as long as the caterpillar when fresh. The specimens of *C. sinensis* are very rare, occurring in the far reaches of high cold and arid hills at an altitude between 10,000 fts and 14,000 fts above the sea level. The fungus is sold by the local people at the rate of Rs. 30 to 35 per specimen or Rs. 70,000 to 80,000 per kg. The healing capacity of the mushroom can be summarized as a tonic to enhance vitality. Its main TCM applications are for treating exhaustion, TB, asthma, back pain and reproductive issues (i.e. premature ejaculation). Its use as an aphrodisiac seems to be the driving force with male consumers.

INTRODUCTION

The importance of *Cordyceps sinensis* for rural populations, especially nomads in the Tibetan areas, cannot be overstated. In past centuries it was an important bartering good to obtain tea from China, a mainstay in the Tibetan diet. Nowadays, it provides a cash income to households, which are still carrying out most of their pastoral and, where possible, agricultural activities on a traditional subsistence basis. It enables otherwise impoverished households to purchase goods, and to pay school fees, hospital bills or taxes. Collection of medicinal plants has a long-standing history in Tibetan culture, be it for personal use or for trade. With the continuous advance of a cash economy onto the Tibetan Plateau, collection of tradable plants and fungi, as medicine, as food or aroma source, has gained increasing importance, especially for rural populations who are otherwise participating only marginally in the new cash economy.

MEDICINAL PROPERTIES

Cordyceps sinensis contains crude protein cordycepin, an antibiotic (3'-deoxyadenosine) ($C_{10}H_{13}0_3N_6$) and *d*-mannitol. Owing to its high efficacy and potency in curing various diseases, this fungus was recommended by the ancient medicinal practitioners as the "*Panacea of all ills*". However, it caught the attention of the people since 1993 when a group of nine Chinese women athletes who have been taking *Cordyceps*, shattered nine world records in world outdoor track field championships in Germany. Having high efficacy and potency in curing various diseases, this fungus is in wide popularity as an important nourishing tonic and adaptogen, which increases the physical stamina. The research findings have indicated that this fungus enhances the oxygen supply to the brain and heart and thus improves the resistance to hypoxia.

In China *Cordyceps* has been in traditional use for the treatment of various diseases like chronic bronchitis, insomnia, hypertension, pneumonia, tuberculosis, pulmonary emphysema, anemia, night sweats and cough (Xie *et al.*, 1988; Zhu *et al.*, 1998). Based on the recent advaces in literature, the major pharmacological actions by *Cordyceps sinensis* on hepatic, renal, cardiovascular and endocrine systems, anticancer activity, immunomodulation and hypoglycaemic activity, and effects on erythropoiesis and haematopoiesis have been reviewed (Bao *et al.*, 1988; Wang-Sheng *et al.*, 2000). *Cordyceps sinensis* works as an immunomodulatory agent (Zhu, 1990; Kuo *et al.*, 1996). The various products of *Cordyceps* regulate the normal functioning of different body parts, strengthen the immune system and promote the overall vitality and longevity. This fungus increases IL-1, interferon and TNF production in *in vitro* Kupffer cells of rats and elevate serum level of these substances (Liu *et al.*, 1996). *Cordyceps* exhibits anti tumour and antiviral activities (Bok *et al.*, 1999). This fungus has a diuretic effect and thus is good for the prevention of nephralgia (Chiu *et al.*, 2000).

Cordyceps is found beneficial in case of climatic age illness, impotence, emission, neurasthemia, rheumatoid arthritis, cirrhosis, flabby waist and knee. Being a hormone stimulator, *Cordyceps* is an important anti-aging medicine. It also inhibits the formation of active monoamine oxidaze, an enzyme responsible for aging in man. The frequent use of this fungus may prevent the senile disorder. *Cordyceps* is useful in preventing arteriosclerosis, coronary heart diseases and other diseases related to the blood vessels of brain.

Cordyceps contains various bioactive components which regulate and ensure the various physiological activities in the human body. A nucleoside derivative called cordycepin was isolated from *C. militaris* (Cuningham *et al.*, 1950; Kredichand Guarino, 1961). The various biological activities regulated by this nucleoside derivative includes inhibition of DNA and RNA synthesis in cancer cells (Plageman and Erabe, 1971), enhancement of cell differentiation (Mathew *et al.*, 1989), restructuring of cytoskeleton in cells (Deitchand, 1979; Zeive and Roemer; 1988), inhibition of protein kinase activity (Glazer and Juo; 1977), antitumour activity in bladder, kidney, colon, lungs carcinoma as well as fibroblastoma (Hubell *et al.*, 1985), inhibition of infection and reverse transcriptase activity of human immunodeficiency virus type I (Montefiori *et al.*, 1989; Muller *et al.*, 1991), inhibition of methylation of nucleic acids (Noval-Fernandez and Leory, 1980) and inhibition of chemotaxis and specific protein synthesis of the macrophage cell line (Aksamit *et al.*, 1983). The water-soluble solution of *Cordyceps sinensis* is found to have the stimulatory effect in the production of corticosteroids (Wang *et al.*, 1998). *Cordyceps sinensis* has shown the outstanding activity against the human leukemia (Chen *et al.*, 1997). This fungus works as an immunomodulatory agent and thus confirms the presence of immunosuppressive ingredients in it (Zhu, 1990; Kuo *et al.*, 1996). Restoration of cellular immunological function is observed during the trials on patients suffering from advanced stage of cancer (Zhou and Lin, 1995). *Cordyceps sinensis* is also beneficial to the patients suffering from hyperglycemia and hypertension since it has the potential to reduce the blood sugar as well as the blood cholesterol (Kiho *et al.*, 1996).

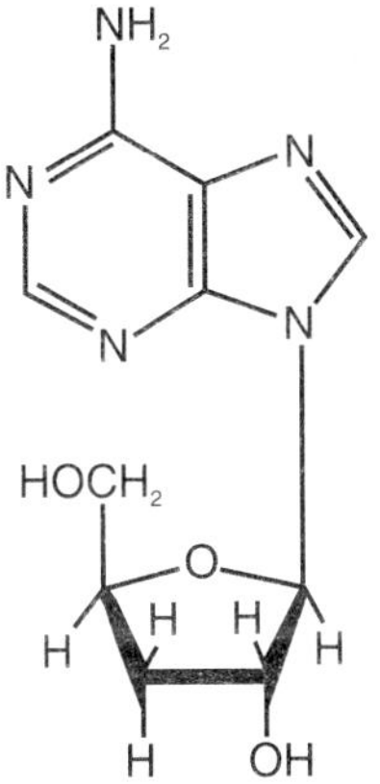

Molecular structure of Cordycepin (3-deoxyadenosine)

ECONOMY, ECOLOGY & ETHNO-MYCOLOGY

Cordyceps sinensis is a fungus parasitizing the larvae of a moth of the genus *Thitarodes* (*Hepialus*), which lives in alpine grasslands of the Tibetan Plateau. Collection and trade of the caterpillar fungus is one of the most important sources of income for pastoral Tibetan communities. Tibetans know the fungus as *Yarsa gamboo*, "summer grass-winter worm", and often derive over 50% of their annual cash income from its collection in spring and summer. Written records in Tibet date back at least 500 years, but the market is driven by Chinese consumers, who know it as chongcao (dongchong xiacao), a highly valued tonic in Traditional Chinese Medicine. A field study in Lithang County and other sites in Ganzi Tibetan Autonomous Prefecture, Sichuan (China), analyzed the local market, and the quantity and value of the harvest. Results are contrasted to Chengdu and Lhasa market prices, which peaked at 40,000 Renminbi/kg in 2003/4. Such information as is available regarding *Cordyceps* collection in other plateau regions, such as Golok TAP and Tibet Autonomous Region as well as the Himalayas, is integrated. In addition, the mycology (esp. reproduction, taxonomy and recognition of new species), ethno-mycology, distribution and geo-ecology of *Cordyceps* are presented, based on Western, Tibetan and Chinese sources to lay a foundation for further research on the issue of the sustainability of current harvesting rates.

Ethno-mycological Summary

The Tibetan name Yarsa gamboo (dbyar rtswa dgun 'bu) means "summer grass winter worm". Apparently "grass" (rtswa) is also used to denote other mushrooms such as *Ganoderma lutescens*, which is also collected as a medicinal mushroom for the Chinese market in some Tibetan areas. Boesi (2003) notes that this term describes the life stages of *Cordyceps*; Tibetans "believe that during winter the Yarsa gamboo lives as a worm and that, after a metamorphosis occurring at the beginning of spring, it transforms into a kind of grass. Tibetans recognise two distinct phases in the transformation process. At first, from the head of the larvae the "grass" starts growing. At this stage the worm, whitish in colour, is still alive and it is possible to see it moving over the ground with a short horn protruding from its head. Subsequently, as the season progresses, the horn continues to grow until the worm dies. When the metamorphosis is completed, the worm, [now] brownish-yellow in colour, is transformed into the root of a kind of grass".

The first mentioning of Yarsa gamboo in Tibetan Medicine can be traced to the 15th century according to Yonten Gyatso's ongoing research (com. 2004). Yonten points out that Yarsa gamboo is mentioned in Zurkhar Nyamnyi Dorje's (Zur mKhar mNyam Nyid rDo rJe's [1439-1475]) text: Man nGag bYe Ba Ring bSrel (Oral Instructions on a Myriad of Medicines). Thus, Tibetan Medicinal use of *Cordyceps sinensis* is clearly documented for over 500 years, predating the first reference in Chinese medicinal texts by three centuries. In Gawä Dorje's (1995) contemporary compendium on Tibetan Materia Medica Yarsa is placed in the category of "medicinal essences" (rTsi sMan), which includes several tonics. It is used for general strengthening, boosting the immune system, vitality, and is prescribed for kidney and heart problems. It is also used for treatment of Hepatitis B. In Tibetan Medicine Yarsa is mostly prescribed in composite remedies, which contain a variety of ingredients to balance each other, thus optimizing their efficiency and minimizing side effects.

It is interesting to note that Yarsa is collected by a majority of Tibetans who base their subsistence on traditional land use techniques, although there used to be a traditional taboo placed on digging Yarsa and other medicinal plant roots. Namkhai Norbu (1997), who grew up in Derge (Ganzi TAP), reports that "Rigya [ri rgya], the general laws that govern relations with the environment", forbid the digging of certain tubers and Yarsa gamboo. Nomads of Dzachukha and Sertar regard Yartsa as a treasure of "the earth spirits". Digging Yarsa gamboo provokes these earth spirits who will strike the offender, his family and clan with sickness and punish his livestock with ill health. This Tibetan belief probably predates Buddhist thought. However, in Lithang Boesi (2003) did not encounter such beliefs. He speculates that "this may be related either to the loss of that belief in the area or to a different local tradition". It is very plausible that the taboo has been lost in Lithang. It can be speculated that throughout the centuries Tibetans observed Chinese collectors digging in Lithang without suffering too many ill effects. Also pointing in this direction is the fact that in Golok Yarsa gamboo was not really exploited by Tibetans in the past, but large-scale exploitation was ushered in after PRC took control in the 1950s. Before that Golokpas were famous for keeping all outsiders, Chinese or Western, out of their territory, thus very successfully protecting their traditions, while Lithang was much earlier impacted by Chinese imperial policy.

Chongcao in Traditional Chinese Medicine

In Traditional Chinese Medicine (TCM) *Cordyceps sinensis* is known as dôngchóng xiàcão, "winter-worm summer-grass", clearly the literal translation of the Tibetan name. Commonly it is just referred to as "chóngcão", "worm-grass". In TCM it was first recorded in Wu Yiluo's "Ben cao cong xin" ["New compilation of materia medica"] in 1757 (Zhu, Halpern and Jones, 1998). However, Chinese doctors had learned about the healing power of *Cordyceps sinensis* earlier. Already in 1736 the Jesuit Du Halde describes how the Emperor's physician at the Court in China treated him successfully with "Hia Tsao Tong Tchong". Du Halde mentions the fact that it is hard to "procure" and that its value is four times its weight in silver. It can be taken as a food ingredient, traditionally as a stuffing of duck, chicken or pork, as a soup ingredient, or ground up. *Cordyceps* tastes agreeable and I would describe it as primarily fungal without much difference in taste between the fruiting body and the former body of the larva. The recommended daily dosage is 3-9 g (Yao and Zhang, 1996), which would cost about 75-350 RMB [$9-$43] in 2003 retail prices, making it an extremely costly treatment. In general chongcao is used as a tonic in TCM to strengthen the system and regain energy, somewhat similar to the tonic functions of ginseng (*Panax* spp.). Liu (1994) summarizes chongcao's healing capacity as a tonic to enhance vitality. Its main TCM applications are for treating exhaustion, TB, asthma, back pain and reproductive issues (i.e. premature ejaculation). Its use as an aphrodisiac seems to be the driving force with male consumers in China, a fact Tibetan informants liked to comment on, pointing out that they have no need to take *Cordyceps*. The development of new drugs addressing erectile dysfunction might reduce the demand for *Cordyceps* (Huber, 1958).

Mycology - *Cordyceps* and its hosts

Cordyceps is an entomophagous (feeding on insect) flask fungus (Pyrenomycetes, Ascomycotina) in the family Clavicipitaceae (Dube, 1983). There are 300 to 400 species of *Cordyceps* (Sung, 2004), of which 68 species have been recorded in China (Wang, 1999); 33 species have been recognized in the Tibetan Plateau and Himalayan region (Zang and Kinjo, 1998). *Cordyceps* species parasitize insects, spiders, and sometimes deer truffles (Elaphomyces, a hypogeous ascomycetous fungus). Only a few are collected for their medicinal properties. Most commonly used among these is *Cordyceps sinensis*. Several other species are also referred to as chongcao in TCM, such as *Cordyceps militaris*, which is widely distributed in North America and Eurasia (Arora, 1986), as well as *Cordyceps ophioglossoides, C. sobolifera, C. liangshanensis and C. cicadicola* (UNW, 1996). However, no other species is considered as powerful as *Cordyceps sinensis*, or is as costly (Chen, Yin, *et al.*, 2002). Zang and Kinjo (1998) have described several distinct species so far clustered under *C. sinensis: C. gansuensis, C. kangdingensis* and *C. nepalensis*. In this article *C. sinensis* will be used sensu lato. There will be no further differentiation of *C. sinensis* from the apparently very similar newly recognized alpine species.

Cordyceps sinensis (Berk.) Sacc. parasitizes a range of grass root-boring caterpillars, which would hatch as whitish ghost moths, when not attacked by *Cordyceps*. The normal reproductive cycle for *Thitarodes (Hepialus)* takes up to five years; most of the life cycle is lived as a caterpillar, the moth itself living for only a short time, 2-5 days in the case of *H. biruensis* (Chen *et al*., 2002). As many other ghost moths, the adult *Thitarodes* moth is not able to eat. So far, the host for *C. sinensis* most commonly reported is *Thitarodes armoricanus* (Oberthür) Ueda, Hepialidae, Lepidoptera. Other host larvae have been identified, such as *Hepialus oblifurcus* Chu & Wang *H. baimaensis* Liang and *H. biruensis* Fu Huang and Chen (Chen *et al*., 2002). All in all, nearly 40 species of *Thitarodes (Hepialus)* moths are recognized in the Tibetan Plateau region and, according to Chen *et al*., 30 species can be infected by *Cordyceps sinensis*. Chinese entomologists are still using the generic name Hepialus, although it has been restricted to a single European species – *Hepialus humuli* – for some thirty years. The genus Thitarodes was erected in 1968 to accommodate *Hepialus armoricanus* and other related species placed originally in *Hepialus* (Nielsen *et al*., 2000), many of them the host for *Cordyceps sinensis*.

Kendrick (1992) reports that *Cordyceps* fungi have developed a special adaptation to improve their chances of reproductive success. Since reproduction is dependent on a very specific host, each spore fragments into 100 or more part-spores so that each fungal fructification produces 32 million propagules, thus increasing the odds of landing on a larva. Li *et al*. (1998) reported that *Cordyceps sinensis* produces 30-60 propagules, and that usually 15 days pass between spore dispersion, their breaking up into propagules and larval infection. The propagules usually attach themselves to the larval state of the insect, but they also can attach themselves to mature moths as well. Apparently the larvae is forced by the fungus to move into its final position before being immobilized, since the fungus needs proximity to the surface to grow its fruiting body (stroma) above ground. No infected larvae will hibernate close to the surface. The mycelium – a cotton-like mesh composed of white threads (hyphae) – develops inside the body of the insect, first feeding on less vital parts, until it has taken over the complete organism filling the caterpillar with its hyphae. After the insect is completely mummified and emptied of nutrients, leaving behind the larval exoskeleton filled with the *Cordyceps* mycelium, the fungus will develop a fruiting body out of the head above the eyes, where the larvae has a horn-like protuberance in early spring. The 5-10 cm long, brown, club-shaped fruiting body grows above ground to have its propagules dispersed by the wind in order to find a new host insect. The stroma is nearly twice as long as the caterpillar when fresh (Fig. 1b). It takes several weeks for the spores to mature. As the insect is the sole source of food for the fungus, the size of its stroma is dependent on the size of the host caterpillar (Arora, 1986). Fruiting of *Cordyceps sinensis,* according to collectors in Lithang, is enhanced by high precipitation – plateau winters are arid and frigid – and by mild temperatures in spring. However, as Boesi (2003) reports, if too much snow accumulates during the harvest season the stromata will rot and much of the harvest is lost. While most other commonly collected mushrooms, many of them living in ectomycorrhizal symbiosis with trees, fruit in clusters, *Cordyceps* does not fruit in groups, but is highly dispersed, since each individual fungus is feeding on a different larva. However, specific site conditions might contribute to a concentration of larvae and thus of fruiting bodies.

DISTRIBUTION AND ECOLOGY

The specimens of *C. sinensis* are very rare, occurring in the far reaches of high cold and arid hills at an altitude between 10,000 fts and 14,000 fts above the sea level. The local people gather specimens of this fungus during the month of May-June just after the melting of snow. Whole of the families residing in these remote localities are found busy in collection of this fungus throughout its period of occurrence. Before moving for collection, people generally worship the local god by lightening scent stick (agarbatti) and pray him to get more number of specimens. It has been observed that each family having 5 to 6 members engaged in collection of this fungus for 35 to 45 days, are able to collect around 1 kg of dried fungus specimens in one season. After collection, people take the specimens to their tent (temporary residence nearer to the areas of occurrence of fungus) and wash with water to remove the adhering soil to the surface of caterpillar portion (Fig. 1b). Since the caterpillar portion is also surrounded with a net like structure (cocoon), people normally use toothbrush to remove it. The cleaned specimens are spread over a plain surface of stones for drying. The complete drying of specimens takes approximately one-week time since it is cold and humid weather prevailing all around. Less dried specimens containing considerable amount of moisture are spoiled by the contamination of other fungi. Hence, people are very cautious to dry their specimens completely and collect them in some self-made cotton bags and wrap with some polythene sheet. The fungus is sold by the local people at the rate of Rs. 30 to 35 per specimen or Rs. 70,000 to 80,000 per kg.

Few locations, namely Chhipalakedar, Brahamkot, Ultapara, Ghwardhap, Chhipalakot and Najari, which are spread in about 8 kilometres of area in Dharchula region of Uttarakhand, were initially identified as a place of occurrence of *C. sinensis* (Arif and Kumar, 2003; Garbyal *et al.*, 2004; Das *et al.*, 2005) . But looking to its possibility of occurrence in several other places having the similar type of climatic conditions in the Indo-Tibet Himalayan ridges, local people started searching the neighbouring places for its occurrence. Till date many more places in the areas namely Lashpa, Darti, Milam, Burfu, Mapa, Tola, Ralam in Johar velley and Nagindhura, Golfa, Bona, Chhipalakedar in the basement of Panchachuli in Munsyari were also identified as places of its occurrence (Fig. 1a).

Cordyceps sinensis is endemic to the Tibetan Plateau including the adjoining high altitude areas of the Central and East Himalayas (Nepal, Bhutan and India's Uttarakhand, Sikkim, Himachal and Arunachal Pradesh). In addition, Zang and Kinjo (1998) also report *C. sinensis* from the Tian Shan and Altai Shan in Xinjiang, NW China. However, there is also abundant misleading information available regarding the distribution of *Cordyceps sinensis*, which is explained by the fact that TCM's "chongcao" includes a range of other *Cordyceps* species. Countless webpages list Guizhou, Hubei, Zhejiang and some other Chinese provinces; this might be quoted from "Icones of Medicinal Fungi from China" (Ying *et al.*, 1987), which also misreports *Cordyceps sinensis* growing in marshlands. *Cordyceps sinensis* is common in grasslands and shrublands of the Tibetan Plateau including West Sichuan, North Yunnan and vast areas of Tibet AR, Qinghai and West Gansu (Fig. 1). Its distribution is limited to areas with an average annual precipitation above 350-400 mm. In general, *Cordyceps* is not found in areas where precipitation is below 300 mm/a, such as the Changtang and other arid areas of the north-western plateau.

The grasslands providing habitat for Thitarodes (Hepialus) moths and thus for *Cordyceps sinensis* consist predominantly of Kobresia sedges. Kobresia can cover up 80-90% of the subalpine

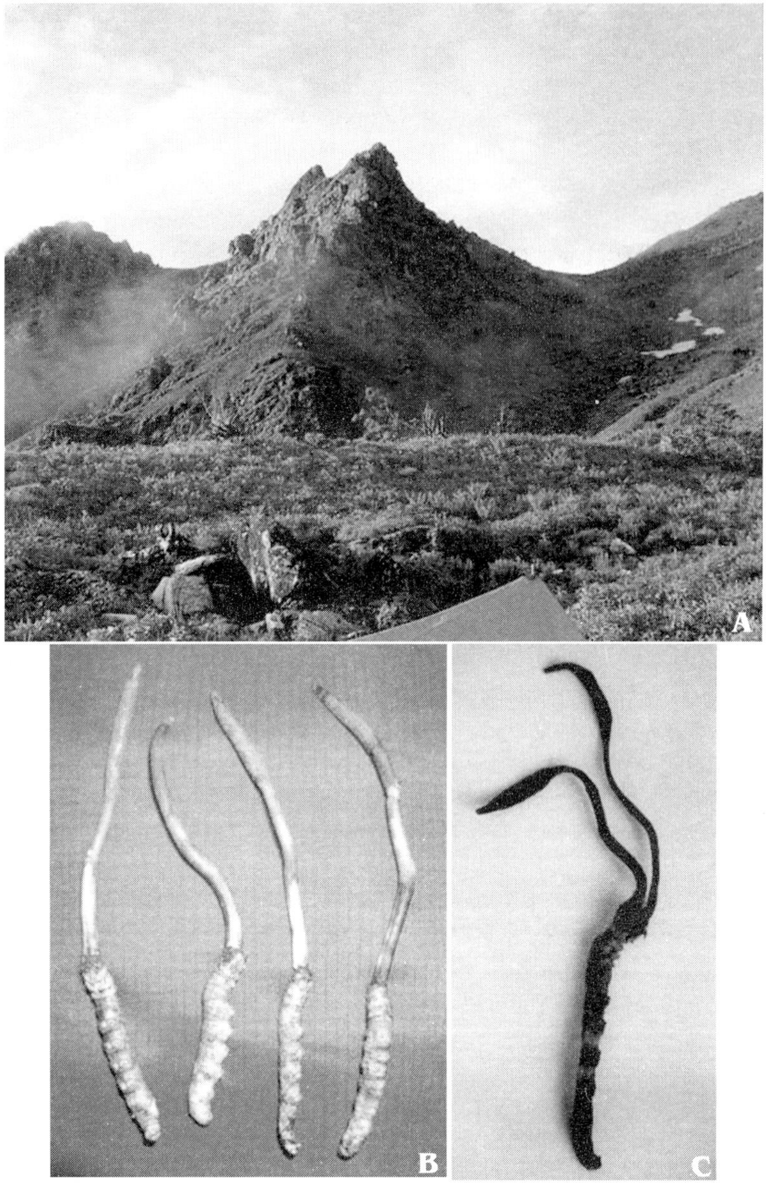

Fig. 1. A. Himalayan region where *C. sinensis* is growing, B. The photograph showing the stromata of *Cordyceps sinensis* emerging out of the caterpillar larvae, C. Two stromata of the fungus emerging out of the larva.

grasslands (Wu, 1997a). According to Miehe *et al.* (2004) Kobresia pastures of the Tibetan Plateau are the world's largest alpine ecosystem covering 450.000 km^2 with an altitudinal range from 3000 m to nearly 6000 m gradually rising from the south-east to the north-west of the Plateau. In general, *Kobresia* pastures are dominated by these *Kobresia* species: *K. capillifolia, K. humilis, K. pygmaea* and *K. schoenoides*. The dominance of *Kobresia* in plateau grasslands seems to be a result of intensive grazing (ibid.). Aside from Kobresia sedges, a range of grasses and a great variety of forbs are common. For dominant species in Lithang (see Tab. 1). Zang and Kinjo (1998) reported that alpine *Cordyceps* species associate with *Polygonum affine, P. viviparum, P. macrophyllum, P. glaciale, Astragalus balfourianus and A. craibianus*. Chen *et al.* (2000) stated that Thitarodes (Hepialus) prefers to feed on young roots of plant species of the families of Polygonaceae, Fabaceae, Cyperaceae, Poaceae and Liliaceae. Its favourite fodder species in Nagchu are *Polygonum viviparum, Rheum pumilum, Astragalus yunnanensis* and *Salix lindleyana*. Here in good habitat an average density of 0.42 specimen per m^2 is reported, but can reach 4 specimens per m^2.

Caterpillar fungus thrives in subalpine and alpine grasslands or meadows as well as open dwarf scrublands around the potential treeline. Based on the potential treeline, a relatively well-researched geo-ecological indicator (i.e. Li, 1994; Miehe, 2004) assessment can be made regarding the altitudinal distribution range of *C. sinensis*. It is distributed within a range of 400m - 500m of altitude to the potential treeline. Basing the distribution on the potential treeline is helpful, since the current actual treeline has been strongly influenced by human activities. Wide areas of the Tibetan Plateau forests have been replaced by pastures through the millennia, (Winkler, 1998, 2003).

Area Distribution of *Cordyceps sinensis*, Potential Treeline and Lower Limit Upper Limit Source

Ganzi TAP (Sichuan) 3500m 4700m Liu J.B. (1994) 3800m-4400m
Lithang, Ganzi TAP (Sichuan) 3800m 4800m Winkler 4200m-4400m
Qamdo Prefecture (Tibet AR) 4300m 5000m Gawä Dorje (1995) 4400m-4600m
Naqu Prefecture (Tibet AR) 4100m 5000m Chen, Yin *et al.* (2000) 4400m-4600m
Bhutan 4070m 4800m Namgyel (2003) 3800m-4000m
Kumaon Hills, Uttarakhand, India 3300m 4300m Arif and Kumar (2003) 3600m-3800m

In Lithang, caterpillar fungus is mostly found on north-facing slopes according to Boesi (2003). Pastures on north-facing slopes – all grasslands are grazed at some point through the year – usually retain more moisture through the arid winter months, since snow will not melt and/or evaporate as quickly as on south-facing slopes. In Machen (Maqen) County (Golok/ Goluo Prefecture) and other counties of South Qinghai Province and NW Aba TAP, Cordyceps proliferates on well-drained sunny slopes with lush grass vegetation. In Nagchu Prefecture (Tibet AR) *Cordyceps* is confined to rich pastures and absent from marginal areas. Also, sites which are too wet or waterlogged, such as bogs and tussock grass areas, do not have *Cordyceps* populations.

Yarsa Gamboo Harvest

Each year, when spring arrives in the grasslands, Tibetan nomads, farmers and also some city dwellers, venture on foot or on horseback out into the grasslands to look for "bu", as Khampas and other Tibetans call it for short. At lower elevations the season can start as early as mid-April, but in many places not before the beginning of May. In colder high altitude areas, such as in Bachen or Biru County, Nagchu Prefecture, it lasts into July. The collection peak phase should not last longer than 6 weeks in a given area. The present decentralized grassroots gathering (no pun intended) is very efficient. Some people are concerned it might be too efficient and there are concerns regarding its sustainability. The harvest procedure itself is simple; collectors usually carry a small hoe or use their knife to lift the caterpillar fungus out of the fertile topsoil. This needs to be done carefully, since the Kobresia turfs are very dense and breaking off the stroma from the head of the larva reduces the value.

The actual gathering of yartsa requires an intensive but extensive search. The grasslands and shrublands are vast and the visible part of the fungus is very small, often not longer than 2-5 cm. It can be spotted only at close range (Bao et al., 1988) and individual yartsa are dispersed, not like many other fungi whose mycelia may produce several fruiting bodies close together. A daily harvest in Lithang's grasslands is anywhere from a few specimens to several dozen specimens, sometimes up to 100 specimens per collector. In high season a common average seems to be around 20 to 40 specimens per collector per day.

During Yarsa season Tibetans focus on gathering, and keep other activities to a minimum, since the income generated is very important. Consequently, during peak season, it is difficult to carry out labour-intensive projects. Community or government infrastructure projects, such as local road construction, come to a halt. Finding local workers for reforestation can become difficult. To attract workers wages need to exceed the potential income from gathering, often a multiple of common wages. However, since the mushroom hunt also includes the risk of finding very few specimens, some people prefer a guaranteed income, as I was told by local working in reforestation in Lithang. In some areas schools give vacations, so that students can join in collecting. For example, in adjoining Nyarong County (Pinyin: Xinlong, Kandze/Ganzi TAP) most schools close for 20 days in late May and also in Beizha, Nangchen County (Nangqian, Yushu TAP, Qinghai), the Gar Monastery school scheduled their month-long vacation period to coincide with the Yarsa gathering season. Even attendance at traditional festivals, such as Cham dances in Songpan (Aba, TAP), suffers from people collecting "bu" in the grasslands instead of joining the celebration.

In Lithang's Junba District communities live and farm in deep, forested valleys (Tibetan: rong), but graze their livestock above treeline on the grasslands (thang) in summer. In spring often most members of a household camp out in the grasslands without livestock, turning the hunt into a joyful event. Also, some people set up tents around *collectors*' camps to serve as shops or bars, thus taking their share from the freshly made *Cordyceps* money. In Chungba (Junba) I was told that collectors are confined in collecting to their legal grazing grounds. Collectors from outside reported that they had to pay a flat fee of 25 RMB ($3) per collector to the administration for the 2001 season in Chungba District. However, in Lithang, grasslands are, in general, perceived as openly accessible for collection of plants and fungi. Local herders do not own their pastures according to Chinese law, but have usage rights. Exploitation of resources on common property often causes disputes. Although in Lithang and Sershul (Serxu/ Shiqu) many collectors complained that there are too many people collecting nowadays, but no violent conflicts were reported during Yarsa collection. Also Boesi (2003), who carried out a detailed study of the yarsa harvest in Lithang, reports that "cropping activities are not controlled by any particular local rules or traditions, there is no competition between the gatherers of the produce (`Bu rKo Ba), either they are recurrent or occasional labourers or they come from different villages to exploit the produce. Nobody claims the right to pluck over a certain area".

HISTORY OF COLLECTION AND TRADE

Having wide medicinal properties of *C. sinensis,* it is highly prized in the markets world over. Traditionally the fungus is traded in China for its weight in silver and gold and the market prize is astronomically high as of December 1999, it was sold at a wholesale price of rupees one lakh per kilogram. The trade and market channel in India is not a transparent one. It is being smuggled out to China and Tibet across the boarder lines. In these boarder lines the cost of per kg mushroom varies from rupees 80,000 to 1, 30,000 while in international market this fungus may fetch a price between rupees 1.5 to 2.5 million per kg. However, the cost of this mushroom varies among the trade channels starting from field gatherers to broker and their agents who collect the material from different locations and sell it at a higher rate. The current prevailing price per kg of *C. sinensis* in the markets of Tibet ranges between Rs. 68,000 and Rs. 80,000, in Nepal between Rs. 80,000 and Rs. 90,000 and in India between Rs. 1,25000 to Rs. 1,30,000 according to the survey done by Sharma (2004). But, presently it is being smuggled at a price between Rs.12-20 lac per kg.

Many early western explorers of Eastern Tibet (Rockhill, 1891, Pratt, 1892, Bacot, 1912, Wilson, 1913; and Bailey, 1945) reported on the collection and the trade of caterpillar fungus from Tibet to China often in exchange for tea or luxury goods like silk, confirming its economic importance. Bailey (1945) brought the first photo of Yartsa collection taken in Lithang in 1911 to the west. Shelton (1921), a missionary who spent 17 years in Bathang wrote: "Wonderful and awe-inspiring concoctions of Chinese medicine contribute much to the industries of the Tibetans. In the spring and summer months Tibetans also dig plants and collect fungi and other articles of supposed medicinal value for export to the Chinese market. One of the ingredients is the grass worm. This seeming combination of animal and vegetable life is not only used as a medicine, but is also eaten as a delicacy". Coales (1919) reported on trade in "Tachienlu" (Pinyin now: Kangding, Tibetan: Dardsendo) "the medicines exported are mainly rhubarb and other vegetable drugs for the Chinese market. The most interesting is the curious Chungtsao or insect grass, a dried caterpillar about 2 inches long, which has been killed by a fungus of about the same length growing out of one of its segments. It is supposed to be an excellent restorative to weak constitutions". Yarsa traditionally and still today functions as currency in many areas. It is a favourite trading good, since it is small and light-weight (<1 g), it does not readily deteriorate, and is very valuable.

Historically 60% of China's chongcao supply came from the Tibetan areas of Sichuan (Liu, 1994). Already during the late Qing dynasty (1796-1874) 10 t of yartsa gunbu were exported annually from the Kham (Bhattarai *et al.*, 1995) region when trade between China and Tibet blossomed (Wang, 2000). Statistics for Sikang (Pinyin: Xikang), a former province (1920s-1961) comprising most of the Tibetan cultural area of Kham list chongcao production at 15 t in 1939; back than 1 kg was worth 8 Yuan. In (1953) in Kandze/Ganzi Prefecture alone, there was an output of 10 t. According to the Ganzi Mushroom Institute in Kangding the highest annual production in Ganzi TAP was 20 t (Boesi, 2003). During the Cultural Revolution and the Commune phase the chongcao market cooled off, but it resumed with the economic liberalization in the early 1980s. In 1982 prices were still very low: one specimen cost 0.1 RMB or 200 to 250 RMB per pound in Lithang. In Chengdu 1 jin (500 g) was traded for 400 to 500 RMB. The overall production of *Cordyceps sinensis* is estimated at nearly 100 t per year in all of China according to an unpublished Chinese industry report.

The present *Cordyceps* market

Since economic liberalization in the early 1980s, caterpillar fungus has developed into one of the most important "cash crops" on the south-eastern Tibetan Plateau. Since caterpillar fungus is consumed dried and not fresh, there is no need for rapid marketing as with culinary mushrooms. Its small size, its easy preservation, and its high value make transport very easy. Furthermore, dried specimens often fetch much higher prices in winter. Lack of cash and fluctuating prices deter most people from holding back their harvest and speculating on profits to be made by selling in winter. However, Yarsa is also used as a way of keeping savings at home.

Cordyceps is traded in several categories; the main criteria are water content (fresh, dried), state of processing (uncleaned, cleaned) as well as size, the main retail price factor. The fungus becomes more valuable with increased cleanness and dryness. Freshly harvested caterpillar is encased in a black topsoil layer. The coating is first opened by hand or commonly with a toothbrush. Careful brushing exposes the caterpillar's body; the stroma itself is not encased in soil. It is important not to break off the stroma during cleaning. Broken specimens lose value; sometime a piece of match is used to reconnect the broken-off stroma. After cleaning, specimens are dried again, often on paved sidewalks or side streets in Lithang. Also of importance is the origination of the *Cordyceps*. Dealers in Chengdu's Hehuazi market distinguish between origin from Tibet AR, Qinghai and West-Sichuan. However, the differences seem to be based mostly on size, not medicinal qualities. Tibet AR, especially Biru and Tengchen, is known for growing the biggest specimens. Those from Qinghai and West-Sichuan fungi are both smaller and regarded as being similar in quality.

After collection most caterpillar fungus is usually first marketed in the local administrative center or directly in the county seats or other important market towns. In remote areas in Lithang, prices are 5% to 10% lower than in the County seat. For example, in 1999 in Chungba (Junba) town 1 jin (= 1 metric pound = 500g) was sold for 1300 RMB. In Lithang town, 6 hours away the same bunch would fetch 1400 RMB. Asked about selling their harvest in Kangding or Chengdu informants from Lithang said that the prices are fluctuating and the extra input of time does not justify taking the risk.

Some collectors clean their mushrooms themselves; others sell their uncleaned fungi to local dealers, who will hire locals for cleaning before reselling. Prices nearly double for the same amount of specimen after cleaning due to the weight loss incurred during cleaning. Cleaned fungi are bought up by middlemen, whose average acquisition from collectors is 250 g to 1 kg at around 1500 RMB per jin. Local dealers often do 10 to 20 such transactions per day, which would require a capital of 3,000 to 30,000 RMB. Many of these dealers also trade other medicinal plants or matsutake mushrooms in the summer. All of these transactions take place on Lithang's main streets. The whole area is buzzing from the fungi deals. Regular shopkeepers are not involved in the trade. Sellers from the most remote areas in Lithang and adjacent counties are in town to sell off their fresh harvest, be it from a couple of days hunting or from 6 weeks of work. Dealers with scales, which signal their interest, walk up and down the streets to attract the next deal or they simply hang out on a chosen spot waiting for sellers to come by. Prices are being discussed constantly, but are relatively stable. However, prices usually change through the season on the basis of supply and demand. Sellers, flush with cash, use their visit to town to buy staples or luxury goods and enjoy their windfall in fancy restaurants.

The daily quota of the middlemen, mostly Tibetans and also Hui (Bok *et al.*, 1999), was estimated by informants at 1 to 5 kg, is later bought up by big dealers, many of them lowland Chinese or Tibetans, for South Chinese phytopharma companies. These transactions do not take place on the street. Several middlemen estimated that there are about 100 dealers buying up small quantities from local collectors. This information allows a rough estimate of a daily transaction of 100 kg to 500 kg in Lithang Town. For a 6-week caterpillar fungus season this would indicate 4.5 t-22.5 t of *Cordyceps sinensis* is dealt annually in Lithang. However, the higher end seems clearly too high a guess etimate. Dealers and officials estimated the annual harvest at between 5 t and 10 t (value: 7.5–15 million RMB), however one Yartsa dealer was convinced that Lithang annually produces only about 1t, but local trades included another 1-2t collected in neighbouring counties. All in all, these figures compare well with official figures from Kandze TAP's Derge County, which reports an annual production of Cordyceps of 3t (Derge County statistics, 1997). Tengchen County (Denqen/Dingchen, NE Qamdo Prefecture, Tibet AR) is reported to annually trade up to 7.5t, which provides 60% of Denqen's GDP (CTI 2003). Chen *et al.* (2000) report from Naqu Prefecture (Tibet AR) that the annual yield of *Cordyceps sinensis* is 7t. There, Yartsa mostly grows in the relatively warmer and moister southern counties, such as Diru (Biru), Drachen (Bachen/Baqin), Sok (Suo), Lari (Jiali), Nyanrong (Nierong), and Nagchu (Naqu) of Nagchu/Naqu Prefecture.

Market Prices

During survey in Lithang in May and June 1999, prices were relatively stable. Individual specimens were traded for 1 to 5 RMB depending on size, the bigger the fungus, the higher the price. The price of the biggest specimens increased from 4 RMB to 5 RMB through this season. 1 Jin (500 g) unsorted wet and cleaned yarsa was sold for 1400 to 1500 RMB, which indicates about 800 to 1000 specimens per kg. Prices in 1999 were substantially higher than in 1998, when 1 jin wet and cleaned Yarsa sold for only 900 RMB. In 1997, the price was much higher at 2000 RMB/jin for cleaned fungi.

The degree of dryness of yarsa is an important factor when basing the price on weight. Dealers assess the moisture content by the flexibility of the fungus; the less pliable the fungus, the dryer it is. At Chengdu's Hehuazi market, China's biggest medicinal plant market, informed dealers stated that 3.5 jin of fresh wet fungi will weigh 1 jin when completely dry. This would indicate a weight reduction of 71.5% (Chen *et al.*, 2000). Using this formula, if 1 jin of fresh, wet and clean chongcao costs 1,500 RMB in Lithang, 1 jin of dried chongcao would elsewhere cost 5,250 RMB plus transportation cost and profit margins. However, even relatively fresh fungi will have lost some moisture when coming onto the market, especially when collectors sell off the harvest of a month's search. Dealers at Chengdu's TCM market add about 1,000 RMB to each jin. Overall, what seems like a steep increase in prices from Lithang's 1,300-1,500 RMB/jin to Chengdu at 7,000-8,500 RMB/jin mostly evaporates once moisture content is taken into account.

In Hong Kong in October 1999, 37.5 g of packed bigger specimen cost 528 HK$, small ones sold for 480 HK$. This indicates a retail price of 7,125 to 7,750 RMB/jin. A former chongcao dealer reported that in 1992, 1 jin of dried and cleaned fungus cost 1,800 RMB in

Kandze Prefecture. In Guangdong, the same jin was worth 2,250 RMB, which is an increase of value of 25%.

Prices clearly fluctuate through the years, especially depending on the annual harvest. Seasonally, the prices are at their lowest in harvest season in spring and peak in winter, since there is much less product available, but this is also partly caused by the much lower water content of *Cordyceps* so long after harvest season. Thus, winter prices are better compared to the markets in the lowlands, which trade mostly in completely dried fungi. In the winter of 97/98 the price was at 6,000 RMB/jin for fully dried caterpillar fungus. In Chengdu 1 jin of big chongcao can cost up to 18,000 RMB, small ones to over 10,000 RMB/jin. Here seasonal price changes have a range of 30% to 40%, the price of bigger specimens being more susceptible to fluctuation. Dealers would like to buy several hundred pounds in Chengdu, but going to the production areas is too risky. The fluctuation of prices makes it too uncertain. Some dealers go to Kangding to purchase fungi, but there are not enough fungi available. One has to go further, to the counties of origin. Overall, the inflation of the price of *Cordyceps sinensis* is very impressive. While one jin (500 g) of average yarsa was available for 200-250 RMB in Lithang in 1982, 21 years later one jin fetched from 8,000-20,000 RMB, which is a price increase of 40 to 100 times. Recent dramatic price increases are reported by some as a result of the SARS epidemic in China in 2002, since people apparently believed that *Cordyceps* would help prevent infection with SARS.

Economic Importance

The income derived from yarsa gamboo is very important for Lithang (Chen *et al.*, 2002) and all the other areas of Tibet where *Cordyceps* occurs. Resource control, be it usage rights or ownership, are crucial factors in accessing and developing economic plant and fungus resources. Unlike many other natural resources in the region, such as timber, gold and also increasingly hydroelectricity, where the profits are captured by the state sector, fungal income (and other income from wild collected plants) goes directly to rural households. There are three main factors that enable rural households to participate successfully in accessing this resource; open access of the grasslands, availability of harvesting knowledge, and no capital required for participation. Thus, resource access is assured to the people, who are otherwise marginalized by government control of local resources such as timber, by lack of formal education, lack of access to credit etc. Thus, almost all rural households, who base their survival on traditional subsistence herding and agriculture, participate in yarsa gamboo collection (Winkler, 2003).

For comparison, in 1993 the gross value of industrial and agricultural output in Lithang County is reported at 67 million RMB, including an agricultural output of 47 million RMB. Checking these figures back, by extrapolating Lithang's Chungba (Junba) District figures from 1998, where *Cordyceps sinensis* and the booming Matsutake mushroom (*Tricholoma matsutake*) collection contributed 60% to the District income of 959 RMB per capita in 1998, in Lithang County this would figure to around 40 million RMB while the official GDP figured at 107.4 million Yuan in Lithang in 1998 (Chen *et al.*, 2002). However, much of the income derived from the mushroom industry apparently is not accounted for in official statistics. Under reporting of earnings reduces tax burdens and secures provincial and/or central subsidies and thus is a widespread phenomenon in rural China.

The importance of yarsa for rural households has been confirmed in many other locations. In connection with rural development project, I have also interviewed local people in several counties in Kandze TAP, such as Nyarong (Xinlong), Sertar's Golok (Seda) area, Sershul (Shiqu) and Derge (Dege). Unless people have access to precious non-timber forest products such as matsutake and morels, caterpillar fungus typically contributes 50% to 80% of the annual household cash income.

Future Marketing

The marketing mechanisms seem vary evolved for a trade in such a remote region. The fact that the price between remote villages and Lithang's county seat differs by only about 8%, without deducting minimal transport cost, indicates an established marketing network. However, the network is informal with fluctuating prices, thus adding some uncertainty, which deters many dealers from getting involved in long distance trading. Only a few people occupy this niche, since trading in bigger quantities requires large amounts of capital. Even local dealers, buying a daily quota of 20 jin require a capital of 30,000 RMB, which exceeds 5 years salary for an average government worker. The fungal trade, especially caterpillar fungus and matsutake, has helped to generate local capital that allowed some local dealers to finance jeeps and other vehicles. The informality of the sector is further illustrated by the fact that there is no real taxation, as one informant put it, "you need to get caught with a truck load at a check point", and that governmental statistics seem to not fully include the caterpillar fungus trade in their statistics.

Furthermore, Lithang and many other Tibetan localities need to establish their own brands of "chongcao" and raise its profile as a distinct geographic origin of medicinal fungi. Preparing a product ready for marketing would surely increase local profit margins and create job opportunities. In Lithang, there is already a distillery that produces liquor (qingke jiu) from regionally grown barley with a few caterpillar fungi or fritillary bulbs (Fritillaria spp., beimu) floating in the bottle. Having processing and packaging companies in Lithang will be an important step in securing sustainability, since reduction in local output would be felt directly by the companies. In addition, higher profit margins are captured locally and employment opportunities created. Moreover, once this process is initiated and gains momentum it can be applied for many other precious medicinal resources.

Artificial and Semi-artificial *Cordyceps* Cultivation

Chinese laboratories pioneered artificial cultivation of *C. sinensis* derived organisms in the early 1980s. The resulting asexual organisms are distinct from C. sinensis and since the fungus is not parasitizing *Thitarodes* caterpillars, but feeding on other substrates, the organism develops differently, and thus is regarded as a different species. There are many new species, known as agamotypes, i.e. *Paecilomyces hepiali, Hirsutella sinensis,* and *Cephalosporium donchongxiacae* (Wang, Chen, Li, Zhou, Zhong 1998). In China, liquid culture or fermentation is widespread. Tanks containing a liquid medium formulated to optimize mycelial growth are inoculated with a *Cordyceps* derived agamotype. Sterilized silkworm residue is commonly used as base for this medium. In America and Japan unconverted grain substrates are frequently used in solid culture (Holliday *et al.*, 2004).

Cultivators like to advertise their product derived from the asexual propagation as "superior", since it is asexually reproduced under controlled laboratory conditions and thus its potency is supposed to be guaranteed. Also, medical research is carried out on the efficiency of these lab strains. A consistent product might be guaranteed, but this does not assure that these products contain the same range of active ingredients as the natural product. Many cultivated products do not even contain cordycepin, a key active ingredient of *Cordyceps sinensis* (Holliday *et al.*, 2004). Thus, these widespread products need quality control. In addition, Asian consumers do value the fact that medicines are gathered from the wild. The plateau origin adds to the lure of the caterpillar fungus besides the curiosity of this "symbiosis" of insect and fungal world.

Wild fungi, as long as they are available, will probably not be replaced by laboratory produced fungal strains, but prices might be negatively affected. Artificially grown *Cordyceps sinensis* mycelium so far has not impacted the market of wild *Cordyceps* in Asia at all, as clearly documented in recent price increases. Western consumers prefer artificially grown products to the real organism, since Western culture lacks a tradition of appreciating edible insects.

There are also experimental semi-artificial *Cordyceps sinensis* growing programs under way in Tibet AR (Lhasa), Qinghai and West Sichuan's Kangding. In Kangding (administrative centre of Kandze TAP), for example, the Thitarodes (Hepialus) host larvae are bred, building on China's rich knowledge of silkworm breeding. About 100 larvae each are placed into shoe carton-sized plastic containers with lids, which are filled with grassland soil containing the tubers and roots of their favourite natural foods collected from the wild, as well as some other roots from cultivation. After two years spores of *Cordyceps sinensis* are inoculated and about 10% of the larvae are actually taken over by the fungus and grow stromata, reducing the natural growing cycle from 5 to 2 years. In April 2004, longish stromata were observable in many boxes. This was the first large scale fruiting, I was told, the culmination of 25 years of research (Chiou, 2000). In another experiment, in a field station near Kangding, bred larvae were released to their subalpine grassland habitat, but there no *C. sinensis* fruiting has yet taken place.

Artificially and semi-artificially grown fungal substance and fungus might prove to be extremely helpful to reduce future pressure on the natural resource, which surely will increase steadily with the growing availability of disposable income in China and ongoing discovery of *Cordyceps sinensis* by Western consumers and continued medical research with promising results.

Cordyceps in Golok TAP

Although I have not researched *Cordyceps* in Golok TAP myself [I only looked into the issue in Golok areas in the north of Sertar County, Kandze TAP], I have collected information from researchers and informants. Golok TAP, South Qinghai Province, is an important production area for *Cordyceps*. In general, these reports portray a similar situation regarding collection, marketing and the economic importance for local communities, but contain one striking difference: the fact that thousands of non-local migrating Hui and Han would come into pastoral areas for digging. To give an idea of the number of outsiders involved in the collection, Goldstein (1996) reports from Sangrima (Xueshan) Township, Dari County, that 3,000 outsiders would come in to dig fungus for a fee of 700 RMB per person, while the local population numbered 1031. In this case the fees were used by local authorities for pastoral development projects, which considerably improved the local pastoral economy according to Goldstein. Horleman (2002) reported "it seems that [*Cordyceps*] resources are presently mainly exploited by Han seasonal migrants and not by Golok pastoralists. The collectors have to buy a permit from the local authorities, which grants exclusive collection rights for certain sites. However, the pastoralists receive only a small fraction of the collection fees, although they are the ones who suffer most from the grassland degradation caused by extensive digging. Most of the profit from the permits goes to the local authorities, that is, to the townships and to the villages and herders' associations." Horleman was told by officials that the permits can cost up to 10,000 RMB per season and that over 100,000 migrants are coming to Golok TAP, a number comparable to the overall population. How much a traditional taboo placed on digging Yarsa had contributed to the fact that outsiders were encouraged by the administration to exploit these resources needs further research.

Informants and reports also have bitterly complained about the licensing approach, which not only deprived local herders of one of their most lucrative resources, but also left them stuck with degrading holes in their pastures. Outsiders often dig much bigger hoes and thus damage the turf more seriously. Not surprisingly, there are many reports of conflicts between local collectors and outside collectors in Golok. An informant from Jigdril County (Jiuzhi Xian), who financed his high school education with Cordyceps collection, reported frequently scaring away Han and Hui collectors, who had come up to this remote county for chongcao collection. Showing his long Tibetan knife, he cut the encounters short and for that day secured his resource without further discussion.

Overall, many Golokpas were not content with the licensing system and wished to exclude outsiders from collecting. Horleman was told that the outside exploitation of medicinal "plant" resources was banned in 2000, but actual enforcement took several years. Costello (2003) reports that administrations finally halted licence sale to outsiders in 2003. However, many outsiders are still coming for the season. The popular demand for preventing outsiders from collecting is understandable taking Costello's (2003) observation into account: "most pastoral families especially the poorer ones make most of their cash income not from selling livestock products, but from digging Yarsa. The number of yarsa an individual can find each season varies greatly by individual, but averages between 600 and 1,000. The price for an average yarsa varies both yearly and through the year, varying between 2-4 Yuan each (Christiensen *et al.*, 1975). For those who dig with even average success, the income from Yarsa, supplemented with dairy products provides enough for a basic living". In addition, Yarsa also provides income to local dealers. Costello observed that yarsa changes hands 5 times before leaving town. This might explain a common complaint that local collectors do not receive fair prices and that too much profit is made by middlemen.

Cordyceps in the Tibet AR

In Tibet AR *Cordyceps sinensis* in general is distributed in areas with more than 350 mm annual precipitation. In the southern counties of Nagchu Prefecture, such as Diru (Biru) and Bachen, where some of the best fieldwork so far published on *Cordyceps sinensis* and its host insects (Chen *et al.*, 2000, 2002) has been carried out. South Nagchu is known for its large Yarsa specimens. In Biru County Yarsa collection has heavily impacted the local economy according to Ben Jiao. Now many locals earn enough money through yartsa collection that they can afford to hire Tibetans from poor areas in Shigatse Prefecture to carry out their agricultural work. During collection season tent cities develop in the grasslands complete with bars, brothels and shops. The county government tries to keep outside pickers away by charging high collection fees; also, there is a tax on trade of *Cordyceps sinensis* in Biru. Yarsa riches have been spent on new houses and trucks. Also Nyingchi County (Pinyin: Linzhi, Nyingchi Prefecture) blocked access for outsiders in 2003 and started to sell licences only to locals. However, some locals figured out that selling their licences to outsiders for good money is easier than collecting themselves. Licences were partially introduced, since the previous harvest was much lower than previous years. This was attributed to over-collection according to Luorong Zhandui.

Cordyceps in the Himalayas

In recent years the *Cordyceps* collection, Zang and Kinjo (1998) recognized the caterpillar fungus from Nepal as a distinct species, *C. nepalensis* that has increased dramatically in the Himalayan region. Most of the product is destined for export to Tibet AR. While Shresta *et al.* (1998) reported from Dolpo, Nepal that locals receive below 15% of the price on the Chinese market; prices now have sharply risen in Nepal and also in India and Bhutan in the last years. In Bhutan, collection of *Cordyceps* is illegal in order to give them full protection. However, Turkelboom *et al.* (2001) rightly remark that outlawing harvesting of very valuable species without proper monitoring leads to poaching and all sustainability issues will be disregarded.

Similarly, in Nepal Yarsa gamboo is under legal protection and has been banned from export without a government permit since 1996. In 2001 it was opened for trading with the royalty of NR 500 (US$ 6.60) per piece, an irrational fee, which surely will encourage smuggling (Das, 2001). It is interesting to note that in connection with the insurgency in Nepal, where government control has ceased in many alpine areas, the insurgents have established a more realistic and thus more profitable taxation. Marietta Kind reported that the insurgents "control all the Yarsagumba business in Dolpo now. They collect "revolution tax" on each kilogram (5,000 NR) traded, combined with a tax per picker (500 NR) and a tax per trader (20,000 NR). They also encourage collectors from other regions of Nepal to join the picking to optimize their "tax base". So in 2003 the area with its formerly hidden valleys was crowded with thousands of pickers digging holes everywhere, using up scarce firewood and leaving behind piles of garbage.

Overall, *Cordyceps* harvest has increased dramatically in Nepal and other Himalayan areas. In addition Yarsa gamboo is also being collected in India's Arunachal Pradesh and Uttarakhand (Costello, 2003), where it has only recently come to the attention of researchers (Garbyal, 2001; Arif and Kumar, 2003), while its occurrence in the high altitude areas of Sikkim has been known for long time.

Sustainability

It is known that Yarsa has been collected for centuries. Some sources indicated harvest rates totaling at least 100 million specimens per year since the 1950s. As reported, collectors in Lithang and other production areas did not mention reduced fruiting, but acknowledged the fact that there are more and more collectors each year. Dealers did not report a reduction in harvest quantities either. In short, no widespread disappearance of *Cordyceps sinensis* has been reported from the Tibetan Plateau. However, it was reported to have historically disappeared from Xishan near Kunming (CAC, 1990:185), but this could have been caused by a variety of reasons, such as local climate change or any other cause impacting the host moths or the fungus; In addition Xishan is a relatively small disjunct mountain area.

Still, in recent years the issue of the sustainability of current harvest rates has been raised frequently (He and Sheng, 1995; Bhattarai, 1995; Shei, Wang and Xie, 2001; Zhang *et al.*, 2001). An assessment regarding sustainability of current harvesting levels of *Cordyceps sinensis* is not possible based on the research currently available in the West. Accurate baseline data is missing [as with most mushrooms collected from the wild]. In addition, annual fluctuations are very common as with all mushroom fruiting, thus further complicating the interpretation of available data. However, a clear trend toward increased harvesting pressure is recognizable over its entire distribution, which is further accelerating with increasing prices.

Pastoralists in many areas of the plateau complain about the destruction of their grasslands by diggers of *Cordyceps* (and other medicinal plants), who make unnecessarily big holes in the turf by using relatively big hoes and by not closing the hole after removal of the fungus. Such careless extraction, often attributed to outsiders poaching Yarsa gamboo, not only impacts pasture quality, but also disturbs the habitat of the *Thitarodes (Hepialus)* moths, thus potentially undermining sustainability. In this context, the hotly debated issue of overgrazing and grassland degradation also needs to be considered. Although there is no data available on the impact of grassland degradation through overgrazing or desiccation and the abundance of *Thitarodes (Hepialus)*, it can be speculated that a healthy grassland environment is favorable to larval development. Also, no informant reported on more abundant fruiting of the fungus in degraded areas (Dai and Li, 1994).

The long-term impact of intensive collection on caterpillar fungus and its reproduction is still unknown. Although there are a variety of entomophagous fungi (Das, 2001), *Cordyceps sinensis* being economically by far the most important, not much research was carried out on these organisms and to my knowledge no research on the long-term impact of recent intense harvesting has yet been published. From a mycological point of view, it can be speculated that, if the fungus is collected after it had time to release enough of its spores, there might be no negative impact at all. To the contrary, handling the fungus might even result in wider "spore" dispersal. As described above *Cordyceps* has developed the rare capacity to divide each spore into 50-100 propagules producing 32 million part-spores to increase its chances of finding its host organism. However, this astounding capacity is of no help if the fungus is removed before "spores" are released. This is the case with *Cordyceps* gathered early in the season. In late April 2004, at the onset of the collection season, none of the specimen offered for sale in Kangding showed signs of maturity of the asci, the spore producing organs. Collection of these specimens will negatively impact reproduction. Also Boesi (2003) observed in Lithang, that most of the fungi were collected "before its spores have been abundantly spread by the wind". It is evident that more research is crucial to assess the question of sufficient "spore" dispersal in order to secure sustainability. Also a reduction of *Cordyceps* might seriously impact the ecological balance of the grasslands, since *Thitarodes (Hepialus)* larvae feed on roots of the forage species.

Outreach programmes and public education campaigns need to be initiated to ensure that fungi are harvested only after sufficient spore dispersal has occurred. A simple measure facilitating post-collection spore dispersal could be the introduction of collection containers that allow spores to disperse, such as baskets or netted bags. Hopefully the question of sustainability will be answered in time before *Cordyceps sinensis* populations actually dwindle, an assumption that is not far fetched looking at the track record of *Homo sapiens* unwisely exploiting natural resources. Everything possible needs to be done to ensure that a healthy environment is sustained and sustainable harvesting carried out, so that medicinal fungi and plants will be able to provide a basic income to rural households and healing for generations to come.

REFERENCES

Adshead, S.A.M. (1984). Province and Politics in Late Imperial China-Viceregal Government in Szechwan, 1898-1911, Curzon Press, London, 1-139.

Arif, M. and Kumar N. (2003). *J. Expt. Zoo India,* **6.1:** 45-55.

Arora, D. (1986). Mushrooms Demystified, Ten Speed Press, Berkeley, 1-959.

Atlas of Tibet AR [Xizang Zizhiqu Dituce] (1996). Zhongguo Ditu Chubanshe Chuban (China Cartographic Publishing House), Beijing, 1-167.

Bacot J. (1912). *In:* Revue d'Ethnographie et de Sociologie, 203-210.

Bailey, F.M. (1945). China-Tibet-Assam: A journey, 1911. Jonathan Cape, London, 1-175.

Bao, T.T., Wang, G.F. and Yang, J.L. (1988). Chang His I Chieh Ho Tsa Chih. **8(6):** 352-354.

Bhattarai, N.K. (1995). *In:* Non-Wood Forest Products 11 - Medicinal Plants for Conservation and Health Care - FAO, Rome.

Boesi, A. (2003). The dbyar rtswa dgun 'bu (Cordyceps sinensis Berk.): An important trade item for the Tibetan population of the Lithang County, Sichuan Province, China. The Tibet J. **28.3:** 29-42.

Bok, J.W., Lermer, L., Chilton, J., Klingeman, H.G. and Towerrs, G.H. (1999). Phytochem. **51(7):** 891-898.

CAC (1990). The Conservation Atlas of China. Sponsored by National Environmental Protection Agency, Compiled by Changchun Institute of Geography CAS, Science Press, Beijing.

Chen, S.J., Yin, D.H., Li, L., Zha, Xi., Shuen, J.H. and Zhama, C. (2000). *In:* Zhong Yao Cai **23.11:** 673-5. [Chin., Engl. Abstract].

Chen, S.J., Yin, D.H., Zhong, G.Y. and Huang, T.F. (2002). *In:* Zhongguo Zhong Yao Za Zhi **27.12:** 893-5. [Chin., Engl. Abstract].

Chen, Y.J., Shiao, M.S., Lee, S.S. and Wang, S.Y. (1997). *Lif. Sci.* **60(25):** 2349-2359.

Chen, Y.Q., Wang, N., Zhou, H. and Qu, L.H. (2002a). *Planta Med.* **68.7:** 635-9.

Chiou. W.F., Chang, P.C., Chou, C.J. and Chen, C.F. (2000). *Life Sci.* **66(14):** 1369-1376.

Christensen, C.M. (1975). Moulds, Mushrooms and Mycotoxins, Christensen, C.M. (Ed.), pp. 164. University of Minnesota Press, Minneapolis.

Coales, O. (1919). *In: Geographic J.* **54:** 242-247.

Costello, S. (2003). The Flow of Wealth in Golok Pastoralist Society: Toward an assessment of local financial resources for economic development. Draft Paper presented at Oxford University, 10th Seminar of the International Association for Tibetan Studies.

CTI (2003). Caterpillar Fungus in Denqen. China's Tibet Information, August 7, http://www.tibetinfor.com.cn/english/news/2003-8-7/News0200387170347.htm

Dai Xiancai and Li Taihui (1994). (Mushroom Key of Ganzi Prefecture, Sichuan Province) Sichuan Zhen Ganzi Zhou Jun Lei Zhi, Sichuan Chishu Chubanshe , Chengdu, 1-330. (Chin.)

Das, B. (2001). Minutes of the 11th NNN Sharing Meeting July 6, 2001 http://www.panasia.org.sg/nepalnet/ansab/11meetingminute.html

Degliantoni , G., Murphy, M., Kobayashi, M., Francis, M.K., Erussia, B.P. and Trinchieri, B. (1985). *J. Exp. Med.* **162:** 1512-1530.

Derge County Statistics (1997). Derge County Administration, unpublished.

Dubey, H.C. (1983). *In:* An introduction to fungi, pp 231-232. Vikas Publishing House, New Delhi.

Du Halde, P. (1736). The General History of China: Of the Empire of China, Chinese-Tartary, Corea & Thibet. Including an Extract & Particular Account of their Customs, Manners, Ceremonies, Religion, Arts and Sciences, J. Watts, London.

Garbyal, S.S. (2001). Occurrence of Cordyceps sinensis in upper Himalaya, Dharchula Subdivision, Pithoragarh District, Uttaranchal, India The Indian Forester, **127.11:** 1229-1231.

Gawä Dorje (dGa' ba'i rDo rje), (1995). 'Khrungs dpe dri med shel gyi me long. Mi rigs dpe skrun khang, Beijing.

G.E.Clarke (Ed.). Development, Society and Environment in Tibet, Proc. 7th Seminar IATS, Vol. 5, Verlag der Österreichischen Akademie der Wissenschaften Wien, Austria, Vienna, 79-96.

Goldstein, M.C. (1996). "Nomads of Golok, Qinghai: A Report" .On-line: http://www.cwru.edu/affil/tibet/nomads.html.

Guzman, G., Allen, J.W. and Gartz, J. (1998). A Worldwide Distribution of Neurotropic Fungi. In: *Ann. Mus. Civ. Rovereto*, vol.14, 189-280.

Halpern, G.M. (1999). Cordyceps, China's Healing Mushroom", Avery Publishing Group, New York.

He, S.A. and Sheng, N. (1995). Utilization and conservation of medicinal plants in China with special reference to Atractylodes lancea. In: Non-Wood Forest Products 11 – Medicinal Plants for Conservation and Health Care - FAO, Rome. http://www.fao.org/docrep/W7261E/W7261e13.htm

Holliday, J.C., Cleaver, B.S., Loomis-Powers, M. and Patel, D. (2004). The hybridization of Cordyceps sinensis strains and the modifications of their culture parameters, in order to optimize the production of target medicinal compounds. http://www.nwbotanicals.org/nwb/lexicon/hybridcordyceps.htm

Horlemann, B. (2002). *In:* Toni Huber (Ed.), Amdo Tibetans in Transition, Brill Academic Publisher, Leiden, **2002:** 241-269.

Huang, K.C. *The Pharmacology of Chinese Herbs,* pp 263-264, CRC Press, Boca Raton, London, New York, Washington D.C, 1999.

Huber, J. (1955). *Arch. Midrobiol.* **29:** 257-276.

Jagdale, S.V. and Patil, M.S. (1983). *Indian J. Mycol. Pl. Pathol.*, 13: 150-153.

Kendrick, B. (1992). The Fifth Kingdom, Focus Publishing, Newburyport, Ma, 1-406.

Kiho, T., Yamane, A., Hui, J., Usui, S. and Ukai, S. (1996). *Biol. Pharm. Bull.* **19(2):** 294-296.

Kuo, Y.C., Tsai, W.J., Shiao, M.S., Chen, C.F. and Lin, C. (1996). *Am. J. Chin. Med.* **24(2):** 111-125.

Li, B.S. (1993). The Alpine Timberline of Tibet. *In:* J. Alden, J.L.Mastrantonio, S. Odum (Eds). Forest Development in Cold Climates, Plenum Press, NYC., 511-527.

Li, Q.S., Zeng, W., Yin, D.H. and Huang, T.F. (1999). *In:* Zhongguo Zhong Yao Za Zhi [China Jour. Chin. Mat. Med.] **23.4:** 210-2. [Chin., Engl. Abstract].

Liu Jianbang (Ed.). (1994). Forestry History of Ganzi Tibetan Autonomous Prefecture [Ganzi Zangzu Zizhizhou Lingyezhi], Sichuan Kexue Jisu Chubanshe , Chengdu, 1-323. (Chin.)

Liu, P., Zhu, J., Huang, Y. and Liu, C. (1996). Chung Kuo Chung Yao Tsa Chih. **21(6):** 367-369.

Lu Rongsen, (1998). Enterprises in Mountain-specific Products in Western Sichuan, China. ICIMOD Discussion Paper No. MEI 98/7, Kathmandu, 51 p.

Miehe, G., Miehe, S., Koch, K. and Will M. (2003). *In:* Mountain Research and Development **23.4:** 324-328.

Namgyel Phuntsho, (2003). Household Income, Property Rights and Sustainable Use of NTFP in Subsistence Mountain Economy: The Case of Cordyceps and Matsutake in Bhutan Himalayas. Paper Presented at the Regional CBNRM Workshop, Nov. 2003, 1-23 unpubl.

Namkhai Norbu, (1997). Journey among the Tibetan Nomads. An Account of a Remote Civilization. Dharamsala: Library of Tibetan Works and Archives, Dharamsala, 88 p.

Nielsen, E.S., Robinson, G.S. and Wagner, D.L. (2000). *J. Natural History* **34:** 823-878.

Pilz, D., Hosford, D., Molina, R. and Amaranthus, M. (1997). Ecology and Management of the Commercially Harvested American Matsutake Mushroom. General Technical Report PNW-GTR-412, Forest Service, Pacific Northwest Research Station, Portland, 1-68.

Pegler, D.N., Yao, Y.J. and Li, Y. (1994). The Chinese "Caterpillar fungus". Mycologist **8:** 3-5.

Pratt, A.E. (1892). To the Snows of Tibet Through China, London, (reprint 2001, AES Delhi), 1-268.

Rockhill, W.W. (1891). The Land of the Lamas – Notes of a journey through China, Mongolia and Tibet, London (reprint 1991, AES Delhi), 1-383.

Schei, P.A., Wang Sung, and Xie Yan. (2001). *In:* Conserving China's Biodiversity (II). China Environmental Science Press. Beijing. 78-100 p.

Shelton, A.L. (1921). *National Geographic Magazine* **40:** 294-326.

Shresta, K.K., Ghimire, S.K., Gurung, T.N. and Lama, Y.C. (1998). Conservation of Plant Resources, Community Development and Training in Applied Ethnobotany at Shey-Phoksundo National Park and its Buffer Zone, Dolpa. Unpubl. report, WWF Nepal & People and Plant Initiative, Kathmandu: 1-69.

Sichuan Zhibei, (1980). Editorial board of Sichuan's Vegetation, Chengdu, Sichuan People's Press, (Vegetation of Sichuan, in Chinese), 1-465.

Sung Jae-Mo, (2004). Cordyceps Diversity and its Preservation in Korea. In: Inoculum – Supplement to Mycologia **55.4:** 1-3.

Turkelboom, F., Gurung, T.R. and Dukpa, D. (2001). Role and Use of Common Property Resources (CPRS) in Bhutan Himalayas: Between Tradition and Globalisation. dlc.dlib.indiana.edu/archive/00000663/00/turkelboom.pdf

UNW (1996). University of Nanjing Webpages - Catalog of outstanding examples in bioresources of food therapy. http://www.nju.edu.cn/foode/jun/b3.htm

Wang, Hong (1999). Discussion of Several Issues Regarding Research and Development of *Cordyceps sinensis.* In: Development of Agriculture and Pastoral Production 1999. 6, Nanchang, 21-22. [Chin.].

Wang, Huailin (2000). Xanzhao Kangba [Searching for Kham], Sichuan Chubanshe, Chengdu, 1-319.

Wang-Sheng, Y., Shiao-Ming, Shi and Wang, S.Y. (2000). *J. Food and Drug Analysis* **8 (4):** 248-257.

Wang S.M., Lee, L.J., Lin, W.W. and Chung, C.M. (1998). *J. Cell Biochem.* 15, **69(4):** 483-489.

Wang, Wei., Chen, Tecan., Li, Qiongying., Zhou, Shining, Zhong, Yingchang, (1998).

Wilson, E.H. (1913). A Naturalist in Western China. (reprint Cadogan Books, 1986), London, 1-480.

Winkler, D. (1998). *In:* Patterns of Forest Distribution and the Impact of Fire and Pastoralism in the Forest Region of the Tibetan Plateau. Ed. Winkler, D. (2000). *In:* Environmental Change in High Asia, Eds. Miehe, G. and Zhang, Yili. Marburger Geographische Schriften **135:** 201-227.

Winkler, D. (2003). *In:* The Ecological Basis And Sustainable Management Of Forest Resources. Informatore Botanico Italiano **35-Sup.1:** 116-125.

Wu, Ning, (1997a). *In:* Rangelands and Pastoral Development in the Hindu Kush-Himalayas, Eds. Miller, D.J. and Craig, S.R. ICIMOD, Kathmandu, 23-40.

Wu, Ning, (1997b). Ecological Situation of High-Frigid Rangeland and its Sustainability – A Case Study on the Constraints and Approaches in Pastoral Western Sichuan/China. Abh. Anthropogeo. 55, Berlin, 1-281.

Xie, Z., Huang, X., Lou, Z., Li, S., Zhou, L., Yang, Z., and Tang, Z. (1988). Dictionary of Traditional Chinese Medicine. The Commercial Press Limited, Hong Kong.

Xu, Jianchu and Wilkes, A. (2002). *Global Environmental Research Journal* **6.1:** 103-110.

Yang Zhanchang (ed.). (1999). A Guide to Investment in Ganzi TAP, Foreign Investment Bureau of Ganzi Prefectural People's Government, Ganzi Chubanshe, Kangding, 1-32 p.

Yao, Damu and Zhang Jingbao, (1996). Coloured Atlas of the Chinese Materia Medica Specified in Pharmacopoeia of the PRC, Pharmacopoeia Commission of the Ministry of Public Health, PRC, Guangdong, 1-519.

Ying, J.Z., Mao, X.L., Ma, Q.M., Zong Y.C. and Wen, H.A. (1987). Icones of Medicinal Fungi from China. Science Press, Beijing, 1-575.

Yu thog Yon tan mGon po, (2002). bDud rtsi snying po yan lag brgyad pa gsang ba man ngag gi rgyud. ("the [4] Tantras [comprising the] Precious Nectar of the 8 Secret and Extraordinary Branches [of Medicine]"). Bod ljongs mi dmangs dpe skrun khang. Lhasa.

Zhang Jinfeng., Wang Wenbing, and Geng Yunfen (2001). *In:* The International Seminar on NTFP.

Zang, M. and Kinjo, N. (1998). Notes on the Alpine Cordyceps of China and Nearby Nations". Mycotaxon **66:** 215-229.

Zhou, D.H. and Lin, L.Z. (1995). *J. Chin. Trad. Med.* 15: 476.

Zhu, J.S., Halpern, G.M. and Jones, K. (1998). *J. Alternative and Complementary Medicine* **4:3:** 289-303 & **4:4:** 429-456.

Zhu, X. (1990). *Chin. J. Modern Devs. Trad. Med.* **10:** 485-487.

Chapter **8**

Biotechnological Application in Textile Industry—Antimicrobial Textiles

R. Rajendran

P.G. and Research Department of Microbiology, PSG College of Arts and Science Coimbatore–641 014, India

E-mail: rrajendranmicro@yahoo.co.in

Textile industry is next biggest to agriculture sector in India. There are a variety of areas in textiles where microorganisms play a significant role. The protective aspects of textile have provided the most textile ground for innovative developments. Hygiene has acquired importance in recent years. The consumers are now increasingly aware of the hygienic life style and there is a necessity and expectation for a wide range of textile products finished with antimicrobial properties. The ability to make textiles resistant to microbial contamination has several advantages in many applications and market segments. The environmentally safe antimicrobial that does not leach or wash off creating permanent antimicrobial property is necessary. The types of antimicrobials used in textiles may be leaching or non leaching type to confer biocidal or biostatic functions depending on the chemical nature of the agent and its reactivity with the type of fabrics including cotton, viscose, wool, nylon, polyester, bamboo and acrylic. The existing commercial antimicrobial agents are mostly chemical in nature and hence search for eco-friendly antimicrobial agents from natural sources such as medicinal plants is the need of the hour. There are various benefits of antimicrobial additives in textiles based on the end use application. The incorporation of antimicrobials into textiles may be carried out in different stages of textile processing such as yarn, fibre, fabric and garment. The major challenging task is to increase the wash durability by applying suitable cross linking agent and evolve the most efficient eco-friendly cross-linking agent. Microencapsulation technology is suitable alternative to increase the wash durability of the antimicrobially finished fabrics. However, such value addition in textiles should not at any cost interfere with the comfort characteristics of the fabric. On a global scenario the antimicrobial textiles can be validated with the advent of several international methods and standards laid down by AATCC. This chapter mainly emphasizes the developments of antimicrobial textiles and its recent trends in the value addition of textile industry.

INTRODUCTION

'Wellness' is a trend in modern society, which opens up a wide range of new possibilities—textiles with antimicrobials. There are many different types of textiles used today in an attempt to either limit or prevent the transmission of hazardous microorganisms. These materials range from solitary layers of non-woven single use products to composites of woven and knitted multiple use products that include film reinforcements (Achwal, 2003). There is a multitude of various fibre types, ways in which fibre assemblies and textile and film structures can be used in textile constructions to impart varying degrees of protection against microorganisms. Often, there are entirely different performance objectives for textiles, depending on the nature of their intended application, which can dictate which types of textiles structures are appropriate for use as microbial barriers (Edward, 2001). It has been estimated that there are at least 193 important biological agents that show infectious, allergenic, toxic, or carcinogenic activities in the working population.

Although bacteria and fungi are normal components of the natural environment, they cause specific problems that we could like to avoid: odours, staining, discolouration and deterioration of the fabric. In the textile industry we have many opportunities to add value and improve products by incorporating novelty finishes (Sakagami, 1995). The trend is now to protect textiles against both microbial infestation and the effects of soil, water, and/or oil. The major advantages of the antimicrobial textile finish or modification are protection to the wearer (health related) against biohazards, control of foul odour (hygiene related) and improvement in the performance and aesthetic characteristics of textiles (Mucha *et al.,* 2002).

Natural fibres are more prone to microbial attack due to inherent characteristics such as moisture regain, chemical composition and residues left after chemical processing, which provide room for infestation by microbes. Besides, the climatic conditions also influence the microbial activity, in terms of relative humidity and temperature (Vandana and Padma, 2001). It is observed that the nuisance caused by microbes is more in tropical and sub tropical regions of earth. Though man-made fibres are somewhat resistance to microbial growth, they act as carriers of pathogens. It is pointed out that penetration characteristics of textiles to air and liquid does not rule out the possibility of transmission of infectious microorganisms. Depending on the intended application, the goal should be to demonstrate that textile materials are effective microbiological barriers tested for definitive test methods as like microbiological challenge.

The inherent properties of the textile fibres provide room for the growth of microorganisms. Besides, the chemical process treatments may induce the growth of microbes. Natural fibres like cotton and wool are more susceptible than synthetics because their porous hydrophilic structure retains water, oxygen and nutrients providing a perfect environment for bacterial growth. Humid and warm environment still aggravate the problem. Nuisance caused by microbes is severe in tropical and subtropical regions. Microbial infestation causes cross infection by pathogens and development odour where the fabric is worn next to skin (HaSoo *et al.,* 1999). In addition, microbes cause staining and loss of the performance properties of textile substrates. Basically, with a view to protect the wearer and the textile substrate itself antimicrobial finishes are inevitable. The applications of antimicrobial textiles are expanding from industrial textiles to even daily use sports or outdoors wear. This development is accompanied by an increase in awareness of hygiene issues and demand for hygienic products on one hand and increasing criticism of the use of chemicals on the other. Currently, a number of research works has been carried out on natural antimicrobial agents originating from natural resources (Payne and Kudner, 1996).

HISTORICAL ACCOUNT

The use of antimicrobial agents, on textiles, dates back to antiquity when the ancient Egyptians used spices and herbs to preserve mummy wraps. In 1935, Domagk, a German scientist, developed an important class of antimicrobial agents based on quaternary ammonium salts (Ha-Soo, 1999). Although people have used natural materials to combat diseases for millennia, and we have known that bacteria and microbes cause diseases for centuries, only in the twentieth century we begin to produce antimicrobial compositions and add them to textile materials. During World War II, cotton fabrics were used extensively for tentage, tarpaulins and truck covers, for the protection of fabrics from rotting. This was particularly a problem in the South Pacific campaigns, where much of the fighting took place under jungle like conditions. During the early 1940s, the U.S. army Quartermaster Corps collected and compiled data on fungi, yeast and algae isolated from textiles in tropical and subtropical areas throughout the world. Cotton duck, webbing and other military fabrics were treated with mixtures of chlorinated waxes, copper and antimony salts that stiffened the fabrics and gave them a peculiar odour. At the time, potential polluting effects of the application of these materials and toxicity-related issues were not a major consideration (Ramachandran *et al.*, 2004).

After World War II, and as late as the mid-to late-1950s, fungicides used on cotton fabrics were compounds such as 8-hydroxyquinoline salts, copper naphthenate, copper ammonium fluoride and chlorinated phenols. As the government and industrial firms became more aware of the environmental and work place hazards that these compounds caused, alternative products were sought. A considerable amount of work was done by the Southern Regional Research Laboratory of the U.S. Department of Agriculture (USDA), the Institute of Textile Technology (ITT) and some of the ITT's member mills to chemically modify cotton to improve its resistance to rotting and improve other properties by acetylation and cyanoethylation of cotton. These treatments had limited industry acceptance because of relatively high cost and loss of fabric strength. In addition, the growing use of man-made fibres such as nylon, acrylics and polyester, which have inherent resistance to microbial decomposition, came into wider use to replace cotton in many industrial fabrics.

FACTORS INFLUENCING THE GROWTH OF MICROBES IN TEXTILES

Cotton textiles in close proximity to the human body provide an ideal living environment for bacteria, yeasts and fungi (Payne and Kudner, 1996). All the conditions required for the growth of these organisms are satisfied:

a) **Nutrients:** Soil, dust and some textile finishes can all be sources of nutrients for microorganisms. Perspiration contains salts, amino acids, carboxylic acids and other essential nutrients. Dead skin cells or oils secreted from the skin are also a potential source of carbon. Cellulose itself is not a nutrient, but many bacteria produce extracellular cellulase enzymes, which convert cellulose into the readily metabolized glucose. This has the dual effect of promoting further growth and degrading the fibre.

b) Water: Cotton consists of porous fibres of hydrophilic cellulose and has a high affinity for water. Perspiring human beings have been estimated to give off an average of 100 g/hour of water, which accumulates in clothing and bedding. A humid environment will provide enough water to support fungal growth. Bacteria need more water and require the fabric to be damp.

c) Oxygen: The atmosphere provides a ready source of oxygen.

d) Warmth: Most fungi and bacteria will grow at ambient temperatures of 10-20°C. Certain bacteria prefer the slightly warmer conditions of clothing or bedding in close proximity to the skin. Bacteria such as *Staphylococcus aureus, Staphylococcus epidermidis, and Cornybacterium* sp. are most commonly found on the skin. (McNeil *et al.*, 1994).

In addition to providing these conditions, textiles are continuously exposed to microbial contamination from the skin, dust and airborne particles and other sources. It is not surprising that large populations of microorganisms have been isolated from textiles. Bacterial counts reported range from 1.3×10^4 organisms per gram of pillows, up to 4×10^5 per 5 cm^2 of socks and 5×10^6 per square inch of T-shirts (McNeil *et al.*, 1994). Synthetic fibres such as polyester, nylon do not provide such ideal living conditions for microorganisms. The fibres have a lower capacity for retaining water and are more resistant to degradation by enzymes. Nevertheless, these fibres will also support microbial growth (Mairal and Patel, 2002). In the majority of cases, the presence of microorganisms on cotton passes unnoticed. In certain circumstances, however, undesirable and possibly serious effects can occur (Table 1).

Table 1. Some common microbes and their influence

	Kind of microbes	Influence on human body		Garments most stained		
		Odour	Hosiery	Underwear		Trousers
Bacteria	*Staphylococcus*	+	Acute purulence	+	+	
	Bacillus subtilis		Conjunctivitis	+	+	+
	Escherichia coli	+	Ulcers	+	+	+
	Pseudomonas aeruginosa		Phthorax, Otitis media	+		
	Proteus vulgaris	+		+	+	
	Klebsiella pneumoniae				+	+
Yeast Fungi	*Candida albicans*		Thrush	+	+	
Mould	*Trichophyton interdigital*		Cause of athlete's foot	+		
	Aspergillus niger		Weakness and De-colouring of clothing	+	+	

+ : Positive for influence/staining.

ANTIMICROBIALS

Bactericidal, bacteriostatic, fungicidal, fungistatic or biocidal and biostatic are just a few examples of commonly used terms to define the nature of antimicrobial activity. Figure 1 is used to clarify and distinguish between these terms (Huber, 2001). If an active principle has a negative effect

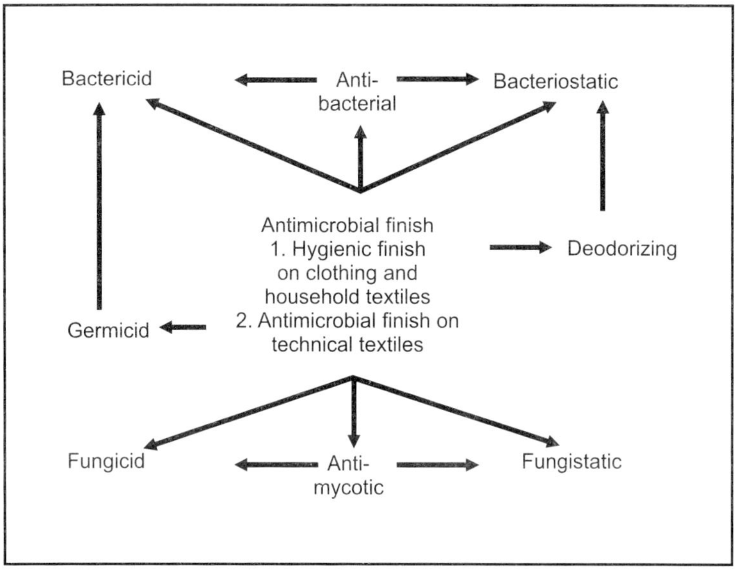

Fig. 1. Differentiation of antimicrobial activity.

on the vitality of microorganisms, this is generally referred to as antimicrobial activity. If the active principle affects only bacteria or fungi, this is referred to as antibacterial or antimycotic activity respectively. The degree of the effect is denoted by –cidal(lethal) where there is significant germicidal activity, or –static where the active substance serves to inhibit the growth of bacteria.

Unbound Antimicrobials

An unbound antimicrobial agent cannot be bonded to a textile surface in order to function properly. It must diffuse from the treated substrate and be consumed by the microorganism in order to be effective. Once inside the organism, the chemical agent will act like a poison interrupting some key metabolic or life sustaining process of the cell, causing it to die. They can contact with the skin and potentially effect the normal skin bacteria, cross the skin barrier and/or have the potential to cause rashes and other skin irritations. Once the antimicrobials is depleted or is washed away during regular maintenance, protection vanishes. After application, an unbound antimicrobial continues to diffuse or leach from the substrate on which it has been applied. As this diffusion continues, the concentration of the active ingredient becomes diluted below effective levels. Under these conditions, microorganisms have the ability to adapt or build up a tolerance to these antimicrobials. Conventional (unbound) antimicrobials can often be very effective against specific types of microorganisms, but are generally limited in their ability to offer broad-spectrum control. In other words, they may be effective against specific bacteria, but not all or they may destroy all bacteria, but be ineffective against fungi, yeasts or algae. The safety and toxicity of 'unbound' antimicrobial treatments vary considerably depending on the specific chemistry involved. For example, many organotin antimicrobials must be handled with great care for the workers and consumer health and the effects of disposal on environment. In fact, some of these organotin compounds are currently under close scrutiny by the EPA (Environmental Protection agency) and have been restricted in Japan and in some European countries (Hales *et al.,* 1986).

Bound Antimicrobials

Bound antimicrobial agents remain chemically attached to the surface on which they are applied. They function by interrupting the organism's delicate cell membrane. This prevents microorganisms from carrying on vital life process. These antimicrobials kill organisms on contact and can do so again and again. Since a 'bound' antimicrobial is fixed to the fibre and is not continually diffusing off the surface leaving behind a level that is less and less effective, the adaptation process described earlier cannot and does not occur. Since they are chemically bonded to the surface and are not washed away during regular maintenance, protection continues. The unique mechanism by which bound antimicrobials exhibit their activity permits them to effectively control a broad spectrum of microorganisms. Bacteria, molds, mildew, fungi, yeast and algae can be controlled with this type of antimicrobial (Gettings, 1985). All of the bound antimicrobials on the market today have a very favourable toxicologic and safe profile. Most commonly used in the market place today, quaternary ammonium compounds have undergone extensive toxicological testing to satisfy EPA requirements prior to it becoming registered for commercial use. And due to its unique chemistry, this particular compound can be rendered inactive in solution by simply adding an anionic detergent. This permits safe disposal of any extra treatment solution (Hales *et al.,* 1986).

Biocidal and Biostatic Functions

Antimicrobial materials can be divided into two categories based on the abilities against microorganisms (Anita *et al.,* 2002).

- Biocidal functions: Inactivation of microorganisms on the materials of total kill.
- Biostatic functions: Inhibition of the growth of microorganisms on the materials or partial kill.

Antimicrobial Textiles

In principle, during an antimicrobial treatment of textiles the growth of the microorganisms should be inhibited. This also reduces the undesired by-products. There are different chemical and physical possibilities for the production of antimicrobial textiles. In practice the antibacterial effects are achieved through application of antibacterial chemicals in the finishing or through incorporating antibacterial agents during the spinning process of the synthetic fibre (Wei and Karen, 2000).

Possibilities are:

- Addition of microbicides to the spinning solution.
- Modification through grafting or other chemical reactions.

- Fibre blends.
- Finishing of the fabric. After thermal treatment (drying polymerization) by incorporating these agents in polymer or resin containing finishing chemicals, they are fixed on the fabric.

Protection against biological attack may be subdivided into three categories (Table 2).

a) Protection of the wearer or user of a textile material against microorganisms for aesthetic (suppressing or killing odour causing bacteria), hygienic (preventing skin and related infections usually caused by dermatophytic fungi) or medical purposes (suppressing or killing pathogenic and/or parasitic microorganisms problematic in hospitals and institutions).

b) Protection of the textile itself from biodeterioration caused by mold, mildew and rot producing fungi.

c) Protection of textiles from insects and other pests for preservation of the fibre and/or protection of persons wearing clothing from insects and pests (Deborah, 2001).

Table 2. General strategies for protecting textiles and their users against biohazards

Technical approach	Relevant features
Physiochemical techniques	Physical barriers impermeable to both liquid and vapour and/micro porous coatings permeable only to vapour but not to liquids. Microencapsulation of bioactive agents that operate by controlled release
Chemical methods	Controlled release of active biocide or biostat applied by finishing methods Subclasses including chelating agents, oxidizing agents and reducing agents and various types of antibiotics
Multifunctional property fabrics	Flame retardant / antibacterial fabrics Fabrics with multifunctional properties including antimicrobial resistance Clothing resistant to both chemical and biohazards

Physical barrier or blocking action is achieved by either using inert films or coatings for physically blocking bacteria or by films and coatings having direct surface contact activity against bacterial growth. Coating can be universally applied to all fibre types and other surfaces to produce activity against a broad range of microorganisms and durable to normal washings. Hydrolysis product of trialkoxysilyl quaternary ammonium salt for example is applied through surface bondings to give such protection (Gangsun, 1997).

Specific Applications of Antibacterial and Antimycotic Agents for Textiles

Practically, every class of chemical compound has been utilized to impart antibacterial activity to textiles. Earlier efforts were primarily based on insolubilization of inorganic compounds like copper and other organometallic salts. Later organic compounds like polyhalogenated phenols, their esters and bisphenols were used. Cellulosics are best treated with various wrinkle recovery, nitrogenous resins which also impart rot and mildew resistance. More recent approaches include synthesis of reactive dyes containing antimicrobial substituents on cotton fabrics to simultaneously dye and rotproof fabric. Many commercial formulations are available today. Dow corning used its expertise in silicone chemical technology to incorporate a standard antimicrobial substance (a quaternary amine) into a silane. With its silane backbone, the cored antimicrobial polymer becomes an integral part of any surface to which it is applied. That means it cannot volatilize into the air or leach into water or other liquids; and it is not washed away by normal cleaning.

Actigard from sanitized AG, Switzerland, is designed for items that are seldom or never washed such as carpets, furnishing fabrics and also filling such as polyurethane foam. Applied to textile fibres, it inhibits fungal growth and reduces mould spores. It is UV stable fast to dry cleaning but not machine washing. A product in the hyfresh range from Japanese manufacturer Daiwa, and supplied by LJ specialties takes a different approach to dust mite control. This product acts as a repellent, driving the mites into areas where they can be more readily controlled by cleaning and vacuuming. The product retains its effectiveness even after 50 washes. Another member of the HyFresh range is designed for odour control. The active ingredient in the finish inhibits both bacteria and fungi. Applied by either pad or exhaust methods, it can be used on synthetic or synthetic/natural blends. Still active after 40 washes, it has been used in household textile and clothing. Tests have also shown that it is capable of alleviating the symptoms of athletes foot when applied to socks. Biosil from Toyobo is a well-known finishing agent based on organic silicones with tertiary ammonium where the cations block the cell division. Commercial products for PET with tertiary ammonium salts are sanitan from kuray and peach fresh from Nisshinbo.

With a view to develop such textile materials, considerable research has been carried out by making use of organic and inorganic compounds, antibiotics, heterocyclics, quaternary ammonium compounds, and so on. Some recent developments in antibacterial products include a process involving the preparation of antibacterial resins containing phenol derivatives. (Nonaka, 1997) The resins exhibited higher antibacterial activity against *Escherichia coli* and *Staphylococcus aureus*. Beliakova *et al.* (1998) discussed the process for preparing an antibacterial formulation by reaction of carboxymethyl starch with trimethylolated melamine in the presence or absence of cupric ions to render cotton fabric antibacterial. Fabrics made from viscose fibres containing polysilicic acid and aluminium silicate have been given urea peroxide treatment to make them antibacterial as well as deodorant. The cellulose has been modified chemically with biocides accompanied by redox reaction to achieve durable and regenerable antibacterial activity on cotton and other cellulosic fabrics (Sun and Xu, 1997). The finish poly (hexamethylene biguanide hydrochloride) imparts antibacterial property to cotton to cotton blended materials that is effective against a broad spectrum of bacteria, fungi and yeasts and also durable to 50 launderings (Payne and Kudner, 1996; Payne *et al.*, 1997). Vigo (1993) studied the antimicrobial effect of polyols on cotton fabrics and found that cross-linked polyols suppressed the growth of microbes as well as odour. A formulation that is hydrolytically stable at ambient temperature for at least two months and thermally stable below 300°C has been prepared by making use of magnesium hydroperoxy acetate and magnesium dihydroperoxide to render cellulosics, chemical fibres and their blends bactericidal up to 50 launderings (Vigo and Danna, 1996). Chitosan treatment on cotton fabrics imparts antibacterial activity against *Staphylococcus aureus*, but the effect decreases on subsequent launderings (Lee *et al.,* 1999). It is to be mentioned that synthetic fabrics can also be made antibacterial by treating them with antimicrobial agents (Anon, 1992; Buchenska, 1996; Buchenska, 1996; Kawata, 1998; Zeibneva, 1998). In addition to imparting antibacterial activity to cotton and other cellulosics materials, research has also been carried out to protect them from fungus and mildew attack. The fungi *Aspergillus niger* and *Chaetominum globosum* normally cease the growth of mildew on textiles especially under high humid condition that leads to deterioration, staining, and discoloration of fabrics (Turner, 1989; Andrasic, 1993). Studies proved that octhilinone controls the growth of mildew on cotton fabrics (Meeks and Greenley, 1998) and chlorine bleach removes the stain produced by mildew.

TYPES OF ANTIMICROBIALS IN TEXTILES

Antimicrobials are of two types (Edward Menezes, 2002):

1. Leaching type (conventional antimicrobials)
2. Non-leaching type

1. Leaching Type

Diffuses from the garment to come in contact with the microbe and products migrate off of the garments, forming a sphere of activity, and any microbes coming into the sphere are destroyed. But in the course of time the strength decreases and thus it just 'hurts' the microbes, giving them a chance to form a stronger strain by mutation. The microbes consume the antimicrobials as it acts on them. The product is eventually used up by the bacteria and slowly loses its effectiveness.

2. Non-leaching Type

Bound to the product, allowing control of the microbes and products do not migrate off of the garments and destroy the bacteria coming in contact with the surface of the garment. The microbes do not consume the antimicrobials, which destroy them by acting on the cell membrane. These products do not lose their effectiveness; the finishing will be permanent and will remain functional throughout the life of the fabric and will withstand more than 40 laundry washes.

MODE OF ACTION OF ANTIMICROBIALS AND ADAPTATION

Migration and Zone of Inhibition

Conventional antimicrobials are often effective, but they are used up in the process of working or wasted in random misses. Some companies incorporate leaching technologies into fibres and slow the release rate to extend the useful life of the antimicrobial or even add them to chemical binders and claim they are now 'bound'. Whether leaching antimicrobials are extruded into the fibre, placed in the binder or simply added as finish to fabrics or finished goods, they all function the same. In all cases leaching antimicrobial technologies provide a killing field or 'zone of inhibition'. The zone of inhibition is the area around the treated substrate into which the antimicrobial chemistry leaches or moves to, killing or inhibiting microorganisms. This killing or inhibiting action of a leaching antimicrobial is witnessed when an AATCC 147 test or other zone of inhibition test is run. These tests measure the zone of inhibition created by leaching antimicrobial and clearly defines the area where the antimicrobial has come off the substrate and killed the microorganisms in the agar. This shows the difference between the leaching and the non-leaching antimicrobial treatments on textile both as first treated and then after five household launderings (Curtis *et al.,* 1997).

Adaptation Mechanism

Microbes are living organisms and like any living organism will take extreme measures to survive. Microorganisms can be genetically mutated or enzymatically induced into tougher 'super-strains' if they are exposed to sublethal doses of antimicrobial agents. The exposure of the microbe to a sublethal dose of an antimicrobial can cause mutation of their genetic materials allowing for resistance i.e., then replicated through the reproductive process creating generations of microorganisms that are no longer affected by the chemistry. This phenomenon is of serious concern to the medical community and food processing industries and should be considered for the textile industry as it chooses the antimicrobials to which it will be exposing the public and their workers. A leaching antimicrobial is strongest at the fabric surface and the weakest the farther it travels from the reservoir. The outermost edge of the zone of inhibition is where the sublethal dose can be found. This is where resistant microbes are found that have been produced by leaching antimicrobials.

Bonding Mechanism

A significant different and much unique antimicrobial technology used in the textile industry does not leach but instead remains permanently affixed to the surface applied to. Applied in a single stage of the wet finish process, the attachment of this technology to surfaces involves two means. First and the most important is a very rapid process, which coats the substrate (fabric, fibre etc), with the cationic species. The second mechanism is unique to materials such as silane quaternary ammonium compounds. In this case, the silanol allows for covalent bonding to receptive surfaces to occur. The bonding is then made even more durable by the silanol functionality, which enables them to homopolymerize. After they have coated the surface in this manner, they become virtually irremovable (Fig. 2).

Once polymerized, the treatment does not migrate or create a zone of inhibition, so it does not set up conditions that allow for adapted organisms. Since the antimicrobial stays on the substrate, it does not cross the skin barrier and does not affect the normal skin bacteria, causing rashes or skin irritation. It does not poison the microorganism and is not consumed by the microorganism. When a microbe contacts the oragnofunctional silane treated surface of the fabric, the cell is physically ruptured by a sword like action and then electrocuted by a positively charged nitrogen molecule. This technology has been used for nearly twenty-five years without

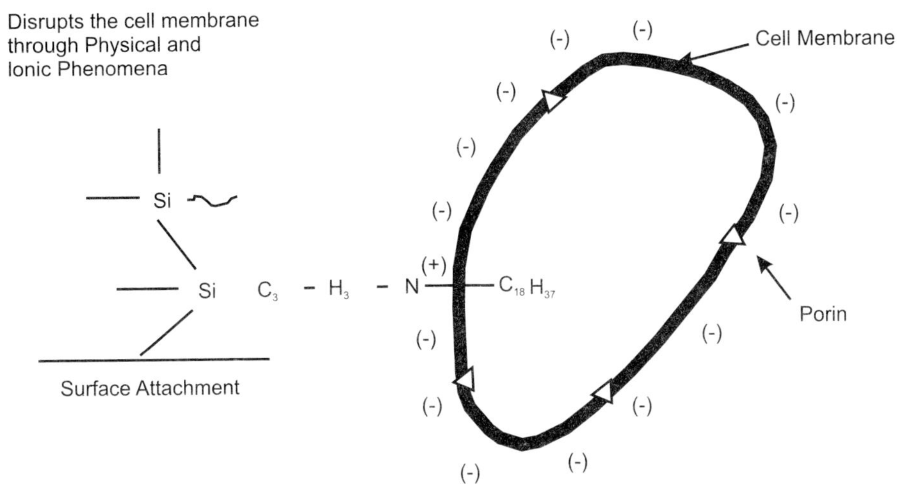

Fig. 2. Mechanism of microbial cell disruption by antimicrobial agents

any human health or environmental problems in manufacturing facilities or in actual end use situations (Curtis *et al.,* 1997).

ANTIMICROBIAL TREATMENT OF VARIOUS FABRICS

Cotton and Viscose

Cotton and viscose being most widely acceptable textile fibres for apparels due to their comfort properties should be made antibacterial as they are widely used for production of surgical gowns and baby clothes. Human body produces sweat, which contains moisture, urea and other waste materials, providing a suitable shelter for microbial growth. The moisture content further accelerates the microbial growth. At lower moisture content, a cotton sample contains 1.4 million bacterial organisms per gram. This number ceases to 9000 million at higher moisture level (Cletus and Clark, 1983). Quaternary ammonium compounds (QAC) and their water soluble halides have been used extensively on cotton and viscose fibre materials for effective control over microbial growth. Quaternary ammonium halides are more soluble than stearates. The advantages of halides are that they are easily absorbed in the fabric due to their better solubility and a higher antimicrobial property. However, their wash fastness properties are very poor. Although stearates are less absorbed onto the fabric, their properties are stable to repeated washings, thus, tetra alkyl ammonium and alkyl pyridinium halides, which are soluble in water, are suitable only for the preparation of one time action (microbicidal articles), while the less soluble stearates give stable antimicrobial property for staple fibres based on cellulose (Young *et al.*, 1998).

In an interesting study on cotton material, the cotton fabric was printed with a composition containing 1% aqueous konnyaku powder (thickener) 900 g, glyoxal resin 100 g, a catalyst 40 g, albumin 50 g, clove 20 g, and rosin 30 g. This fabric was then calendered at 130°C under a 150-500 kg, cm^{-1}pressure, cured for 2-7 min at 120-180°C. The fabric was claimed to be having very good antimicrobial property even after 50 washings. In another study, a cotton fabric was treated with an aqueous dispersion containing 1% silver and silica-alumina-mica powder for 30 min at 90°C. The fabric with 100% bacteria extinction capacity was obtained when it was exposed to a medium containing *Staphylococcus aureus* for 24 hours at 28°C. Kanebo ltd, has come up with an antibacterial and fungicidal fabric by padding the mercerized cotton woven fabric with an aqueous solution containing 50 g 5% N-[(flurodichloromethyl) thio] phthalmide, 100g of 30% Ashiguard AG710, and 50 g/l of 80% polyester polyurethane (Vondic 140). The fabric was then dried, heat treated at 160 °C for 3 mins. It was reported that this fabric exhibited good retention of properties even after 10 washings (Seventekin and Ucarci, 1993).

It may be understood that mere treating with an antimicrobial chemical will not serve the purpose but the fixation of the chemical is more important. Chemical and physical entrapment of the antimicrobial agent in the fabric is to be facilitated. Thus the slow release of the agent would ensure no microbial growth. Although various chemicals are being used all over the world for antimicrobial activity, the exact reaction mechanism is not understood yet (Murugesh, 2003).

Wool

Wool, having higher moisture regain (14%), is more prone to microbial attack. However, in the tropical countries, woolen fabrics are seldom used and microbial attack on wool is not an issue. Research on insect resistant wool has picked up momentum. It is important to know that the consumption of wool in European countries is relatively high in the form of sweaters, overcoats, carpets etc. As these clothings are usually subjected to limited washings antimicrobial treatment is very important (Davidson and Xue, 1994). Meiji milk products Co. Ltd have disclosed a method for the manufacture of washfast antimicrobial wool fibres. Wool yarns were first treated with 2% aqueous 2-mercaptoethanol. They were washed and then treated with 0.1M $AgNO_3$ for 45 minutes at room temperature. After washing and drying the fabrics made from the yarns were found to possess very good antimicrobial property and wash fastness. Another patent claims the production of odour absorbing antimicrobial wool fibres. Wool slivers were immersed in a solution of 40 ppm of Ni^{2+}. The slivers were repeatedly stretched, relaxed and further chlorinated using aqueous solution of HOCl. The fibres thus obtained were reported to allow a bacterial growth of 0.04% of the untreated fibres. The stretching and relaxing of the wool fibres in water containing transition metal ion may cause the breakage of cystine bonds and absorption of metal ions, which causes the oxidation and removal of exocuticle proteins and exposure of endocuticle. Endocuticle, upon chlorination, shows antimicrobial property (Murugesh, 2003).

Polyester and Nylon

Polyester being an important fibre has taken the driver's seat in the man-made fibre industry. It has very low moisture absorption capacity and thus the microbial growth over polyester might be comparatively low (Bide *et al.,* 2002). However, it has become necessary to produce antimicrobial polyester, as the consumption of polyester in apparels is considerably high. Some researchers (Mairal and Patel, 2002) have reported a very effective antimicrobial agent for polyester and nylon. In addition, it was claimed that antimicrobial polyester fibres, namely Bactekiller® fibre, have been developed. The antimicrobial agent is a zeolite powder containing antimicrobial metal ion. Zeolite, an inorganic substance with network structure, shows a peculiar phenomenon in the ion exchange function, and was found to be safe and thermally stable. The antimicrobial metal ions may get entrapped in the zeolite network. The finely powdered zeolite containing antimicrobial metal ions Bactekiller® may be spun along with polyester to produce antimicrobial polyester. The spinning and dyeing properties were reported to be approximately similar to the regular PET fibres. Due to the high adsorption capacity of the zeolite, the fibres containing the zeolite would be deodorant also (Angelina, 2003).

A patent of Japan Ester Company describes the production of antimicrobial and deodorant fibres. The stretched PET fibre was coated with an aqueous emulsion containing 10 parts of 2,4,4 – trichloro – 2 hydroxy diphenyl ether, 5 parts of lauryl benzalkonium chloride and 28.5% non-ionic and anionic surfactants at 50°C to give PET fibre having 98.9% reduction of *Staphylococcus aureus* content (Angelina, 2003). Socks and underwear, which are made up of nylon, polyester or acrylic fibres act as good shelters for microbial attack due to sweat. In synthetic fibres, which have lower fluid transport properties, the escape of the sweat to the atmosphere through the undergarments should be made easy to maintain the dryness (Young and Gang, 2000). Foul smell of socks is due to microbial attack in the presence of sweat. The irritation between foot fingers is a common problem associated with microbial growth. Toray industries introduced antimicrobial underwear and socks from nylon, polyester and acrylic fibres. Nylon 66 knit was grafted with an aqueous mixture containing acrylic acid and methacrylic acid to give a fabric with carboxylic acid group. The grafted fabric was treated with an aqueous composition containing 7 g/L benzyl dimethyl stearyl ammonium chloride for 30 minutes at 100°C. This technique seems to be more lasting compared to other finishing, as a quaternary ammonium chloride will be attached to the grafted carboxylic groups (Murugesh, 2003).

Acrylic

Acrylic fibres have been made antimicrobial through chemical modifications. Old Russian communications of 70's dealt with inclusion of mercury salts at the as-spun stage. An acrylic as-spun fibre based on acrylonitrile (92.5%), methyl acrylate (6%) and itaconic acid (1.5%) was imparted with antimicrobial property by treating it with aqueous solution of mercury nitrate, air dried, heat-set at 120°C and then thoroughly washed to remove the salt and dried at 80°C (Stefan, 2003). These modified fibres were washed for one hour in boiling distilled water. The resulting fibres were subjected to biological activity. In their designed experiments infected agar-agar method was used, where the microbial growth in the presence and absence of the treated fibres was compared at 37°C for 24 hour and at 24°C for 48 hour with bacteria and fungoid flora respectively. It was reported that the modified polyacrylonitrile fibre containing 1.9% mercury inhibits the growth of certain pathogenic flora. However, mercuric compounds are not advisable, as they are poisonous to human body (Stefan, 2003).

Quaternary ammonium chlorides are also widely used for antimicrobial property in acrylonitrile-based fibres. The best-known preparations of this group include Ketamin AB, and alkylryltrimethyl ammonium chloride. Acrylic fibres with phenol derivatives such as 3-methyl-4-isopropylphenol and 2,4,4,'trichloro hydroxy diphenyl ether also reported to have very good antimicrobial activity (Young *et al.,* 1998). A European patent disclosed the idea to bond silver sulphonate to the acrylic fibres without aggregation. Hundred parts by weight of cashmilon acrylic fibre was soaked in 3000 parts by weight of water that contains 0.126 parts by weight of sulphonate. This was then boiled for 30 minutes, allowed to cool until 50°C and thoroughly washed in water and dried. 0.2 g of this cashmilon coated with a silver sulphonate group and 0.2 ml of standard bouillon suspension of *Staphylococcus aureus* were mixed at 37°C for 18 hours and washed or extracted with 20 ml of physiological saline buffer containing phosphate. Recently, Courtaulds Co. Ltd. reported antimicrobial acrylic fibres in the trade name "Amico" acrylic fibres. These fibres can be blended in various rations with virgin fibre in desired level as per the requirements (Murugesh, 2003).

Antimicrobial with Active and Passive Effect

Antimicrobial active principles can be divided into 2 categories: materials with passive effects and materials with active effects. Passive materials do not have specific bioactive substances. The microbial colonization of the textiles is only prevented by the surface structure of the fibre (e.g. the lotus effect or micro-domain structured surfaces). In other words, the bacterial cells themselves are not affected; instead the cells are prevented from adhering to the fibre surface. This has a negative effect on the living conditions for microorganisms (anti-adhesive effect). Materials with active finishes contain specific active antimicrobial substances which act upon microorganisms either on the cell membrane during metabolism or within the core substance (genome). The majority of bioactive substances with the exception of antibiotics (Beumer *et al.*, 2000; Russell and Chopra, 1996) have different effects depending upon where the antimicrobial activity occurs, either in or on the cell. Wallhauber *et al.,* (1995) have dedicated an extensive chapter to antimicrobial substances, giving details of the range of antimicrobial effects, toxicity and the various fields of application. Many antimicrobial substances, which are used on textiles, are already known from the foodstuff and cosmetics sectors (Roy *et al*., 1999), where, in some instances, they have been in use for decades. They are used under the assumption that they have been sufficiently tested and are toxically safe and that they also tend to be used in lower concentration in textile finishing than in the cosmetics industry. This assumption has unfortunately not always proved true, however, as was demonstrated in the controversial discussions concerning tributyl tinoxide and triclosan (U.S.E.P.A-toxicological review, 1997; Virnich, 2000; Heath and Rock, 2000; Beumer *et al.,* 2000).

Antibiotics are also classified as antimicrobial substance. However due to the very specific nature of their effect, it is important to make a clear distinction between antibiotics and other substances which have a broad range of uses (Hyung-Min, 2004). Oxidizing agents include aldehydes, halogens and peroxo compounds. They primarily affect the cell membrane of the microorganisms. If these substances get into the cytoplasm, they affect the enzymes (WallhauBer, 1995; Beumer *et al.*, 2000; Russell and Chopra, 1996).

Coagulants are predominantly alcohols. They irreversibly denature the protein structures (WallhauBer, 1995; Russell and Chopra, 1996; Beumer *et al.,* 2000). Radical formers include halogens, isothiazones and peroxo-compounds. These are highly reactive substances due to the presence of a free electron, and they react with virtually all-organic structures, in particular oxidizing thioles in amino acids (WallhauBer, 1995; Beumer *et al.*, *2000*; Russell and Chopra, 1996). Radical formers pose a particular risk to nucleic acids as they trigger mutations and dimerisation even when they are only present in extremely small concentrations. Quaternary ammonium compounds, biguanides, amines and glucoprotamine display polycationic, porous and absorbent properties. Fibres finished with these substances bind microorganisms to their cell membranes and disrupt the lipopolysaccharide structures, eventually resulting in the penetration of the membrane and breakdown of the cell (WallhauBer, 1995; Beumer *et al.*, *2000*; Russell and Chopra, 1996; Tiller, 2001). A number of metallic compounds are also used: various silver compounds are currently very popular. These are already successfully being used in the antimicrobial preparation of drinking water (WallhauBer, 1995). In general, complexing metallic compounds cause an inhibition of the active enzyme centers, i.e. they inhibit metabolism. This is also referred to as oligodynamic action or effect, i.e. it is effective in minute quantities (WallhauBer, 1995). The effect is greatest with cadmium and reduces gradually with silver, brass, copper and mercury. With gold, platinum, iron, aluminium and zinc, the effect does not occur at all. Silver is significantly more effective with Gram-negative bacteria than Gram-positive and than with fungi: It is also 3-4 times more effective at pH 8 than pH 6. (WallhauBer, 1995). Natural products such as Chitosan are also being used more frequently. This is a chitin derivative, which is obtained from crab shells. In the case of Chitosan, the protonized amino group binds to the glucose ring with the negatively charged cell surface of the bacterium (Bohringer *et al.*, 2000). It has not yet been clarified whether this is then destroyed or merely remains bound. The polycationic effect on the bacterial cell membrane is also known from n-alkylated polyvinyl pyridine (Tiller, 2001) In many cases, the exact active principles have still not been sufficiently researched. Materials with active finishes contain specific active antimicrobial substances, which act upon microorganisms either on the cell, during the metabolism or within the core substance (genome). However, due to the very specific nature of their effect, it is important to make a clear distinction between antibiotics and other active substances, which have a broad range of uses (Youngsook *et al.*, 2000).

One of the most durable type of antimicrobial products is based on, a diphenyl ether (bis-phenyl) derivative, known as either 2, 4, 4'-trichloro-2'hydroxy dipenyl ether or 5-chloro-2-(2, 4-dichloro phenoxyl) phenol. Triclosan products have been used for more than 25 years in hospitals and personal care products such as antimicrobial soap, toothpaste and deodorants. Triclosan inhibits growth of microorganisms by using an electrochemical mode of action to penetrate and disrupt their cell walls. When the cell walls are penetrated, leakage of metabolites occurs and other cell functions are disabled, thereby preventing the organism from functioning or reproducing. Triclosan when incorporated within a polymer, migrates to the surface where it is bound. Because it is not water-soluble, it does not leach out, and it continuously inhibits the growth of bacteria in contact with the surface using barrier or blocking action (Tiwari, 2001).

EXISTING COMMERCIAL ANTIMICROBIAL AGENTS AND FIBRES

Thomsan Research Associates markets a range of antimicrobial under the trade name 'Ultrafresh' for the textile and polymer industry. Ultrafresh products were developed for normal textile processes. Most Ultrafresh treatments are non-ionic and are compatible with a wide range of binders and finishes. To incorporate antibacterial into high temperature fibres like polyester and nylon, it is necessary to use an inorganic antimicrobial like Ultrafresh CA-16 or PA 42.These must be added as a special master batch to the polymer mixture before the extrusion process. Fibres such as polypropylene are extruded at lower temperatures, it is possible to use organic antimicrobial such as Ultrafresh Nm 100, Dm-50 or XQ-32 (Ian, 2002). Rossari's Fabshield with AEGIS microbe shield programme is based on rupturing the cell membrane of the bacteria when the microbes come in contact with the treated surface, thus, preventing the consumption of antimicrobial over a period of time and remain functional throughout the life of the product. The active substance 3-Trimethoxy silyl propyl dimethyl octadecyl ammonium chloride gets attached to the substrate either through bond formation on the surface or by micropolymersing and forming a layer on the treated surface; the antimicrobial agent disrupts the cell membrane of the microbes through physical and ionic phenomena (Ian, 2002).

Avecia's Purista-branded products treated with Reputex 20, which is based on poly (hexamethylene) biguanide hydrochloride (PHMB), claimed to possess a low mammalian toxicity and broad spectrum of antimicrobial activity. PHMB is particularly suitable for cotton and cellulosic textiles and can be applied to blends of cotton with polyester and nylon. Another product, Ciba Tinosan NW200 is designed for spray application to textiles that do not require durability to washing. Both Tinosan AM110 and NW200 contain as an active antimicrobial Ciba Irgasan DP300, generically termed triclosan, which has been used for over 30 years successfully in skin and body-care products (Franz, 1978; Ochs, 1998; Cox, 1999) and it is effective against most Gram-positive and Gram-negative bacteria, as well as some fungi and yeasts (Regös *et al.,* 1979). Tinosan AM110 behaves like a colourless disperse dye and can be exhausted at a very high exhaustion rate onto polyester and polyamide fibres when added to the dye. The toxicity and allergy risk to mammals are extremely low and triclosan has also been shown to have an anti-inflammatory efficacy. In addition to the aforesaid antimicrobial agents, the textile fibres derived from synthetic polymers with built-in antimicrobial properties are shown in Table 3.

Table 3. Antimicrobial fibre on the basis of synthetic polymers.

Polymer	Company	Brand
Polyester	Trevira	Trevira Bioactive
	Montefibre	Terital SANIWEAR
	Brilen	Bacterbril
Polyacryl	Accordis	Amicor
	Sterling	Biofresh
Polyamide	Kaneba	Livefresh
	R-STAT	R-STAT
	Nylstar	Meryl Skinlife
Polypropylene	Asota	Asota AM Sanitary
Polyvinyl chloride	Rhovyl	Rhovyl's As antibacterial
Regenerated cellulose	Zimmer AG	Sea Cell Activated

ANTIMICROBIAL SUBSTANCES OF PLANT ORIGIN

An innovative approach to make the cloth antimicrobial is to apply the plant extracts containing active substances. These natural antimicrobial substances are not only ecofriendly but also renewable. Earlier studies conducted on fabrics finished with some medicinal plant extracts also confirmed the presence of antimicrobial activity. Naturally available plant species have been traditionally known for their ability in curing diseases. Botanical research (Osato *et al.*, 1993) shows that some plant species exhibit antibacterial, antiviral, and antifungal properties. Neem (*Azadirachta indica*) added as a preservative in the size bath yielded better results (Sarkar *et al.*, 2003), and Tulsi (*Ocimum sanctum*) applied to textile fabric by ultrasound energy method exhibited antimicrobial activity (Bhawana *et al.*, 2000). Extraction, isolation and structure elucidation of the active principles of plants are done (Schaaf *et al.*, 2000).

BENEFITS OF ANTIMICROBIAL ADDITIVES

A wide range of textile product is now available for the benefit of the consumer. Initially, the primary objective of the finish was to protect textiles from being affected by microbes, particularly fungi. Uniforms, tents, defense textiles and technical textiles such as geo-textiles have therefore all been finished using antimicrobial agents. Later, the home textiles such as curtains coverings, and bathmats came with antimicrobial finish. The application of the finish is now extended to textiles used for outdoor, healthcare sector, sports and leisure. Novel technologies in antimicrobial finishing are successfully employed in non-woven sector, especially in medical textiles. Textile fibres with built-in antimicrobial properties will also serve the purpose alone or in blends with other fibres. Bioactive fibre is a modified form of the finish, which includes chemotherapeutics in their structure, i.e., synthetic drugs of bactericidal and fungicidal qualities. These fibres are not only used in medicine and health prophylaxis applications, but also for manufacturing textile products of daily use and technical textiles. The field of application of the bioactive fibres includes sanitary materials, dressing materials, surgical threads, materials for filtration of gases and liquids, air conditioning and ventilation, constructional materials, special materials for food industry, pharmaceutical industry, footwear industry, clothing industry, automotive industry, etc. Textiles and filling materials with antimicrobial finishes have been used for many years in a wide variety of sectors.

The benefits that can be claimed from the use of antimicrobial additives can be placed into three categories—hygiene factors, increased product lifetime and health benefits.

Hygiene Factors

Bacteria and fungi require three things in order to proliferate—nutrients, moisture and favourable temperature. The human body and most of the environments in which we operate provide these three basic requirements. If we therefore consider an article of clothing or bedding that is newly laundered, we would consider this to be clean and hygienic by everyday standards. We start the day therefore with a clean body and clothing. The body and the clothing and the air that we breathe are not sterile but contain a healthy background flora of bacteria/fungi. During the next

8-12 hours the body is generating moisture (which contains nutrients) and feed on a constant basis. Most of the moisture and heat will pass through the clothing that we wear and then is lost to the atmosphere. Our skin and clothing and other fabrics that we use next to the body provide an ideal breeding ground for the bacteria and fungi to grow, and during the 8-12 hours of working day the background levels of bacteria/fungi will increase dramatically. This can lead to the development of odours and of course a high bioburden. The use of antimicrobial products will greatly reduce this malodour/bioburden issue by controlling the growth of bacteria and fungi. We can also consider hygienic factor in the food preparation and food delivery areas. The major concern in these environments is not so much malodours but cross contamination that can lead to food poisoning, etc. Good examples of this hazard are wipes, towels and the clothing worn by workers preparing or delivering the food. If you consider a kitchen towel, this may be used for a few hours or upto a week between laundry cycles. This product is almost constantly moist and in a warm environment. It will be used dozens of times during the day by a number of different people. It will therefore be constantly in contact with a wide range of bacteria and fungi that will be transferred to any surface on which the towel is used. The use of antimicrobial additives can minimize this risk even against bacteria such as *Salmonella sp., E.coli* and *Listeria sp.*

Increased Lifetime

Many products contain cellulosic fibres for very good results. However, the duration of the lifetime of the product may be reduced to the slow but sure deterioration of the tensile properties of these cellulosic fibres by biodegradation. It has been demonstrated in a number of products, which have been engineered using antimicrobial fibres in blends with cellulosic fibres, that this deterioration is greatly reduced. This results in an increased product lifetime that can be quite significant (upto 30%).

Health Benefits

Antimicrobial products are finding an increase in use in end uses in which health benefits form part of the product claims. A good example of this is in the area of foam textiles in which anti-allergy claims are made. These claims are based upon the ability of antimicrobial products to control the number of dust mites that live within the bedding environment. It is the dust mite that is the main source of the Algerian Derp 1, which is known to trigger asthma in the people sensitized to this allergen. Amicor pure is a blend of two antimicrobial fibres, one having very good antibacterial performance, and one having very good, but quite specific antifungal activity. This fibre is patented and the use of Amicor pure to control dust mites is also patented. Amicor pure is able to prevent the *Aspergillus repens* from developing and hence prevents the breakdown of the human skin. The allergen DerP1 is acknowledged as the main trigger of asthma within the home environment. What is less widely known is that the house dustmite and its associated allergen are also the major cause of sinusitis, rhinitis and eczema.

Environmental Concerns

Bioactive textiles should be environment-friendly. They should not be harmful to the environment during manufacture. For example, during the production of the bioactive fibres, all water and sludge waste should be recycled, and during the use and disposal of the final product, no heavy metals should accumulate in solution either in the manufacturing process or during post manufacturing washings of the bioactive textile.

END USE APPLICATION (EDWARD, 2001)

a) **Socks**: The antimicrobial finishing makes the socks free from unpleasant odour by controlling the microbial growth.

b) **Undergarments**: These are directly in contact with the body and get enough moisture and temperature favouring growth of microbes, which can lead to several skin diseases. The antimicrobial treated undergarments will control the growth of microbes in turn inhibiting the chance of diseases.

c) **Carpet/Car upholstery**: The fibres used in carpets mainly wool tend to be degraded because of the growth of microbes and consequently reduce the life of the product. Treatment with an antimicrobial product will improve the life of the product. Also these microbes leave stains on the carpets.

d) **Pillow covers, pillow fills**: To improve the life of the pillow covers/fills, it is very important to control the growth of the microbes on it. Hence the requirement treating these with antimicrobials.

e) **Apparels**: The bad odour due to perspiration is one of the major problems with the apparels and the major cause for this is the growth of microbes.

f) **Shoes:** Microbes leave several stains on the shoelaces and shoes during rainy season and when the shoes are kept unused for longer time.

g) **Medical industry**: Medical industry is the area where hygiene is a must and microbes are the real cause of unhygienic conditions. Hence, it is very important to make all the textiles used in the medical industry free from microbes. Apart from the above there are several areas in textiles where this treatment needs to be given (Table 4).

Table 4. Textiles with antimicrobial finishes or modifications and their field of application

Medicine	Medicine and Leisure	Outdoor	Technical	Household
Stockings	Shoes	Jacket	Carpets	Curtains
Protectives	Socks	Tents	Roof covers	Covers
Underlays	T-shirts	Uniforms	Facade covers	Dish-cloths
Encasings	Cycling dresses	Personal overalls	Air filters	Bath-mats
Bed fillings	Team dresses	Artificial	Auto-textiles	Sanitary ware
Pillows	Joggings-wear	Sun Umbrellas	Geo-textiles	Under wares
Implants and dressings		Awnings		Carpets

Types of Antimicrobial Finish

Antimicrobial finishes and modifications must now comply with the requirements governing their stability and hence, should not pose a risk to the wearer's health. This means that the active substance should not be removable by washing and has to be resistant to dry cleaning, nor should it be possible for the substance to come into contact with the skin through perspiration. Three fundamental procedures therefore have been established—the active substance is applied directly to the spinning mass, the active substance is applied to the fibre surface, e.g., with the aid of spacer molecules, the manufactured fabric is coated with a resinogenic-finishing agent.

Application Methods

Three different ways of application of biocides are identified.

- Polymer modification
- Spun-in additives
- Post-treatment

Polymer modification: Biocides can be incorporated in the fibres during the polymerization process. This is called polymer modification. A polymer modification can give fibre inherent antibacterial characteristics. The polymer modification is known for synthetic fibres such as acrylics, as well as polyethylene and polypropylene fibres. Examples of fibres used for polymer modification are cationic amines or quaternary ammonium salts.

Spun-in additives: In this application a biocide is incorporated into the yarn during the spinning operation, which makes the agent an integrated and permanent part of the yarn (Teresa and Marianna, 2000). Like polymer modifications, spun-in additives are used for man-made fibres e.g., acetates and acrylics. No literature has been found on the use of spun-in additives for natural fibres. Examples of biocides used as spun-in additive are triclosan, chlorinated biphenyls and natural terpenes.

Post-treatment: Post treatment biocides can be applied on the fibre after fibre production or in the finishing process of the fabrics, where biocides are added together with antistatistics, water repellants, dyes, etc. For natural fibres, biocides can be added at different phases of the life cycle e.g., on the fibre, after spinning, on the fabric, etc. The biocides are added in order to prevent proliferation of rot and mildew when transported or stored under warm and humid conditions (Edward, 2001). Biocides can also be added for man-made fibres such as polyesters, acrylic and viscose products as post treatment to the fibres. These biocides can be metal complexes of copper or zinc. Other post-treated biocides can be quaternary ammonium compounds, polysiloxanes, sulphur compounds and organic salts of metals like zinc, silver and mercury (Hoon and Sung, 2004).

Durable Press Finishing

Chemical treatments applied to cellulosic textiles to impart wrinkle resistance, permanent creases, shrinkage resistance, and smooth drying properties are often referred to as durable press (DP), easy care, or wash-ware finishes. Such dimensional stabilization or shape fixation processes are applied to yarns, fabrics, or entire garments made of cotton or its blends with polyester. DP finishes are also applied to textiles of rayon or other forms of regenerated wood cellulose. Occasionally, fibres of linen and ramie are treated. Chemical finishes for wool, a protein fibre, have made launderable wool garments a commercial reality. The DP finishing of silk, likewise a protein material, has been the subject of promising recent research and development (Yang and Li., 1994).

Cross-linking agents: Methods used in DP finishing of cellulosic textiles is to apply a cross-linking agent that reacts with hydroxyl group of cellulose in the presence of heat and catalysts to form covalent cross-links between adjacent cellulose molecular chains. Since cotton cellulose is a high polymer with molecular weights exceeding one million, multiple cross-linking creates a three-dimensional network within each fibre, flexed, or otherwise deformed. During garment use or laundering, the fabric returns to the flat or creased configuration it possessed at the time the cross-links were put in placed.

Finishing with 1,3-dimethylol-4, 5-dihydroxy ethylene urea (DMDHEU)

DMDHEU was introduced in 1964 as a promising new durable press finish. Unlike the UF and MelF finishes, DMDHEU did not react with itself to form a network polymer, but formed nearly monomeric cross-links between cellulose molecules. Synthesis of DMDHEU is accomplished by reacting urea, glyoxal and formaldehyde. Reductions in the amount of formaldehyde released by fabrics finished with DMDHEU have been accomplished through improved control of curing conditions, better catalysts and by reacting DMDHEU with alcohols or polyols to convert one or more of the N-hemiacetal groups to an acetal. The resulting "capped" finishes are the dominant cross-linking finishes in today's textile market. Because of the concern over formaldehyde released from DP-finished fabrics, a search for formaldehyde-free cross-linkers has been in progress for several years. One solution to this problem has been to utilize etherified DMDHEU that do not release formaldehyde.

Finishing with 1,2,3,4 –Butanetetracarboxylic acid (BTCA)

The commercial preparation of BTCA is a two-step synthesis. The Diels-Alder addition of butadiene to maleic anhydride yields 1, 2, 3, 6-tetrahydrophthalic anhydride. This is oxidized to cleave the carbon-carbon double bond and form BTCA (Franz *et al.*, 1965; Lynn and Roberts, 1961). Currently produced on a pilot-scale, BTCA is at present too expensive to use as a DP finishing agent. The projected price of BTCA for every large-scale production would be about double the present price of DMDHEU. BTCA is the most effective of the polycarboxylic acids which have been studied thus for, in terms of DP performance imparted, speed of curing, laundering durability of the finish and whiteness of the treated fabric. The agent is normally applied at 4–5.5% levels, based on the weight of the fabric. A detailed and comprehensive comparison between BTCA and DMDHEU has been made with respective DP appearance rating, tensile and tearing strengths, shrinkage resistance, and flexed abrasion resistance in treated fabric. These properties were measured initially and after 30 launderings (Brodman, 1990). With sodium hypophosphite as the catalyst the DP performance of BTCA was equal to that of DMDHEU catalyzed by magnesium chloride. BTCA finishing afforded a higher attention of tensile strength, tearing strength, and flex abrasion resistance. The shrinkage resistance with BTCA was moderately less than for DMDHEU. The two agents were compared for any tendency to cause colour shade changes in dyed fabrics. With most dyes, the agent gave similar results. Towards some sulphur dyes, however, the sodium hypophosphite used as a catalyst in BTCA finishing acted as reducing agent. As non-reducing catalysts, disodium phosphite (Brodman, 1990) and potassium dihydrogen phosphate (Brotherton *et al.,* 1989) have been recommended for fabrics dyed with sulphur dyes or turquoise reactive dyes. BTCA finishes contain unesterified carboxyl groups and readily take up basic dyes (Andrews and Welch, 1989; Brodman, 1990; Andrews, 1990). The increased dye absorption is directionally proportional to the concentration of BTCA applied and can be used as a quantitative measure of the level of the treatment (Brodman, 1990). The after dying of poly carboxylic acid finishes with cationic dyes is of interest, as it offers a possible means of dyeing garments which have already been DP finished in fabric or garment form (Andrews *et al.,* 1990). Increased affinity for anionic dyes can be imparted to cotton dye BTCA finishes by including in the treating formulation a tertiary amine possessing one or more hydroxyl group per molecule. The alkanolamine becomes bonded to cotton cellulose by the cross–linking agent and furnishes cationic centres to which the anionic dye quickly become affixed (Blanchard, 1991).

Finishing with Citric Acid

Citric acid is familiar to the cotton textile industry as a widely available, low cost chelating agent and catalyst activator. This agent has been approved as safe for use in food and beverages and offers few environmental concerns (Young-Sick *et al.,* 1998). Several studies have shown that citric acid is considerably less effective and less durable than BTCA as a DP finishing agent (Welch and Andrews, 1989; Andrews *et al.,* 1989). The alpha hydroxyl group of citric acid interferes to some extent with the desired esterification and cross-linking of cellulose by this agent. The interference was shown by direct comparison of citric acid and 1,2,3-propanetetracarboxylic acid (PCA) in the finishing of cotton print cloth (Welch and Peters, 1997). The later agent imparted a DP appearance rating of 4.6, a wrinkled recovery angle of 278°, warp+ fill, and the finish was durable to over 70 launderings. Citric acid at the same molality (0.36) imparted a DP rating of 4.2 and a wrinkled recovery angle of 278° and the finish was durable to 22 launderings. If 1% PCA was added to 7% CA formulation, the DP rating increased to 4.7 and the finish was durable to over 80 launderings. The only difference between citric acid and PCA is the alpha hydroxyl group in the citric acid molecules. In these treatments, the catalyst was 4.8% sodium hypophosphite monohydrate, curing was at 180°C for 90s, and 0.5% polyethylene was the fabric softener used (Andrews *et al.,* 1993). The adverse effect of the alpha hydroxyl group of citric acid on the esterification and cross-linking of cellulose has been measured by FTIR spectroscopy (Yang *et al.,* 1997).

The degree of fabric whiteness in the most favorable case was comparable to that obtained in DMDHEU finishing. With regard to DP performance, fabric breaking strength, and fabric whiteness, the treatments were found to be an improvement over DMDHEU finishing, which used magnesium chloride catalyst activated with citric acid. Selecting the citric acid concentration, catalyst concentration, cure temperature, and cure time that gave the best compromise between

DP performance and fabric whiteness was difficult. By contrast, if a 6% concentration of a 3:1 mixture by weight of citric acid and BTCA were used as the cross-linking agent with 6.2% hypophosphite catalyst, the whiteness index reached or surpassed that afforded by DMDHEU finishing, with DP ratings of 4.0–4.3 for a number of curing temperatures-time combinations (Andrews, 1990). The citric acid was serving as an inexpensive extender for BTCA. Citric acid has also been used as an extender for cyclopentane tetra carboxylic acid at a weight ratio of 2:1. In this instance, the curing catalyst was monosodium phosphate. Other curing additives, which improve the fabric whiteness obtained in citric acid finishing of the cotton, include triethanolamine and several of its salts (Andrews *et al.,* 1993; Choi, 1993), triisopropanolamine, N-methyl diethanolamine, and polyethylene glycols (Choi, 1993) of 2-12 monomer units. Glycerol, pentaerythritol, sorbitol and other polyhydric alcohols are also effective (Choi, 1993). Boric acid and borates suppress yellowing (Choi, 1993; Fung *et al.*, 1993) but can also lower the DP performance (Choi, 1993). The whiteness index of fabric just removed from the curing increases as the fabric regains moisture from the air. Additives that serve as humectants often improve the whiteness. It is probable that moisture causes partial reversal dehydration reactions responsible for the yellowing produced by heat curing.

MICROENCAPSULATION

The concept of encapsulation has been considered and practiced to some degree by several industries for many years. Mechanical encapsulation technology dates back to the late 1800s. Microencapsulation provides a means of packaging, separating, and storing materials on a microscopic scale for later release under controlled conditions. Minute particles or droplets of almost any material can be encased by an impervious capsule wall and thus isolated from reactive, corrosive, or otherwise hostile surroundings. The content of a capsule can be made available by mechanical rupture of the capsule wall, its disintegration by electrical or chemical means, or by a leaching action carried out in an appropriate liquid environment. Since in most applications the particles are very small, the term microencapsulation as used here is appropriate (Achwal, 2003). The terms "encapsulation" and "microencapsulation" have been applied rather loosely to describe any particle-coating process that produces capsules varying in size from a few microns in diameter to 4000 μ and even larger. As to the size range at which the term "micro" becomes applicable, there has been much variation, but it may be considered to be about 200 μ. Research by The National Cash Register Company led to the development of chemical microencapsulation processes. The first significant application was NCR (No Carbon Required) Paper introduced in 1954. By encapsulating one of two colourless components, which react to form a color, the components may be separately carried by one or more pieces of paper in colourless form. The pressure of type striking the paper ruptures the capsules and allows the reaction to take place, producing the resulting image.

Purpose of Encapsulation

Encapsulation of active ingredients for a wide range of industries is carried out for one or more of the following purposes (Mecatta, 1989; Ramesh and Rane, 2004):

1. Rendering liquids into powders.
2. To prevent clumping and improve mixing.
3. To protect active ingredients from oxidation, heat, acidity, moisture or evaporation.
4. To prevent ingredients from interacting with other compounds in the system, which may result in degradation.
5. To mask the taste of unpleasant flavours or odours.
6. To improve handling of an ingredient before processing.
7. To release active chemicals in a controlled or targeted fashion.
8. To protect workers or end users from exposure to hazardous substances.

Microencapsulation in Textile Finishing

Microencapsulation was developed as a technique over sixty years ago, but it is only over the last decade that it has become of interest for textile dyeing, printing finishing (Ian, 2003). Broadly speaking, microencapsulation may be defined as a micro packaging wherein active core material is encapsulated in a polymer shell of limited permeability. The objective of this technology is either to protect the active core material from the external environment till required or to affect the controlled release of the active core to achieve desire delay until the right stimulus is encountered. Interestingly, this process was initially developed for the carbonless copy industry. However, it has now attracted the attention of a wide number of industries including pharmaceutical, agricultural, bulk chemical, food processing, cosmetics, as also textiles. The textile industry has been late and slow to react to the possible opportunities this technique could offer. Efforts are still seen more at the research and development stage. Interesting possibilities do exist in medical and technical textiles where this technology could be used for applying finishes and for imparting desirable properties that either are not possible through conventional application procedures or are not cost-effective. Some other examples include antimicrobial, moth repellency, flame retardancy, etc. Use of phase change materials in the core has also been suggested for thermal regulation functions in clothing for sports and extreme weather. Although there is a great potential of the microencapsulation technology to succeed in textile arena, most applications, as of now, are very specific and therefore have had limited commercial success.

The following characteristics of microcapsules do interest the researchers (Cordon, 2001):

a) **Size and size distribution** – Capsules can be made in sizes ranging from few microns to 4000 microns in diameter and even larger. Low size increases the mechanical strength and also ease of application.

b) **Loading fraction** – This is the weight ratio of core to wall of the microcapsule; the higher is the ratio, the better is the efficiency but poorer would be the stability.

c) **Release properties** – Rate of release from microcapsules depends largely on the structure of the polymer wall, which in turn is influenced by the conditions employed in the preparation. Wall characteristics like crystallinity, cross-link density and porosity play a big role in determining the release rate. As the crystallinity and cross-link density of the wall increase, the release rate reduces substantially. The other important factor is the outside environment; if it is of the same type as that of the core material, the rate of release will be high. The core ingredient may be released by mechanical stimulus (Fig. 2), chemical stimulus and thermal stimulus. The resultant release rate can normally be expressed as a first order rate process, i.e., –dc/dt = kc, k is the diffusion constant and c is the concentration gradient.

d) Thermal stability – It is very important when field of application is at a high temperature.

Materials for Microencapsulation

There are various materials that have been encapsulated like the solvents, acids, bases, etc. (Table 5). These are called the core materials. The wall material may be natural, semi-synthetic or synthetic (Table 6).

Table 5. Wall Materials (Ramesh and Rane, 2004)

Material type	Example
Natural	Gum acacia, agar, agarose, maltodextrin, sodium alginate, calcium alginate, dextran, fats and fatty acids, acetyl alcohol, milk solids, molasses, gelatin, gluten, albumin, shellac, starches, casinates, sucrose and waxes.
Semi-synthetic	Cellulose acetate, cellulose acetate butyrates, cellulose acetate phthalate, cellulose nitrate, methyl cellulose, sodium carboxymethyl cellulose.
Synthetic	Acrylic polymers and copolymers, aluminum monostearate, carboxyl vinyl polymers, polyamides, polyamides, polystyrene, polyvinyl alcohol, poly lysine, polycarbonates, poly urea.

Table 6. Materials for microencapsulation (Ramesh and Rane, 2004)
Core Materials:

Material type	Example
Solvents	Benzene, toluene, cyclohexane, chlorinated Phenyls, paraffins, esters, ethers, alcohols and water
Plasticisers	Phthalate, adipate, phosphate, Silicones and chlorinated hydrocarbons
Acids and bases	Boric acid, caustic alkali and amines
Catalysts	Curing agents, oxidants, free radical initiators and reducing agents
Colorants	Pigments and dyes
Adhesives	Polysulphides, cyanoacrylates, isocyanates, epoxy resins, thermally sensitive adhesive materials
Fragrances	Menthol, essences and speciality compounds
Foods	Oils, fats, spices and flavours
Agricultural chemicals	Herbicides, insecticides and pesticides
Pharmaceuticals	Aspirin, vitamins and amino acids
Recording materials	Reprographic toners, coupling agents, developers, silver halides, fixing agents, photo chromatic compound an liquid crystals
Rust inhibitors	Zinc chromate
Others	Detergents, bleaches and fire retardants

Microencapsulation Methods

The methods that can be used for microencapsulating solids, liquids and gases are many and varied. The technique of manufacture depends upon many considerations, and especially upon factors such as the core (or nucleus) i.e., the material to be encapsulated and the choice of polymer coating (wall). Other factors include the relative solubilities of the core and wall polymer, the range of particle sizes, the wall thickness and permeability and particularly the type and release mode of the contents of the core (Nelson, 1991). In addition, the economics of manufacturing the microencapsulated finish must also be considered. Of the many forms of microencapsulation available, not all are as yet used for textile end uses. However, the potential manufacturing methods for microencapsulation may be broadly divided into two groups, namely, chemical microencapsulation and mechanical microencapsulation (Ian, 2003).

Chemical microencapsulation: Chemical microencapsulation processes are generally characterized by batch operation, high capacity, conventionality of equipment and diversity in physical state and size of nucleus particle. Here the production and isolation of capsules is done in liquid medium and involves chemical or phase change separation (Gomes and Genovez, 1997).

Mechanical microencapsulation: Mechanical microencapsulation processes are generally characterized by continuous operation, lower capacity, restricted diversity in physical state and size of the nucleus particles and need for specialized auxiliary mechanical equipment for the formation and isolation of capsules, e.g. spray drying (Gomes and Baptista, 2001).

Based on these two principles, there are several types of microencapsulation techniques (Aggarwal *et al.,* 1998) such as coacervation process, interfacial polymerisation, heat melt microencapsulation, spray drying, centrifugal extrusion process, air suspension coating and pan coating.

Coacervation technique: Coacervation method is also known as phase separation method. Here the core material is dispersed in a polymer solution. While continuing to uniformly suspend the material through timing, the polymer's solubility is decreased by addition of a non-solvent for the polymer, lowering the temperature, addition of second polymer, changing the pH of the solution or by producing an ion gradient (Lazko *et al*., 2004). Depending on the solubility of polymer, the polymer either precipitates or phase separates into a polymer rich and polymer poor phase. The polymer migrates to the interface to form the wall (Gokhan and Merih, 2004).

Interfacial polymerization: Interfacial polymerization was discovered in 1951 by Ziegle and co-workers. This involves a polycondensation reaction, which is a simple technique to obtain capsule wall. It consists in making an emulsion of two non-miscible phases; each one containing dissolved monomers able to react with other monomers present in the other phase. As a consequence, the polymer formation is the result of an interfacial reaction between the monomers, leading to the formation of capsule like particles. The stirring speed affects the microcapsule dimension.

Hot-melt microencapsulation: In this method, a core material is added to molten polymer. The molten droplets solidify and entrap the core material, while the temperature of the polymer is rapidly reduced below the glass transition of the polymer (Gokhan and Merih, 2004).

Spray-drying: This is a common method of microencapsulation. The microspheres that include the core material are dispersed or dissolved in a polymer solution. This solution or dispersion is pumped through a nozzle into a chamber, which contains heated air, to form an aerosol. Then the solvent evaporates from the micro droplets. The core:wall material ratio, concentration, viscosity and temperature of starting solution affect the properties of the microcapsules (Mauriello *et al.,* 1999).

Centrifugal extrusion process: This is a mechanical process of forming capsules containing either liquids or solids. This process involves the use of a multiple-orifice, centrifugal apparatus which forms a continuous film of the wall material across the orifice of a nozzle, upon which the nucleus material impinges. It is based on the principle of forming a membrane of fluid film and enveloping a tiny droplet as it is forced through the fluid film (Gokhan and Merih, 2004).

Air suspension coating: In air suspension coating technique, particles are coated by dissolved molten polymers that are suspended in upward-moving air stream in unstable equilibrium. Equipment for this type of coating is made by several firms.

Pan coating methods: This method has been used in pharmaceutical industry for many years to produce small-coated particles. Solid particles are tumbled in a rotating pan, whilst suitable coating material is slowly added, with a controlled temperature profile. Additional coatings of film, forming polymers may be added in successive stages (Aggarwal *et al.,* 2004).

Commercial microencapsulated products for textile use (Gokhan and Merih, 2004). Microcapsules have been developed and manufactured by RT Dodge for several years. These microcapsules survive 8-20 washing cycles and are strength to conventional tumble dryers. Celessence International offered Celessence TXT capsule systems, which can be applied by conventional textile application methods to a variety of textile substrates. LJ specialities has developed microencapsulated fragrance product for textiles. The mechanism release depends on light abrasion. Euracil has introduced microcapsules containing aromas, moisturizers, creams, and antibacterial agents. Eldorado Int. Co. has produced fabrics that contain encapsulated aromas. These capsules break on movement of the wearer. Courteulds Jersey underwear has introduced scented fabrics created with essential oils of lemon and orange. Speciality Textile Products Ltd has offered Biocap series. These products are microcapsules containing vitamins such as vitamins A, D and E and *Aloe vera.* Cognis has introduced microcapsules for textiles, which is called skintex. These contain chitosan as wall material and various cosmetics as core material. Hong and Park prepared microcapsules using melamine-formaldehyde systems containing migrin oil. The microcapsules were attached to the fabric by printing using acrylic binder.

ADVANTAGES AND DISADVANTAGES OF MICROENCAPSULATION

Microencapsulation has found a wide range of applications in textile industry due to the advantages such as controlled release of the active material, eco-friendliness, energy saving, easy application, protection of the core material from the reactive or hostile environments, does not affect the existing textiles, has a shelf-life on a garment that allows normal care processes to take place and it is a micro-packaging technique, which enables the supplier to provide new opportunities for a variety of end use applications. Although microencapsulation serves as a desirable technique, it is relatively expensive when compared with the conventional methods. Another major disadvantage of the technique is the possible cross-reaction that may occur between the core and wall material selected.

EMERGING TECHNOLOGIES

Traditionally, chemical finishes are applied by padding or low wet pick up techniques such as lick roll, porous bowls, vacuum extraction or foam. Alternative methods include coating and lamination technologies. New application technologies are emerging. Some of these are briefly mentioned below.

Water Resistant Breathable Finish: Present trend is to impart water repellency without affecting water vapour permeability of fabrics and garments for use in outdoor activities, foul weather clothing, etc. Improving the transpiration rate of perspiration through the textile materials is particularly important in sports activities, where the relative metabolic rate is high. Water resistant breathable finishes are permeable to water droplets but allow the escape of water vapour. It is possible to achieve this property because of enormous size difference between liquid water molecule (100 microns in diameter) and water vapour molecule (0.0004 micron) i.e., there is factor of 2,50,000 between two sizes. Such type of garments is designed for sports wear, skin wear, tracksuits, rainwear, clothing for mountaineering, etc (Achwal, 2003). Lightweight fabrics coated with polyvinyl chloride and polyurethane has become increasingly popular for foul weather clothing. Such fabrics provide exceptional protection against, rain but considerable build-up of moisture vapour inside the clothing may cause discomfort. This problem arises because of water vapour impermeability of the polymer coating. To overcome this problem, breathable polymer coatings are now available which have greatly enhanced the comfort properties of these clothing (Jeong-sook, 1997).

Fluorochemicals: Fluorochemicals or fluorocarbons are increasing in use because of greater ability to engineer the chemical structure to provide good water repellancy and soil and stain release properties (Wei and Karen, 2000). Hybrid fluorochemical finishes have been introduced that contain per-fluoro alkyl containing polymer segments and also hydrophilic segments. In air the fibre-air interface is dominated by closely packed dominated by per fluoro alkyl moieties, which promote soil repellancy. In aqueous wash liquor, molecular reorientation occurs, leading to the hydrophilic segments being oriented towards fibre-liquid interface, thereby promoting soil release. The use of per fluoro alkyl acrylate and methacrylate based copolymers modified to suit particular end uses is well-established (Cho and Cho, 1997).

Nextec Process: It is patented by Nextec applications Inc. USA. It is a novel process by means of which individual fibres within the fabric are encapsulated or wrapped with an ultra thin film of polymer. This can be utilized to provide a breathable barrier film within the fabric, preventing liquid water from penetrating while allowing moisture vapour to pass through at as close to the same as original fabric. The handle of the fabric is relatively unaffected. It is also possible to introduce chemicals for imparting wrinkle resistance, soil release, and stain resistance. Chemical and biochemical additives may be positioned on the fabric in order to provide water repellancy, UV protection, flame resistance or viral barriers. So far this unique fibre encapsulation process has been applied using silicone-based polymers, polyurethane, polyacrylate and other polymers. This process would open up many opportunities for providing multifunctional finishes on cotton fabrics (Achwal, 2003).

Plasma Treatment: The plasma treatment is based on ionized gases produced by electrical discharges. The plasma treatment is highly surface specific and does not affect the bulk properties. The surface properties enhanced include wettability, adhesion, cross-linking, biocompatibility, chemical affinity or inertness. Relative free radical intensity of the plasma treated fibres measured by ESR increases in the following order:

Cotton > Wool > Silk > Nylon = PET

Plasma treatment may enable to avoid precoating and prelamination preparation of the substrate, replacing some costly and environmentally polluting processes. The dry treatment and environment friendliness for improved wetting, addition and bondability offer scope for improvement in functional finishes for cotton (Gu *et al.*, 2002).

Use of Biotechnology: The use of biotechnology for the modification of cotton is at an exciting stage and the greater applicability of the enzyme treatments as chemical pretreatments and finishing treatments are worth exploiting. The next decade should bring forward some interesting developments in this field, and the genetic manipulation of cotton gene may usher in new cotton varictics with improved properties and performance that can be further exploited in chemical finishing to produce innovative effects.

VALIDATION OF ANTIMICROBIAL TECHNOLOGY

Safety aspects: Antimicrobials have been used for thousands of years and have an important role to play in the control of troublesome organisms. In general terms, those agents that declare a "broad spectrum of activity" are more toxic to higher order species including humans. An exception to this generalization is based on an organofunctional silane. Effective levels of this technology do not leach or diminish over time, and in twenty years of use has no known negative effects. This type of antimicrobial technology is being preferred for textiles that are likely to have human contact or where durability is of value. Besides affecting durability and useful life, products dependent on leaching have the potential to cause a variety of other problems when used in garments. These include their negative effects because they can contact the skin and potentially effect the normal skin bacteria, cross the skin barrier, and/or have the potential to cause rashes and other skin irritations. A more serious concern with leaching technologies is the potential for cell mutation and resistance. Bacteria and fungi have a role to play in our ecosystem being responsible for returning dead and discarded material back into the environment. The total eradication of all bacteria and fungi is not the objective but the control of over population and this can be achieved with appropriate application levels and product selection.

Antimicrobial Treatment Verification

Another important property of a useful antimicrobial is that its presence should be verifiable. There is no easy way to tell whether leaching antimicrobials are present on a product. The only verification technique for a leaching chemistry is to use exacting laboratory tests, which take days or weeks to perform. With the bound antimicrobial technology though, a simple staining test can be performed in a matter of minutes at the mill or in a store to verify proper treatment of a fabric or other surface. This is a very important part of a quality assurance programme that gives the manufacturer, the retailer and the customer confidence that a feature, normally invisible to the senses, can be seen (Curtis *et al.*, 1997).

Antimicrobial Regulation

As mentioned before, not all antimicrobials are alike. There are technical differences between antimicrobials that affect their life, performance, safety and cost. But one thing is true for all antimicrobials and sometimes treated products. All antimicrobials must be regulated by the Environmental Protection agency, the European Union or other regulatory agencies around the world for their specific uses (Curtis *et al.,* 1997).

MODERN TEXTILE CHARACTERIZATION METHODS

Standard Testing Methods: Textile characterization must take into consideration an in-depth understanding of the nature of fibre forming materials (polymers), fibre structure, its physical, mechanical, and chemical properties, and how these properties relate to further engineering operations that result in fabrics/textiles and finished products. The end-use performance of finished products will depend upon all these factors, and can be predicted on the basis of fundamental theories of fibre science and sound characterization methods. There are numerous well-known organizations, such as the International Standard Organizations (ISO), the American Society for Testing and Materials (ASTM), the American Association of Textile Chemists and Colorists (AATCC), the European Standardization Committee (CEN), and various others that develop standards test methods for evaluating and predicting performance of fibrous systems. However, generally, there is a significant time lag between the developments in textile characterization methods and their acceptance as standards methods.

Most textile fibres, both natural and synthetic, are organic polymers that manifest some characteristic morphological structure with easily discernable features. In natural fibres such as cotton and wool, this structure is genetically determined and may be influenced by variety or growing conditions. Synthetic fibres or filaments are designed and engineered to a predetermined structure. These special structural features may be used for fibre identification, for prediction of fibre characteristics, or for determination of changes that have occurred in the fibre due to chemical reaction and finishing, or damage by insects, fungi, heat, wear, or chemical attack. Study of deformations or rapture due to deformation can provide useful information on internal morphology as well as reaction deformation.

Objective Measurement of Fabric Hand

There are two types of clothing fabric performance. One type is utility performance, such as strength, colour durability, shrinkage resistance, etc. While this type of performance is, of course, very important for clothing materials, consumers are generally satisfied with fabrics that meet these criteria to a certain extent. Beyond this the consumers' attention turns to higher-level performance factors, such as improved quality from the standpoint of garment appearance and comfort. This second factor of fabric quality-type performance factors is related to the idea of "better fit" to the human body, and is also an essential requirement in clothing material. The evaluation of fabric quality performance is, however, more difficult than the evaluation of utility performance. The quality of clothing fabric with regard to the second type of performance has been evaluated by consumers and textile producers subjectively by means of the hand touch of fabric from the mechanical-comfort viewpoint. This evaluation is called hand evaluation and the fabric property relating to this evaluation is fabric hand ("handle" in England). The subjective judgment of fabric hand is based on human sensitivity and experience. It is true that this subjective method is the most direct method for evaluating fabric mechanical comfort, as the human body and sensitivity feel the comfort of clothing.

Kawabata (1980) and his co-workers began researching fabric hand around 1969 based on their concepts of fabric hand and on the work of many of their predecessors in this field. The research focused on the analysis of the judgment of fabric hand as carried out by experts in textile mills, especially finishing mills for wool textiles. The first step of research was to standardize the fabric hand expressions that were traditionally used by the experts in wool textile mills. Based on this standard, a numerical expression of fabric hand became possible; then subjective hand judgment was transferred to an objective evaluation system based on fabric mechanical properties (Billmeyer and Saltzman, 1981).

GENERAL CHARACTERIZATION OF TEXTILES

End-use Requirements

The overall performance requirements of textiles in each personal protection and infection control end-use application can be quite different. Understanding the performance requirements for each and end-use application is the key to developing a successful strategy of experimentation and defining risk reduction decision logic. In each end-use application, all the technical attributes that are required for the textile to perform adequately should be identified. Two different end-uses may require the same protection against the penetration of microorganisms; however, there may be completely different requirements for physical, chemical and thermal properties. In some cases, one attribute may have to be sacrificed in order to obtain the goal for another attribute. As an example, strength objectives might dictate that the weight of the textile has to be increased; however, increasing the weight can negatively impact the hand (stiffness) of the textile. When considering the performance objectives for textiles in just a few end uses, such as firefighter turnout clothing, emergency medical response clothing, and surgical apparel, it becomes obvious that the bio barrier demands for the textiles used in this applications will have to be integrated and balanced with the multitude of other technical attributes. Other significant influences, which are not technical performance attributes but can have a direct impact on deciding which textile to use, are cost and environmental impact.

PHYSICAL PROPERTIES

Four of the easiest and most objective analyses for textiles are weight, thickness, bulk density and microscopy. Together, these tools can serve to benchmark different textiles for comparison and help to predict other important microbial barrier and physical attributes.

Weight: The weight of textile is often determined in order to be used with thickness measurements to obtain a bulk density. Weight measurements for similar textile structures can also be related to other physical attributes, such as strength, abrasion resistance, and stiffness. The international units of measurements for weight are grams per square metre.

Thickness: As stated earlier, thickness determinations can be used with weight measurements to calculate bulk density, which can be used for a variety of purposes. Thickness measurements for similar textile structures can also be related to other physical attributes, such as strength, abrasion resistance, and stiffness. The compressibility of various textile structures may be different and need to be taken into account when making this measurement for comparative purposes. The international units of measurement for thickness are millimetres.

Bulk Density: Bulk density calculations for textiles can be related to the insulative properties and may also be useful in helping to understand the liquid, air, moisture vapour, and microbial penetration resistance characteristics. Generally speaking, for textiles with similar physical and chemical structures, as the bulk density increases the penetration resistance increases. The international units of measurement for density are kilograms per cubic metre.

Microscopy: Visually observing the magnified images of textiles is perhaps the most interesting and informative analysis that can be made in order to better understand the physical structures. Virtually all of the physical elements can be investigated, including the type of structure (type of weave, knit, or nonwoven), additional features of the structures (calendared, texturized, entangled, point bonded), the complexity of the structure (multiple layers, fibre blends, film reinforcements), the porosity of the structure (define pathways, approximate the size of yarn and fibre interstices, approximate the pore size of some films), yarn characteristics (filaments/yarn), fibre characteristics (diameter/cross-sectional shape, classification of some types), film bonding techniques (adhesives laminated, direct coated/extruded, point bonded), and film types (monolithic, bicomponent, microporous). Many of these futures can be determined using either a stereomicroscope or mono-objective compound microscope; however, scanning electron microscope (SEM) has the flexibility to determine all of them and more. In order to illustrate the diversity of some of the structures, products that represent the most common types of textile structures currently being used in personal protective clothing and infection control applications were chosen and SEM was employed to view their outside surface and cross-section.

Surface Characterization of Textiles Using SEM: In modern material sciences, structural characterization is basic to a thorough understanding of the nature of material. The microscope has played an important part in formulating structural concepts in polymeric materials. The light microscope (LM) provides information of an external structure of bulk fibrillar polymers and the transmission electron microscope (TEM) and the scanning electron microscope (SEM) produce surface images that are realistic, three-dimensional reproductions of the sample, and its depth of focus and range of magnifications fill the gap between the LM and the TEM. It has thus, opened a whole new field in fibre structure studies.

ANTIMICROBIAL PROPERTIES

Textiles that include antimicrobial additives require special considerations when assessing the microbiological barrier properties. These textiles may confound normal microbial challenge testing by eliminating the challenging organisms. Most of the standard microbial challenge tests evaluate the ability of a textile to prevent penetration and assume that the challenging organism remains viable throughout the process. If the mechanism by which the textile acts to protect against the penetration of microorganisms is to inactivate the microorganisms, then this fact should be fully understood and appropriate means to evaluate the protective properties of that textile should be utilized in the laboratory. Textiles incorporating antimicrobial agents might not prevent the penetration of or direct contact with viable pathogenic microbes during use, as the effectiveness of antimicrobial agents requires direct contact with the microbes for a specific period of time under defined conditions. The antimicrobial agents can be ineffective if penetration of the microbe through the textile occurs quickly or if the activity of the antimicrobial agent is diluted or inactivated by the vehicles carrying the microbe (such as with blood or body fluids) (Yuke, 1989).

CHEMICAL RESISTANCE PROPERTIES

In virtually every end-use application for textiles being used as microbiological barriers, there exists the potential for exposure to various types of chemical agents. A complete list of all chemical agents that are likely to come into contact with textiles in each personal protection and infection control end-use application should be made. Some of these agents will be hazardous and require that the textile also act as a barrier to them. This list could include many different chemical types, such as disinfectants, acids, bases, solvents, fuels and lubricants. If dual protection to both microorganisms and chemicals is required, then both types of laboratory analyses for textiles in these applications will need to be performed. Depending on the type of chemical that the textile is exposed to and the chemical resistance properties of the textile, the chemical may seriously compromise the ability of the textile to effectively limit or prevent microbial penetration. Whether the chemical is hazardous or non-hazardous, some types of textiles and film reinforcements, particularly those that are porous in nature, can be made to allow microbial penetration to occur after contact with chemical pre wetting and contaminating agents. If dual exposure situations can be clearly defined, then preconditioning textile samples with the chemical agents of concerned prior to microbial challenge testing would be appropriate.

THERMAL PROPERTIES

Most of the personal protection and infection control end-use applications have standards and regulations governing flammability requirements for textile based products. These requirements should be recognized and appropriate flammability testing should be performed. In each end-use application, situations involving possible exposures to heat and ignition sources should be recognized and appropriate strategies should be developed in an effort to avoid over exposure. The thermal stability of textiles, including flame resistance, dimensional stability, insulative properties, etc., should be well characterized for the end-use applications requiring thermal protection. The microbiological barrier properties of some textile materials may degrade with repeated or prolonged exposure to hot environments. For the applications requiring both thermal protection and microbiological barrier properties, preconditioning textiles samples with thermal exposures prior to assessing the microbial barrier properties should be considered.

MICROBIOLOGICAL TEST METHODS AND THEIR APPLICATIONS FOR TEXTILES

The most important recognized international test systems are summarized in Table 7. When testing for the specific antimicrobial activity of an active substance, reference samples and control materials, which are not finished, are necessary. They must be of the identical structure and chemical composition as the test material, but without the antimicrobial finish.

Table 7. Overview of the methods used to test textiles, fibres, yarns and polymers for antimicrobial effectiveness

Designation	Title	Origin	Principle
SN 195920-1992	Textile fabrics: Determination of the antibacterial activity: Agar diffusion plate test	Switzerland	Agar diffusion test Semi-quantitative
SN 195921-1992	Textile fabrics: Determination of the antimycotic activity: Agar diffusion plate test	Switzerland	Agar diffusion test Semi-quantitative
AATCC 147-1993	Antibacterial activity: Assessment of Textile materials: Parallel Streak Method	USA	Agar diffusion test Semi-quantitative
AATCC 90-1982	Antibacterial activity: Assessment of textile materials Agar Plate Method	USA	Agar diffusion test Semi-quantitative
AATCC 174-1993	Antibacterial activity: Assessment of carpets	USA	Agar diffusion test Semi–quantitative
JIS L 1902 1998	Testing method for antibacterial of textiles	Japan	Agar diffusion test Semi-quantitative and Challenge test Quantitative
AATCC 100-1993	Antibacterial finishes on textile materials: Assessment of carpets	USA	Challenge test Quantitative
SN 195924-1983	Textile fabrics: Determination of antibacterial activity-Germ count method	Switzerland	Challenge test Quantitative
XP G 39.010-2000	Properties of Textiles: Textiles and polymeric surfaces having antibacterial properties – Characterization and measurement of antibacterial activity	France	Challenge test Quantitative
AATCC 30-1993	Antifungal activity: Assessment of textile materials: mildew and rot resistance of textile materials	USA	Agar diffusion test Soil burial test and humidity chamber test
JIS Z 2911-1992	Methods of test for fungus resistance	Japan	Fouling tests
DIN EN ISO 846-1997	Plastics – Evaluation of the action of microorganisms	International	Fouling tests Soil burial test
DIN EN ISO 11721-1-2001	Textiles: Determination of resistance of cellulose containing textiles to microorganisms: Soil burial test – Part 1: Assessment of rot retardant finishing	International	Soil burial test

CONCLUSION

In this chapter the antimicrobial finish for textiles and its significance in the development of eco-friendly antimicrobial textile finishes for various textile products have been discussed. This development can be extended to a number of other end-uses of textile products. Manufacturers who do not currently treat fabrics with a durable antimicrobial finish should consider shielding their products from eroding value by incorporating antimicrobial technology. As manufacturers look to enhance the value of their products, they should recognize antimicrobial finishes as a feature with a future and the future is now.

REFERENCES

AATCC Technical Manual, (1995). *Research Triangle Park,* **73:** 148-149.

AATCC Technical Manual, (1995). *Research Triangle Park,* **73:** 261-262.

AATCC Technical Manual, (2001). *Research Triangle Park,* **73:** 205-208.

AATCC Technical Manual, (2001). *Research Triangle Park,* **73:** 63-65.

AATCC Technical Manual, (2001). *Research Triangle Park,* **73:** 85-86.

AATCC Technical Manual, (2003). *Research Triangle Park,* **73:** 176-178.

AATCC Technical Manual, (2004). *Research Triangle Park,* **73:** 81-84.

Achwal, W.B. (2003). *Colourage,* **Jan.:** 58-59.

Aggarwal, A.K., Dayal A. and Kumar, N. (1998). *Colourage,* **Aug.:** 15-24.

Andrasic, T. (1993). *IFI Fabricare news,* **22:** 13.

Andrew, B.A.K., Blanchard, E.J. and Reinhardt, R.M. (1993). *Text Chem. Color* **25(3):** 52-54.

Andrews, B.A.K. (1990). *Text Chem. Color.* **22(9):** 63.

Andrews, B.A.K., Blenchard, E.J. and Reinhardt, R.M. (1990). *Am. Dyest. Rep.* **79(9):** 48.

Andrews, B.A.K., Welch, C.M. and Trask-Morrell, B.J. (1989). *Am. Dyest. Rep.* **78(6):** 15.

Angelina Georgieva. (2003). *ITB international Textile Bulletin,* **4:** 60-61.

Anita, A. Desai., Chitra Joshi Ajmeri. and Ajmeri J.R. (2002). *The Tex. Int. and Trade J.,* September-October: 43-47.

Anon, (1992). *Medical Textiles,* **7:** 1.

Anon, (1993). *Medical Textiles,* **10:** 3.

Beliakova, M.K. and Hebeish, A. (1998). *Am. Dyest. Rep.* **87:** 46-49.

Beumer, R., Bloomfield, S.F., Exner, M., Fara, G.M., Nath, K.J. and Scott, E. (2000). A Review by the International Scientific Forum on Home Hygiene (IFH), September.

Bhawana Ghorpade., Vandana Tiwari., Anuradha Mishra. and Padma, S. Vanka. (2000). *Asian Textile J.,* May: 56-57.

Bide, M., Zhong, T., Ukponmwan, J., Phaneuf, M., Quist, W. and LoGerfo, F.W. (2002). AATCC International Conference and Exhibition, Charlotte, Sc, October.

Billmeyer. and Saltzman's, (1981). (ed. Berns, R.S.) New York: John Wiley and Sons, 140.

Blanchard, E.J., Reinhardt, R.M. and Andrews, B.A.K. (1991). *Text Chem. Color.* **23(5):** 25.

Bohringer, A., Rupp, J. and Yonenaga, A. (2000). *Int. Text. Bull.* **46(**5**):** 12-26.

Brodman, G.L. (1990). *Text Chem. Color,* **22(11):** 13.

Buchenska, J. (1996). *Fib. And Text in Eastern Europe,* **4:** 80-84.

Buchenska, J. (1996). *Fib. And Text in Eastern Europe,* **4:** 53-59.

Cho, J.S. and Cho, G. (1997). *Textile Res. J.,* **67(12):** 875-880.

Choi, H.M. (1993). *Text Chem. Color,* **25(5):** 19-24.

Cordon Nelson. (2001). *Rev. Prog. Color.,* **31:** 57-64.

Cox, D.M. (1999). *In:* Nanostructure Science and Technology: High Surface Area (eds. Sigel, R.W., Hu, E. and Roco, M.C.) International Technology Research Institute, VA; 49-66.

Curtis White M.S., Robert A. Monticello. and James W. Krueger. (1997). Patrice Vandendaele, AEGIS Environments, file://A:\cati.htm.

Davidson, R.S. and Xue, Y. (1994). *J. Soc. Dyers Colour,* **110:** 24-29.

Deborah K. Lickfield, (2001). *International Finer J.*, August: 42.
Edward Menezes. (2002). *International Dyer,* **12:** 13-14.
Edward Menezes. (2001). *Textile Industry and Trade J.,* **1-2:** 35-38.
Franz, J.E., Knowles, W.S. and Osuch, C. (1965). *J. Org. Chem.,* **30:** 4328.
Frick, J.G., Kottes Andrews, B.A. and Reid, J.G. (1960). *Textile Res. J.* **30:** 495-504.
Fung, K.W., Wong, K.H. and Brotherton, D.L. (1993). U.S. Patent 5; 199,953 to ortec, Inc.
Gang Sun, (1997). *AATCC International Conference and Exhibition,* 181.
Gettings, R.L. (1985). *Dow Corning Corporation.*
Gettings, R.L. and Triplett, J.L. (1978). *AATCC Book of Papers,* 259-261.
Gokhan Erkan. and Merih Sariisik. (2004). *Colourage*, Annual: 61-64.
Gomes, J.I.N.R. and Baptista, A.L.F. (2001). *Textile Res. J.,* **77(2):** 153-156.
Gomes, J.I.N.R., Genovez, M.C. and Hrdina, R. (1997). *Textile Res. J.*, **67:** 537-540.
Gu, J., Yang, X. and Zhu, H. (2002). *Mater. Sci. Eng. C.*, **20(1-2):** 199-202.
Ha, N.J., Jung, Y.J., Lee. J.S., Kim, Y.T. and Kim, Y.M. (1998). *Arch. Pharm. Res.*, **21(5):** 570.
Hales, M.G., Sarkin, M.E. and White, W.C. (1986). *Cleaning and Restoration Magazine.*
Heath, R.J. and Rock, C. (2000). *Nature* **406:** 145.
Hoon Joo Lee. and Sung Hoon Jeong. (2004). Bacteriostasis of Nanosized Colloidal Silver on Polyester Nonwovens, *Textile Res. J.*, **74(5):** 442-447.
Huber, B. (2001). SOFW–J. 127, **8:** 42-45.
Hyung-Min Choi, (2004). *Textile Res. J.,* **74(4):** 333-342.
Ian Holme, (2002). International Dyer, December: 9-11.
Ian Holme, (2003). *International dyer,* August: 9-12.
Jeong- Sook Cho. and Gilsoo Cho. (1997). Textile Res. J., **67(12):** 875-880.
Kawabata, S. (1980). The Hand Evaluation and Standardization Committee, Tokyo. 28-57.
Kawata, T. (1998). *Chem. Fibres Int.*, **48:** 38-43.
Lazko, J., Popineau, Y. and Legrand, J. (2004). *Colloids and Surfaces B: Biointerfaces*, **37:** 1-8.
Lee, S., Cho, J.S. and Cho, G. (1999). *Text Res. J.*, **69:** 104-112.
Lynn. J.W. and Roberts, R. (1961). *J. Org. Chem.,* **26:** 4303.
Mairal, A.K. and Patel, P.G. (2002). *The Text. Industry and Trade J.,* **9-10:** 19-28.
Mauriello, G., Aponte, M., Andolfi, R., Moschetti, G. and Villani, F. (1999). *J. Food Protection*, **62:** 773-777.
McNeil, E. (1994). *American Dyestuffs Reporter,* 87-90.
Mecatta, J.J. (1989). Encyclopedia of Chemical Processing and Design, **66(3):** 30,165.
Meeks, J.K.A. and Greenley, D.E. (1998). *Text Rental,* **71:** 68-70.
Morris, C.E. and Welchi, C.M. (1983). *Textile Res. J.,* **53(3):** 143-147.
Mucha, H., Hofer, D., Abfalg, M. and Swerev, (2002). *Melliand International,* **8:** 148-151.
Murugesh Babu, K. (2003). *Asian Textile J.,* **4:** 64-68.
Nelson G. (2002). *Intern. J. Pharma.* **242:** n 1-2, 55-62.
Nelson, G. (1991). *Rev. Progr. Colo* **21:** 72.
Nonaka, Y., Uemura, Y., Ohse, K., Jyono, K. and Kurihara, S. (1997). *J. Appl. Poly. Sci,* **66:** 1621-1630.
Ochs. (1998). *Speciality Chemicals,* **18(9):** 389-391.
Osato, J.A., Santiago, L.A., Remo, G.M., Cuadra, M.S. and Mori, A. (1993). *Life Sci.,* **53:** 1383-1389.
Payne, J.D. and Kudner, D.W. (1996). *Text Chem. Color,* **28(5):** 28-30.
Payne, J.D. and Kudner, D.W. (1996). *Am. Dyest. Rep.,* **85:** 26-30.
Payne, J.D., Kudner, D.W., Pierce, and Lee, S.K. (1996). *International Nonwovens conference,* 225-237.
Ramachandran, T., Rajendrakumar, K. and Rajendran, R. (2004). *Institute of Engineers (India) J.,* **84:** 42-47.
Ramesh Kumar, A. and Rane, Y.N. (2004). *International Dyer,* **7:** 14-21.
Roy M., Broughton, Jr., Worley, S.D., Unchin Cho, Jian Lin, and Gang Sun. (1999). Incorporation of Antimicrobial Materials in Fabric.
Russell, A.D. and Chopra, I. (1996). *Ellis Horwood.*
Sakagami, Y. (1995). *J. Soc. Fibre Sci. Technol.* **51(6):** 256-261.
Sarkar, R.K., Purushottam De. and Chauhan, P.D. (2003). *Indian J. Fibre & Textiles Research,* **28:** 322-331.
Seventekin, N. and Ucarci, O. (1993). *J. Text. Inst.,* **84(3):** 304-314.
Stefan Zikeli, (2003). *Textile Asia*, **1:** 35-39.

Sun, G. and Xu, X. (1997). *Am. Chem. Soci. Abst.,* Part I CELL 047.
Teresa Wierzbowska. and Marianna Stachowiak-Nogacka. (2000). *Asian Textile J.,* **12:** 95-98.
Tiller, C. et al., (2001). *Proc. Natl. Acad. Sci.,* USA **98:** 5981-5985.
Tiwari, S.K., Prajapati, K.G. and Gharia, M.M. (2001). *Colourage,* **9:** 17-20.
Turner, A.J. (1989). *Text Rental,* **72:** 82-88.
U.S. Environmental Protection Agency: Toxicological Review, Tributilin oxide: In Support of Summary Information on the Integrated Risk Information System (IRIS), July 1997.
Vandana Tiwari and Padma, S. Wanker. (2001). *Asia Textile J.,* **8:** 82-84.
Vigo, T.L. (1993). *Proceedings of the beltwide production conference.*
Vigo, T.L. and Danna, G.F. (1996). *Am. Chem. Soc. Abst, Part 1 CELL 090.*
Virnich, A. (2000). *Melliand Textilber,* **81:** 408-410.
Wallhaußer, K.H. and Praxis der Sterilisation, (1995). Konservierrung. Georg Thieme Verlag Stuttgart New York.
Wei Huang. and Karen K. Leonas. (2000). *Textile Res. J.,* **70(9):** 774-782.
Welch, C.M. and Peters, J.G. (1997). *Text Chem. Color,* **29(3):** 22.
Yang, Y. and Li, S. (1994). *Text Chem. Color,* **26(5):** 25.
Yang, C.Q. Wang, X. and Kang, I.S. (1997). *Text. Res. J.,* **67(5):** 334-342.
Young Hee Kim. and Gang Sun. (2000). *Text. Res. J.,* **70(8):** 728-733.
Young Ho Kim., Hyung-Min Choi. and Jung Hee Yoon. (1998). *Textile Res. J.,* **68(6):** 428-434.
Young-Sik Chung., Kwang- Keun Lee, and Jin-Woo Kim. (1998). *Textile Res. J.,* **68(10):** 772-775.
Yuke, O. (1989). *Antimicrofinish,* Textiles Co., 182-184.
Zgibneva, Zh. A. (1998). *Mater Resp. Konf. Tekst. Khim,* **3:** 24-25.

Chapter 9

Applications of *Streptomyces* sp. in Pharmaceutical Industry

D. DHANASEKARAN
Department of Microbiology, School of Life Sciences
Bharathidasan University, Tirichirapalli-620024
E-mail : dhansdd@yahoo.co.in

Streptomyces is an actinomycetes, which can produce unique bioactive compounds. These microorganisms are found in a wide range of aquatic and terrestrial environments, and when faced with harsh conditions, have the ability to activate the metabolic pathways that ensure their survival. The unique bioactive compounds that may be produced through this process have commercial applications. Historically, a large percentage of the world's antibiotics have been found initially in a group of microorganisms known collectively as actinomycetes. Anti-cancer agents and other interesting bioactive compounds have also been detected in actinomycetes. Currently throughout the world there are a series of institutes and a number of companies devoted to the commercialization of bioactive compounds from actinomycetes. The fields of application for these bioactive compounds or the actinomycetes range from human and veterinary antibiotics to bio-remedial processes. Recently with the advent of wide ranging antibiotic resistance to presently available antibiotics, there is resurgence in the search for new antibiotics in clinically important bacteria, especially in actinomycetes. This is of particular importance due to the increasing occurrence of virulent infections in hospitals worldwide. In addition to the specific activities described above, there are numerous potential commercial applications for the bioactive compounds that may be identified by the company through selective screening techniques. Other potential applications for actinomycetes and the compounds, which they produce, may include the concentration and extraction of various metals in the mining industry, the discovery of new fluorescent dyes for paints and colour directed pH indicators, the discovery of new methods of production of amino acids and the discovery of new sunscreens. Recent developments by other researchers unrelated to the company have led to the discovery of bioactive compounds produced by actinomycetes, which are capable of alleviating clinical conditions such as blood clots and obesity in humans. Those applications include other antibiotics and anti-fungal, anti-viral and anticancer agents. It is anticipated that the company's screening techniques may be extended in identifying microorganisms that have application for bio-remedial control of environmental and mining pollutants. Examples include microorganisms that may digest various sources of cellulose, domestic wastes, toxic chemicals in polluted ground or water, plastics and the treatment of salt affected soils. This chapter specially describes the applications of *Streptomyces* in pharmaceutical industry.

INTRODUCTION

Streptomycetes are Gram-positive, filamentous soil bacteria that undergo morphological differentiation during their life cycle. They normally occur as spores, but in the presence of sufficient moisture and nutrients, the spores can germinate and form vegetative mycelium. In response to environmental signals, such as a shortage of nutrients or water, the process of differentiation is set in motion, and spores resistant to desiccation and starvation are formed again. At the same time, the production of pigments, antibiotics and other secondary metabolites is initiated (Kutzner 1986, Williams *et al.*, 1989). Streptomycetes are common in soil, but are also found in composts, fodder and aquatic habitats. Due to their characteristic life cycle, they are good survivors under the fluctuating growth conditions predominating in nature (Kutzner 1986). Streptomycetes have also been isolated from indoor environments, air, building material and dust samples (Nevalainen *et al.*, 1991, Andersson *et al.*, 1999, Hyvärinen *et al.*, 2002). They are considered as indicators of moisture conditions in buildings that are favourable for microbial growth.

The genus *Streptomyces*

The streptomycetes are aerobic, Gram-positive bacteria, which produce extensive branching vegetative (substrate) mycelium and aerial mycelium bearing chains of arthrospores. The substrate mycelium and spores can be pigmented, but diffusible pigments are also produced. On agar plates, they form lichenoid, leathery or butyrous colonies. The G+C-content of the DNA is 69-78% (Williams *et al.*, 1989). L-diaminopimelic acid is the characteristic compound present in the cell wall peptidoglycan of streptomycetes (Lechevalier and Lechevalier 1970). The streptomycetes are able to utilise a wide range of organic compounds as a carbon source, including complex biological materials, such as cellulose and lignin, and can also utilise an inorganic nitrogen source (Kutzner 1986). They are also known as producers of many secondary metabolites, such as antibiotics and other bioactive compounds (Williams *et al.*, 1989, Kutzner 1986). These bacteria are widely distributed in various habitats, including man-made environments.

CLASSIFICATION

The number of species in the genus *Streptomyces* is increasing continually. In 1997, 464 validly described species and 45 subspecies were reported (Hain *et al.*, 1997), in September 2002 there were over 650 species listed in the German Collection of Microorganisms and Cell Cultures (DSMZ). Thus, the genus is the largest of the order *Actinomycetales* within the class *Actinobacteria* (Stackebrandt *et al.*, 1997). The genera *Streptoverticillium* and *Kitasatospora* have been included in the genus *Streptomyces* (Witt and Stackebrandt 1990, Wellington *et al.*, 1992), although the taxonomic position of *Kitasatospora* is unclear (Zhang *et al.*, 1997). The classification of streptomycetes was originally based on morphological and biochemical characterisation, later on physiological tests (Kutzner 1986, Williams *et al.*, 1983, Goodfellow *et al.*, 1987, Kämpfer *et al.*, 1991, Goodfellow *et al.*, 1992). Serological methods (Ridell *et al.*, 1986), phage typing (Korn-Wendish and Schneider 1992) and protein profiling (Ochi 1995, Taguchi *et al.*, 1997) have also been used in their classification. The application of genetic methods, such as DNA-DNA reassociation (Labeda 1992, Kim *et al.*, 1999) and 16S rRNA gene sequence analysis (Gladek *et al.*, 1985, Stackebrandt *et al.*, 1992, Kim *et al.*, 1996, Takeuchi *et al.*, 1996, Hain *et al.*, 1997, Kataoka *et al.*, 1997, Kim *et al.*, 1999) has partly confirmed the phenotypic classification, but this approach has also provided new information.

IDENTIFICATION

A probabilistic identification matrix for streptomycetes, based on 41 characters like spore chain and spore morphology, pigmentation, physiological abilities, antibiosis and resistance to antibiotics, was developed by Williams *et al.* (1983). This scheme was applied to unknown isolates, 80 % of which could be assigned to a cluster. Kämpfer *et al.* (1991) revised the data using more characters and testing more strains and comparing the data with published genetic and chemotaxonomic data. Goodfellow *et al.* (1992) supplemented the method with rapid enzyme tests using fluorogenic substrates. Phage sensitivity and serological as well as cellular fatty acid characterisation can be useful for assigning an unknown isolate to the genus *Streptomyces*, however, these methods have difficulties to discriminate the species at the intragenus level (Anderson and Wellington 2001).

DNA-based molecular methods have been used for species differentiation and the identification of streptomycetes. DNA-DNA reassociation is a method measuring the DNA relatedness of two organisms and has proved to be suitable for the investigation of relationships between closely related taxa, such as species (Stackebrandt and Goebel, 1994). Strains belonging to the same species will generally have greater than 70% DNA-DNA relatedness (Stackebrandt and Goebel 1994). The method has been used in numerous studies dealing with streptomycetes; in some cases it has shown good correlation with the identification based on morphology and physiology, but in other cases there was no apparent correlation (Healy and Lambert 1991, Doering-Saad *et al.*, 1992, Labeda 1992, Labeda and Lyons 1992, Kim *et al.*, 1999, Bouchek-Mechiche *et al.*, 2000). While DNA-DNA reassociation has shown to be useful in the identification of *Streptomyces* species, because of the instability of the genome, it should not be used alone, but in connection with other tests (Anderson and Wellington 2001).

The sequence analysis of the genes coding for the ribosomal subunits (16S, 23S, and 5S rRNA), in particular the 16S rRNA gene, has become an important tool in bacterial identification, since it provides information about the phylogenetic placement of species (Woese 1987, Brenner *et al.*, 2001). The DNA sequences of the ribosomal genes are highly conserved, but the genes also contain variable regions, which sometimes can be useful for species discrimination (Stackebrandt and Goebel 1994, Rosselló-Mora and Amann, 2001). Kataoka *et al.* (1997) utilised the hypervariable region of the 16S rRNA gene to create an index for *Streptomyces* species identification. However, the 16S rDNA sequence information alone is not sufficient for species identification (Rosselló-Mora and Amann, 2001). Also repetitive intergenic DNA sequences (rep-PCR) (Sadowsky *et al.*, 1996), and PCR-RFLP of the 65-kDa heat shock protein gene (Steingrube *et al.*, 1997) have been used for the classification and identification of pathogenic and other clinically important *Streptomyces* species.

ISOLATION AND ENUMERATION

In order to isolate and enumerate streptomycetes from natural habitats it is necessary to eliminate or reduce fungal and bacterial growth on isolation media without adversely affecting the target organisms. This is usually achieved by applying appropriate selective pressures at various stages of the dilution procedure. The numerous strategies evolved for the selective isolation of streptomycetes have recently been reviewed. Fungal contaminants can be virtually eliminated by adding antifungal antibiotics, such as actidione, nystatin and pimaricin to isolation media.

Selectivity may be enhanced by pretreatment of the material or propagules in suspension prior to plating onto selective media. Examples of pretreatments applied in the isolation of streptomycetes are shown in Table 1. Heat pretreatments have often been used to decrease the numbers of bacteria on isolation plates. The basis of this effect has still to be explained but streptomycetes spores are more resistant than Gram-negative bacteria which cause most of the problems on isolation plates. Membrane filtration and centrifugation have been used to concentrate streptomycetes propagules in soil, water and sediment samples. After filtration, each filter is placed on the surface of isolation medium, where it remains or is removed after a few hours. The number of colonies developing on isolation plates is influenced by the type of filter used. A further departure from the dilution plate procedure is used to isolate thermophilic streptomycetes from dry, self-heated plant material. The latter is shaken in a wind tunnel or sedimentation chamber and the spore cloud impacted onto plates of surface dried medium in an Andersen sampler. Nutrient enrichment of the material has been used to increase the numbers of streptomycetes prior to isolation. Baiting techniques have occasionally been employed to the same end. This has generally involved the introduction of chitinous substrates into the soil or water and their subsequent removal for examination and plating once they have been colonized.

The selectivity of isolation media is influenced primarily by nutrient composition, the presence of selective inhibitors, pH, temperature and time of incubation. Many formulations have been developed for the isolation of streptomycetes (Table 2), but as the various ingredients have usually been chosen empirically, the basis of their selectivity is not always clear. Colloidal chitin and starch casein agars have been employed with particular success for the isolation of streptomycetes. The selectivity of chitin agar was considered to rest on the virtual universal ability of strains to hydrolyze this compound, yet only 25% of representative sample of the genus *Streptomyces* were found to be strongly chitinolytic. Similarly, starch casein agar supplemented with sodium chloride (4.6%, w/v) to reduce the growth of spreading Gram-negative bacteria will not support the growth of streptomycetes unable to tolerate salt at this concentration. Specific fractions of the streptomycetes community have been isolated by supplementing media with antibacterial antibiotics.

It is now known that the pH of the medium is an important selective factor. Neutrophilic streptomycetes have been the subject of most ecologists surveys, even though Jensen isolated acidophilic streptomycetes from acid soils over fifty years ago. The former are isolated on media with a reaction around neutrality, grow between pH 5.0 and 9.0, and occur as populations of over one million per gram in most soils. Acidophilic streptomycetes, which grow between pH 3.5 and 6.5, but not at pH 7.0, are wide spread in natural and man-made acid soils, but are only detected when acidified selective media are used. Indeed, in acid soils they are found in greater numbers than neutrophilic streptomycetes. In turn, alkaline soils harbor alkalophilic acid-sensitive streptomycetes. Mesophilic streptomycetes develop within 14 days at 25°C to 30°C and thermophilic streptomycetes in about 5 days at 45°C to 60°C.

The inability of workers to identify streptomycetes below the genus level prevented detailed studies on the activities of specific groups in nature. It is now possible to identify streptomycetes and to formulate media selective for fractions of streptomycetes communities found in natural habitats. The formulation media has been assisted by the application of the DIACHAR programme to the streptomycete data base of Williams *et.al.* (1983) to highlight the most distinctive growth requirements and tolerances of selected groups. The use of objectively formulated media, and the subsequent computer identification of isolates led to the detection of clear qualitative differences in streptomycetes isolated from the sample. It is evident that there is no such thing as a non-selective medium for streptomycetes and that several selective media must be used if the qualitative composition of streptomycete communities is to be unravelled.

Table 1. Examples of pretreatment for isolation of streptomycetes.

Treatments	Material	Predominant isolates
Physical:		
a. Agitation	Mould hay	Thermophilic streptomycetes
b. Centrifugation	Sea water and mud	*Streptomyces spp*
Heating:		
a. 40°C for 2 to 6 h	Soil, roots	*Streptomyces spp*
b. 50°C for 10 min	Soil and sediment suspension	*Streptomyces spp*
c. 65°C	Air-dried soil	*Micromonospora and Streptomyces spp*
d. 75-80°C for 10 min	Water	*Micromonospora and Streptomyces spp*
e. 100°C for 15 min	Air-dried soil	*Micromonospora and Streptomyces spp*
f. 120°C	Soil	*Streptomyces spp*
Membrane filtration:	Water	*Micromonospora and Streptomyces spp*
	Soil	*Streptomyces spp*
Chemical:		
a. Ammonia and sodium hypochlorite	Water	*Streptomyces spp*
b. Chloramine	Water	*Micromonospora and Streptomyces spp*
c. Calcium carbonate to increase pH with incubation	Soil	*Streptomyces spp*
d. Nutrient enrichment with chitin and incubation	Soil	*Streptomyces spp*
e. Phenol	Soil and water	*Streptomyces spp*

Table 2. Some media used for selective isolation of streptomycetes

Major constituents	Antibiotics	Predominant isolates
Arginine, glycerol, mineral salts	None	*Streptomyces* spp
Colloidal chitin, mineral salts	None	*Micromonospora and Streptomyces* spp
Glucose, yeast extract, malt extract	Actidione	Thermophilic streptomycetes, *Micropolyspora faeni*
Half strength nutrient agar	Actidione	Thermophilic streptomycetes, *Thermoactinomyces vulgaris*
Half strength nutrient agar	Actidione and rifampicin	*Streptomyces albus, Saccharomyces viridis*
Paraffin, mineral salts	None	*Streptomyces* spp
Starch casein, mineral salts	Actidione, nystatin, penicillin and polymyxin	*Micromonospora and Streptomyces* spp
Starch, mineral salts	Bacitracin, penicillin and polymyxin	*Streptomyces* spp F1
Starch, casein and sea water	Actidione	*Streptomyces* spp
Starch, casein, sodium chloride and mineral salts	Actidione, aerosporin, penicillin	*Streptomyces* spp

Isolation Methods

There is no single correct way to selectively isolate streptomycetes from environmental samples. Some pre-treatment methods have been used for the enrichment, such as heating or treating with ammonia and sodium hypochlorite, chloramines or $CaCO_3$, these being based on the fact that streptomycete spores are more resistant to these treatments than Gram-negative bacteria (Kutzner 1986, Goodfellow and Simpson 1987). Further selectivity can be achieved by using selective nutrient sources in the cultivation media. In contrast to many other bacteria, streptomycetes are able to utilize many biopolymers and are satisfied with an inorganic nitrogen source like nitrate (Kutzner 1986, Goodfellow and Simpson 1987, Williams *et al.*, 1989). Isolation media containing starch or glycerol as the carbon source and nitrate, casein or arginine as the nitrogen source have proven to be the most effective growth media for selective isolation of streptomycetes (Kutzner 1986). The addition of antifungal agents to the isolation media suppresses the growth of fungal species on the plates, however antibacterial antibiotics cannot be used to reduce the amount of other bacterial isolates because they will also inhibit the growth of many streptomycetes. Cycloheximide (50-100 μg/ml), as well as pimaricin and nystatin (10-50 μg/ml) have been used (Kutzner 1986). Mesophilic streptomycetes are usually cultivated at temperatures from 22 to 37°C for 14 days and thermophilic species at 40 to 55°C for 5 days (Kutzner 1986, Goodfellow and Simpson 1987, Williams *et al.*, 1989). Most streptomycetes are neutrophilic, and the isolation media commonly have a neutral pH, thus, if acidophilic strains are to be isolated, the pH of the medium can be adjusted to 4.5, and for alkalophilic strains to pH 10-11. However, some species may also show remarkable adaptation to a wide pH range (Suutari *et al.*, 2000).

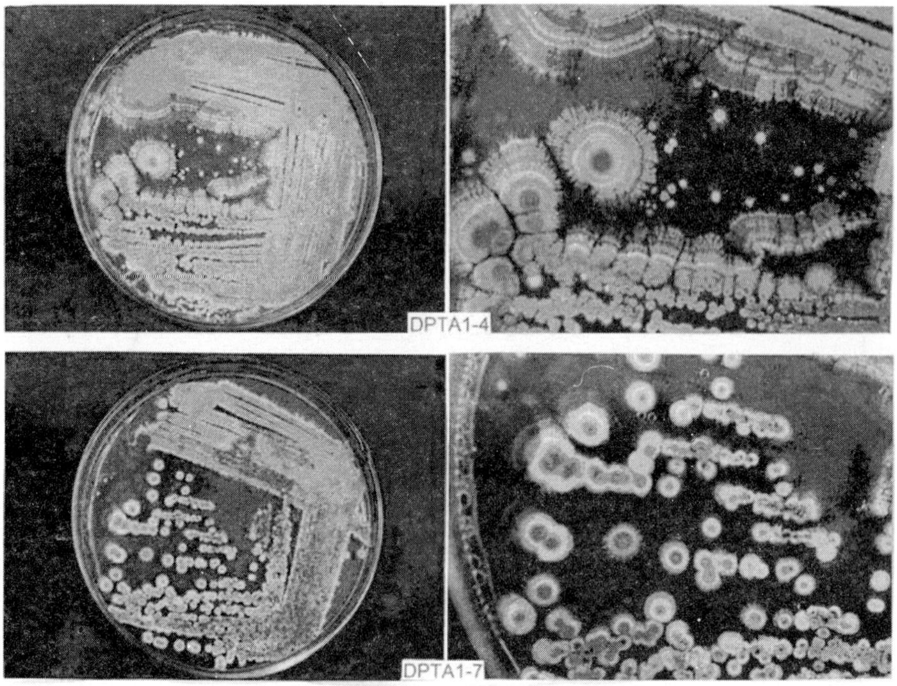

Plate 1. *Streptomyces* isolates in starch casein agar.

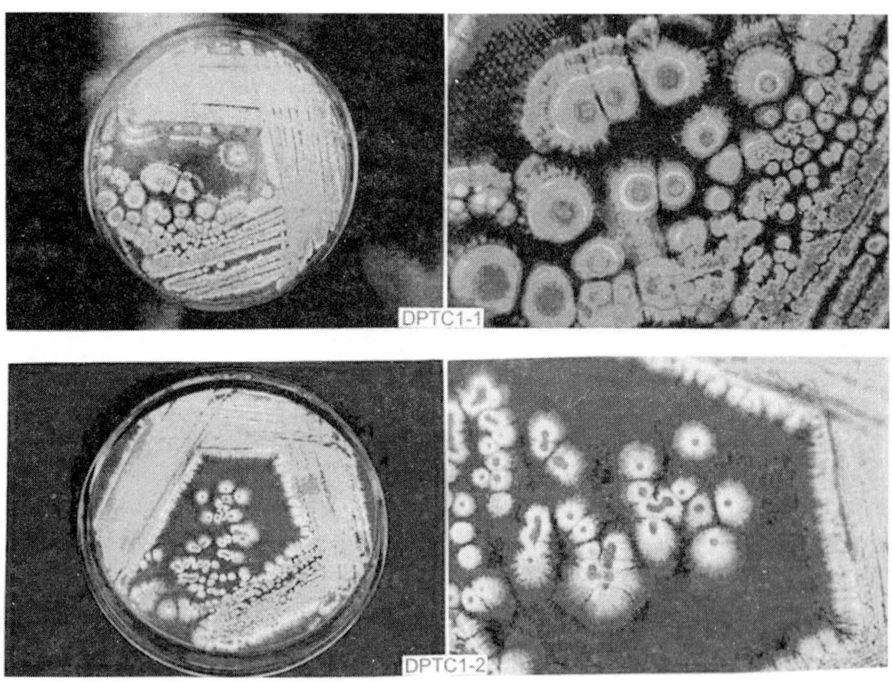

Plate 2. *Streptomyces* isolates in starch casein agar.

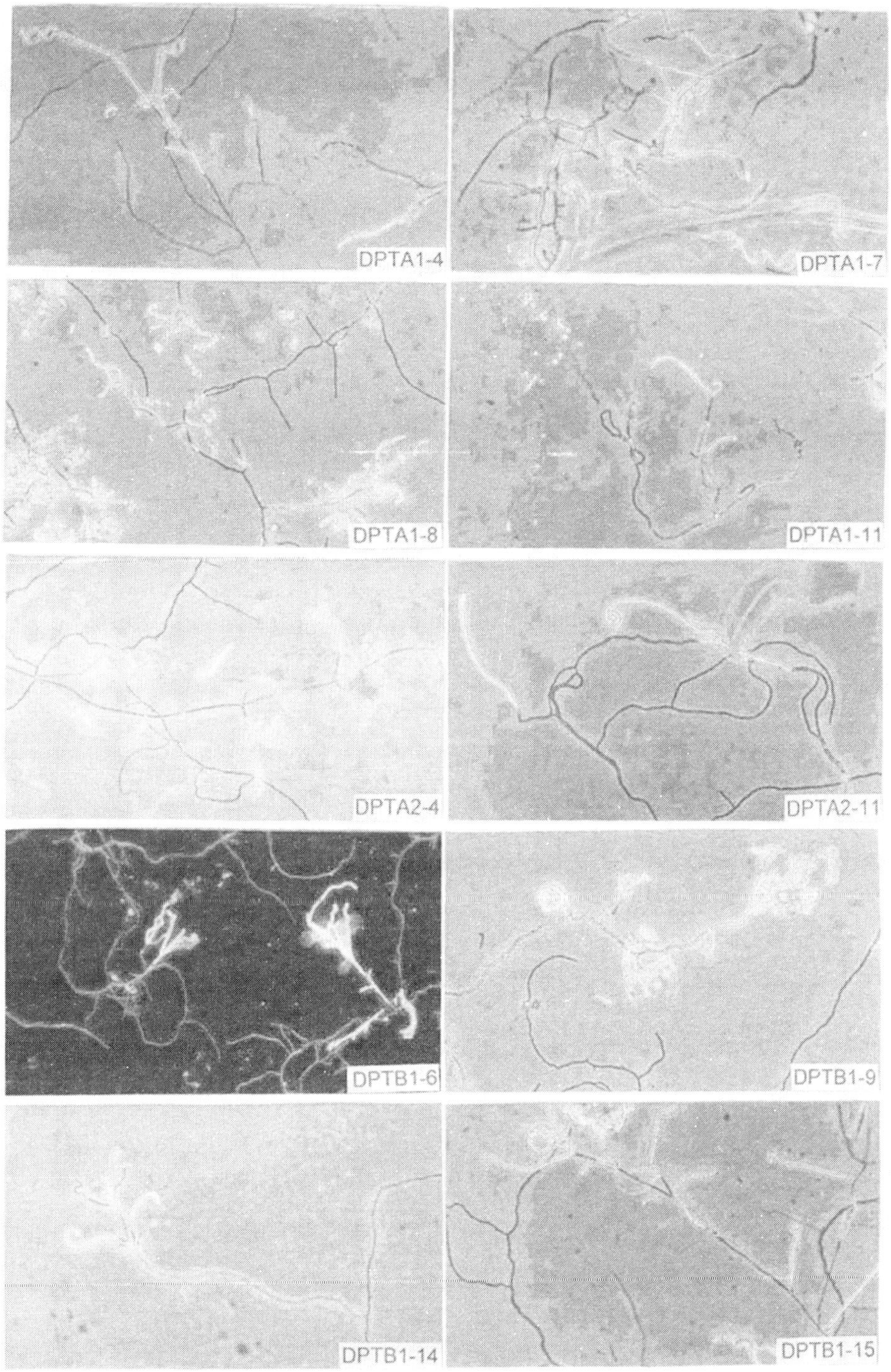

Plate 3. Light microscopic view of some *Streptomyces* isolates

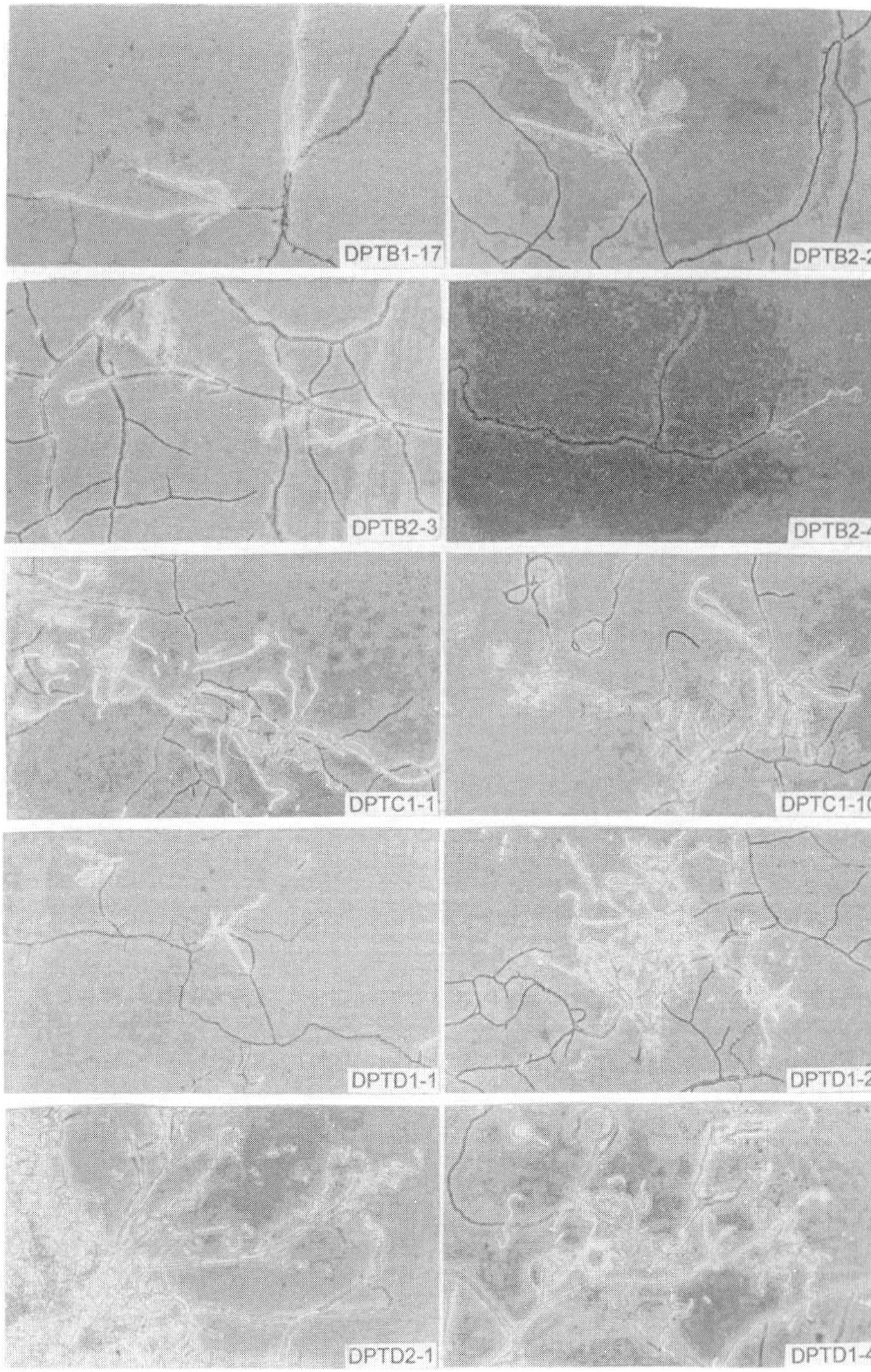

Plate 4. Light microscopic view of some *Streptomyces* isolates

Molecular Detection Methods

Streptomyces coelicolour A3 and *S. lividans* 66 were examined for their morphological and cultural characteristics as well as DNA-DNA relatedness in comparison with those of *S. coelicolour*, *S. lividans* ISP5434 and *S.violaceoruber* ISP5049 in order to clarify their taxonomic status (Hatano, 1994).

A neighbour joining method for reconstructing phylogenetic trees from evolutionary distance data was reported by Saitou and Nei (1987). Czerwlinska *et al.* (1990) indicated that restriction endonuclease analysis patterns could be used both to distinguish between species of *Streptomyces* and to recognize variation within the species. Stackebrandt *et al.* (1991) studied the designation of streptomycetes 16S and 23S rRNA based target regions for oligonucleotide probes. Weisburg *et al.* (1991) reported the 16S ribosomal DNA amplification for phylogenetic study. Ikeda and Omura (1991) studied the strategic strain improvement of antibiotic producer. The molecular variations among *Streptomyces* strains causing potato scab disease were analysed by Saad *et al.* (1992). Taxonomic status of *Streptomyces* hosts for gene manipulation in *Streptomyces coelicolour* A3, *S. kasugaensis* and *S. lividans* was studied (Nimi and Hotta, 1994). SLP2, pHG, pBL1, pSCL and other linear plasmids in β-lactam producers and plasmids in *S. rimosus* were also described (Kinashi, 1994). Thompson *et al.* (1994) used Clustal W: improving the sensitivity of progressive multiple sequence alignment programme for sequence weighting, position-specific gap penalties and weight matrix choice of *Streptomyces*.

Chun and Goodfellow (1995) reported a phylogenetic analysis of the genus *Nocardia* with 16S rRNA gene sequences. The restriction analysis of actinomycetes chromosomal DNA was made and described by Novella *et al.* (1996). The use of repetitive intergenic DNA sequences to classify pathogenic and disease-suppressive *Streptomyces* strains was assessed (Sadowsky *et al.*, 1996). Takeuchi *et al.* (1996) studied phylogentic analysis of *Streptomyces* spp. causing potato scab disease based on 16S rRNA sequences (*S. scabies, S. diastatochromogens, S. bottropenesis* and *S. neyagawaensis*). The characterization of a polyketide synthase genes encoding involved in the biosynthesis of mithramycin from *Streptomyces argillaceus* was reported by Lombo *et al.* (1996). Molecular detection of streptomycin producing streptomycetes in Brazilian soil has been reported by Huddleston *et al.* (1997). Complete nucleotide sequence of the *S. lividans* plasmid P[IJ101] and correlation of the sequence with genetic properties were analysed by Kendall and Cohen (1998). Molecular identification of *Streptomyces* species in pure culture and in environmental samples was performed (Stackebrandt and Pukall, 1999). Clark *et al.* (1998) studied the diversity within *S. ipomoeae* based on inhibitory interactions, repetitive element based polymerase chain reaction (rep PCR) and plasmid profile.

Mao *et al.* (1999) observed the molecular characterization and analysis of biosynthetic gene cluster for antitumor antibiotic mitomycin 'C' from *Streptomyces lavendulae* NRRL 2564. The results of phylogenetic analysis of 16S rRNA genes and PCR analysis of the *nec1* gene from *Streptomyces* spp. causing common scab, pitted scab and netted scab disease in Finland were reported by Kreuze *et al.* (1999). Phylogenetic analysis of the genus *Thermoactinomyces* based on 16S rDNA sequences was made (Yoon and Park, 2000). Zotchev *et al.* (2000) reported the identification of a gene cluster for antibacterial polyketide-derived antibiotic biosynthesis in the nystatin producer *Streptomyces* noursei (ATCC 1455). Global analysis of growth responsive gene expression and regulation of antibiotic biosynthetic pathways in *Streptomyces coelicolour* using DNA microarrays was reported by Huang *et al.* (2001). PCR-RFLP method could serve as a rapid tool for estimating the approximate phylogenetic relationship of *Streptomyces* isolates without the need for interspacer of DNA or 16s rDNA sequencing (Atalan 2001). Song *et al.* (2001) reported the phylogenetic diversity of thermophilic actinomycetes and *Thermoactinomyces* species, isolated from mushroom composts in Korea, based on 16S rRNA gene sequence analysis. The cultivation and independent population analysis of bacterial endophytes in three potato varieties based on eubacterial and actinomycetes specific PCR of 16S rRNA genes have been reported (Sessitsch *et al.*, 2002). Studies have been carried out on phylogentic analysis based on 16S rDNA sequence data in a novel genus and species *Longispora albida* nov., in the family Micromonosporaceae (Matsumota *et al.,* 2003) and actinobacteria diversity in a deep-sea sediment by comparing phylogenetic diversity and estimating species number using a number statistical method (Stach *et al.,* 2003). The genetic and phenotypic traits of *Streptomyces* were used to characterize antibiotic activities of microbes collected from agricultural field (Davelos *et al.*, 2004). Kim *et al.* (2004) reported the rpo B sequence similarity of genera *Streptomyces* and *Kitasatospora* based on partial RNA polymerase -subunit gene sequence. The relationship of the marine actinomycetes *Salinospora* was confirmed by phylogenetic analysis of 16S rRNA gene sequencing (Kim *et al.,* 2005).

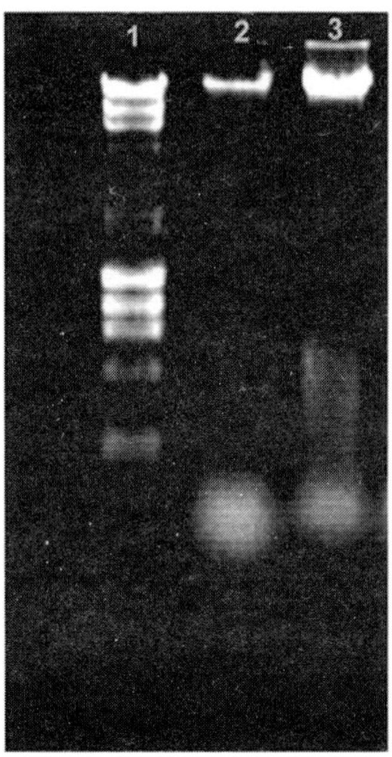

Lanes

1. Marker lambda DNA, Hind III digest
2. *Streptomyces* sp. (DPTB16) DNA
3. *Streptomyces noursei* (DPTD21) DNA

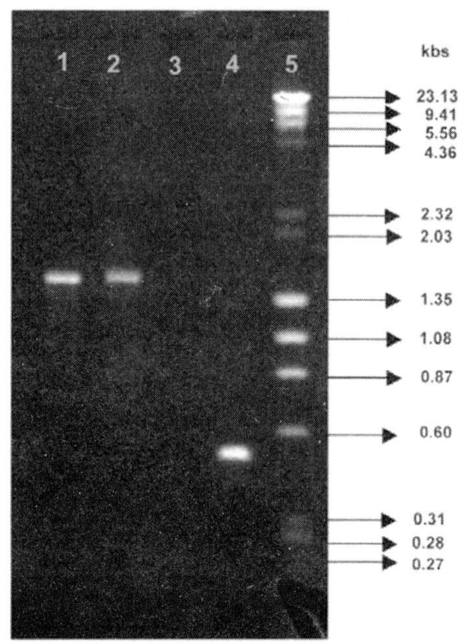

Lanes

1. *Streptomyces* sp. (DPTB16) amplified DNA
2. *Streptomyces noursei* (DPTD21) amplified DNA
3. Negative control
4. Positive control
5. Marker lambda DNA, Hind III digest

Plate 5. (a) Agarose gel electrophoresis of genomic DNA of *Streptomyces* spp.
(b) Agarose gel electrophoresis of amplified 16s rDNA of *Streptomyces* spp.

Free Energy of Structure = –152.0 kkal/mol

(a)

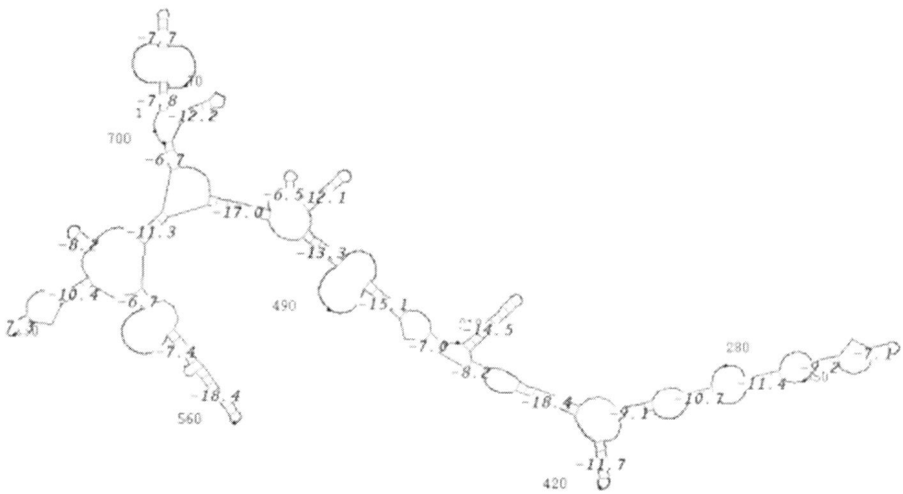

Free Energy of Structure = –187.2 kkal/mol

(b)

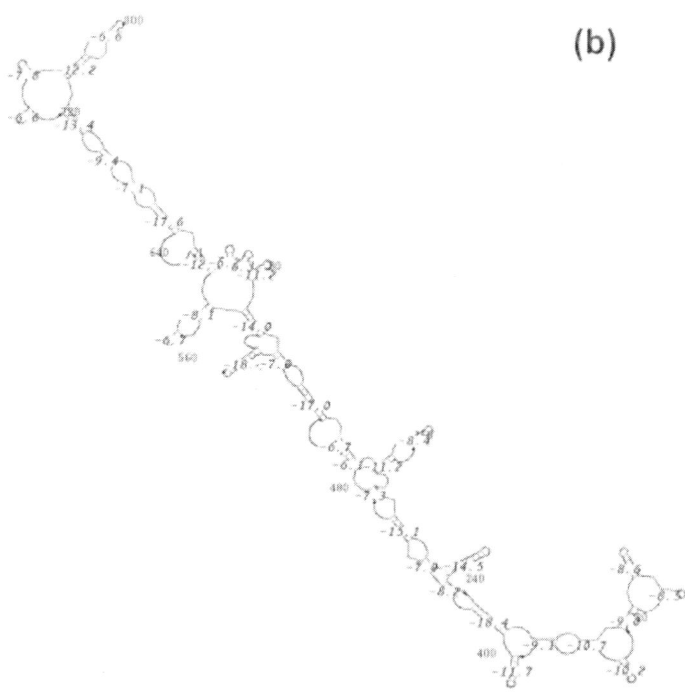

Plate 6. (a & b) Secondary structure 16s rDNA of *Streptomyces noursei* (DPTD21) and *Streptomyces* spp (DPTD16)

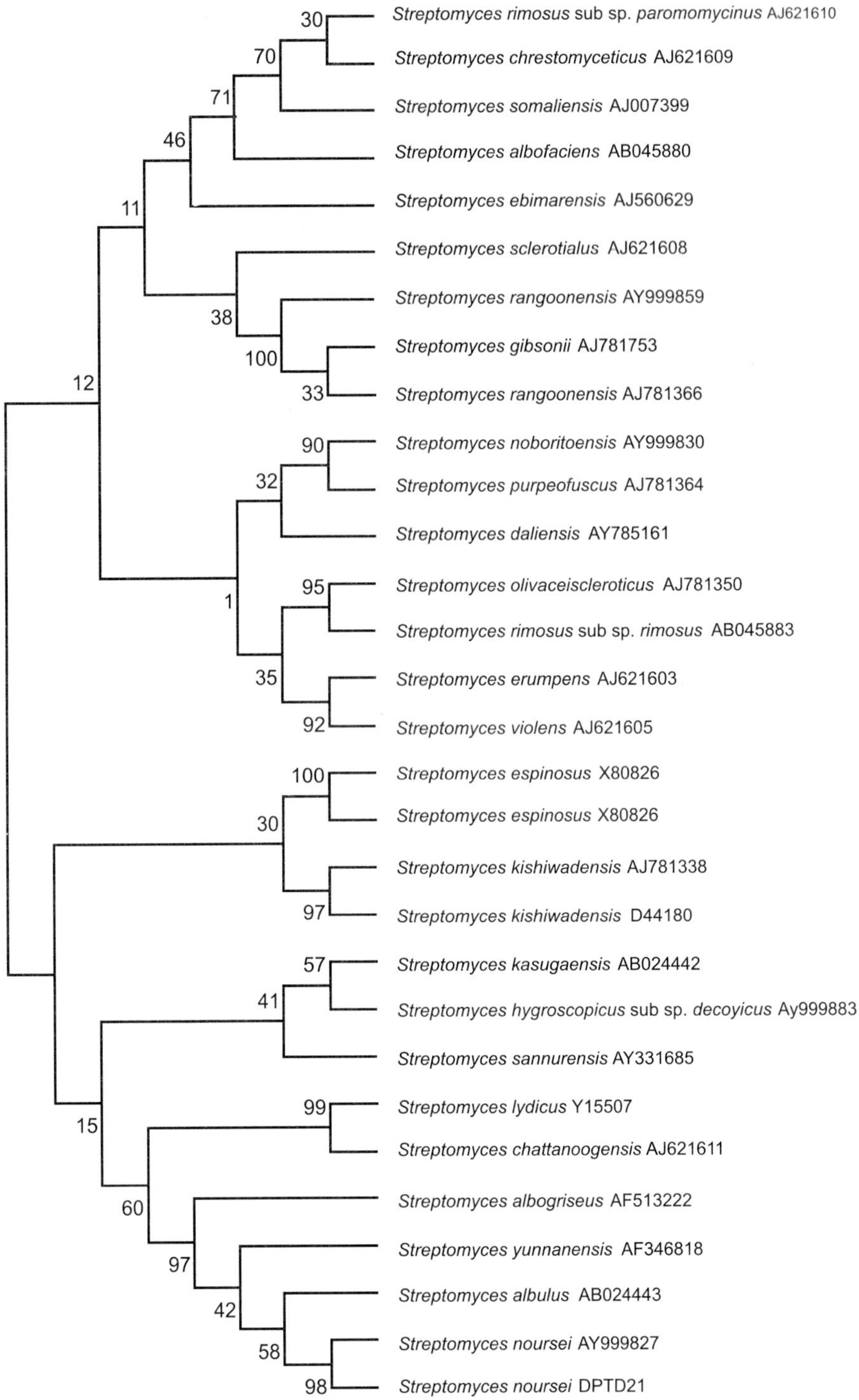

Plate 7. Phylogenetic analysis of *Streptomyces noursei* DPTD 21

OCCURRENCE OF STREPTOMYCETES IN NATURAL AND MAN-MADE ENVIRONMENT

Streptomycetes are able to produce extracellular enzymes (Morosoli *et al.*, 1997), which can decompose various materials. As reviewed by Williams (1985), streptomycetes have been implicated in the biodeterioration of silicone and natural rubber, wool, PVC plastic, cotton, jute fibre, wood, hay and grain. Jendrossek *et al.* (1997) isolated 50 rubber-degrading actinomycetes, from which 33 were identified as streptomycetes. Mesophilic actinomycetes, many of which belong to the genus *Streptomyces*, have been frequently isolated from all kinds of building materials, especially from ceramic materials, paints and glues (Hyvärinen *et al.*, 2002). They are able to grow and proliferate on plasterboards (Murtoniemi *et al.*, 2002).

Composts and Fodder

Streptomycetes are found in composts and fodder, especially in self-heated hay or grain (Lacey and Crook, 1988). During the early stages of composting or self-heating, mesophilic species are present, but these are replaced by thermotolerant species like *Streptomyces albus* or *Streptomyces griseus*, and with increasing temperature, the real thermophilic species take their place (Goodfellow and Simpson, 1987). Thermophilic strains were isolated from composts in numbers ranging from 103 to 105 cfu/g dry weight, and in mouldy hay from 105 to 107 cfu/g dry weight (Goodfellow *et al.*, 1987). The strains were assigned to clusters *Streptomyces thermovulgaris, Streptomyces thermoviolaceus, Streptomyces macrosporus, Streptomyces megasporus* and *Streptomyces thermolineatus* (Goodfellow *et al.* 1987).

Habitat

There seems to be no doubt that soil is the natural habitat of most members of the family Streptomycetaceae. They find the suitable conditions for growth and proliferation. These may not be optimal *per se* but allow successful competition with other microorganisms: (i) streptomycetes are able to degrade constituents of plant and animal residues, not as easily attacked by other bacteria and fungi, such as polysaccharides (starch, pectin, chitin- but cellulose only to a limited extent), proteins (e.g., keratin, elastine), and- although not extensively- aromatic compounds; (ii) streptomycetes are non-fastidious and satisfied with an inorganic nitrogen source and do not require vitamins or growth factors; (iii) in soil, streptomycetes find plenty of surfaces to support their mycelial growth; by their growth cycle "spore- mycelium- spore" they also are adapted to the other physical conditions of this habitat, such as moisture and aeration, which may undergo dramatic changes within a short time according to weather and season; (iv) although the "spores" (arthrospores, conidia) are not as resistant toward unfavorable conditions as bacterial endospores, they certainly contribute to the survival over longer periods of drought, freeze, and anaerobic conditions produced by water saturation.

Other biotopes may be infected or contaminated from soil. Few streptomycetes have been isolated from animal or human sources. Pathogenicity to these hosts is certainly not a typical property of these organisms. The same is true in regard to infestation of plants: only a few species cause bacteriosis (e.g., *Streptomyces scabies* scab of potatoes and other plants). Fodders and grain may contain streptomycetes. Their number depends on the degree of contamination by soil particles or dust and conditions of storage; however, these organisms seldom contribute significantly to the spoilage of fodders and grain (thermophilic actinomycetes of other families are of much greater importance in this respect). By drainage after rainfalls, creeks and rivers become contaminated with soil streptomycetes that find their way into the sediments of freshwater lakes and eventually to marine biotopes. Streptomycetes may exist which are strongly adapted to these sediments but these are usually the habitat of other actinomycetes. By the same route, drinking water supplies are also contaminated with streptomycetes; some of them produce odorous compounds leading to spoilage of the water.

Soil as Habitat

Soil is the most common habitat of streptomycetes; those bacteria can be found in all kinds of soils (Kutzner 1986, Williams *et al.*, 1989). It has been observed that the streptomycetes in soil grow attached to surfaces, such as plant residues or fungal hyphae (Mayfield *et al.*, 1972), and they may have an important ecological role in the degradation of litter in soil (Khan and Williams 1975, Kutzner 1986, Goodfellow and Simpson 1987). The largest concentrations of streptomycetes can be found in the organic horizon (Hagedorn 1976). In most soils, streptomycetes comprise 1-20% of the total viable count (Kutzner *et al.* 1986), and 64-97% of the cultivable actinomycetes (Xu *et al.*, 1996, Wang *et al.*, 1999). The more arid the soil and cooler the climate, the higher the percentage of the streptomycetes in the total count of actinomycetes (Xu *et al.*, 1996). In coniferous forest soil in Finland, streptomycetes represented 5% of the viable count (Elo *et al.* 2000). Plate counts of streptomycetes in acidic soils, peat and mine waste ranged from 103-106 cfu/g dry wt. soil (Khan and Williams 1975).

Streptomycetes may play a role in promoting plant growth through control of root pathogens or in some indirect way, since some species are able to produce antifungal compounds (Yuan and Crawford 1995, Hamby and Crawford 2000). Streptomycetes are commercially used as biocontrol agents (Raatikainen 1995). On the other hand, *Streptomyces* spp. are the causative agents of potato scab, a plant disease, which causes significant financial losses to the agricultural community (Takeuchi *et al.*, 1996).

Growth in Soil

By direct observation of the soil microflora, several authors could show that streptomycetes perform their typical life cycle in this natural habitat. Pfennig (1958) demonstrated the following sequence of events when glass slides were buried into the soil (Cholodny's method, 1930): Organic particles are colonized by "substrate" (= vegetative) mycelium; from there, hyphae grow away to soil particles (or the free glass surface supplied by the investigator), and chains of arthrospores are finally produced (terminally or as side branches), their morphology (straight, spiral, etc.) corresponding to what we know of these organisms in pure culture. The hyphae growing on glass slides either undergo autolysis or they fragment to cylindrical elements which were called by this author "chlamydospores". Thus, under these conditions two types of propagules result from the growth, chains of arthrospores probably dominating. Pfennig's studies answered by direct observation the question often asked whether streptomycetes exist as mycelium or spores in soil? Vegetative growth is limited to organic particles (plant and animal residues including hyphae of fungi and protozoa) and in most cases is of short duration (In these experiments just a few days). The number of propagules produced by these microcolonies depends on the quantity and quality of nutrients supplied; if these are very scanty, the germ tube of an arthrospore itself may be transformed into sporogenous hyphae and produces a short chain of arthrospores.

Similar observations were described by Pfennig (1958), and by other authors, either by direct observation (Cholodny's, 1930; Glathe, 1955; Llyod, 1969; Szabo, Marton, and Partal, 1964), or by a combination of this technique with colony counts (Jensen, 1943) or only by colony counts from samples treated in various ways (Skinner, 1951). Mayfield *et al.* (1972), using scanning electron microscopy, could demonstrate even more clearly this growth behavior of streptomycetes in soil.

Of interest is the observation by Pfennig (1958), Llyod (1969), and Mayfield *et al.* (1972) that spores of streptomycetes produced on agar media and inoculated into unsterile soil did not germinate (or only to a limited extent). Instead the spores supplied on glass slides were "grazed" by soil amoebas. The inability of added spores to germinate may be due to inhibitory effects of other microorganisms and/or shortage of nutrients. In sterile soil (γ-irradiated), however, added spores showed a high germination rate (Llyod, 1969); further, both unsterile and sterile soil supported growth of germinated spores (germlings). Although Lloyd (1969) did not definitely mention the two types of propagules observed by Pfennig (1958) and only mentions "sporulation" or "fragmentation into arthrospores", he writes: "Occasionally, a germ tube failed to grow and instead fragmented into rectangular arthrospores whose linear pattern preserved the general shape of the original germling". This appears to be rather reminiscent of Pfennig's "chlamydospores."

Colony Count

From what has been discussed in the previous paragraph, it becomes clear that colony counts reflect the number of propagules, especially arthrospores, at the time of sampling rather than actively growing vegetative mycelium. Nevertheless, we are inclined to relate the colony count with metabolic activities exhibited by these organisms in the soil under study—although this activity may have been in the recent past.

The determination of colony counts was of high concern to soil microbiologists during the first half of this century. The result of 18 papers published between 1903 and 1956 have been summarized by Flaig and Kutzner (1960b). In most soils, a count of colonies between 10^4 and about10^7 per gram can be expected; exceptionally high numbers of 200 million and 2,000 million have been reported by Jensen (1943) and Jagow (1956), respectively. In the 18 studies evaluated, streptomycetes accounted for about 1-20% of the total viable count, but in some soils these organisms may dominate.

Interest in general counts has now declined; it is nowadays centered around questions such as (i) influence of certain conditions, (ii) occurrence of streptomycetes at special micro sites and in habitats which were out of the scope of the general soil microbiologist, (iii) species composition of the flora of different soils, (iv) disturbution of species with special properties in certain biotopes.

Influence of Various Factors

Although not one single condition such as water and air circulation, pH, content of organic matter, vegetation, soil type, season, and climate by itself determines the occurrence of streptomycetes in soil quantitatively or qualitatively (and the influence of a single factor can hardly be determined), some tendencies can be observed.

Moisture Tension and pH

As shown in an investigation by Williams *et al.* (1972b), streptomycetes not only survive quite well in dry soils due to the formation of arthrospores but also need a lower water tension for growth than other bacteria. On the other hand, these organisms are very sensitive to water-logged conditions. In this case, shortage of oxygen is less effective than the increase of carbon dioxide above 10%. This study not only confirms but also by its detailed experiments explains numerous observations of earlier authors that well-drained soils (sandy loam, soil covering limestone) harbor more streptomycetes than heavy clay soils.

In accordance with the pH requirement isolates in pure culture, streptomycetes prefer neutral to alkaline soils in a natural habitat. This has been shown, by Flaig and Kutzner (1960b) who investigated a series soil of different pH. However, this seems now to be an over-simplified view caused by the use of neutral media for counting and isolating soil actinomycetes. As a result of more detailed studies of acid and alkaline soils that employed media adapted to the pH of the habitat (and supplemented with antifungal agents), it has emerged that there are acidophilic as well as alkalophilic streptomycetes. *Streptomyces acidophilus* has been isolated by Jensen (1928) from acid forest soil. Corke and Chase (1964) reported that an acid podzol contained actinomycetes adapted to the low pH of this habitat, and Williams *et.al.* (1971) and Khan and Williams (1975) could prove that acid soils, peat, and mine waters harbor a specific flora of streptomycetes. This comprised a number of different species (not named by the authors), none of which could be definitely identified as *S. acidophilus* Jensen (the type culture of which is no longer available). Further, the isolates were distinct from neutrophilic species in a number of properties, which suggests that they were not merely acidophilic variants of neutrophilic streptomycetes. At the opposite extreme, alkaline soils contain alkalophilic, acid-sensitive streptomycetes, e.g., *S. caeruleus* (Taber, 1959, 1960).

Organic Matter; Vegetation

As would be expected, more streptomycetes were found in a lot receiving organic manure than supplied only with mineral fertilizer or none in a 70-year-old field trial on "eternal growing of rye" (Flaig and Kutzner, 1960b). When soil samples are taken from arable land and nearby grass vegetation, the latter often yielded more streptomycetes (Conn, 1916; Flaig and Kutzner, 1960b). Although Conn (1916) expressed the view that actinomycetes may be active in the decomposition of grass roots, the difference as compared with the adjacent field soil may instead be due to more favourable water and air circulation (Krzywy, *et al.*, 1961).

Occurrence at Special Micro Sites

From direct observations it becomes obvious that streptomycetes–like many other soil bacteria–are not evenly distributed in soil but rather occur as micro colonies on particles of organic material which serve as food. Besides this heterogeneous distribution, which does not need to be considered when counting or isolating these organisms, there are two micro sites that have attracted the attention of soil microbiologists: the rhizosphere and earthworm casts. Studies on the occurrence of streptomycetes in the rhizosphere as compared with the surrounding soil have been carried out by several investigators. Often a significantly higher number of these organisms were found at this micro site-depending, however, very much on the species and age of the plant. For example, soy and maize harbored 10 to 18 times as many "actinomycetes" in their rhizosphere (Abraham and Herr, 1964). Kaunat (1963) found an increase of actinomycetes in the rhizosphere of maize, beans, sunflower, poppy, and spinach, whereas in 15 other plant species this plant-site was free of actinomycetes. Analogous observations were made by Rouatt, *et al.* (1951) and Glathe, *et al.* (1954). The intestinal tract of the earthworm has been regarded by Brusewitz (1959) as a "breeding place" for actinomycetes, provided that appropriate food is supplied. Similar results were reported by Ruschmann (1953), Schutz and Felber (1956), and Parle (1963a,b).

Streptomyces Flora of Different Soils

Two questions can be raised: (i) How many different "species" do we find in a soil when plate counts are performed without enrichment or other special treatments. (ii) Do the various species of streptomycetes prefer particular soils or in other words, do we find flora of varying species composition in different soils?

The number of "species" (or types) to be observed is to a certain extent dependent on the number of isolates made from the soil sample: further on the fact whether the investigator applies a narrow or wide species concept. However, if we disregard these points, we can make the general statement that habitats with a low colony count (and therefore also supplying fewer isolates) also yield a low number of different types. The second question stated above is difficult to answer. Because of the past chaotic state of the taxonomy of these organisms, an exact identification of these species–the prerequisite of ecological studies–was almost impossible. One, therefore, wonders what value species names such as *Streptomyces albus*, *S. griseus*, and *S. coeliocolor* have in earlier publications. Also, the differences with regard to the species concept have to be mentioned again. To arrive at a meaningful interpretation of the findings, the present author believes that–at least in ecology–a narrow species (or subspecies) has to be applied. There are only a few publications that can be evaluated in this respect.

Species with rather extreme requirements for pH occur only at respective habitats. A flora adapted to high salt concentration and alkaline reaction in Hungarian soils was observed by Szabo *et al.* (1959). A special geographical distribution has been claimed to exist for *S. malachiticus*, which has been mainly found in tropical or sub-tropical zones (Kuster, 1970),

Misiek (1955) reported that samples of the same soil type differed in the occurrence of "subgroups"; however, variation was observed between different soil types. Some streptomycetes were found in many of the 47 samples, others only in few. Flaig and Kutzner (1960b) isolated 1,500 strains of *Streptomyces* from various soils. The samples were taken from nearby areas of different vegetation (arable land, grass, forest). Seventy-six of the 382 "subgroups" differentiated by these authors preferred certain vegetation: forest, 14; grass, 5; arable land, 11 (belong *S. violaceoruber* [from 27 soil samples] and *S. hirsutus* [4 samples]); arable land or grass, 42 (belongs *S. griseinus* [grisein producing strain of *S. griseus* from 12 samples], *S. coelicolor* [34 samples], and *S. phaeochromogenes* [17 samples]); 4 subgroups were found only in a compost of manure and in a field trial receiving only manure, among them were *S. prasinus* (4 samples). Of course, in the soil type nor the vegetation alone but rather the combination of conditions prevailing at these habitats has to be regarded as responsible for the occurrence of particular organisms.

Some species were mentioned above as occurring in many soils, others being found rather seldom – an observation made by all ecologists. This may be due to various factors: (i) species with high frequency may be metabolically more versatile, as observed by Davies and Williams (1970) and Kuster (1976), or more successful in competition with other soil microorganisms (ii) these species may be found more often because they produce more propagules in soil (iii) the result may simply reflect the methods and media used for isolation.

In the course of screening programmes for producers of antibiotics, the question has been raised whether certain habitats are especially rich sources of antagonistic streptomycetes. Numerous papers on this subject have been revieved by Routien and Finlay (1952). The general tendency is that habitats with a quantitative and qualitative rich flora of streptomycetes harbour more antibiotic producers. It is also obvious that produces differ widely in their distribution, some occurring rather frequently (e.g., producers of actinomycins), others rather seldom. However, since no prognosis can be made where to find producers of new antibiotics, the search for further strains must proceed by empirically screening various habitats and employing more sophisticated methods of isolation. The occurrence of antibiotic-producing streptomycetes in soil has also been of concern to the soil microbiologist and plant pathologist.

Interaction with Other Soil Microorganisms

The phenomenon of antibiosis *in vitro* is now well known: it was first observed by soil microbiologists many decades ago when dilution plates for isolating bacteria and fungi were prepared. Its outcome has been the finding of more than two thousand antibiotic substances from bacteria and fungi, many of which have found application in medicine. Most of the antibiotics in use in chemotherapy are produced by streptomycetes. This fact led to the speculation that production of antibiotics may also take place in soil, conferring an advantage to its producer in competition with other microbes and thus playing a role in the establishment of an equilibrium among the members of soil flora. The enormous body of information available in this field can be summarized as follows:

(i) Production of antibiotics has been demonstrated unequivocally only in sterilized soil supplemented with nutrients and inoculated with a potent producer.
(ii) Although direct evidence of antibiotic production in natural soil is lacking, most soil microbiologists share the view that antibiosis could be important in micro habitats of the soil.
(iii) Several observations interpreted as being due to antibiotic effects may rather be due to other kinds of interaction such as competition and induction of lysis or autolysis.

The phenomenon of antibiosis in soil has especially attracted the attention of plant pathologists. Numerous studies have shown that soil-borne diseases can be controlled by amending soil with nutrients such as chitin or fungal cell wall, which leads to an increase of the flora of antibiotic-producing actinomycetes. However, further investigations have suggested that it is not the potential of the enriched flora to produce antibiotics that is important, but rather their capacity to exhibit lytic effects on fungi, cell wall of which also contains chitin.

Aquatic Habitats

Streptomycetes have been isolated from fresh water as well as marine environments, although, it has been a subject of debate, whether they are indigenous, or have been washed off from the surrounding soils (Cross 1981, Good fellow and Simpson, 1987). Moran *et al.* (1995) showed that in coastal marsh sediment, streptomycetes accounted for 2-5 % of the microbial community, and were an indigenous population. Occasionally, streptomycetes also grow in drinking water reservoirs affecting the water quality by causing earthy odors, which are due to their production of volatile secondary metabolites, such as geosmin and methyl *iso*-borneol (Bentley and Meganathan 1981, Cross 1981, Wood *et al.*, 1985).

Secondary Metabolites

According to one definition, microbial secondary metabolites are substances that are not needed for the growth or other essential processes in the cell (Vining, 1990). Secondary metabolites are mainly produced by microbial genera inhabiting soil and undergoing morphological differentiation, such as actinobacteria, bacilli and fungi (Vining, 1990). There are over 23,000 known microbial secondary metabolites, 42 % of which are produced by actinobacteria, 42% by fungi, and 16% by other bacteria (Lazzarini *et al.*, 2000). The streptomycetes are very potent producers of secondary metabolites. Out of the approximately 10,000 known antibiotics, 45-55 % are produced by streptomycetes (Demain 1999, Lazzarini *et al.*, 2000). The secondary metabolites produced by them have a broad spectrum of biological activities, e.g., antibacterial (streptomycin, tetracycline, chloramphenicol), antifungal (nystatin), antiviral (tunicamycin), antiparasitic (avermectin), immunosuppressive (rapamycin), antitumor (actinomycin, mitomycin C, anthracyclines), enzyme inhibitory (clavulanic acid), diabetogenic (bafilomycin, streptozotocin) etc. The pigments also have structures similar to those of other secondary metabolites and are synthesised by the same kinds of mechanisms (Metsä-Ketelä *et al.*, 1999, Nakano *et al.*, 2000). Some of the important secondary metabolites are listed in Table 3.

Table 3. Examples of secondary metabolite produced by Streptomycetes, their biological activities and the producing species known so far.

Compound	Biological Activity	Species	Reference
Avermectin	Antiparasitic	*S. avermitilis*	Burg *et al.* 1974
Anthracyclines	Antitumor	*S. galileus*	Fuji and Ebizuka 1997
Bafilomycin	ATPase-inhibitor of microorganisms, plant and animal cells	*S. griseus* *S. halstedii*	Werner *et al.* 1984 Frandberg *et al.* 2000
Chloramphenicol	Antibacterial, inhibitor of protein biosynthesis	*S. venezulae*	Bewick *et al.* 1976
Hygromycin	Antimicrobial, immunosuppressive	*S. hygroscopicus*	Omura *et al.* 1987 Uyeda *et al.* 2001
Lincomycin	Antibacterial, inhibitor of protein biosynthesis	*S. lincolnesis*	Peschke *et al.* 1995
Mitomycin C	Antitumor, binds to double-stranded DNA	*S. lavendulae*	Mao *et al.* 1999
Rapamycin	Immunosuppressive, antifungal	*S. hygroscopicus*	Vezina *et al.* 1975
Streptomycin	Antimicrobial	*S. griseus*	Egan *et al.* 1998
Streptozotocin	Diabetogenic	*S. achromogenes*	Herr *et al.* 1967
Tetracyclines	Antimicrobial	*S. aureofaciens* *S. rimosus*	Saleh *et al.* 1985 Hansen *et al.* 2001
Valinomycin	Ionophor, toxic for pro and eukaryotes	*S. griseus*	Andersson *et al.* 1998

APPLICATIONS OF STREPTOMYCES IN PHARMACEUTICAL INDUSTRIES

Streptomyces are Gram-positive, soil-dwelling bacteria. The well known property of many *Streptomyces* spp. to produce multiple secondary metabolites has attracted much attention (Gregory *et al.,* 2003). The secondary metabolism of *Streptomyces* to produce numerous antibiotics makes them interesting objects in pharmaceutical research (Andreas, 2006). *Streptomyces* are well known as a source of :

- Antibiotics: Antifungal and Antibacterial
- Anti-tumor agents
- Bioactive compounds: Immunosuppressants
- Enzymes
- Enzyme inhibitors
- Pigments
- Miscellaneous compounds

Antibiotics

Streptomyces is the largest antibiotic producing genus, producing both antibacterials and antifungals. Indeed, it was only with the discovery of the antibiotic Actinomycin from *Streptomyces antibioticus* by Waksman and Woodruff in the early 1940s that an interest was taken in a group of bacteria that were considered the preserve of a few soil microbiologists.

Antifungal Antibiotics

The fungal diseases have become very common as secondary infections to various predisposing factors as well as agents of primary diseases. The field of antifungal chemotherapy is presently rapidly moving, which began in 1903 with the successful use of potassium iodide. There was little progress for the next half-century when nystatin was introduced in 1951, the first polyene antibiotic from *Streptomyces*. The other antifungal antibiotics from *Streptomyces* include the following (Table 4).

Table 4: Antifungal antibiotics from actinomycetes

Organism	Antifungal agent	Reference
S. noursei	Nystatin	Akaike and Harata. 1994
S. nodosus	Amphotericin B	Fanos and Cataldi. 2000
S. natalensis	Pimaricin (Natamycin)	Sweetman, 2004
S. pimprina	Hamycin	Shervani and Bachhawat, 1993

Antibacterial Antibiotics

Streptomyces have provided about two-thirds of the naturally occurring antibiotics discovered, including many of those important in medicine such as aminoglycosides, anthracyclines, ß-lactams, tetracyclines, chloramphenicol and macrolides. Antibiotics are synthesized and accumulated at the idiophase (from the late log phase to the stationary phase). The various antibacterial antibiotics from *Streptomyces* under different parameters taken into consideration are:

Table 5: Antibacterial antibiotics from Streptomycetes

Unusual cultivation conditions	Organism	Antibacterial antibiotic	Reference
High inorganic phosphate (100-200 Mm)	*S. sapporonensis* *S. tateyamensis*	Bicyclomycin Thiopeptin	Miyoshi *et al.*, 1972 Miyairi *et al.*, 1970.
Poor nutrition	*S. griseus*	Aplasmomycin	Okami *et al.*, 1976
Low temperature (0-18°C)	*S. griseus*	Cryomycin	Yoshida *et al.*, 1972
High aeration (550 rpm)	*S. lavendulae*	Chlorocarcin	
		Mimosamycin	Arai *et al.*, 1977
		Saframycin	
Mixed culture (with *Pseudomonas maltophila*)	*S. griseorubiginosus*	Biphenomycin	Ezaki *et al.*, 1985; Imanaka *et al.*, 1986
S. tenjimariensis	Istamycin		Morioka *et al.*, 1988

Bioconversion of Foreign Antibiotics

The low substrate specificity of the enzymes present in antibiotic-producing microorganisms can be used as the basis for converting known antibiotics to new molecules. For example, Istamycin producer, *Streptomyces tenjimariensis* converts Fotimicin A to a new antibiotic 1-epi-dactimicin (Morika *et al.*, 1988).

Mutasynthesis

Blocked mutants that grow without special media supplementation but need supplementations for antibiotic biosynthesis are called idiotrophs, with the process being known as "mutasynthesis" or "mutational biosynthesis". Here an analogue of the major supplement is incorporated, thereby leading to the biosynthesis of a new antibiotic (Table 6).

Table 6: Some of the newly synthesized antibiotics.

Organism	Usual antibiotic	New antibiotic	Reference
S. erythraeus	Erythromycin	Unidentified	Martin *et al.*, 1974
S. fradiae	Neomycin	Hybrimycins A and B	Shier *et al.*, 1969
S. galilaeus	Aclacinomycin	11-hydroxyaclacinomycin A	Oki *et al.*, 1980
S. griseus	Streptomycin	Streptomutin A	Nagaoka and Demain, 1975
S. platensis	Platenomycin	5-O-mycaminosylnarbonolide	Maezawa *et al.*, 1976
S. kanamyceticus	Kanamycin	1-N-methyl Kanamycin	Kojima and Satoh, 1973
S. niveus	Novobiocin	Unidentified	Sebek, 1976
S. ribosidificus	Ribostamycin	1-N-methyl Ribostamycin	Kojima and Satoh, 1973
S. tendae	Nikkomycin	Nikkomycin Z	Delzer *et al.*, 1984
S. rimosus var. paramomycinus	Paramomycin	Hybrimycin C	Shier *et al.*, 1974

Changed Biosynthetic Pathways

Mutants blocked or modified in biosynthetic pathways involved in antibiotic production can be another source of antibiotic intermediates or antibiotic metabolites that are not accumulated by the original strains. For example, blocked mutants of *Streptomyces platensis* produce demycarosyl platenomycin and 9-dehydromycarosyl platenomycin (Furumai and Suzuki, 1975). Other examples include Doxorubicin (Arcamone *et al.*, 1969), Cirerubin (Nakagawa *et al.*, 1986) and Oxaunomycin (Yoshimoto *et al.*, 1986) from blocked mutants of Daumycin producing strain of *S. peucetius*, Cinerubin producing strain of *S. violaceochromogenes* and Baumycin producing strain of *Streptomyces* spp. respectively.

Many reports on antagonistic actinomycetes against the pathogenic bacteria and fungi are given below:

Table 7: Antagonistic actinomycetes against the pathogenic bacteria and fungi.

Organism	Activity against	Investigator(s) and year
Actinomycetes	*Fusarium oxysporum* *F. cubense* and *Penicillium graminicolum*	Meredith, 1946
Actinomycetes	*Mycobacterium tuberculosis*	Johnstone, 1947
Streptomyces spp.	*Trichophyton* sp., *Fusarium* sp., *Penicillium* sp. and certain bacteria	Leben *et al.*, 1952
Actinomycetes	Yeast	Takahashi, 1952 Takahashi *et al.*, 1993

Table 7: Contd.

Organism	Activity against	Investigator(s) and year
Actinomycetes	*S. aureus* and *E. coli*	Krasilnikov *et al.*, 1953
Streptomyces antibioticus	*Helminthosporium sativum*	Stevenson, 1956
Streptomyces sp.	Soil fungi and bacteria	Mukerjee and Chandra, 1975 Zeeck *et al.*, 1987; Kobinata *et al.*, 1993; Elson *et al.*, 1994
Actinomycetes	*Candida albicans* and some bacteria	Mukhopadhyay *et al.*, 1999 Hwang *et al.*, 2001
Streptomyces aureus	*H. oryzae, C. lunata* and *T. mentagrophytes*	Chakrabarthy and Chandra, 1979
S. hydroscopicus	*T. mentagrophytes* and *C. albicans*	Gurusiddaiah *et al.*, 1979
S. violaceoniger	*Macrophomia phaseolina* and *C. albicans*	Hussain and El-Gammal, 1980 Ivanova *et al.*, 1998
S. globus	*Alternaria solani, A. niger, C. pallescens, T. rubrum, T. mentagrophytes, C. albicans* and *Phytophthora* sp.	Nair *et al.*, 1994 Hwang *et al.*, 1996.
S. arabicus	*A. brassicae*	El-Shahed, 1994
S. anandii	Bacteria	Christopher Franco *et al.*, 1987
Streptomyces sp.	*Botrytis* sp., *Helminthosporium* sp., *Fusarium* sp. and *Pyricularia* sp.	Tanaka *et al.*, 1987 De and Gupta, 1991
Streptomyces sp.	Pathogenic Gram-positive bacteria	Theriault *et al.*, 1987
Streptomyces sp.	Gram-positive and negative bacteria	Saitoh *et al.*, 1988; Oscarson *et al.*, 1989 Fukuda *et al.*, 1990 Anwar *et al.*, 1994
Streptomyces sp.	*Phytophthrora* sp.	Omura *et al.*, 1988; Omura, 1990; Hwang *et al.*, 1994; Tang *et al.*, 2000
S. aerocolonigens	*C. albicans*	Nishio *et al.*, 1989
S. griseochromogenes	Phytopathogenic fungi; *S. scabies* and *Botrytis* sp.	Cheng *et al.*, 1989 Philips and Mc Closkey, 1990 Eckwall and Schottel, 1997 Xiao *et al.*, 2002
S. parasinopilosus	*Trichophyton* sp., *Candida* sp. and *E. coli*	Masanito *et al.*, 1989
Streptomyces sp.,	*Trichophyton* and Phytopathogenic fungi	Nair *et al.*, 1989
Streptomyces sp. (altromycin and bioxlomycins)	*Staphylococcus* sp., *Streptococcus* sp. and *Enterococcus* sp.	Jackson *et al.*, 1990 Bernan *et al.*, 1994
S. violaceus	Gram-positive and negative bacteria and *C. albicans*	Johdo *et al.*, 1991a Malviya *et al.*, 1994 Fiedler *et al.*, 1996 Hwung *et al.*, 1996 Siu *et al.*, 1997
S. roseiscleroticus (sultriecin) and *S. hygroscopicus* (yatakemycin)	Antitumor and antifungal *C. albicans, Cryptococcus neoformans, A. fumigatus*	Ohkuma *et al.*, 1992 Satoh *et al.*, 1993 Imamura *et al.*,1993

Table 7: Contd.

Organism	Activity against	Investigator(s) and year
	F. moniliforme, T. mentagrophytes, Blastomyces dermatitides and *Petriellidium boydii*	Atalan, 1997; Zheng *et al.*, 2000 Datta *et al.*, 2001
S. auermitilis (Avermectins)	Antiparasitic – nematodes and arthropods	Ikeda and Omura, 1993
Streptomyces sp. (non-polyene)	*Candida albicans, C. tropicalis, B. cinerea, A. fumigatus, F. solani, F. oxysporum, P. irregulare,* A. mellec, *T. mentachrophytes, Pseudomonas* sp., *Salmonella* sp. and *Legionella* sp.	Hacene *et al.*, 1994 Yon *et al.*, 1995 Ouhdouch *et al.*, 1996
S. violaceusniger (new-macrolide)	*C. neoformans, C. albicans, C. tropicalis, C. parapsiolis, C. glabrata, A. fumigatus, A. flavus, T. mentagrophytes, T. rubrum, M. canis, M. gypseum, P. capsici, M. grisea* and *C. gloeosporides*	Ubukata *et al.*, 1995 a Kook and Kim, 1995 Woo and Kamei, 2003 Fulguerir *et al.*, 2004
S. auroefaciens (sub sp. *viridulaus*)	*B subtilis, B. megaterium, B. cereus, Arthrobacter simples* and *E. coli*	Mansour *et al.*, 1996
Streptomyces sp.	*S. aureus, B. cereus, S. cerevisiae, F. moniliforme, C. albicans, E. coli, P. aeruginosa, Salmonella typhi, Botrytis pumilis, F. oxysporum, A. macrosporum, Shigella* sp. and *Klebsiella* sp.	Haque *et al.*, 1992 Abussaud, 1996 Malviya *et al.*, 1994 Saadoun *et al.*, 1999
Streptomyces sp.	*B. subtilis, S.aureus, M. luteus, A.fumigatus, A. niger, M. hiemalis, P. roqueforti, P. rariotii, C. albicans, Cryptococcus humicolus S. cerevisiae P. cinnamoni, Pestalotiopsis sydowiana, B. cinerea, Sclerotinia homoeocarpa, A. alternata, M. circinellodies, P. citirum, P. niotinanae, P. aphanidermatum, P. oligandrum, P. porphyrae, P. ultimo, T. rubram* and *Ustilogo maydei,*	Frandberg *et al.*, 2000 Shimizu *et al.*, 2000 Ndonde and Semu, 2000 Woo and Kamei., 2003 Moncheva *et al.*, 2002 Narayana *et al.*, 2004 Augustine *et al.*, 2004 Augustine and Kapadnis, 2005 Dhanasekaran *et al.*, 2005 b
Marine actinomycetes	*Salmonella* sp., *E. coli, Klebsiella* sp., *R. solani, pyricularia oryzae, H. oryzae* and *Colletotrichum falcatum*	Kathiresan *et al.*, 2005 Dhanasekaran *et al.*, 2005 c, d
S. albovinaceus	*Bacillus subtilis, B. pumilus, S. aureus, E. coli* and *P. aeruginosa*	Ellaiah *et al.*, 2005

Antiviral Property of Marine Actinomycetes

Aquaculture can provide an effective solution to the world's increasing demand for proteinacious food, but the sustained production is being hampered primarily by the outbreak of diseases. The shrimp culture industry is currently going through a period of severe crisis due to the outbreak and recurrence of a viral infection, commonly referred to as the White Spot Syndrome Virus (WSSV). A well-formulated strategy is needed to control this disease. Use of aquatic plants and animals for biomedical research and the potential of lower marine organisms as sources of pharmaceuticals have opened up new vistas to the whole scenario of aquaculture activities.

Recently, Achuthankutty and Desai (2004) have described a patented leaf extract formulation that is effective as a prophylactic and therapeutic agent against WSSV in penaeid shrimps. Although actinomycetes constitute only <0.05% of the culturable microbial community (Choi *et al.*, 1999), they have provided many important bioactive compounds (Okami *et al.*, 1988) and many of them have exhibited antiviral activity (Hasobe *et al.*,1985 and Nagawa *et al.*, 1985). However, no studies on the anti-WSSV properties of marine actinomycetes have been made so far. Pentalactones isolated from the fermentation broth of *Streptomyces* sp. M-2718 has been reported to be active against several DNA viruses (Nagawa *et al.*, 1985). The antiviral activities of pentalactones and pyrrole-2-carboxylic acid against herpes simplex virus had already been assayed and described (Mc Laren, *et al.*, 1983). Researchers have reported that guanine-7-N-oxide produced by *Streptococcus* sp. was found to inhibit *in vitro* replication of the fish herpes virus (Onchorhynchus Masou Virus), rhabdovirus (Infectious Hematopoitic Virus) and a bisegmented double-stranded fish virus (Infectious Pancreatic Necrosis Virus) (Hasobe *et al.*, 1985). All these studies indicate that actinomycetes function as antiviral antibiotics. Thus, it leads to the obvious conclusion that isolates of actinomycetes in the culture broth may have produced bioactive compounds that possess potent antiviral activities. These isolates, when incorporated in the feeds, lowered WSSV infection in shrimps. Isolation of these compounds and their characterization are essential which would lead to the possibility of developing antiviral agents effective against white spot disease in shrimps.

Antitumor Agents

Streptomyces, the largest antibiotic producing genus, also produces valuable compounds that are currently being used in the treatment of cancer/tumors. The antitumor or anticancer agents derived from *Streptomyces* spp are listed in Table 8.

Table 8: Showing different anticancer agents.

S.No	Organism	Antitumor agent	Reference
1.	*S.antibioticus*	Pentostatin	Gralla *et al.*, 2005
2.	*S.platensis*	Migrastatin	
		Isomigrastatin	Shan, 2005
3.	*S.lavendulae*	Mitomycin	Fukuyama and Yang,1987
4.	*S.peucetius*	Doxorubicin (Adriamycin)	Laginha, 2005
5.	*Streptomyces spp*	Echinomycin (Quinomycins)	Yoshida *et al.*, 1968
6.	*S.verticillius*	Bleomycin	Fujii *et al.*, 1974

Bioactive Compounds

Pharmacological or biological activity is an expression describing the beneficial or adverse effects of a drug on living matter. When the drug is a complex chemical mixture, this activity is exerted by the substance's active ingredient or pharmacophore but can be modified by the other constituents. The main kind of biological activity is a substance's toxicity. *Streptomyces* is the largest antibiotic producing genus, producing both antibacterials and antifungals, and also a wide range of other bioactive compounds such as immunosuppressants. Examples of immunosuppressant drugs include the following:

Organism	Immunosuppressant drug	Reference
S. tsukubaensis	Tacrolimus	Pritchard, 2005
S. hygroscopicus	Sirolimus	Pritchard, 2005

Enzymes

Enzymes, after antibiotics, are the most important products of *Streptomyces*. At present, microbial enzymes are used widely in food processing, detergent manufacturing, textile and pharmaceutical industries, medical therapy, bioorganic chemistry and molecular biology. The various enzymes derived from *Streptomyces* and their uses are given in Table 9.

Table 9: Various *streptomyces* species with their enzymes.

Organism	Enzyme	Uses	Reference
S. nigrifaciens	Choline oxidase	Test Phospholipid level in blood serum	Manyu Seijaku, 1982
S. violascens	Cholesterol oxidase	Cholesterol level in blood serum	Tomioka *et al.*, 1976
S. cyanogenus	Urate oxidase	Treat hyperuricaemia and primary gout	Ohe and Watanabe, 1981
S. albus	SalI	Restriction enzyme	
S. aureofaciens	SauI	Restriction enzyme	
S. caespitosus	ScaI	Restriction enzyme	
S. fimbriatus	SfiI	Restriction enzyme	Roberts, 1984;
S. phaeochromogenes	SphI	Restriction enzyme	Kessler *et al.*, 1985
S. stanfordii	SstI, SstII	Restriction enzyme	
S. tubercidicus	StuI	Restriction enzyme	
S. achromogenes	SacI, SacII	Restriction enzyme	
Streptomyces spp	Penicillin amidase	Production of 6-Aminopenicillanic acid	Nara *et al.*, 1971
S. olivaceus	Aminocyclases	Production of optically active	Batchelor *et al.*, 1961;
S. roseiscleroticus		amino acids	Sugie and Suzuki, 1978
S. tuirus			
S. sparsogenes			
S. karnatakensis	L-Asparaginase	Antineoplastic against murine neoplasms and human acute lymphatic leukaemia	Mostafa, 1982
S. griseus	Pronase	Liquifies hardened material present in age-related cataracts when injected into lens capsule prior to their surgical removal	Awadr *et al.*, 1972a

Table 9: Contd.

Organism	Enzyme	Uses	Reference
S. avidinii	Streptavidin	Strong affinity for biotin	Zimmermann *et al.*, 1994
S. lavandulae	Cholesterol esterase	Lipolytic activity	Toshio *et.al.*, 1979
S. scabies	Phospholipase	Estimation of choline	Kyowa Hakko Kogo Co.Ltd., 1983
S. cinnamoneus			Okawa and Yamaguchi,
S. hiroshimensis			1976a and b;
S. hygroscopicus	Amylase	Amylolytic activity	McKillop *et al.*, 1986
S. limosus			
S. praecox			Fairbairn *et. al.*, 1986
S. antibioticus	Cellulase	Cellulolytic activity	Enger and Sleeper, 1965
S. flavogriseus	Xylanse	Degradation of terrestial biomass	Kleupfel and Ishaque, 1982
S. griseus	Chitinase	Chitinolytic activity	Berger and Reynolds, 1958
S. antibioticus			Jeuniax, 1969
S. orientalis			Tominaga and Tsujisaka, 1976

Enzyme Inhibitors

Apart from enzymes, enzyme inhibitors are also derived from *Streptomyces*. For example, Orlistat, a saturated derivative of lipstatin, isolated from *Streptomyces toxytricini*, is a potent natural inhibitor of pancreatic lipases (Barbier and Schneider, 1987). Cycloheximide from *S. griseus*, which is a potent inhibitor of proteins in eukaryotes, is another example for enzyme inhibitor derived from *Streptomyces* (Berg *et al.*, 2002).

Pigments

Numerous *Streptomyces* possess a strikingly colored substrate mycelium (Lechevalier, 1971; Blinov, 1970; Habermehl, 1977; Hussein, 1970; Krassilinikov, 1970). They may be (i) Endopigments, bound to certain cell structures or (ii) Exopigments, excreted into the surrounding medium. Some pigments are produced as colourless leuco-pigments, which after excretion are transformed into recognizable pigment. In addition to bacteriochlorophylls, there are other pigments which are listed below:

Table 10: Showing different pigments.

Chemical class	Pigment	Organism
Prodigiosin	III Undecylprodigiosin IV Metacycloprodigiosin	*S. longispororuber DSM 40599* *S. longispororuber* DSM 40599
Naphthoquinone	Proto-actinorhodin Granaticin	*S. violaceoruber DSM 40049* *S. litmocidin DSM 40164*
Anthracyclin-glycoside	Rhodomycin	*S. violaceus/S. purpurascens DSM 40704*
Phenoxazinone	Actinomycin	*Numerous Streptomycetes*
Diaza-indophenol	Indochrome	*S. coelicolor DSM 40233*

Miscellaneous Applications

Avermectin (Burg *et al.*, 1974), the drug used to treat tapeworm infection and the drug Fosinopril (Okumura *et al.*, 2001) used to treat hypertension and chronic heart-failures are derived from *Streptomyces avermitilis* and *Streptomyces tanashiensis* respectively.

REFERENCES

Akaike, N. and Harata N. (1994). *Jpn. J. Physiol.*, **44(5):** 433-73.

Anderson, A.S. and Wellington, E.M.H. (2001). *Inter. J. Syste and Evol. Microbio* **51:** 797-814.

Andersson, M.A., Mikkola, R., Kroppenstedt, R.M., Rainey, F.A., Peltola, J. and Helin, J. *et al.* (1998). *Appl. and Environ. Microbiol.* **64:** 4767-4773.

Andersson, M.A., Weiss, N., Rainey, F. and Salkinoja-Salonen, M.S. (1999). *J. Applied Microbiology* **86:** 622-634.

Anwar, M.N., Rahman, M., Chowdhury, J.U. and Begum, J. (1994). *Indus. Research*, **29(3):** 71-81.

Arai, T., Takahashi, K. and Kubo, A. (1977). *J. Antibiotics* **30:** 1015-1018.

Arcamone, F., Cassinelli, G., Fantini, G., Gein, A., Orezzi, P., Pol, C. and Spalla, C. (1969). *Biotech. and Bioengine.* **11:** 1101-1110.

Argoudelis, A.D., Coats, J.H. and Johnson, L.E. (1974). *J. Antibi.* **27:** 738-743.

Awad, W.M. Jr., Ochoa, M.S. and Toomey, T.P. (1972). Proceedings of National Academy of Sciences USA **69:** 2561-2565.

Abussaud. (1996). *Acta Microbiol. Immunol. Hungarica*, **43(1):** 47-53.

Augustine, S.K., Bhavsar, S.P., Baserisallhi, and Kapadnis, B.P. (2004). *Indian J. Exp. Biol.,* **42:** 928-932.

Augustine, S.K. and Kapadnis, B.P. (2005). *J. Microb. World,* **7(2):** 328-331.

Barbier, P. and Schneider, F. (1987). *Helvetica Chimica Acta* **70(1):** 196-202.

Batchelor, F.R., Chain, E.B., Richards, M. and Rolinson, G.N. (1961). Proceedings of Royal Society Series B **154:** 552-531.

Bentley, R. and Meganathan, R. (1981). *FEBS letters* **125:** 220-222.

Berg, J., Tymoczko, J. and Stryer, L. (2002). *Biochemistry.* W.H. Freeman and Company.

Bewick, M.W., Williams, S.T. and Veltkamp, C. (1976). *Microbios* **16:** 191-199.

Blinov, N.O. and Khokhlov, A.S. (1970). *In:* (Ed. Prauser, H.) The Actinomycetales. The Jena International Symposium on Taxonomy. September 1968. Jena: Gustav Fischer Verlag.

Bouchek-Mechiche, K., Gardan, L, Normand, P. and Jouan, B. (2000). *Intern. J. Syste. and Evolut. Microb.* **50:** 91-99.

Brenner, D.J., Staley, J.T. and Krieg, N.R. (2001). *In:* Bergey's Manual of Systematic Bacteriology. Second edition. (Eds. Boone, D.R., Castenholz, R.W.), Springer-Verlag, New York Berlin Heidelberg, Volume 1, pp. 27-48.

Burg, R.W., Miller, B.M., Baker, E.E., Birnbaum, J., Currie, S.A. and Hartman, R., *et al.* (1979). *Antimicro Agents and Chemoth,* **15:** 361-367.

Bernan, V.S., Montenegro, D.A., Korshalla, J.D., Maiese, W.M., Steinberg, D.A. and Greenstein, M. (1994). *J. Antibiot.,* **47(12):** 1417-1424.

Choi, J.D. and Park, U.Y. (1999). *Bull. Kor. Fish Soc.*, **26:** 305-311.

Cross, T. (1981). *J. Applied Bacteriology,* **50:** 397-423.

Czerwinska, J.A., Kuczek, K., Mastalarz, A.G. and Mordarski, M. (1990). *Actinomycetes,* **1(3):** 70-82.

Cheng, X.C., Kihara, T., Ying, S., Uramoto, M., Osada, H., Kusakabe, H., Wang, B.N., Kobayashi, Y., Ko, K., Yamaguchi, I., Shen, Y.C. and Isono, K. (1989). *J. Antibiot.*, **42(1):** 141-144.

Delzer, J., Fiedler, H.P., Muller, H., Zahner, H., Rathmann, R., Ernst, K. and Konig, W.A. (1984). *J. Antibiot.*, **37:** 80-82.

Demain, A.L. (1999). *Applied Microbiology and Biotechnology* **52:** 455-463.

Doering-Saad, C., Kämpfer, P., Manulis, S., Kritzman, G., Schneider, J. and Zakrzewska-Czerwinska J, *et al.* (1992). *Applied and Environmental Microbiology* **58:** 3932-3940.

Datta, I., Banerjee, M., Mukherjee, S.K. and Majumdar, S. K. (2001). *Indian J. Exp. Biol.* **39(6):** 604-606.

Dhanasekaran, D., Sivamani, P., Arunagirinathan, N., Panneerselvam, A. and Thajuddin, N., (2005a). *J. Microbial World,* **7(1):** 62-66.

Dhanasekaran, D., Sivamani, P., Panneerselvam, A., Thajuddin, N., Rajakumar, G. and Selvamani, S., (2005b). *Pl. Pathol. J.,* **4(2):** 91-95.

Dhanasekaran, D., Rajakumar, G., Sivamani, P., Selvamani, S., Panneerselvam, A. and Thajuddin, N. (2005c). *Internet J. Microbiol.,* **1(2)**.

Egan, S., Wiener, P., Kallifidas, D. and Wellington, E.M.H. (1998). *Applied and Environmental Microbiology* **64(12):** 5061-5063.

Elo, S., Maunuksela, L., Salkinoja-Salonen, M., Smolander, A. and Haahtela, K. (2000). *FEMS Microbiology Ecology* **31**: 143-152.

Ezaki, M., Iwami, M., Yamshita, M., Hashimoto, S., Komori, T., Umehara, K., Mine, Y., Kohsaka, M., Aoki, H. and Imanaka, H. (1985). *J. Antibio.* **38:** 1453-1461.

Eckwall, E.C. and Schottel, J.L. (1997). *J. Indus. Microbiol. Biotechnol.,* **19(3):** 220-225.

Ellaiah, P., Adhinarayana, K., Saisha, V. and Vasu, P. (2005). *Indian J. Microbiol.,* **45(1):** 33-36.

Fanos. V. and Cataldi. L. (2000). *J. Chemother* **12:** 463-70.

Floss, H.G. and Yu, T. (2005). *Chem. Rev.* **105:** 621.

Frändberg, E., Petersson, C., Lundgrenm L.N. and Schnurer, J. (2000). *Canadian J. Microbiology,* **46:** 753-758.

Fujii, A., Takita, T., Shimada, N. and Umezawa, H. (1974). *J. Antibio* **27:** 73-77.

Fujii, I. and Ebizuka, Y. (1997). *Chemical Reviews* **97:** 2511-2523.

Fukuyama, T. and Yang, L. (1987). *J. Am. Chem. Soc.,* **109:** 7881-7882.

Fukuda, D.S., Mynderse, J.S., Baker, P.J., Berry, D.M., Boeck, L.D., Yao, R.C., Mertz, F.P., Nakatsukasa, W.M., Mabe, J., Ott, J., Counter, F.T., Ensminger, P.W., Allen, N.E., Alborn, W. and Hobbs, J.N., (1990). *J. Antibiot.,* **43(6):** 623-633.

Furumai, T. and Suzuki, M. (1975). *J. Antibio.* **28:** 775-782.

Fiedler, H.P., Nega, M., Pfefferle, C., Groth, I., Kempter, C., Stephan, H. and Metzger, J.W., (1996). *J. Antibiot.,* **49(8):** 758-764.

Fulgueira, C.L., Amigot, S.L. and Madni, C. (2004). *Curr. Microbiol.,* **48(2):** 135-139.

Frandberg, E., Peterson, C., Lundgren, L.N. and Schnurer, J. (2000). *Canadian J. Microbiol.,* **46:** 753-758.

Gladek, A., Mordarski, M., Goodfellow, M. and Williams, S.T. (1985). *FEMS Microbiology Letters,* **26:** 175-180.

Goodfellow, M., Lacey, J. and Todd, C. (1987). *J. General Microbiology* **133:** 3135-3149.

Goodfellow, M. and Simpson, K.E. (1987). *Frontiers in Applied Microbiology* **2:** 97-125.

Gralla, R., de Wit, R., Herrstedt, J., Carides, A., Ianus, J., Guoguang-Ma, J., Evans, J. and Horgan, K. (2005). *Cancer* **104(4):** 864-8.

Goodfellow, M., Ferguson, E.V. and Sanglier, J.J. (1992). *Gene,* **115:** 225-233.

Hagedorn, C. (1976). *Applied and Environmental Microbiology,* **32:** 368-375.

Hain, T., Ward-Rainey, N., Kroppenstedt, R.M., Stackebrandt, E. and Rainey, F.A. (1997). *International J. Systematic Bacteriology,* **47:** 202-206.

Hamby, M.K. and Crawford, D.L. (2000). Abstracts of the 100th General Meeting of the ASM, American Society for Microbiology, Washington DC, pp. 567.

Hansen, L.H., Ferrari, B., Sørensen, A.H., Veal, D. and Sørensen, S. (2001). *Applied and Environmental Microbiology,* **67:** 239-244.

Hacene, H., Sabaou, N., Bounaga, N. and Lefebrvre, G. (1994). *Microbios,* **79(319):** 81-85.

Hasobe, M., Saneyoshi, M. and Isono, K. (1985). *J. Antibiot.,* **38:** 1581-1587.

Hatano, K. (1994). *Actinomycetologica,* **8:** 115.

Healy, F.G. and Lambert, D.H. (1991). *International J. Systematic Bacteriology* **41:** 479-482.

Herr, R.R., Jahnke, H.K. and Argoudelis, A.D. (1967). *J. the American Chemical Society* **89:** 4808-4809.

Horinouchi, S. and Beppu, T. (1992). *Gene,* **115:** 167-172.

Hussein, A.M. (1970). *In:* (Ed. Prauser, H.). The Actinomycetales. The Jena International Symposium on Taxonomy. September 1968. Jena: Gustav Fischer Verlag pp. 133-141.

Hyvärinen, A., Meklin, T., Vepsäläinen, A. and Nevalainen, A. (2002). *International Biodeterioration and Biodegradation* **49:** 27-37.

Hwang, K.K., Ahn, S.J. and Moon, S. S. (1994). *Canadian J. Bot.,* **72(4):** 480-485.

Haque, S.K.F., Sen, S.K. and Pal, S.C. (1992). *Hindustan Antibiot. Bull.* **34(3-4):** 77-83.

Imanaka, H. (1986). *Biseibutus* **2:** 407-411.

Isono, K. and Suhadolnik, R.J. (1976). *Archives of Biochemistry and Biophysics* **173:** 141-153.

Imamura, N., Nishijima, M., Adachi, K. and Sano, H. (1993). *J. Antibiot.,* **46(2):** 241-246.

Ikeda, H. and Omura, S. (1993). *Actinomycetologica,* **7:** 133-144.

Jendrossek, D., Tomasi, G. and Kroppenstedt, R.M. (1997). *FEMS Microbiology Letters* **150:** 179-188.

Jackson, M., Karwowski, J.P., Theriault, R.J., Hardy, D.J., Swanson, S.J., Barlow, G.J, Tillis, P.M. and Mc Alpine, J.B. (1990). *J. Antibiot.,* **43(3):** 223-228.

Johdo, O., Ishikura, T. and Yoshimoto, A. (1991a). *J. Antibiot.,* **44(10):** 1110-1120.

Kataoka, M., Ueda, K., Kudo, T., Seki, T. and Yoshida, T. (1997). *FEMS Microbiology Letters* **151:** 249-255.

Katz, E. (1974). *Cancer Chemotherapy Reports,* **58:** 83-91.

Kessler, C., Neumaier, P.S. and Wolf, W. (1985). *Gene,* **33:** 1-102.

Khan, M.R. and Williams, S.T. (1975). *Soil Biology and Biochemistry,* **7:** 345-348.

Kim, D., Chun, J., Sahin, N., Hah, Y.C. and Goodfellow, M. (1996). *International J. Systematic Bacteriology,* **46:** 581-587.

Kim, B., Sahin, N. and Minnikin, D.E., Zakrzewska-Czerwinska, J., Mordarski, M. and Goodfellow, M. (1999). *International J. Systematic Bacteriology,* **49:** 7-17.

Kojima, M. and Satoh, A. (1973). *J. Antibiot.* **26:** 784-786.

Korn-Wendish, F. and Schneider J. (1992). *Gene,* **115:** 243-247.

Kutzner, KJ. (1986). *In:* (Eds. Starr, M.P., Stolp, H., Trper, H.G., Balows, A. and Schlegel, H.G.). The prokaryotes, A Handbook on Habitats, Isolation, and Identification of Bacteria, vol. 2, Springer-Verlag, New York, pp 2028-2090.

Kämpfer, P., Kroppenstedt, R.M. and Dott, W. (1991). *J. General Microbiology,* **137:** 1831-1891.

Krassilinikov, N.A. (1970). *In:* (Ed. Prauser, H.). The Actinomycetales. The Jena International Symposium on Taxonomy. September 1968. Jena: Gustav Fischer Verlag.

Kook, H.B. and Kim, B.S. (1995). *Pest. Sci.* **44(3)**: 255-260.

Kathiresan, K., Balagurunathan, R. and Masilamani Selvam, M. (2005). *Indian J. Biotechnol.*, **4:** 271-276.

Labeda, D.P. (1992). *Gene,* **115:** 249-253.

Labeda, D.P. and Lyons, A.J. (1992). *Applied and Environmental Microbiology,* **58:** 532-535.

Lacey, J. and Crook, B. (1988). *Annals of Occupational Hygiene,* **32:** 515-533.

Laginha, K.M. (2005). *Clinical Cancer Research.* October 1. Vol. 11 (19).

Lechevalier, M.P. and Lechevalier, H. (1970). *International J. System Bacteriology,* **20:** 435-443.

Lechevalier, H.A., Lechevalier, M.P. and Gerber, N.N. (1971). *Advances in Applied Microbiology,* **14:** 47-72.

Maezawa, I., Kinumaki, A. and Suzuki, M. (1976). *J. Antibiot,* **29:** 1203-1208.

Manyu Seijaku, K.K. (1982). Choline oxidase production by fermentation. Patent Japanese Kokai Tokkyo Koho IP 57 132 880.

Mao, Y., Varoglu, M. and Sherman, D.H. (1999). *Chemistry & Biology,* **6:** 251-263.

Martin, J.R., Egan, R.S., Goldstein, A.W., Mueller, S.L., Hirner, E.A. and Stanaszek, R.S. (1974). *J. Antibiot,* **27:** 570-572.

Mayfield, C.I., Williams, S.T., Ruddick, S.M. and Hatfield, H.L. (1972). *Soil Biology and Biochemistry,* **4:** 79-91.

McDowall, K.J., Thamchaipenet, A. and Hunter, I.S. (1999). *J. Bacteriology* **181:** 3025-3032.

Mc Laren, C., Ellis, M.N. and Hunter, G. A. (1983). *Antiviral Res.*, **3:** 223-234.

Metsä-Ketelä, M., Salo, V., Halo, L., Hautala, A., Hakala, J., Mäntsälä, P. and Ylihonko, K. (1999). *FEMS Microbiology Letters,* **180:** 1-6.

Miyairi, N., Miyoshi, T., Aoki, H., Kohsaka, M., Ikushima, H., Kunugita, K., Sakai, H. and Imanaka, H. (1970). *J. Antibiot,* **23:** 113-119.

Miyoshi, Y., Miyairi, N., Aoki, H., Kohsaka, M., Sakai, H. and Imanaka, H. (1972). *J. Antibiot,* **25:** 569-575.

Moellering, R.C. Jr. (2006). *Clin. Infect Dis.,* **42:** S3-S4.

Moran, M.A., Rutherford, L.T. and Hodson, R.E. (1995). *Applied and Environmental Microbiology,* **61(10):** 3695-3700.

Morika, M., Hotta, K., Naganawa, H., Okami, Y. and Umezawa, H. (1988). *J. Antibiot.* **41:** In press.

Morosoli, R., Shareck, F. and Kluepfel, D. (1997). *FEMS Microbiology Letters* **146:** 167-174.

Mostafa, S.A. (1982). *Zentralblatt fur Mikrobiologie,* **137:** 63-71.

Murtoniemi, T., Nevalainen, A., Suutari, M., Toivola, M., Komulainen, H. and Hirvonen, M.R. (2001). *Inhalation Toxicology,* **13:** 233-247.

Masanito, N., Takahashi, Y., Itoh, H., Kamiya, K., Shiratsuchi, M. and Otani, G. (1989). *J. Antibiotics,* **32(11):** 1535-1540.

Malviya, H.K., Golwalkar, M.P., Narayanan, M.S. and Tandon, G.G. (1994). *Hindustan Antibiot. Bull.,* **36(1-2):** 21-29.

Mansour, F.A., Elshirbiny, S.A. and El-Metwaly, N.A. (1996). *Egyptian J. Microbiol.,* **31(2):** 237-255.

Moncheva, P., Tishkov, S., Dimitrova, N., Chipeva, V., Nikolova, S.A. and Bogatzevska, N. (2002). *J. Culture Collections,* **3:** 3-14.

Nagaoka, K. and Demain, A.L. (1975). *J. Antibiot,* **28:** 627-635.

Nakagawa, A., Tomoda, H., Hao, V.M., Iwai, Y. and Omura, S. (1985). *J. Antibiot.,* **8:** 1114-1115.

Nakagawa, M., Furihata, K., Adachi, K., Seto, H. and Otake, N. (1986). *J. Antibiot.* **39:** 1178-1179.

Nakano, T., Miyake, K., Ikeda, M., Mizukami, T. and Katsumata, R. (2000). *Applied and Environmental Microbiology,* **66:** 1400-1404.

Nara, T., Misawa, M., Okachi, T. and Yamamoto, M. (1971). *Agricultural and Biological Chemistry,* **35:** 1676.
Nevalainen, A., Pasanen, A.L., Niininen, M., Reponen, T., Kalliokoski, P. and Jantunen, M.J. (1991). *Environment International,* **17:** 299-302.
Nishio, M., Tomatsu, K., Konishi, M., Tomita, K., Oki, T. and Kawaguchi, H. (1989). *J. Antibiot.,* **XLII(6):** 852-868.
Nair, M.G., Putnam, A.R., Mishra, S.K., Mulks, M.G., Taft, W.H., Keller, J.E., Miller, J.R., Zhu, P.P., Meinhart, J.D. and Lynn, D.G. (1989). *J. Nat. Prod.,* **52(4):** 797-809.
Ndonde, M.J.M. and Semu, E. (2000). *World J. Microbiol. Biotechnol.,* **16:** 595-599.
Narayana, K.J.P., Ravikiran, D. and Vijayalakshmi, M. (2004). *Indian J. Microbiol.,* **44(2):** 147-148.
Ochi, K. (1995). *International J. Systematic Bacteriology,* **45:** 507-514.
Ohe, T. and Watanabe, Y. (1978). *J. Fermentation and Technology,* **56:** 477-483.
Okami, Y., Okazaki, T., Kitahara, T. and Umezawa, H. (1976). *J. Antibiot* **29:** 1019-1025.
Okami, Y. and Hotta, (1988). *J. Mar. Biotechnol.* **1:** 59-65.
Okamura, H. and Nishimura, E. (2001). **121(3):** 253-7.
Oki, T., Yoshimoto, A., Matsuzawa, Y., Takeuchi, T. and Umezawa, H. (1980). *J. Antibiot.* **33:** 1331-1340.
Ohkuma, H., Naruse, N., Nishiyama, Y., Tsuno, T., Hoshino, Y., Sawada, Y., Konishi, M. and Oki, T. (1992). *J. Antibiot.,* **45(8):** 1239-1249.
Omura, S., Sadakane, N., Tanaka, Y. and Matsubara, H. (1983). *J. Antibiot,* **36:** 927-930.
Omura, S., Nakagawa, A., Fujimoto, T., Saito, K., Otoguro, K. and Walsh, J.C. (1987). *J. Antibiot,* **40:** 1619-1626.
Omura, S. (1990). *J. Antibiot.,* **43(8):** 1034-1036.
Ouhdouch, Y., Jana, M., Imziln, B., Boussaid, A. and Finance, C. (1996). *Actinomycetes,* **7(1):** 12-22.
Peschke, U., Schmidt, H., Zhang, N.Z. and Piepersberg, W. (1995). *Molecular Microbiology,* **16:** 1137-1156.
Pritchard, D. (2005). *Drug Discov Today,* **10(10):** 688-91.
Philips, D.R. and McCloskey, J.A. (1990). *J. Antibiot.,* **44(4):** 375-381.
Raatikainen, O. (1995). University Publications C/36. Kuopio University Printing Office.
Ridell, M., Wallerström, G. and Williams, S.T. (1986). *Systematic and Applied Microbiology,* **8:** 24-27.
Roberts, R.J. (1984). *Nucleic Acids Research,* **12:** 167-204.
Sadowsky, M.J., Kinkel, L.L. Bowers, J.H. and Schottel, J.L. (1996). *Applied and Environmental Microbiology,* **62:** 3489-3493.
Saleh, E.A., Mahmoud, S.A., el-Haddad, M.E. and Abdel-Fatah, M.K. (1985). *Zentralblatt für Mikrobiologie.* **140:** 325-332.
Saitou, N. and Nei, M. (1987). *Mol. Biol. Evol.* **4(4):** 406-425.
Saitoh, K., Tsunakawa, M., Tomita, K., Miyaki, T., Konishi, M. and Kawaguchi, H. (1988). *J. Antibiot.,* **41(7):** 855-861.
Sebek, O.K. (1976). *In:* American Society of Microbiology, Washington DC. Microbiology-1976. pp. 522-525.
Selman Waksman and Antibiotics - ACS National Historical Chemical Landmarks.
Shan, D. (2005). *PNAS.* **102(10):** 3772-3776.
Shier, W.T., Rinehart, K.L., Jr. and Gottlieb, D. (1969). Proceedings of National Academy of Sciences. USA **63:** 198-204.
Shier, W.T., Schaefer, P.C., Gottlieb, D. and Rinehart, K.L., Jr. (1974). *Biochemistry* **13:** 5073-5078.
Stackebrandt, E., Liesack, W. and Witt, D. (1992). *Gene,* **115:** 255-260.
Stackebrandt, E. and Goebel, B.M. (1994). *International J. Systematic Bacteriology* **44:** 846-849.
Stackebrandt, E., Rainey, F.A. and Ward-Rainey, N.L. (1997). *International J. Systematic Bacteriology,* **47:** 479-491.
Steingrube, V.A., Wilson, R.W., Brown, B.A., Jost, Jr. K.C., Blacklock. Z., Gibson, J.L. and Wallace, Jr. R.J. (1997). *J. Clinical Microbiology,* **35:** 817-822.
Sugie, M. and Suzuki, H. (1978). *FEMS Microbiology Reviews,* **25:** 39-67.
Suutari, M., Lignell, U., Hirvonen, M.R. and Nevalainen, A. (2000). *ASM News* **66(10):** 588.
Gregory, L. Challis and David, A. (2003). *PNAS.* **100:** 14555-14561.
Satoh, A., Murakami, T., Takebe, H., Imai, S. and Satoa, H. (1993). *Actinomycetologica,* **7:** 128-132.
Siu, M.C., Wang, S.S., Lu, S.B. and Hsu, Y.H. (1997). *J. Chinese Agri. Chem. Soc.,* **35(3):** 243-251.
Saadoun, I., Al-Momani, F., Malkawi, H.I. and Mohammad, M.J. (1999). *Microbios,* **100(395):** 41-46.
Shimizu, M., Nakagawa, Y., Sato, Y., Furumai, T., Igarashi, Y., Onaka, H., Yoshida, R. and Kunch, H. (2000). *J. Gen. Pl. Pathol.,* **66(4):** 360-366.
Taguchi, S., Kojima, S., Terabe, M., Kumazawa, Y., Kohriyama, H., Suzuki, M., Miura, K.I. and Momose, H. (1997). *J. Molecular Evolution,* **44:** 542-551.
Takeuchi, T., Sawada, H., Tanaka, F. and Matsuda, I. (1996). *International J. Systematic Bacteriology,* **46:** 476-479.

Tatsuta, K., Takahashi, M., Tanaka, N. and Chikauchi, K. (2000). *J. Antibiot.* **53(10):** 1231-4.

Theriault, R.J., Karwowski, J.P., Jackson, M., Girolami, R.L., Suguna, G.N., Vojtko, C.M. and Coen, L.J. (1987). *J. Antibiot.,* **40(5):** 567-574.

Tomioka, H., Kagawa, M. and Nakamura, S. (1976). *J. Biochemistry,* **79:** 903-915.

Tang, Y.Q., Sattler, I., Thiericke, R., Grabley, S. and Feng, X.Z. (2000). *J. Antibiot.,* **53(9):** 934-943.

Uyeda, M., Mizukami, M, Yokomizo, K. and Suzuki, K. (2001). *Bioscience, Biotechnology and Biochemistry,* **65:** 1252-1254.

Ubukata, M., Morita, T.I. and Osada, H. (1995a). *J. Antibiot.,* **48(4):** 293-299.

Vezina, C., Kudelski, A. and Sehgal, S.N. (1975). *J. Antibiot.,* **28:** 721-726.

Vining, L.C. (1990). *Ann. Rev. Microbiology,* **44:** 395-427.

Walton, R.B., Mc Daniel, L.E. and Woodruff, H.B. (1962). *Developments in Industrial Microbiology,* **3:** 370-375.

Wang, Y., Zhang, Z.S., Ruan, J.S., Wang, Y.M. and Ali, S.M. (1999). *J. Industrial Microbiology and Microbiology,* **23:** 178-187.

Wellington, E.M.H., Stackebrandt, E., Sanders, D., Wolstrup, J. and Jorgensen, N.O.G. (1992). *International J. Systematic Bacteriology,* **42:** 156-160.

Werner, G., Hagenmaier, H., Drautz, H., Baumgartner, A. and Zähner, H. (1984). *J. Antibiotics,* **37(2):** 110-117.

Williams, S.T., Goodfellow, M., Wellington, E.M.H., Vickers, J.C., Alderson, G. and Sneath, P.H.A. *et al.* (1983). *J. General Microbiology,* **129:** 1815-1830.

Williams, S.T. (1985). *International Biodeterioration,* **21:** 201-209.

Williams, S.T., Goodfellow, M. and Alderson, G. (1989). *In:* Bergey's manual of systematic Bacteriology, (Eds. Williams, S.T., Sharpe, M.E. and Holt, J.G.) vol. 4, Williams & Wilkins, Baltimore, pp. 2452-2492.

Witt, D. and Stackebrandt, E. (1990). *Systematic and Applied Microbiology,* **13:** 361-371.

Wood, S., Williams, S.T. and White, W.R. (1985). *J. Appl. Microbiology,* **58:** 319-326.

Woo, J.H. and Kamei, Y. (2003). *Appl. Microbiol. Biotechnol.*, **62(4):** 407-413.

Xu, L-H., Li, Q-R. and Jiang, C-L. (1996). *Appl. and Environmental Microbiology* **62:** 244-248.

Xiao, K., Kinkel, L.L. and Samac, D.A. (2002). Biologi. *Cont.*, **23:** 285-295.

Yoshida, T., Kumura, Y. and Katagiri, K. (1968). *J. Antibiot.,* **21:** 465-467.

Yoshida, N., Tani, Y. and Ogata, K. (1972). *J. Antibiot.,* **25:** 653-659.

Yoshimoto, A., Fujii, S., Johdo, O., Kubo, K., Ishikura, T., Naganawa, H., Sawa, T., Takeuchi, T. and Umezawa, H. (1986). *J. Antibiot.,* **39:** 902-909.

Yuan, W.M. and Crawford, D.L. (1995). *Appl. and Environmental Microbiology* **61:** 3119-3128.

Yon, C, Suh, J.W., Chang, J.H., Lim, Y., Lee, C.H., Lee, Y.S. and Lee, Y.W. (1995). *J. Antibiot.,* **48(8):** 773-779.

Zhang, Z., Wang, Y. and Ruan, J. (1997). *International J. Systematic Bacteriology* **47:** 1048-1054.

Zheng, Z., Zeng, W., Huang, Y., Yang, Z., Li, J., Cai, H. and Su, W. (2000). *FEMS Microbiol. Lett.,* **188(1):** 87-91.

Zimmermann, R. and Cox, E. (1994). *Nucleic Acids Res.* **22(3):** 492-7.

Chapter 10

Microbial Proteases and their Applications

A. MEHTA

Microbial Technology Laboratory, Department of Botany,
Dr. H.S. Gour University, Sagar–470 003 (India)
E-mail: pradeepunisagar@rediffmail.com

Enzymes are macromolecules, with highiy specialized catalytic functions produced by all living organisms. Enzymes are responsible for many essential biochemical reactions in microorganisms, plants, animals and human beings. The role of microbial enzymes in many processes has been known for a long time, although enzymes have been among us since time immemorial. Enzymes are essential for all metabolic processes. This catalytic capability makes enzymes unique. Enzymes play a diversified role in many aspects of everyday life including aiding in digestion, the production of food and several industrial applications. Enzymes are used for thousands of years to carry out important chemical reactions for making products such as cheese, bear, and wine. Bread and yogurt also owe their flavor and texture to a range of enzyme producing organisms that were domesticated many years ago. Temperature has profound influence on the production of protease by microorganisms. The pH of culture strongly affects many enzymatic processes and transport of compounds across the cell membrane. Enzyme inhibitors are also very useful in elucidating metabolic pathways in cells. Moreover, some drugs useful in medicine appear to function as they can inhibit certain enzymes in malfunctioning of the cells. Inhibition studies primarily give an insight into the nature of an enzyme, its co-factor requirement and the nature of active center. The extracellular proteases are of commercial value and find multiple applications in various industrial sectors.

INTRODUCTION

Enzymes are macromolecules with highly specialized catalytic functions, produced by all living organisms. Enzymes are responsible for many essential biochemical reactions in microorganisms, plants, animals and human beings. The role of enzymes in many processes has been known for long. Their existence was associated with the history of ancient Greece where they used enzymes from microorganisms in various processes. Ever since antiquity, mankind has made use of the catalytic, or "modifying", property of food. Humankind has used them for thousands of years to carry out important chemical reactions for making products such as cheese, beer, and wine. Bread and yogurt also owe their flavor and texture to a range of enzyme producing organisms that were domesticated many years ago. Enzymes are essential for all metabolic processes. Although like all other proteins, enzymes are composed of amino acids, they differ in a function in that they have the unique ability to facilitate biochemical reaction without undergoing change themselves. This catalytic capability makes enzymes unique. Enzymes play diversified role in many aspects of everyday life including aiding in digestion, production of food and several industrial applications. Enzymes are nature's catalysts.

Microorganisms represent an excellent source of enzymes owing to their broad biochemical diversity and their susceptibility to genetic manipulation. The inability of the plant and animal proteases to meet current world demands has led to an increased interest in microbial proteases. Microbial proteases account for approximately 40% of the total worldwide enzyme sales (Godfrey and West 1996). Proteases from microbial sources are preferred than plant and animal sources since they possess almost all the characteristics desired for their biotechnological applications. Proteinases are found in several microorganisms such as viruses, protozoa, bacteria, yeast and fungi.

Bacteria

Organisms belonging to the genus *Bacillus* produce most commercial proteases, mainly neutral and alkaline. Bacterial neutral proteases are active in a narrow pH range (pH 5 to 8) and have relatively low thermotolerance. Due to their intermediate rate of reaction, neutral proteases generate less bitterness in hydrolyzed food proteins than do the animal proteinases and hence are valuable for use in the food industry. Neutrase, a neutral protease, is insensitive to the natural plant proteinase inhibitors and is therefore useful in the brewing industry. The bacterial neutral proteases are characterized by their high affinity for hydrophobic amino acid pairs. Their low thermotolerance is advantageous for controlling their reactivity during the production of food hydrolysates with a low degree of hydrolysis. Some of the neutral proteases belong to the metalloprotease type and require divalent metal ions for their activity, while others are serine proteinases, which are not affected by chelating agents. Bacterial alkaline proteases are characterized by their high activity at alkaline pH (10) and their broad substrate specificity. Their optimal temperature is around 60°C. These properties of bacterial alkaline proteases make them suitable for use in the detergent industry.

Fungi

Fungi elaborate a wider variety of enzymes than bacteria. For example, *Aspergillus oryzae* produces acid, neutral, and alkaline proteases. The fungal proteases are active over a wide pH range (4 to 11) and exhibit broad substrate specificity. However, they have a lower reaction rate and worse heat tolerance than the bacterial enzymes. Fungal enzymes can be conveniently produced in a solid-state fermentation process. Fungal acid proteases have an optimal pH between 4 and 4.5 and are stable between pH 2.5 and 6. They are particularly useful in the cheese making industry due to their narrow pH and temperature specificities. Fungal neutral proteases are metalloproteases that are active at pH 7 and are inhibited by chelating agents. In view of the accompanying peptidase activity and their specific function in hydrolyzing hydrophobic amino acid bonds, fungal neutral proteases supplement the action of plant, animal, and bacterial proteases in reducing the bitterness of food protein hydrolysates. Fungal alkaline proteases are also used in food protein modification.

Viruses

Viral proteases have gained importance due to their functional involvement in the processing of proteins of viruses that cause certain fatal diseases such as AIDS and cancer. Serine, aspartic acid, and cysteine peptidases are found in various viruses (Rawlings and Barrett, 1993). All of the virus-encoded peptidases are endopeptidases. Retroviral aspartyl proteases that are required for viral assembly and replication are homodimers and are expressed as a part of the polyprotein precursor. The mature protease is released by autolysis of the precursor. An extensive literature is available on the expression, purification, and enzymatic analysis of retroviral aspartic protease and its mutants (Kuo and Shafer, 1994). Extensive research has focused on the three-dimensional structure of viral proteases and their interaction with synthetic inhibitors with a view to design potent inhibitors that can combat the relentlessly spreading and devastating epidemic of AIDS. Although proteases are widespread in nature, microbes serve as a preferred source of these enzymes because of their rapid growth, the limited space required for their cultivation, and the ease with which they can be genetically manipulated to generate new enzymes with altered properties that are desirable for their various applications.

Enzymes are proteins with highly specialized catalytic functions produced by all living organisms. Functionally and quantitatively, proteins represent one of the most abundant classes of biomolecules. Protein degrading (proteolytic) enzymes constitute the single most important group of industrial enzymes currently in use. Microorganisms produce proteolytic enzymes which are capable of hydrolyzing peptide bonds into amino acids (Hankin and Anagnstakis, 1976) based upon the size of the molecules that they can attack or preferably attack. These enzymes have generally been treated as consisting of two distinct groups, the proteinases and the peptidases. The distinction between proteinases and peptidases is not very realistic. Enzymes not only work efficiently and rapidly, but they are also biodegradable. Enzymes are highly efficient in increasing the reaction rate of biochemical processes that otherwise proceed very slowly or in some case, not at all. Over 700 enzymes are known and 100 have been crystallized. Enzymes are globular proteins with MW ranging from 10,000 to millions Daltons. Microbial enzymes may be divided into two groups depending upon whether they are secreted into the surrounding medium or remain confined within the cell. The enzymes that belong to the first group are known as extracellular enzymes which are capable of producing their specific actions in the complete absence of living cell. The later group is called intracellular enzymes or endoenzymes as they are produced within the cell. Proteases represent the class of enzyme, which occupies a pivotal position with respect to their physiological role as well as their commercial applications. They perform both degradative and synthetic functions and they are physiologically necessary for living organisms.

Proteases and proteolysis play a vital role in biology of all cells. The action of protease is divided into:

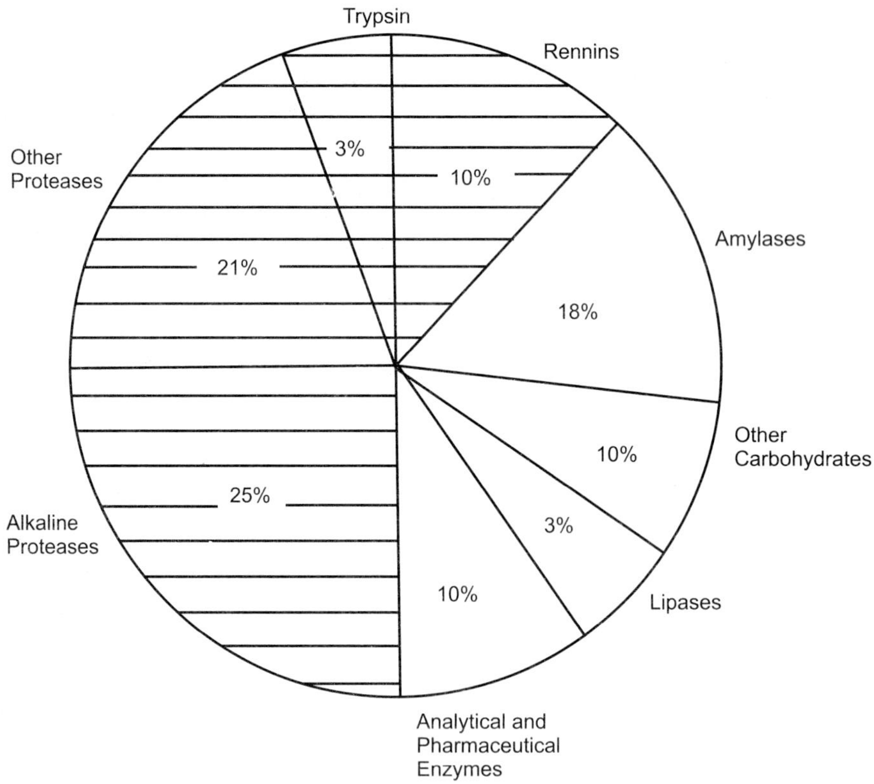

Fig. 1. Distribution of enzymes sale. The contribution of different enzymes to the total sale of enzymes is indicated. The shaded portion indicates the total sale proteases.

1. **Limited proteolysis**, in which they split only a small number of peptide bonds of a target protein.
2. **Non-limited proteolysis**, those which show complete degradation of a target protein.

The role of protease/proteinase enzymes is to digest nutrient proteins or to participate in the turn over of cellular proteins (North, 1982). Proteolysis has a key role in a wide range of cellular processes (Holzer and Tschenesche 1979). According to the nomenclature committee of the International Union of Biochemistry and Molecular Biology (1992), proteases are classified on the basis of their huge diversity of action and structure into three major criteria:

(i) Type of reaction catalyzed
(ii) Chemical nature of the catalytic site
(iii) Evolutionary relationship with reference to structure (Barett 1994)

Protease enzyme is grossly classified into two major groups, i.e., exopeptidases and endopeptidases depending on their site of action.

Table 1. Classification of proteases.

Protease	Mode of Action*	EC No.
Exopeptidases		
Aminopeptidases	●↓○-○-○-○---	3.4.11
Dipeptidyl peptidase	●-●↓○-○-○---	3.4.14
Tripeptidyl peptidase	●-●-●↓○-○---	3.4.14
Carboxypeptidase	---○-○-○-○-○↓●	3.4.16–3.4.18
Serine type protease		3.4.16
Metalloprotease		3.4.17
Cysteine type protease		3.4.18
Peptidyl dipeptidase	---○-○-○-○↓●-●	3.4.15
Dipeptidases	●↓●	3.4.13
Omega peptidases	+-●↓○-○---	3.4.19
	---○-○-○↓●-+	3.4.19
Endopeptidases	---○-○-○↓○-○-○---	3.4.21–3.4.34
Serine protease		3.4.21
Cysteine protease		3.4.22
Aspartic protease		3.4.23
Metalloprotease		3.4.24
Endopeptidases of unknown catalytic mechanism		3.4.99

*Open circles represent the amino acid residues in the polypeptide chain. Solid circles indicate the terminal amino acids, and stars signify the blocked termini. Arrows show the sites of action of the enzyme.

1. Peptidases or Exopeptidases

Peptidases or exopeptidases cleave the peptide bond proximal to the amino or carboxyl termini of the substrates. They are further classified as amino peptidases and carboxy peptidases.

(i) Aminopeptidases: Aminopeptidases act at a free N terminus of polypeptide chain and liberate a single amino acid residue, a dipeptide, or a tripeptide. Aminopeptidases occur in a wide variety of microbial species including bacteria and fungi (Ballinger *et al.,* 1996).

(ii) Carboxypeptidases: The carboxypeptidases act at C terminals of the polypeptide chain and liberate a single amino acid or dipeptide. Carboxyl can be divided into three major groups, serine carboxyl peptidases, metal carboxyl peptidases and cysteine carboxypeptidases, based on the nature of the amino acid residues at the active site of the enzyme. The enzyme can also hydrolyze the peptides in which the peptide group is replaced by an acyl group.

2. Proteinases or Endopeptidases

Endopeptidases are characterized by their prefential action at the peptide bond in the inner regions of the polypeptide chain away from the N and C termini. The presence of free amino or carboxyl group has a negative influence on enzyme activity. The endopeptidases are divided into four subgroups based on their catalytic mechanisms. The four major groups of protease are classified on the basis of active sites and sensitivity to various inhibitors (Hase and Finkelstein 1993).

(a) Serine protease (EC 3.4.21): Serine proteases are characterized by the presence of a serine group in their active site. They are numerous and widespread among viruses, bacteria and eukaryotes, suggesting that they are vital to the organisms. Serine proteases are found in the exopeptidases, endopeptidase oligopeptidase, and omega peptidase groups. They may be divided into a number of subfamilies, based upon structural similarities. Serine proteases are generally active at neutral and alkaline pH, with an optimum pH 7 and 11. The isoelectric points of serine proteases are generally between pH 4 and 6. Serine alkaline proteases that are active and highly alkaline represent the largest sub group of serine proteases. From an applied perspective, the bacterial subtilisins are subgroup of serine proteases of greatest industrial significance. Subtilisins, particularly those produced by selected bacilli, have found widespread application as detergent additives. Serine proteases are also recognised by their irreversible inhibition by 3, 4 dichloro isocoumarine (3, 4-DCI), di-isopropyl fluorophosphate (DFD) and phenyl methyl sulfonyl fluoride (PMSF). They are resistant to denaturating agents (50% ethanol and anionic detergents).

(b) Aspartic protease (EC 3.4.23): Aspartic proteases are a group of acidic protease that contain an essential aspartic acid residue at the catalytic site. Most aspartic proteases display maximum activity at pH values between 3 and 4, and have isoelectric points between pH 3 and 4.5. They generally exhibit molecular masses of 30-45 kDa. The aspartic proteases are inhibited by pepstatin (Fitzgerald *et al.*, 1990). Microbial aspartic proteases can be broadly divided into two groups: (i) pepsin-like enzymes produced by *Aspergillus*, *Penicillium*, *Rhizopus* and *Neurospora* and (ii) renin-like enzyme produced by *Mucor* sp. They are also sensitive to diazoketone compounds such as diazoacetyl–DL-norleucine methyl ester (DAN) and 1, 2-epoxy-3-(p-nitrophenoxy) propane (EPNP) in the presence of copper ion.

(c) Cysteine/Thiol protease (EC 3.4.22): Cysteine proteases are widely distributed in nature and are characterized by the presence of a cysteine and histidine residue at the active site, which forms a catalytic dyad essential for biological activity. Generally, cysteine protease is active only in the presence of reducing agents such as HCN or cysteine. Based on their side chain specificity, they are broadly divided into four groups – (i) papain-like, (ii) trypsin-like with preference for cleavage at the arginine residue (iii) specific to glutamic acid and (iv) others. Papain is the best known cysteine protease. Cysteine proteases have neutral pH, and optimal inhibitors of these protease are p-chloromercuri benzoate indoacetamide. These proteases have limited industrial applications. They have molecular mass between 30,000-45,000 Daltons. The active sites of these enzymes are cysteine residues.

(d) Metallo-protease (EC 3.4.24): Metallo-proteases are the most diverse of the catalytic types of proteases (Barett 1995). They are characterized by their requirement for (divalent) metal ions to sustain biological activity. They include enzymes from a variety of origins such as collagenases from higher organisms and thermolysin from bacteria, a very stable neutral protease (Shanon *et al.,* 1989). These are maximally active at neutrality having molecular mass between 19–37 kDa. Most of the metallo-proteases display maximum activity at neutral to alkaline pH. Zn^{2+} is essential for its activity and Ca^{2+} is important for maintaining stability. Neutral metalloproteases are specific towards hydrophobic bulky amino acid. Metalloproteases are produced by several *Bacillus* species like *B. polymyxa,* (syn. *Paenibanttus polymyxa*) *B. amyloliquefaciens, B. cereus*, while *B. thermoproteolyticus* produces a remarkable protease "thermolysis" with high thermostability upto 80°C. Metalloprotease from *Bacillus* sp. have been developed industrially for use as rennet substitutes, but it is abandoned due to excessive casein hydrolysis. At present these proteases are used for bating in tanneries to hydrolyse barley protein in breweries and for proteolysis in the food industry. They are inhibited by chelating agents like EDTA, phosphoramide, etc.

Microbial proteases can be divided into three groups based on pH range in which their activity is high, namely acid, neutral or alkaline proteases, which are briefly summarised below:

1. Acidic Protease

These are mainly of fungal origin exhibiting maximum stability at pH 2.5-4.5 and having a molecular weight of around 35,000 daltons. They are low in basic amino acid content and have low isoelectric points, but can hydrolyze a wide range of peptide bonds. These are not inhibited by di-isopropylflurophosphate, ethylene diaminetetraacetic acid or orthophenanthroline. These are much like animal enzymes and can be split into two subgroups, pepsin and renin-like acid proteases. Aspartic acid is present in their active centers.

2. Neutral Protease

They are widely spread both in bacteria and fungi having molecular weight in the range of 40,000–45,000 Da and several have been shown to be zinc containing metallo enzymes (Keay *et al.*, 1970). They are inhibited by orthophenanthroline and EDTA but not by diisopropyl-flurophosphate. They exhibit some specificity towards bonds adjacent to leucine or phenyl alanine amino groups. These are used for food processing and have been shown to be active components in dental plague reaction, e.g., *Bacillus subtilis* and *Bacillus thermoproteolyticus,* which produce neutral protease montase and thermoase, respectively.

3. Alkaline Protease

These are frequent among bacteria and fungi. They resemble the animal enzymes, trypsin and chymotrypsin. The alkaline proteases have molecular weight in the range of 25,000-30,000 Daltons. The most studied are subtilisin carlsberg and subtilisin novo, produced by various *Bacilli* (Keay *et al.,* 1970). The alkaline proteases have found extensive uses in detergents, leather tanning and food industries. Alkaline proteases are the thermostable protease of *Streptomyces rectus* (Mizisawa *et al.*, 1964). "Enzyme thermolysin" produced by *B. thermoproteolyticus* can retain 90% activity after 30 min at 70°C and thus find extensive usage in food industries.

The vast diversity of proteases in contrast to the specificity of their action has attracted worldwide attention in attempt to exploit their physiological and biotechnological applications (Fox *et al.*, 1991; Poldermans 1990). In view of the recent trend of development, they play diversified role in many aspects of everyday life. Proteases are envisaged to have extensive applications in several bioremediation processes; however, the worldwide requirement of enzymes for individual application varies considerably.

MODE OF ACTION OF PROTEASES

The mechanism of action of proteases has been a subject of great interest. Purification of proteases to homogeneity is a prerequisite for studying their mechanism of action. Vast numbers of purification procedures for proteases, involving affinity chromatography, ion-exchange chromatography, and gel filtration techniques have been well documented. Preparative polyacrylamide gel electrophoresis has been used for the purification of proteases from *Conidiobolus coronatus* (Phadatare *et al.*, 1992). Purification of staphylocoagulase to homogeneity was carried out from culture filtrates of *Staphylococcus aureus* by affinity chromatography with a bovine prothrombin-Sepharose 4B column (Igarashi *et al.*, 1979) and gel filtration (Zen-Yoji *et al.*, 1961). A number of peptide hydrolases have been isolated and purified from *E. coli* by DEAE-cellulose chromatography (Perlmann and Lorand, 1970).

The catalytic site of proteases is flanked on one or both sides by specificity subsites, each able to accommodate the side chain of a single amino acid residue from the substrate. These sites are numbered from the catalytic site S_1 through Sn toward the N terminus of the structure and S_1' through Sn' toward the C terminus. The residues which they accommodate from the substrate are numbered P_1 through Pn and P_1' through Pn', respectively (Fig. 2).

Protease : N Sn ---- $S_3 - S_2 - S_1$ * $S'_1 - S'_2 - S'_3$ ---- S′n C

Substrate : N Pn ---- $P_3 - P_2 - P_1$ ---+--- $P'_1 - P'_2 - P'_3$ ---- P′n C

Fig. 2. Active sites of proteases. The catalytic site of proteases is indicated by * and the scissile bond is indicated by + ; S_1 through Sn and S_1' through Sn' are the specificity subsites on the enzyme, while P_1 through Pn and P_1' through Pn' are the residues on the substrate accommodated by the subsites on the enzyme.

Serine Proteases

Serine proteases usually follow a two-step reaction for hydrolysis in which a covalently linked enzyme-peptide intermediate is formed with the loss of the amino acid or peptide fragment (Fastrez and Fersht, 1973). This acylation step is followed by a deacylation process which occurs by a nucleophilic attack on the intermediate by water, resulting in hydrolysis of the peptide. Serine endopeptidases can be classified into three groups based mainly on their primary substrate preference: (i) trypsin-like, which cleave after positively charged residues; (ii) chymotrypsin-like, which cleave after large hydrophobic residues; and (iii) elastase-like, which cleave after small hydrophobic residues. The P_1 residue exclusively dictates the site of peptide bond cleavage. The primary specificity is affected only by the P_1 residues; the residues at other positions affect the rate of cleavage. The subsite interactions are localized to specific amino acids around the P_1 residue to a unique set of sequences on the enzyme. Some of the serine peptidases from *Achromobacter* sp. are lysine-specific enzymes (Masaki *et al.*, 1978), whereas those from *Clostridium* sp are arginine specific (Gilles *et al.*, 1979), and those from *Flavobacterium* sp are post proline-specific (Yoshimoto *et al.*, 1980). Endopeptidases that are specific to glutamic acid and aspartic acid residues have also been found in *B. licheniformis* and *S. aureus* (Drapeau *et al.*, 1972). The recent studies based on the three-dimensional structures of proteases and comparisons of amino acid sequences near the primary substrate-binding site in trypsin-like proteases of viral and bacterial origin suggest a putative general substrate binding scheme for proteases with specificity towards glutamic acid involving a histidine residue and a hydroxyl function. However, a few other serine proteases such as peptidase A from *E. coli* and the repressor LexA show distinctly different mechanism of action without the classic Ser-His-Asp triad (Barett, 1994). Some of the glycine residues are conserved in the vicinity of the catalytic serine residue, but their exact positions are variable (Brenner, 1988). The chymotrypsin-like enzymes are confined almost entirely to animals, the exceptions being trypsin-like enzymes from actinomycetes, *Saccharopolyspora* sp and from the fungus *Fusarium oxysporum*. A few of the serine proteases belonging to the subtilisin family show a catalytic triad composed of the same residues as in the chymotrypsin family; however, the residues occur in a different order (Asp-His-Ser). Some members of the subtilisin family from the yeasts *Tritirachium* and *Metarhizium* sp require thiol for their activity. The thiol dependance is attributable to Cys173 near the active-site histidine (Jany *et al.*, 1986).The carboxypeptidases are unusual among the serine-dependent enzymes in that they are maximally active at acidic pH. These enzymes are known to possess a Glu residue preceding the catalytic Ser, which is believed to be responsible for their acidic pH optimum. Although the majority of the serine proteases contain the catalytic triad Ser-His-Asp, a few use the Ser-base catalytic dyad. The Glu-specific proteases display a pronounced preference for Glu-Xaa bonds over Asp-Xaa bonds (Austew and Smith, 1976).

Aspartic Proteases

Aspartic endopeptidases depend on the aspartic acid residues for their catalytic activity. A general base catalytic mechanism has been proposed for the hydrolysis of proteins by aspartic proteases such as penicillopepsin (James *et al.*, 1977) and endothiapepsin (Pearl 1987). Crystallographic studies have shown that the enzymes of the pepsin family are bilobed molecules with the active-site cleft located between the lobes and each lobe contributing one of the pair of aspartic acid residues that is essential for the catalytic activity (Blundell *et al.*, 1991; Sielecki *et al.*, 1991). The lobes are homologous to one another, having arisen by gene duplication. The retropepsin molecule has only one lobe, which carries only one aspartic residue, and the activity requires the formation of a noncovalent homodimer (Miller *et al.*, 1989). In most of the enzymes from the pepsin family, the catalytic Asp residues are contained in an Asp-Thr-Gly-Xaa motif in both the N- and C-terminal lobes of the enzyme, where Xaa is Ser or Thr, whose side chains can hydrogen bond to Asp. However, Xaa is Ala in most of the retropepsins. A marked conservation of cysteine residue is also evident in aspartic proteases. The pepsins and the majority of other members of the family show specificity for the cleavage of bonds in peptides of at least six residues with hydrophobic amino acids in both the Pl and Pl' positions (Keil, 1992).

The specificity of the catalysis has been explained on the basis of available crystal structures (Liu *et al.*, 1996). The structural and kinetic studies have also suggested that the mechanism involves general acid-base catalysis with lytic water molecule that directly participates in the reaction (Fig. 3A). This is supported by the crystal structures of various aspartic protease-inhibitor complexes and by the thiol inhibitors mimicking a tetrahedral intermediate formed after the attack by the lytic water molecule (James *et al.*, 1992).

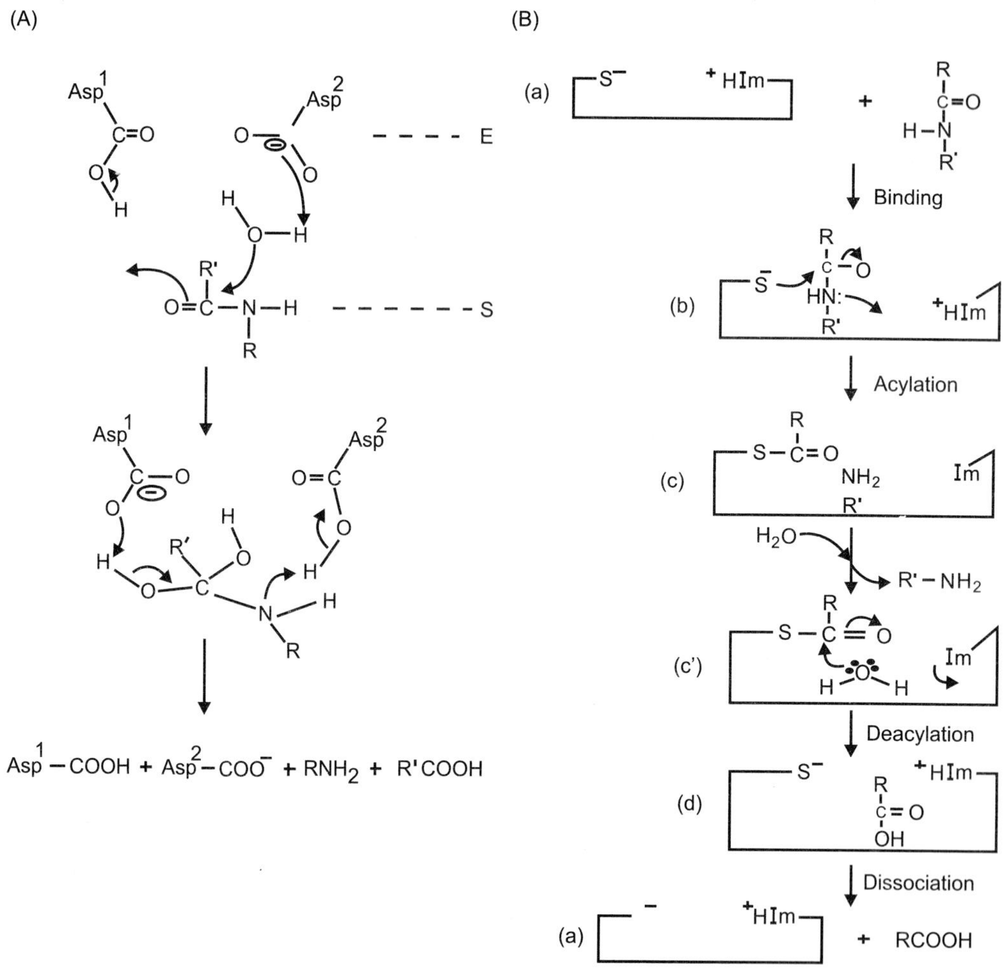

Fig. 3. Mechanism of action of proteases. (A) Aspartic proteases. (B) Cysteine proteases. Im and ⁺Hlm refer to the imidazole and protonated imidazole, respectively.

Metalloproteases

The mechanism of action of metalloproteases is slightly different from that of the above-described proteases. These enzymes depend on the presence of bound divalent cations and can be inactivated by dialysis or by the addition of chelating agents. For thermolysin, based on the X-ray studies of the complex with a hydroxamic acid inhibitor, it has been proposed that Glu143 assists the nucleophilic attack of a water molecule on the carbonyl carbon of the scissile peptide bond, which is polarized by the Zn^{2+} ion (Holmes and Matthews, 1981). Most of the metalloproteases are enzymes containing the His-Glu-Xaa-Xaa-His (HEXXH) motif, which has been shown by X-ray crystallography to form a part of the site for binding of the metal, usually zinc.

Cysteine Proteases

Cysteine proteases catalyze the hydrolysis of carboxylic acid derivatives through a double-displacement pathway involving general acid-base formation and hydrolysis of an acyl-thiol intermediate. The mechanism of action of cysteine proteases is thus very similar to that of serine proteases.

A striking similarity is also observed in the reaction mechanism for several peptidases of different evolutionary origins. The plant peptidase papain can be considered the archetype of cysteine peptidases and constitutes a good model for this family of enzymes. They catalyze the hydrolysis of peptide, amide ester, thiol ester, and thiono ester bonds (Phadatare *et al.*, 1992). The initial step in the catalytic process (Fig. 3B) involves the noncovalent binding of the free enzyme (structure a) and the substrate to form the complex (structure b). This is followed by the acylation of the enzyme (structure c), with the formation and release of the first product, the amine R'-NH_2. In the next deacylation step, the acyl-enzyme reacts with a water molecule to release the second product, with the regeneration of free enzyme.

The enzyme papain consists of a single protein chain folded to form two domains containing a cleft for the substrate to bind. The crystal structure of papain confirmed the Cys25-His159 pairing (Baker and Drenth, 1987). The presence of a conserved aspargine residue (Asn175) in the proximity of catalytic histidine (His159) creating a Cys-His-Asn triad in cysteine peptidases is considered analogous to the Ser-His-Asp arrangement found in serine proteases.

Studies on the mechanism of action of proteases have revealed that they exhibit different types of mechanism based on their active-site configuration. The serine proteases contain a Ser-His-Asp catalytic triad, and the hydrolysis of the peptide bond involves an acylation step followed by a deacylation step. Aspartic proteases are characterized by an Asp-Thr-Gly motif in their active site and by acid-base catalysis as their mechanisms of action. The activity of metalloproteases depends on the binding of a divalent metal ion to a His-Glu-Xaa-Xaa-His motif. Cysteine proteases adopt a hydrolysis mechanism involving a general acid-base formation followed by hydrolysis of an acyl-thiol intermediate.

PHYSIOLOGICAL FUNCTIONS OF PROTEASES

Proteases execute a large variety of complex physiological functions. Their importance in conducting the essential metabolic and regulatory functions is evident from their occurrence in all forms of living organisms. Proteases play a critical role in many physiological and pathological processes such as protein catabolism, blood coagulation, cell growth and migration, tissue arrangement, morphogenesis in development, inflammation, tumor growth and metastasis, activation of zymogens, release of hormones and pharmacologically active peptides from precursor proteins, and transport of secretory proteins across membranes. In general, extracellular proteases catalyze the hydrolysis of large proteins to smaller molecules for subsequent absorption by the cell, whereas intracellular proteases play a critical role in the regulation of metabolism.

CHARACTERIZATION OF PROTEASES

Proteases are difficult to characterize because of their diversity of action and structures. Originally proteases were classified based on molecular size, charge or substrate specificity. However, with the advent of molecular biology, proteases are now grouped into families based on chemical nature of the catalytic or active sites, mechanism(s) of action and the evolutionary relationship of their three-dimensional structure (Rao *et al.*, 1998). Although enzymes derived from a range of organism are being used commercially, certain properties of these proteins can be further improved. For example, it may be advantageous to increase the heat stability or alter the pH optimum of an enzyme, thus improving the efficiency of a defined process. Alternatively, it may be desirable to enhance the activity of enzyme towards one particular substrate or potential substrate. The extent to which a particular active site, side chain participates in substrate, product and transition-state, stabilization can only be achieved by using interdisciplinary mutagenic, kinetic and structural approaches (Wanger and Benkovic, 1990). General principles formulated from such studies are being applied to the design and engineering of novel biocatalysts. Recent advances in genetic/protein engineering allow reliable approaches to modification of the primary structure of enzymes, thus altering some of their physico-chemical and catalytic properties (Singh and Hayashi, 1995; Kuhad *et al.,* 1997). Several methods have been developed for the modification of enzyme activity and stability by using chemical additives, immobilization, and protein engineering. Many of the goals of industrial protein engineering are to design and construct novel enzymes by modifying the properties of existing enzymes. For example, it may be desirable to alter the kinetic properties K_{cat} and K_m, the substrate specificity, the pH optimum, the temperature stability, the stability in the presence of chemical reagent and the isoelectric point of an enzyme (Singh, 1999).

These enzymes are important in a number of diverse and crucial biological processes. They are involved in the regulation of metabolism and gene expression, enzyme modification, pathogenicity, and the hydrolysis of large proteins to smaller molecules for transport and metabolism (Rao *et al.,* 1998). Proteases are essential constituents of all forms of life on earth including prokaryotes, fungi, plants and animals. Proteases are highly exploited enzymes in food, leather detergent, pharmaceutical diagnostics, waste management and silver recovery. The extracellular proteases are of commercial value and find multiple applications in various industrial sectors.

Proteins are degraded by microrganisms, and they utilize the degradation products as nutrients for their growth. The degradation is initiated by proteinases (endopeptidases) secreted by microrganisms followed by further hydrolysis by peptidases (exopeptidases) at the extra- or intracellular site. Numerous proteinases are produced by microrganisms depending on the species of the producers or the strains even belonging to the same species. Several proteinases are also produced by the same strain under various cultural conditions. *Candida albicans* and *C. tropicalis* are the medically more important opportunistic pathogens causing infections in immuno-compromised patients. Their extracellular enzyme, an aspartic proteinase is considered to be a major virulence factor. Human immunodeficiency virus (HIV) a causative agent of AIDS, presents an aspartic protease essential for the retroviral life cycle and it has been a good target for the chemotherapy with specific inhibitors. Most commercial serine proteases, mainly neutral and alkaline, are produced by organisms belonging to the genus *Bacillus*. Representatives of this group are subtilisins. Similar enzymes are also produced by other bacteria such as *Thermus caldophilus* and *Desulfurococus mucosus*, *Streptomyces*, *Aeromonas* and *Escherichia* genera. Fungi produce several serine proteinases. Among them, these enzymes are produced by various strains of *Aspergillus oryzae*. Cysteine proteinases are not so widely distributed as was seen with serine and aspartic proteinases. However, in protozoa belonging to the Trypanosomatidae family they are found as well as metaloproteinases. In several genus such as *Crithidia, Phytomonas, Herpetomonas, Trypanosoma, Leishmania, Leptomonas and Endotrypanum* are involved with the nutrition, life cycle and morphological differentiation of these parasites. *Trichomonas vaginalis*, a flagellated protozoan responsible for trichomonosis, one of the most common sexually transmitted diseases, has numerous cysteine- and some metallo-proteinases. The cysteine enzyme is involved in the damage to the host by the parasite. Sufficient data have been gathered to demonstrate unequivocally that proteases are ubiquitous in microrganisms (Rao *et al.*, 1998).

Nutrition

Nutritional factors play an important role for the secretion of enzymes. Synthetic media have a great advantage overs complex media in that consistency of processes and production is enhanced through avoiding the variability of complex substrates. Thus, synthetic media provides better control and monitoring, improved product recovery and quality, and simplified purification system (Zhang and Greasham, 1999). Mao *et al.* (1992) also reported high production of alkaline protease by *Bacillus licheniformis* in fed-batch fermentation using synthetic media. Ferrero *et al.* (1996) isolated and characterized the thermostable alkaline protease of *Bacillus licheniformis* MIR 29. Protease production by *B-licheniformis* S-40 was reduced to half, is maximum level when glucose was present as carbon source, where soyabean meal was the best nitrogen source for protease production (Sonnleitner, 1983). A similar observation was made by many workers (Kalisz 1988; Lee and Chang, 1990) who observed a significant increase in protease production by the addition of yeast extract, corn steep liquor, bonito extract to soyabean meal. Takami *et al.* (1989) reported the production of extremely thermostable alkaline protease from *Bacillus sp* Ah-101. Gajju *et al.* (1996) observed that *Bacillus coagulants* PB-77, a Gram-positive obligate thermophile, efficiently utilized gelatin, casein and starch. This bacterium exhibited the highest proteolytic activity as compared to other isolates from different thermal habitats in Himachal Pradesh. The 4% inoculum of 24 h old culture proved better for enzyme production with 0.75% casein in the production medium. The higher concentration of casein decreases the enzyme production (Koke *et al.*, 1988). Kaur *et al.*, (1998) reported that the enzyme activity of *Bacillus polymyxa* in modified Reese medium was about 3 times more (350 U) compared to that in the seed medium (110 U). This may be due to repression of protease synthesis by high concentration of yeast extract and glucose in seed medium as compared to the modified medium (Battaglino *et al.*, 1991). Sen and Satyanarayana (1993) concluded that among eleven media combinations used, enzyme production of *Bacillus licheniformis* S-40 was high in soyabean meal starch medium. Major area of focus in the future concerning the production of proteases is the optimization of the media (Johnevelsy and Naik, 2001).

All extracellular protease require the presence of organic nitrogen sources. Every organism is very specific with respect to the nature of inducer. Mckeller (1986) reported in *Pseudomonas* that 1.5% peptone in the medium gives maximum synthesis, while higher concentrations of peptone repressed the synthesis which is probably due to the feed back inhibition. He has also reported the inducing ability of low molecular weight fraction of skimmed milk in *Pseudomonas*. Fujiwara and Yamamto (1987) have found organic sources to enhance the production of enzyme. Gelatin, glycine and potassium nitrate (nitrogen sources) decreased the production of enzyme. Similar results have also been obtained with other *Bacillus* species (Razak *et al.*, 1994; Dhandapani and Vijayaragavan, 1994). The production of alkaline protease was high in starch soya meal medium with wheat bran/starch/glucose/dextrin as a carbon source. Inorganic nitrogen sources supported better protease production than the organic sources. A similar observation was made by several workers (Kalisz 1988; Nehete *et al.*, 1986). The best nitrogen source for protease production by *Bacillus coagulans* PB-77 was beef extract, and addition of inorganic nitrogen sources in the medium resulted in lesser enzyme production in comparison to organic nitrogen sources. (Gajju *et al.*, 1996). Ammonium sulphate has been shown to increase the enzyme production in *B. licheniformis* and *B. firmus* (Moon and Parulekar 1990). Lactose, starch, soybean meal and sucrose have been considered well for industrial protease producer (Koke *et al.,* 1988; Castro *et al.*, 1991). Sinha and Satyaranayana (1991) reported that *Bacillus licheniformis* strain N3 produced alkanine protease on starch soyameal medium, which supported better production of protease. The use of soyameal in the production medium for protease has also been reported (Kroll, 1990). The production of an extracellular proteolytic enzyme during the stationary phase of growth is a characteristic of bacterial species (Durham *et al.*, 1987). Purva *et al.*, (1998) observed the synthesis of protease in late log phase (after 24 h), which then increased throughout the stationary phase for the alkalophilic *Bacillus sp*. IS-3. They also observed maximum yield of protease after 72 h of incubation, while further incubation resulted in sharp decline in protease activity which might be due to cessation of enzyme synthesis together with auto proteolysis. The enzyme production by UV mutant *B. polymyxa* in modified Reese medium was investigated by Madan *et al.* (2000). They also reported that stationary and submerged conditions at 50°C supported maximum enzyme production after 72 h of incubation. Submerged conditions proved to be more effective for protease production than stationary conditions. Matta *et al.* (1994) while studying the effect of agitation on the production of extracellular protease enzyme by *Pseudomonas sp*. AFT-36 found better protease production under continuous agitation (180 rpm) than under intermittent or no agitation. Battaglino *et al.* (1991) reported that glucose repressed protease synthesis, while Gomaa *et al.* (1990) reported glucose to be the best carbon source for protease production by *B. subtilis*. Kaur *et al.* (1998) used different inducers (soyameal, gelatin and tryptone 0.5 %) and evaluated protease production in *B. polymyxa* and found that enzyme activity in presence of soyameal, casein, tryptone and gelatin was 739, 827, 510 and 600 U, respectively indicating soyameal and casein to be good inducers of enzyme activity. Casein concentration when varied from 0.25 to 1.0% resulted in enzyme activity ranging from 246 to 827 U, while casein (0.5%) was found to be optimum, higher or lower concentrations severely inhibited enzyme production. It is well known that bacterial proteases are extracellular enzymes, the synthesis of which depends on the environmental conditions. As inductors, in addition to the substrate, the metabolic end products can also be effective (Muller, 2000). Madan *et al.* (2000) investigated the enzyme production of *B. polymyxa* in modified Reese medium under stationary and submerged conditions at 50°C after different time intervals. Maximum enzyme production was observed after 72 h incubation under stationary as well as submerged condition. Better protease production under submerged condition may be due to improved aeration. Donham *et al.* (1988) reported 17 times higher bacterial density and 7 times higher concentration of exoprotein under aerobic conditions than under anaerobic conditions. Similarly, Kaur *et al.* (1998) reported that incubation of isolated *B. polymyxa* at 70°C for 10 min in modified Reese medium gave about 3 times more protease enzyme activity compared to that in the seed medium. This may be due to repression of protease by high concentration of yeast extract and glucose in seed medium as compared to modified media (Battaglino *et al.,* 1991). Gajju *et al.* (1996) characterized protease thermophile *Bacillus coagulans* PB-77 at different time intervals (10-50 min) and found that increase in incubation time denaturates the enzyme.

pH

Although enzymes derived from a range of organisms are being used commercially, certain properties of these proteins can be further improved. For example, it may be advantageous to increase the heat stability or alter the pH optimum of an enzyme, thus improving the efficiency of a defined process. Alternatively, it may be desirable to enhance the activity of enzyme towards one particular substrate or potential substrate. The extent to which a particular active site, side chain participates in substrate, product and transition-state, stabilization can only be achieved by using interdisciplinary mutagenic, kinetic and structural approaches (Wanger and Benkovic, 1990). General principles formulated from such studies are being applied to the design and engineering of novel biocatalysts. Recent advances in genetic/protein engineering allow reliable approaches to modification of the primary structure of enzymes, thus altering some of their physico-chemical and catalytic properties (Dambrosky, 1998). Several methods have been developed for the modification of enzyme activity and stability by using chemical additives, immobilization, and protein engineering. Many of the goals of industrial protein engineering are to design and construct novel enzymes by modifying the properties of existing enzymes. For example, it may be desirable to alter the kinetic properties Kcat and Km, the substrate specificity, the pH optimum, the temperature stability, the stability in the presence of chemical reagent and the isoelectric point of an enzyme (Singh, 1999).

The stability of an enzyme can be defined as its ability to retain its activity under various conditions (Missel, 1993). The enzymes are inherently unstable. Their activity depends upon structural and functional integrity of the active domain, requiring proper folding of the protein

(Shortle and Sandek, 1995). Stability is not only important during storage or application of the enzyme, but comprises the entire life cycle. While stability during the course of fermentation and downstream processing of the enzyme is expressed in the final yield of the enzyme product, stability during storage and application relates to the quality, sale price and profitability in the market. Proteases form a large group of enzymes, ubiquitous in nature and found in a wide variety of microorganisms. They are molecules of relatively small size and are compact, spherical structures that catalyze the peptide bond cleavage in proteins (Polgar, 1990). Dhandapani and Vijayaragavan (1994) observed the production of thermophilic extracellular alkaline protease by *Bacillus stearothermophilus* up to pH 9, the enzyme was quite stable and retained full activity. Decline in activity was observed with an increase of pH to 11. The pH of culture strongly affects many enzymatic processes and transport of compounds across the cell membrane. Optimum pH (6.5–7.5) has been reported for neutral protease of *Pseudomonas fluorescens* and *Bacillus megaterium* (Fartima *et al.*, 1989), likewise pH 9 to 13 has been reported to be optimum for alkaline protease (Borriss, 1987). *Bacillus brevis II* isolated from kilns was a best protease producer at pH 10 (Joshi and Ball, 1993). Klingeberg *et al.* (1991) characterized the hyperthermophillic thermostable proteases, and they observed that the pH optima for *Thermococcus strain AN1* and *T. celer* were in neutral range and measured 7 and 7.5, respectively. The pH optima for *Thermococcus litoralis* and *Thermobacteroides proteoliticus* were in the alkaline range and measured 9 and 9.5 respectively. Kumar *et al.* (2002) observed maximum protease production in case of *Bacillus sp strain S-4* and *Pseudomonas strain S-22* at pH 7 and 9 respectively and hence indicated that S-4 produced neutral protease and S-22 produced alkaline protease. Mehta *et al.* (2004) observed optimum protease enzyme activity at pH range of 7-8 in case of *Planococcus I, Planococcus II and Micrococcus lylae*. Peek *et al.* (1992) observed that the protease from a *Thermus* sp. *strain Rt41A* exhibited stability for at least for 4 h over a pH 5-10. Park *et al.* (2003) found optimum pH 8 to 8.5 for *Bacillus vallismortis*, *Bacillus subtilis* and *Aeromonas hydrophila* and *Bacillus amyloliquefaciens*. Naidu and Devi (2005) reported that for *Bacillus* sp K-30 the optimum condition for proteolytic activity was at pH 9 and at 55°C. Chaia *et al.* (2000) reported that the optimum pH was 7-11 for the proteolytic activity of *Brevundimonas diminuta*. Each enzyme reduces its activity outside its own optimum pH range. The hydrogen ion concentration may change the ionization and solubility of enzymes, the substrate inhibitors and even the activators. It may also alter enzyme susceptibility to heat. This may be the possible explanation for different optimum pH values for different enzymes. It has been observed that in most enzymes, reaction rate is low at strongly acidic and basic solutions as these are known to cause denaturation of enzymes. Usually, the enzymes are maximally active at or near the isoelectric point, (pH at which the net charge of solution is zero). The effect of pH on the stability of protease enzyme activity was maximum at pH 9 for *Pasteurella pneumotrop* and *Staphylococcus hyicus* and the enzyme was quite stable at this pH, while *Micrococcus* sp, *Micrococcus luteus-1*, *Micrococcus luteus-4* showed maximum activity of enzyme at pH 8 and it was also quite stable at this pH. (Sangeeta, 2006). Protease from *Bacillus* sp. is reported to be stable over a wide alkaline pH range (9-9.5) and up to 50°C (Madan *et al.*, 2000). Broader pH stability in the range of pH 6-13 was reported for protease from *Bacillus amyloliquefaciens*, *Bacillus licheniformis* and *Bacillus sp.* AH-101 (Sinha and Satyanarayana, 1991; El-Beih *et al.*, 1991; Takami *et al.*, 1992). Alkaline proteases generally have broad pH optima in the range of pH 8-12 (Durham *et al.*, 1987). Protease from *B. lentus* exhibited stability in the pH range 7-10

(Betzel *et al.*, 1992). Optimum pH of 9.5 has been recorded for the protease enzyme from *Bacillus steareothermophilus* (Dhandapani and Vijayaraghavan, 1994). Chantawannakul *et al.* (2002) reported that the optimum pH of the proteases from *Bacillus subtilis* strain 38 was found to be at 6.5. However, pH 9.5 and 9 has been reported optimum for proteases of *Bacillus sp.* and *B. stearothermophiles* F1, respectively (Atalo and Gashe, 1993; Rahman *et al.*, 1994). Sharma *et al.* (1996) reported that protease enzyme from *B. laterosporus* was stable at pH 6 to 9, while *Flavobacterium* sp. was stable at pH 7-9. Olajuyigbe and Ajele (2005) investigated the proteolytic activity of three *Bacillus* sp. over the entire range of pH (pH 5-10). However, maximum protease production was observed at pH 8, while at pH 10, the protease production was about 60%.

Temperature

Temperature is one of the most important physical factors affecting bacteria. Bacteria are different from higher plants and animals in the lack of homeostatic mechanism and cannot regulate heat generated by metabolism and are therefore, directly and readily affected by temperature. There is a two-fold increase in the rate of enzyme catalysed reactions for every 10°C rise in temperature. Enzymes are temperature specific, as each enzyme is most active at a specific temperature. Any increase or decrease in temperature than the optima temperature decreases the enzyme activity. Maximum temperature of 55°C was recorded for an alkaline protease from *B. stearothermophilus* sp. AP-4 and 60°C for a protease derived from *Bacillus sp* B21-2. Chaia *et al.*, (2000) observed the optimum temperature for the proteolytic activity of *Brevundimonas diminuta* between 40°C and 50°C. Mehta *et al.*, (2004) observed optimum protease enzyme activity between 40-50°C in case of *Planococcus I, Planococcus II* and *Micrococcus lylae.* Enzymes do not function at above critical temperatures. If damaging temperature is maintained for a longer period, the enzymes become completely inactivated. The denaturation of enzymes is a permanent phenomenon and cannot be renatured. A species *Bacillus stearothermophilus* of TP26 that has been isolated produces an extracellular protease having an optimum temperature of 75°C (Gey and Unger, 1995). Lin *et al.*, (1997) found that enzyme synthesis of *Bacillus* sp. TS-23 occurred between 42 and 60°C with an optimum of 55°C. Kumar *et al.* (2002) reported high protease activity at 30°C from *Bacillus* sp S-4 *and Pseudomonas* S-22 and there was continuous decline in enzyme activity with increase in incubation temperature. All proteases from thermophilic bacteria showed a very broad temperature range with an optimal activity of 80°C. The optimum temperature for the proteolytic activity were 80°C for *Thermococcus stetteri,* 85°C for *Thermobacteroids proteolyticus,* 90°C for *Thermococcus strain* AN1 and 95°C for *Thermococcus litoralis, T. celer* and *Staphylothermus marinus* (Klingeberg *et al.,* 1991). Chantawannakul *et al.* (2002) reported that the optimum temperature of the proteases from *Bacillus subtilis* strain 38 was found to be at 47ºC. Naidu and Devi, (2005) reported maximum production of protease enzyme at 50°C after 96 h in *Bacillus* sp K-30. *Bacillus vallismortis, Bacillus subtilis and Aeromonas hydrophila* and *Bacillus amyloliquefaciens* showed optimum proteolytic activity in the range of 40-70°C (Park *et al.,* 2003). A number of alkaline proteases isolated from *Bacillus* sp. have high optimal temperature for their activity. This is very important characteristic for the use of alkaline protease as detergent additive (Purva *et al.*, 1998). Sharma *et al.* (1996) reported that *Bacillus laterosporus* and *Flavobacterium* sp. isolated from gelatin factory effluents showed maximum enzyme activity at 60°C after 90 min of incubation. Sumandeep *et al.* (1999) found that protease of thermostable alkaline protease of *Bacillus* sp NG–27 retained more than 50% activity over a temperature range of 25 to 80°C with an optimum of 40°C. The purified enzyme of thermophilic *Bacillus coagulans* PB-77 was found to be optimally active at temperature 70°C. A temperature of 70°C has also been reported for alkaline protease from some thermophilic *Bacillus* sp. (Takami *et al.*, 1989). Thermophilic *Bacilli* which grow and produce enzyme at 50–55°C have been found to produce enzymes which prefer alkaline pH and higher temperature (Sonnleitner, 1983). Sen and Satyanarayana (1993) observed that *Bacillus licheniformis* S-40 had 45°C as optimum temperature for the production of protease. Protease enzyme of *Micrococcus luteus-1* and *Micrococcus luteus-4* was found stable at 50°C, whereas *Pasteurella pneumotrop*, *Staphylococcus hyicus* and *Micrococcus* sp. were stable up to 60°C (Sangeeta, 2006). Majority of enzymes are irreversibly destroyed at the temperature range of 70 to 90°C. In a chemically defined medium, thermophilic and alkalophilic *Bacillus* sp JB-99 was also reported to produce thermostable alkaline proteases (Johnevelsy and Naik, 2001). Dominant producers of proteases in fact, are *Pyrococcus*, *Thermococcus* and *Staphylothermus.* Extremely thermostable serine proteases are produced by the hyperthermophilic *Archaeum desulfurococcus* strain (Hanazawa *et al.,* 1996), and thermostable metalloproteases are reported from a Gram-negative thermophilic bacterium (Pawinee and Wipapat, 1996). Sharma *et al.* (1996) reported maximum protease enzyme activity of *Bacillus laterosporus* and *Flavobacterium* sp at 60°C. The enzyme from *Bacillus laterosporus* was stable only at least 60 min at 30 to 80°C, whereas that from *Flavobacterium* sp. was stable only at 40 to 70°C for the same period. Madan *et al.* (2002) found that the enzyme of UV mutant *Bacillus polymyxa* was fully stable which showed 100% enzyme activity at 50°C. Thereafter, the activity declined from 61% and 40% of the original activity at 60°C and 70°C, respectively. However, the enzyme was completely inactivated at 80°C. Adinarayana *et al.*, (2003) found that *Bacillus subtilis* PE-11 protease had a half-life of 250 and less than 50 min at 70°C and 80°C, respectively. The enzyme was almost 100% stable at 60°C even after 350 min of incubation. Temperature optima of 70°C have also been reported for alkaline proteases from some thermophilic *Bacillus* sp. (Kaur *et al.*, 1998). Olajuyigbe and Ajele (2005) showed that three *Bacillus* species under study are producers of extracellular protease at high temperature. The three *Bacillus* species had maximum protease production at 60°C; protease production was about 70% even at 65°C. Other workers have also isolated thermophilic bacteria from the hot springs of Thailand, which were reported to produce thermostable proteases (Sookkheo *et al.*, 2000). This might be an indication that the *Bacillus* sp. produced thermostable neutral and alkaline proteases which could find application in industry and biotechnology. Gajju *et al.* (1996) reported that the thermophilic *Bacillus coagulans* PB-77 produced high protease activity at 55°C. There was no loss in enzyme activity after incubating at 60°C for 12 h and 65°C for 5 h without any metal ions in the reaction mixture. At 65°C, stability of enzyme was affected after 5 h of incubation. The appreciably high enzyme activity and stability at higher temperature and pH makes *Bacillus coagulans* PB-77 (MTCC-2457) an industrially promising and of special interest for basic and applied research. Thermostability studies of protease in the published literature show that the enzyme has half-life of 10 min (Chu *et al.*, 1992) and 23 min at 60°C (Kwon *et al.*, 1994). Kobayashi *et al.* (1996) reported that alkaline serine protease of *Bacillus* sp KSM- K16 remained stable at 50 and 60°C for 10 min. Ferrero *et al.* (1996) also obtained thermostable protease from *B. licheniformis* MIR 29, which showed 100% stability up to 70°C for 10 min. Kaur *et al.*

(1998) studied the stability of the enzyme of *B. polymyxa* at three temperatures (4, 25 and 50°C) for different periods of time. There was no loss in enzyme activity upto 4 d at 4°C and 25°C. At 4°C, the enzyme was quite stable for one month while at 50°C, the enzyme was stable only for 24 h which retained 50% of its original activity after 4 d, and after one week, the enzyme activity was lost completely.

Metal Ions and Inhibitors

Inhibition of enzymic activity is subject to both the chemical and physical factors. A large number of reagents that bind themselves to sulphydryl groups of proteins will inhibit the enzyme. Strong acids generally inhibit enzymes by hydrolysing the protein. Besides, many drugs also inhibit enzymes. Most enzymes can be poisoned or inhibited by certain chemical reagents. From the study of enzyme inhibitors, valuable information has been obtained about the substrate specificity of enzymes, the nature of the functional groups at the active site and the mechanism of the catalytic activity. Enzyme inhibitors are also very useful in elucidating metabolic pathways in cells. Moreover, some drugs useful in medicine appear to function as they can inhibit certain enzymes in malfunctioning of the cells. Inhibition studies primarily give an insight into the nature of enzymes in many processes that have been known for a long time.

Heavy metals ions (Cu^{2+}, Hg^{2+}, Ag^{+}) have been known to act as non-competitive inhibitors of enzymes. This fails to form enzyme substrate complex. In non-competitive inhibition, many enzymes can be poisoned, while some enzyme use Mg^{2+}, Ca^{2+}, Na^{2+} and K^{2+} in large number for its activation. Some substances are capable of restoring the activity of enzymes. These are called activators. A large number of enzymes need metal ions for activation. Such enzymes are called metalloenzymes.

There is no competition for active sites of enzyme by inhibitor and it fixes itself with some other site of enzyme. As a result, the physical structure of enzyme is altered. The enzyme activity of the thermostable alkaline protease from *Bacillus coagulans* PB-77 was greatly enhanced in the presence of 2 mM and 5 mM concentration of Fe^{2+}, Fe^{3+} ions, whereas Hg^{2+} severely inhibited enzyme activity (Gajju *et al.*, 1996). EDTA inhibited about 60% of the enzyme activity which indicates the partial requirement of metal ions (North, 1982). Chantawannakul *et al.*, (2002) studied the effect of various protease inhibitors on the activity of protease of *Bacillus subtilis* strain 38, only 1, 10-phenonthroline decreased the enzyme activity indicating the presence of the metalloproteases. Coolbear *et al.* (1992) have studied the effect of Fe^{2+}, Fe^{3+} and CO^{2+} on protease activity of thermophilic *Bacillus* strain EA-1, but they did not find the effect of these ions better than Ca^{2+} and Zn^{2+}. A psychrophilic strain of *B. coagulans* has been reported to require 4 mM of Fe^{2+} to stimulate the protease activity (Sai *et al.*, 1994). Hg^{2+} suppresses the enzyme activity in most of the thermophilic bacteria (Kobayashi *et al.*, 1995). Kaur *et al.*, (1998) studied the influence of a number of metal ions such as Mg^{2+}·Mn^{2+},Co^{2+},Hg^{2+} and various inhibitors EDTA, PMSF and Sodium azide, at concentration of 1mM under standard enzyme assay conditions. EDTA, Cu^{2+} and Hg^{2+} were the most potent inhibitors of the enzyme, which caused complete inhibition of enzyme activity. As the enzyme was completely inhibited by metal chelator (EDTA) and also by an active site inhibitor of serine protease PMSF, it appears to be a metal ion dependent alkaline serine protease. Sharma *et al.* (1996) indicated that the Ba^{+}

and Ca^{2+} stimulated the activity of the *Bacillus laterosporus* protease, whereas inhibition was recorded with Fe^{2+}, Mg^{2+} and Zn^{2+}. For *Flavobacterium* protease, Ba^{2+}, Ca^{2+} and Fe^{2+} were found to stimulate enzyme activity, while Mg^{3+}, Mn^{2+} and Zn^{2+} were inhibitory. The protease activity of *Bacillus laterosporus* was strongly inhibited by KCN, but was not affected by Hg^{2+}, 2-mercaptoethanol and 8-hydroxyquinoline. A protease from *Arthrobacter aureus* (Michotey and Blanco, 1994) also appeared to be metal dependent. Sumantha *et al.* (2005) studied the effect of various metal ions and known protease inhibitors on the enzyme activity of *Aspergillus oryzae* NRRL 2217. By activating the effect of divalent cations (Mg^{2+}, Ca^{2+}, Fe^{2+}) and inhibiting the effect of certain chelating agents (EDTA, EGTA), the enzyme was found to be metalloprotease. Klingeberg *et al.* (1991) found complete inhibition of protease enzyme activity from *Thermococcus stetteri* when PMSF and DFP were used as inhibitor. Mehta *et al.* (2004) observed that protease enzyme activity was reduced when metals ions like Hg^{2+} and Ba^{2+} were used, whereas calcium chloride and magnesium sulfate increased the activity. EDTA inhibited enzyme production much less in *Micrococcus lylae* when compared to both the species of *Planococcus*.

Bacillus alcalophilus sub sp. *halodurans* KP12 protease is a serine protease, as its activity is strongly inhibited by di-isopropyl-fluoro phosphate and PMSF (Takii *et al.*, 1990). Protease of *B. thermoruber* was inhibited by PMSF and EDTA. The stability was considerably increased by addition of Ca^{2+} and the protease exhibited a relatively high thermal stability (Manachini *et al.*, 1988). The two extracellular proteases from *Brevundimonas diminuta* that were inhibited by typical metal chelators (EGTA and 1, 10-phenanthroline) indicates that both enzymes are metallo protease. Inhibitors to cysteine protease (E-64, iodoacetemide), serine protease (PMSF, aprotinin) and aspartic protease (pepstatin A) failed to inhibit the enzymes (Chaia *et al.*, 2000). The effect of metal chelators on *B. diminuta* extracellular proteolytic activity suggests that these enzymes are calcium activated zinc metalloprotease. Previous reports showed that *Pseudomonas aeruginosa* alkaline protease contains one Zn^{+2} atoms in the catalytic center and requires several Ca^{+2} ions for stabilization of its folding (Morihara *et al.*, 1973). Thermophilic *Bacillus* strains showed enhanced proteolytic activity with Fe^{2+}, Fe^{3+} and CO^{2+} metal ions significantly, while Ca^{2+}, Zn^{2+} metal ions could not do so. The addition of divalent cations Cu^{2+} and Co^{2+} enhanced the thermostability and stimulated the enzyme activity. Other divalent cations like Ca^{2+} and Zn^{2+} had slightly inhibitory effect, while Mg^{2+} had no effect. It is also known that if protease activity was not inhibited by EDTA, it suggests that the metal ion as co-factor was not required for its catalytic function (Sinha and Satyanarayana, 1991). The extracellular alkaline proteases are known to be stabilized by Ca^{++} at high temperature (Pritest 1977).

The effect of different inhibitors on the enzyme activity of the purified protease of *Bacillus subtilis* PE-11 was observed with various inhibitors at 5mM concentration. Phenylmethyl sulfonyl fluoride (PMSF) was able to inhibit the protease completely, while diisopropyl fluorophosphates (DFP) exhibited 94% inhibition (Adinarayana *et al.*, 2003). PMSF sulfonates the essential serine residue in the active site of the protease has been reported to result in the complete loss of enzyme activity (Gold and Fahrney, 1964). Alkaline proteinase with PI 2.8 from *Bacillus* sp was completely inhibited by PMSF and indicated that it is a serine alkaline protease (Tsuchida *et al.*, 1986; Yamagata and Ichishima, 1989). Most of the metal ions tested had a stimulatory effect (Ca^{2+}, Mg^{2+} and Mn^{2+}) or a slight inhibitory effect on enzyme activity. Some of the metal ions such as Ca^{2+}, Mg^{2+} and Mn^{2+} increased and stabilized the protease activity of the enzyme. This is possible because of the activation by the metal ions (Adinarayana *et al.*, 2003). These cations have been reported to increase the thermal stability of other *Bacillus* alkaline proteases (Paliwal *et al.*, 1994). Pan and Lin (1991) suggested that concerned metal ions apparently protected the enzyme against thermal denaturation and played a vital role in maintaining the active confirmation of the enzyme at high temperatures. The normal state of living matter is a delicately balanced, spatially and temporally co-ordinated organisation. If a substance causes an adverse effect on this balance, it is usually termed as a poison. Alternatively, if it redresses a pathophysiological imbalance, it is regarded as a drug. An enzyme inhibitor decreases the activity of an enzyme without significantly disrupting its three-dimensional macromolecular structure. Inhibition is therefore distinct from denaturation and is the result of a specific action by a reagent directed or transmitted to the active site region. Studies on enzyme inhibition can yield much information on the mechanisation of enzyme catalysis.

APPLICATIONS OF PROTEASE ENZYME

Detergents

Proteases are on the standard ingredients of all kinds of detergents ranging from those used for household laundering to reagents used for cleaning contact or denatures. The use of proteases in laundry detergents accounts for approximately 25% of the total world wide sales of enzymes. The preparation of the first enzymatic detergent, "Burnus" dates back to 1913. It consisted of sodium carbonate and a crude pancreatic extract. The first detergent containing the bacterial enzyme was introduced in 1956 under the trade name BIO-40. In 1960, Novo Industry A/S introduced alkalase, produced by *Bacillus licheniformis*. Its commercial name was BIOTEX. This was followed by Maxatase, a detergent made by Gist Brocades. The biggest market for detergents is in the laundry industry, amounting a world wide production of 13 billion tons per year. The ideal detergent protease would possess broad substrate specificity to facilitate the removal of a large variety of stains due to food, blood and other body secretions. Activity and stability at high pH 8 and compatibility with other chelating and oxidizing agents added to the detergents are among the major prerequisite parameters for the best performance of a protease in detergents in its pI, hence, they can withstand higher pH ranges.

Leather Industry

Leather processing involves several steps such as soaking, dehairing, bating and tanning. The major building blocks of skin and hair are proteinaceous. The conventional methods of leather processing involve hazardous chemical such as sodium sulfide, which creates problems of pollution and effluent disposal. The use of enzymes as alternatives to chemicals has proved successful in improving leather quality and in reducing environmental pollution. Proteases are used for non-fibrillar proteins such as albumins and globulins. At present, alkaline proteases with hydrated lime and sodium chloride are used for dehairing, resulting in a significant reduction in the amount of waste water generated. Currently, trypsin is used in combination with other *Bacillus* and *Aspergillus* proteases for bating. The selection of the enzyme depends on its specificity for matrix proteins such as elastin and keratin, and the amount of enzyme needed depends on the type of leather (soft and hard) to be produced. Increased usage of enzymes for dehairing and bating not only prevents pollution problems, but is also effective in saving energy. Novo Nordisk manufactures three different proteases, Aquaderm, NUE and Pyrase, for use in soaking, dehairing and bearing, respectively.

Enzymatic Conversion of Protein Waste

The food production and processing industries generate large quantities of waste protein. Such waste includes dead animals and the unedible portions of animals, such as heads, feet, guts and feathers. Such substances are at best of marginal economic value and are often regarded simply as generating waste disposal problems. The facilities process or render dead animals and animal fall, usually by cooking at high pressure. This yields a valuable source of protein which may subsequently be incorporated into live stock feed slurry of ground feathers with sodium sulfite and an enzyme cocktail consisting predominantly of proteolytic enzymes. The chemical environment generated promotes dissolution of the disulfide bonds. The enzyme activities hydrolyse the ground feather protein (*a*-keratin), yielding a peptide rich and more digestible product. The feather digest may be formulated with another protein source such as soya bean meal in order to upgrade the overall nutritional value of the product.

Dairy Industry

The major applications of proteases in the dairy industry are in the manufacture of cheese. The milk coagulating enzymes fall into three main categories (i) animal rennets, (ii) microbial milk coagulants, and (iii) genetically engineered chymosin. Both animal and microbial milk coagulating proteases belong to a class of acid aspartate proteases and have molecular weights between 30,000 to 40,000 kDa. Rennet extracted from the fourth stomach of unweaned calves contains the highest ratio of chymosin to pepsin activity. In cheese making, the primary function of proteases is to hydrolyse the specific peptide bond (the Phe 105–Met 106 bond) to generate para–k–casein and macropeptides. Chymosin is preferred due to its high specificity for casein, which is responsible for its excellent performance in cheese making.

Protease produced by GRAS (genetically regarded as safe) cleaved microbes such as *Mucor michei, Bacillus subtilis*, and *Endothia parasitica* are gradually replacing chymosin in cheese-making. The production of chymosin in *Aspergillus niger* var. *awamor* has been increased to commercial levels in international markets. More than half the cheese produced world wide now employ fungal acid protease. At present, there are three recombinant chymosin products available and awaiting legislative approval for their use in cheese-making (Godfreyand West, 1996).

Baking Industry

Wheat flour is a major component of baking processes. It contains an insoluble protein called gluten, which determines the properties of the bakery doughs. Endo and exo proteinase from *Aspergillus oryzae* have been used to modify wheat gluten by limited proteolysis. The addition of protease reduces the mixing time and results in increased loaf volumes. Bacterial proteases are used to improve the extensibility and strength of the dough.

Brewing Industries

The cooling of beer after brewing often promotes haze formation. The haze is composed largely of protein compounds. Haze formation can be arrested by addition of proteolytic enzyme to the beer. Although various microbial enzymes have been assessed, plant derived proteases, such as papain and bromelian are most commonly used for such purposes.

Manufacture of Soy Products

Soybeans serve as a rich source of food due to their high content of good quality protein; proteases have been used from ancient times to prepare soy sauce and other soy products. The alkaline and neutral proteases of fungal origin play an important role in the processing of soy sauce. Proteolytic modification of soy proteins helps to improve their functional properties.

Debittering of Protein Hydrolysates

Protein hydrolysates have several applications as constituents of dietetic and health products, infant formulae and clinical nutrition supplements and as flavoring agents. The intensity of the bitterness is proportional to the number of hydrophobic amino acids in the hydrolysate. The peptidase that can cleave hydrophobic amino acids and proline are valuable in debittering protein hydrolysate.

Synthesis of Aspartame

The use of aspartame as a non caloric artificial sweetner has been approved by the food and drug administration. Maintenance of the stereospecificity is crucial, but it adds to the cost of production by chemical methods. Enzymatic aspartame is therefore preferred. An immobilized preparation of thermolysin from *Bacillus thermoprotyolyticus* is used for the enzymatic synthesis of aspartame. Toya soda (Japan) and DSM (Netherland) are the major industrial producers of aspartame.

Pharmaceutical Industry

The wide diversity and specificity of protease used to have great advantage in developing effective therapeutic agents:

- Oral administration of proteases (Luizym and Nortase) from *Aspergillus oryzae* has been used as digestive aid to correct certain lytic enzyme deficiency syndromes.
- Clostridial collagenase or subtilisin is used in combination with broad spectrum antibiotics in the treatment of burns and wounds.

- An asparaginase isolated from *E. coli* is used to eliminate asparagine from the blood stream in various forms of lymphocytic leukemia.
- Alkaline protease from *Conidiobolus coronatus* was found to be able to replace trypsin in animal cell cultures (Chiplonkar *et al.*, 1985).
- Matrix metalloproteases play a prominent role in the degradation of the extracellular matrix during tissue morphogenesis, differentiation and wound healing and may be useful in the treatment of diseases such as cancer and arthritis (Browner, 1995).

ADDITIONAL APPLICATIONS

Besides their industrial and medicinal applications, proteases play an important role in basic research. Their selective peptide bond cleavage is used in the elucidation of structure function relationship, in the synthesis of peptides, and in the sequencing of proteins.

Most enzymes used in the industry are extracellular enzymes derived from microorganisms. These enzymes are employed in soluble form. The definition of immobilized enzymes issued by European Federation Biotechnology (1983) includes "all enzymes that can be recycled". Until 1979, high expectations for immobilized enzymes still existed. Immobilizing strategies were suggested for practically all known enzyme reactions. The practice is very expensive and wasteful. In order to slap this wasteful practice and to reduce among the enzyme leaching of products, enzyme immobilization is practiced. Enzymes are immobilized by various means. Most industrial processes involving immobilized enzymes are conducted in fixed bed reactors. Carrier materials having morphological specifications require synthesis because most of the materials of natural origin cannot meet all the morphological requirements. Other method is covalent coupling of the enzyme to a reactive support or matrix. The covalent binding should be accomplished through functional groups in the enzyme that are not involved in the enzyme's catalytic activity. Generally, the binding is achieved by the nucleophilic groups of the enzyme reacting with activated functional groups of the support material. Aminohydroxyl and thiol group on the enzyme may participate in the linkage. Cysteine, lysine, tyrosine, and histidine residues may be considered as the most active. Among 50 immobilized enzymes, as compared to their soluble forms, Mellose (1971) found 30 to be more stable, 8 less stable and 12 to be showing no change. The shielding of the active sites/groups on the enzymes from reactive groups may have increased the enzyme stability.

INDUSTRIAL ENZYMES AND THE ENVIRONMENT

Enzymes can often replace chemicals or processes that present safety or environmental issues. For example, enzymes can:

- replace acids in the starch processing industry and alkalis or oxidizing agents in fabric desizing;
- reduce the use of sulfide in tanneries;
- replace pumice stones for "stonewashing" jeans;
- allow for more complete digestion of animal feed leading to less animal waste; and
- remove stains from fabrics. Clothes can be washed at lower temperatures, thus saving energy. Enzymes can be used instead of chlorine bleach for removing stains on cloth.

The use of enzymes also allows the level of surfactants to be reduced and permits the cleaning of clothes in the absence of phosphates.

Enzymes also contribute to safer working conditions through elimination of chemical treatments during production processes. For example, in starch, paper and textile processing, less hazardous chemicals are required when enzymes are used.

THE ENZYME INDUSTRY

Until quite recently, industrial enzymes were of interest only to scientists, engineers, and others directly involved in enzyme research and development. Now, with the widespread commercial advertisement of the relative efficiency and safety of enzymes in everyday products such as household detergents and foods, enzymes are an increasing concern to the general public. World consumption of industrial extracellular microbial enzymes is difficult to evaluate, but according to an estimate, the worldwide sale of microbial enzymes amounts to about $1 billion per year (Rao *et al.,* 1998).

The extracellular proteases are of commercial value and find multiple applications in various industrial sectors. Although there are many microbial sources available for producing proteases, only few are recognized as commercial producers. Some of the commercial extracellular proteases and their trade names have been depicted below:

Protease source	Trade name
Bacillus subtilis	Neutrase, Rapidermase, Proteinase 18, Protein
Bacillus lentus	Savinase Esperase, Purafect
Bacillus licheniformis	Alcalase, Optimase, Maxatase P,
Bacillus alcalophilus	Maxapem
Bacillus amyloliquefaciens	Maxacal
Bacillus thermoproteolyticus	Thermoase (N.P.), Thermolysin

REFERENCES

Adinarayana, K., Ellaiah, P. and Prasad, D.S. (2003). *AAPS Pharm. Sci. Tech.* **4(4):** 1-9.
Atalo, K. and Gashe, B.A. (1993). *Biotechnol. Lett.* **15:** 1151-1156.
Austew, B.M. and Smith, E.L. (1976). *Biochem. Biophys. Res. Commun.* **72:** 411-417.
Baker, E.N. and Drenth, J. (1987). *In:* Biological macromolecules and assemblies, vol. 3. (Eds. Jurnak, F.A. and McPherson, A.), Active sites of enzymes. John Wiley & Sons, Inc., New York, N.Y pp. 314.
Battiaglino, R.A., Huwrgo, M., Pilosol, A.M.R. and Bartholomai, G.B. (1991). *Appl. Microbiol. Biotechnol.* **35:** 292.
Ballinger, B.M., Tom, J. and Wells, J.A. (1996). *Biochemistry.* **35:** 13579-13585.
Barett, A.J. (1995). *Methods Enzymol.* **244:** 1-15.
Betzel, C., Klupsch, S., Papendorf, G., Hastrup, S., Branner, S. and Wilson, K.S. (1992). *J. Mol. Biol.* **223**: 427-445.
Blundell, T.L., Cooper, J.B.A., Sali, A. and Zhu, Z. (1991). *Adv. Exp. Med. Biol.* **306:** 443-453.
Borriss, R. (1987). *In:* Biotechnology, (Eds. Rehm, H. and Reed, G.) Weinhein, Verlag Chemie. 35-62.
Brenner, S. (1988). *Nature,* **334:** 528-530.
Browner, M.F., Smith, W.W. and Castelhano, A.L. (1995). *Biochemistry.* **34:** 6601-6610.
Castro, I.M., Lima, A.A., Paula, C.A., Nicoli, J.R. and Brandao, R.L. (1991). *Ferment Bioengg.* **72:** 132.

Chaia, A.A., Giovanni-De-Simone, S., Petinate, S.D.G., Cabral, P.A., Lima, A.P.C.A., Branquinha, M.H. and Vermelho, A.B. (2000). *Braz. J. Microbiol.* **31(1):** 18.

Chiplonkar, J.M., Gangodkar, S.V., Wagh, U.V., Ghadge, G.D., Rele, M.V. and Srinivasan, M.C. (1985). *Biotechnol. Lett.* **7:** 665-668.

Chantawannakul, P., Oncharoen, A., Klanbut, K., Chukeatirote, E. and Lumyong, S. (2002). *Science Asia.* **28:** 241-245.

Coolbear, T., Whittaker, M.J. and Daniel, R.M. (1992). *Biochem. J.* **287:** 367.

Cowan, D.A., Smolenski, K., Daniel, R.M. and Morgan, H.W. (1987). *Biochem. J.* **247:** 121-133.

Chu, M.L., Champion, and Tsushun, I. (1992). *Enzyme Microb. Technol.* **14:** 755.

Dambrosky, J. (1998). *Protein Eng.* **11:** 21.

Dhandapani, R. and Vijayaragavan, R. (1994). *World J. Microbiol. Biotechnol.* **10:** 33-35.

Donham, M.C., Heath, H.E., Leblanc, P.A. and Sloan, G.L. (1988). *J. Gen. Microbiol.* **134:** 2615-2621.

Drapeau, G.R., Boily, Y. and Houmard, J.Y. (1972). *J. Biol. Chem.* **247:** 6720-6726.

El-Beih, F.M., Abu, S.M.R., Gamal, R.F. and Abd-El-Rahim, M.K.I. (1991). *Ann. Agric. Sci.* **36:** 363-376.

Fartima, M.G., Pagliarini, E. and Manachini, P.L. (1989). *Italian J Food Sci.* **3:** 63-70.

Fastrez, J. and Fersht, A.R. (1973). *Biochemistry,* **12:** 2025-2034.

Ferosh, S. (2003). Ph.D Thesis, Dr. H.S. Gour University Sagar. pp. 204.

Ferrero, M.A., Castro, G.R., Abate, C.M., Baigori, M.A. and Sineriz, F. (1996). *Appl. Microbial. Biotechnol.* **45:** 327-332.

Fitzgerald, P.M.D., Mckeever, B.M., Middlesworth, J.F.V., Springer, J.P., Heimbach, J.C., Leu, C.T., Herber, W.K., Dixon, R.A.F. and Darke, P.L. (1990). *J. Biol. Chem.* **265:** 14209-14219.

Fujiwara, N. and Yamamoto, K. (1987). *J. Ferment. Technology,* **65:** 345-348.

Fox, J.W., Shannon, J.D. and Bjarnason, J.B. (1991). *ACS Symp. Ser.* **460:** 62-79.

Fusek, M., Lin, X. and Tang, J. (1990). *J. Biol. Chem.* **26(5):** 1496-1501.

Gajju, H.G., Bhalla, T.C. and Agarwal, H.O. (1996). *Indian J. Microbiol.* **36:** 153-155.

Gey, M. and Unger, K. (1995). *J. Chromatogr.* **166:** 188-193.

Gilles, A.M., Imhoff, J.M. and Keil, B. (1979). *J. Biol. Chem.* **254:** 1462-1468.

Gold, A.M. and Fahrney, D. (1964). *Biochem.* **3:** 783-791.

Godfrey, T. and West, S. (1996). Industrial enzymology, 2nd ed., p. 3. Macmillan Publishers Inc., New York, N.Y. 72).

Govind, N.S., Mehta, B., Sharma, M. and Madi, V.V. (1981). *Phytochemistry.* **20:** 2483-2485.

Gomaa, M.A., Mostafa, M.M., Abounied, M.M. and E.I, Habashy, M. (1990). *Egyptian J. Food Sci.* **16:** 9-26.

Gerhartz, W. (1990). Enzymes in Industry, VCH Publications, Weinheim, 78.

Hankin, L. and Anagnostakis, S.C. (1976). *Mycologia.* **67:** 143.

Hanazawa, S., Hoaki, T., Jannasch, H. and Maruyama, T. (1996). *J. Mar. Biotechnol.* **4:** 121-126.

Hase, C.C. and Finkelstein, R.A. (1993). *Microbiol. Rev.* **57**(4): 823-837.

Igarashi, H., Morita, T. and S. Iwanaga, S. (1979). *J. Biochem.* **86:** 1615-1618.

James, M.N.G., Hsu, I.N. and Delbaere, L.T.J. (1977). *Nature,* **267:** 808-813.

James, M.N.G., Sielecki, A.R., Hayakawa, K. and Gelb, M.H. (1992). *Biochemistry* **31:** 3872-3886.

Jany, K.D., Lederer, G. and Mayer, B. (1986). *FEBS Lett.* **199:** 139-144.

Johnevelsy, B. and Naik, G. (2001). *Proc. Biochem.* **37:** 139-144.

Joshi, A. and Ball, B. (1993). *Indian J. Microbiol.* **33:** 179.

Kalisz, H.M. (1988). *Adv. Biotechnol.* **35:** 1-69.

Kaur, M., Dhillon, S., Choudhary, K. and Singh, R. (1998). *Indian J. Microbiology.* **38:** 63-67.

Keil, B. (1992). Specificity of proteolysis. Springer-Verlag KG, Berlin, Germany.

Keay, L., Moses, P.W. and Wildi, B.S. (1970). Protease of the Genus *Bacillus-II*-Alkaline proteases. Biotechnol. Bioeng. **12:** 213.

Kimura, T. and Horikoshi, K. (1988). *App. Environ. Microbiol.* **54:** 1066.

Klingeberg, M., Hashwa, F. and Antranikian, G. (1991). *Appl. Microbiol. Biotechnol.* **34:** 715-719.

Kok, E.M.M., Draper, I. and Gerson, F. (1988). *Appl. Microbiol. Boitechnol.* **28:** 404.

Kroll, R.G. (1990). *In:* Microbiology of extreme environment, (Ed. Edward, C.) McGraw-Hill Pub. Co., New York. 55.

Kobayashi, T., Hakamada, Y., Adachi, S., Hilomi, J., Yoshimatuse, T., Koike, K., Kawai, S. and Ito, S. (1995). *Appl. Microbiol. Biotechnol.* **43:** 473.

Kuo, L.C. and Shafer, J.A. (1994). *Methods Enzymol.* **241:** 3-17.

Kuhad, R.C., Singh, A. and Eriksson, K.E.L. (1997). *Adv. Biochem. Eng. Biotechnol.* **57:** 45.

Kwon, T.Y., Kim, J.O., Moon, S.Y., Lee, H.H. and Rho, H.M. (1994). *Biotechnol. Lett.* **16:** 413-418.

Lin, L.L., Hsu, W.H. and Chu, W.S. (1997). *J. Appl. Microbiol.* **82:** 325-334.

Liu, H., Muller-Plathe, F. and Van Gunsteren, W.F. (1996). *J. Mol. Biol.* **261:** 454-469.

Madan, M., Dhillon, S. and Singh, R. (2000). *Indian J. Microbiology.* **40:** 25-28.

Madan, M., Dhillon, S. and Singh, R. (2002). *Indian J. Microbiology.* **42:** 155-159.

Matta, H., Kalra, M.S. and Punj, V. (1994). *Indian J. Dairy Sci.* **47:** 785-789.

Mao, W., Pan, R. and Freedman, D. (1992). *J. Ind. Microbiol.* **11:** 1-6.

Masaki, T., Nakamura, K., Isono, M. and Soejima, M. (1978). *Agric. Biol. Chem.* **42:** 1443-1445.

Manachini, P.L., Fortina, M.G. and Parini, C. (1988). *Appl. Microbial. Biotechnology* **28:** 409-413.

Mc Kellar, R.C. (1986). *J. App. Bacteriol.* **53:** 305.

Mehta, A., Ferosh, S., Singh, A., Sharma, A. and Mehta, P. (2004). *Indian J. Microbiology.* **44:** 277-280.

Missel, O. (1993). *In:* Stability and Stabilization of Enzymes (Eds. Van den Tweel, W.J.J., Harder, A. and Buitelaar, A.M.). Elsevier Science, Amsterdam. 111.

Michotey, V. and Blanco, C. (1994). *Appl. Environmental Micorbiology.* **60:** 341-343.

Miller, M., Jasko'lski, M., Rao, J.K.M., Leis, J. and Wlodawer, A. (1989). *Nature* **337:** 576-579.

.Mizisawa, K., Ichishina, E. and Yoshida, F. (1964). *Agric. Biochem.* **28**: 884.

Morihara, K., Suzuki, H. and Oka, T. (1973). *Biophys. Acta.* **309:** 414-429.

Mellose, G.J.H. (1971). *Polymers. Rev. Appl. Chem.* **21:** 83-119.

Mozhaev, V.V. (1993). *Trends Biotechnol.* **11:** 88.

Muller, R. (2000). *Enzyme and Microbial Technology.* **27:** 344-337.

Moon , S.H. and Parulekar, S.J. (1994). *Biotechnol Bioengg.* **37:** 467.

Naidu, K.S.B. and Devi, L.K. (2005). *J. Biotechnology,* **4(7):** 724-726.

Nehete, P.N., Shah, V.D. and Kathari, R.M. (1986). *Enzyme Microbiol. Biotechnol.* **8:** 370-372.

North, M.J. (1982). *Microbiol. Rev.* **46:** 308-334.

Olajuyigbe, F.M. and Ajele, J.O. (2005). *African J. Biotechnology.* 4(8): 776-779.

Paliwal, N., Singh, S.P. and Garg, S.K. (1994). *Biores. Technol.* **10:** 209-211.

Pan, T. and Lin, S. (1992). *J. Chinese Biochem. Soc.* **20:** 49-60.

Park, J.I., Yoon, C.J., Park, J.S., Kim, H.E., Cho, J.Y. and Shin, S.K. (2003). *The J. Microbiology.* **41(2):** 73-77.

Pawinee, K. and Wipapat, K. (1996). *In:* Poster presentation, 10th International Biotechnology sumposium, Sydney, Australia. 48.

Perlmann, G.E. and Lorand, A. (1970). *Methods Enzymol.* **19:** 732-735.

Pearl, L.H. (1987). *FEBS Lett.* **214:** 8-12.

Phadatare, S.U., Srinivasan, M.C. and Deshpande, V.V. (1992). *Eur. J. Biochem.* **205:** 679-686.

Polgar, L. (1990). *Biol. Chem. Hoppe-Seyler.* **371:** 327-331.

Poldermans, B. (1990). *In:* Proteolytic enzymes in industry: production and applications. (Ed. W. Gerhartz) VCH Publishers, Weinheim, Germany. 108-123.

Purva, Soni, S.K., Gupta, L.K. and Gupta, J.K. (1998). *Indian J. Microbiology.* **38:** 149-152.

Priest, F.G. (1977). *Bacteriological Reviews.* **41:** 711-753.

Rajak, N., Samad, M.Y., Basri, M., Yunus, M.Z.W.M., Ampoon, K. and Salleh, A.B. (1994). *World J. Microbiol. Biotechnol.* **10:** 260.

Rahman, R.N.Z.A., Razak, C.N., Ampoon, K., Basri, M., Yunus, W.M.Z.W. and Salleh, A.B. (1994). *Appl. Microbiol. Biotechnol.* **40:** 822-827.

Rao, M.B., Tanksale, A.M., Ghatge, M.S. and Deshpande, V.V. (1998). *Microbiol. Mol. Biol. Rev.* **62:** 597-634.

Rawlings, N.D. and Barrett, A.J. (1993). *Biochem. J.* **290:** 205-218.

Sai Ram, M., Singh, L., Allam, S.I. and Agarwal, M.K. (1994). *World J. Microbiol Biotechnol.* **10:** 356.

Sangeeta, S. (2006). Ph.D Thesis, Dr. H.S. Gour University Sagar. pp. 201.

Sielecki, A.R., Fujinaga, M., Read, R.J. and James, M.N.G. (1991). *J. Mol. Biol.* **219:** 671-692.

Shannon, J.D., Baramova, E.N., Bjarnason, J.B. and Fox, J.W. (1989). *J. Biol. Chem.* **264:** 11575-11583.

Sen, S. and Satyanarayana, T. (1993). *Indian J. Microbiology.* **31:** 43-47.

Sharma, A., Rao, C.L.S.N., Ball, B.K. and Hasija, S.K. (1996). *World J. Microbiology and Biotechnology,* **12:** 615-617.

Shortle, D. and Sondek, J. (1995). *J. Curr. Opin Biotechnol.* **6:** 387.

Singh, A. (1999). *Indian J. Microbiology.* **39:** 65-77.

Singh, A. and Hayashi, K. (1995). *J. Biol. Chem.* **270:** 21928.

Sookkheo, B., Sinchaikul, S., Phutrakul, S. and Chen, S.T. (2000). *Protein Expression and Purification.* **20:** 142-151.

Sonnleitner, B. (1983). *Adv. Biochem. Biotechnol.* **28:** 70-138.

Sumandeep, B.B., Qasin Kalil Beg, and Hoondal, G.S. (1999). *Indian J. Microbiology.* **39:** 185-187.

Sumantha, A., Chandran, S., George, S., Soccol, C.R. and Pandey, A. (2005). *Food Technol. Biotechnol.* **43(4):** 313-319.

Takami, H., Akiba, T. and Horikoshi, K. (1989). *Appl. Microbiol. Biotechnol.* **30:** 120-124.

Takami, H., Akiba , T. and Horikoshi, K. (1992). *Biosci. Biolechnol.Biochem.* **56:** 510-511.

Takaii, H., Takahashi, T., Momose, H., Inouye, M., Maeda, Y., Matsuzawa, H. and Ohta, T. (1990). *J. Biol. Chem.* **265:** 6874-6878.

Tsuchida, O., Yamagota, Y., Ishizuka, J., Yamada, J., Takeuchi, M. and Ichishima, E. (1986). *Curr. Microbiol.* **14:** 7-12.

Wagner, C.R. and Benkovic, S. (1990). *Trends Biotechnol.* **8:** 263.

Yamagata, Y. and Ichishima, E. (1989). *Curr. Microbial.* **19:** 259-264.

Zhang, J. and Greasham, R. (1999). *Appl. Microbiol. Biotechnol.* **51:** 407-421.

Zen-Yoji, H., Terayama, T. and Benoshi, M. (1961). *J. Microbiol.* **5:** 237-247.

Chapter **11**

Proteases: Significance and Applications

RICHA JAIN, N. KANGO AND P.C. JAIN

Applied Microbiology and Biotechnology
Dr. Hari Singh Gour Vishwavidyalaya, Sagar–470 003 (India)
E-mail: richajain_198@hotmail.com

Proteases are widespread in nature and can be obtained from plant, animal and microbial sources. Microbes serve as preferred source of these enzymes because of their rapid growth, the limited space required for their cultivation and the ease with which they can be genetically manipulated to generate new enzymes with altered properties that are desirable for their various applications. They differ greatly in their structure and action. Depending on their site of action, proteases are classified as exo- or endo-peptidases, while on the basis of functional group present on active site, they have been classified as serine, aspartic, cystein and metalloproteases. Currently six classes of proteases have been recognized with the addition of threonine and glutamic peptidases (http://enwikipedia.ofg/wiki/proteases). Proteases execute a large variety of functions and have important biotechnological applications. They represent one of the three largest group of enzymes and find applications in detergents, leather, food, and pharmaceutical industries and provide eco-friendly technologies for bioremediation. Proteases are also useful and important components in biopharmaceutical products such as contact lens cleaners, enzymic debriders. Because of their highly selective nature, they have been employed in protein engineering. Proteases have also been employed for enzymatic peptide synthesis by catalyzing reverse reactions under certain kinetically controlled conditions. Peptides differ from protein in their biological activity and thus are of great pharmaceutical and medicinal value. Modern researchers have indicated that variety of peptides including aspartame, kyotorphin, angiotensin, enkephalins, etc. have been successfully synthesized using proteases. However, developing technologies for generation of biofuels can widen the applications of proteases. Protein engineering is offering possibilities for generating proteases with entirely new functions and applications.

INTRODUCTION

Proteases constitute one of the most important group of enzymes. They cleave peptide bonds present in proteins. Earlier these enzymes were regarded as degradative enzymes, which could only catalyze the total hydrolysis of proteins, but the advances in enzyme research and technology have demonstrated that proteolytic enzymes can carry out highly specific and selective modifications in proteins by limited hydrolysis. Because of these properties, application of proteases in industries has drastically increased. Proteases alone account nearly 60% of the total enzyme sale in the world. They find varied applications in different industries e.g., cosmetics, detergent, food, leather, pulp/paper and pharmaceuticals (Rao *et al.*, 1998; Kocher *et al.*, 2003; Gupta *et al.*, 2005). Certain sectors like diagnostics, medicine and surgery, animal cell culture, molecular biology are among those which require proteases for their various applications (Godfrey and West, 1996; Abe *et al.*, 1997; Onifade *et al.*, 1998; Kuriyama *et al.*, 2000; So *et al.*, 2000; Thangam and RajKumar, 2002; Watanabe, 2004). For application, industries require a bulk of these enzymes and hence their production by novel sources need to be reviewed time to time for their better usage. The discoveries of new specific proteases and improved enzyme technology have made the microbial enzymes more attractive in biotechnology. In view of the above, present review gives a comprehensive account of the classification of proteases, their production by microbial sources and recent trends in their applications in various industries and other sectors of human interest.

PROPERTIES AND CLASSIFICATION

According to the Enzyme Commission (E.C.) of the International Union of Biochemistry (now called as the International Union of Biochemistry and Molecular Biology), proteases are classified in subgroup 4 of the group 3-hydrolases (IUB, 1992).

Based on the action and structure, proteases exhibit a huge diversity among them and hence their nomenclature do not comply with the general system of classification. However, the proteases are classified on the basis of following three major criteria : (1) Type of reaction catalyzed, (2) Chemical nature of the catalytic site, and (3) Evolutionary relationship with reference to their structure (Barett, 1994; 1995). Firstly, proteases are divided into two broad categories on the basis of their attack on the proteins i.e., exopeptidases and endopeptidases. These are further sub divided on the basis of mechanism of action and their active site. Peptidases are grouped into clans and families. The clans are groups of families for which there is an evidence of common ancestry. Based on the functional group present on active site, proteases have been classified into four prominent groups, i.e., serine, aspartic, cysteine and metalloprotease (Hartley, 1960).

Exopeptidases (peptidases)

They cleave peptide bond proximal to the carboxy or amino terminal of the protein and are classified on the basis of their action pattern as:

(i) aminopeptidases, cleaving single amino acid from the amino end of a protein,
(ii) carboxypeptidases, cleaving amino acid from the carboxy end of protein molecule, and
(iii) omega peptidases.

Aminopeptidases

These enzymes act at a free N-terminus of the polypeptide chain and liberate a single amino acid residue (3.4.11). The dipeptidyl peptidase liberates a dipeptide and tripeptidyl peptidase liberates tripeptide (3.4.14) from the amino end of the proteins (Table 1). In general, aminopeptidases are intracellular enzymes but they may be produced extracellularly also (Labbe *et al.*, 1974).

Table 1: Classification of proteases and their mode of action.

Protease	Mode of action	EC Number
Exopeptidase		
Aminopeptidase	●↓-o-o-o-o—-	3.4.11
Dipeptidyl peptidase	●-●↓-o-o-o—-	3.4.14
Tripeptidyl peptidase	●-●-●-↓-o-o—-	3.4.14
Carboxypeptidase	—-o-o-o-o-o-↓●	3.4.16-3.4.18
Serine type protease		3.4.16
Metalloprotease		3.4.17
Cysteine type protease		3.4.18
Peptidyl dipeptidase	—-o-o-o-o-↓●-●	3.4.15
Dipeptidases	●-↓-●	3.4.13
Omega peptidases	.-●↓-o-o—-	3.4.19
	—-o-o-o-↓●-.	3.4.19
Endopeptidase	—-o-o-o↓-o-o-o—-	3.4.21-3.4.34
Serine protease		3.4.21
Cysteine protease		3.4.22
Aspartic protease		3.4.23
Metalloprotease		3.4.24
Endopeptidase of unknown catalytic mechanism		3.4.99

o Open circles represent the amino acid residues in the polypeptide chain.
· Solid circles indicate the terminal amino acids,
* Stars signifies the blocked termini.
↓ Arrows show the site of action of the enzyme.

Carboxypeptidases

The carboxypeptidases act at C–terminals of the polypeptide chain and liberate a single amino acid (Table 1). Based on the nature of the amino acid residue at the active site of the enzyme, these are further classified into three major groups, serine carboxypeptidases (3.4.16), metallocarboxypeptidases (3.4.17), and cystein carboxypeptidases (3.4.18). Most serine carboxypeptidases are similar in their substrate specificities, but differ slightly in other properties such as pH optima, stability, molecular weight, and effect of inhibitors. Metallocarboxypeptidases specifically require Zn^{2+} or Co^{2+} for their activity. Peptidal dipeptidase (3.4.15) liberates C terminal dipeptide, and dipeptidases (3.4.13) act on peptide bonds of a dipeptide.

Omega Peptidases (3.4.19)

These are exopeptidases acting on peptide bonds. For example, (i) Acylaminoacyl–peptidase cleave N–acetyl or N–formyl amino acid from the N–terminus of a polypeptide, (ii) Y-Gly-X–carboxy peptidase (3.4.19.9) cause cleavage of a Y-glutamyl bond to release an unsubstituted C terminal amino acid.

Endopeptidases (proteinases)

Endopeptidases are characterized by their preferential action at the peptide bond in the inner region of the polypeptide chain away from the N- and C- termini. The presence of free amino or carboxy group has a negative influence on enzyme activity. Earlier, the proteinases were divided into 4 subgroups, i.e., serine, cysteine, aspartic and metalloproteases. To facilitate quick and unambiguous references to a particular family of peptidases, Rawling and Barett (1993) have assigned a code letter denoting the type of catalysis, i.e., S, C, A, M or U for serine, cysteine, aspartic, metallo or unknown type of proteases, respectively. Currently, six classes of proteinases have been established by addition of glutamic acid proteinase and threonine proteinases (http://en.wikipedia.org/wiki/protease). The serine, threonine and cysteine peptidases utilize the catalytic part of an amino acid as a nucleophile and form an acyl intermediate. These peptidases can also readily act as transferases. In the case of aspartic, glutamic and metallopeptidases, the nucleophile is an active water molecule.

Serine Proteases

They are characterized by the presence of serine group in their active site. Serine proteases are found in the exopeptidase, endopeptidase and omega peptidase groups. Based on their structural similarities and functional evidences, serine proteases have been grouped into 20 families which have been further subdivided into 6 clans (SA, SB, SC, SE, SF and SG) with common ancestors (Barett, 1994). The primary structures of the members of 4 clans , chymotrypsin (SA), subtilisin (SB), carboxypeptidase C (SC) and *Escherichia coli* D-Ala. D-Ala peptidases A (SE) are totally unrelated suggesting separate origin of serine proteases. The three clans, SA, SB and SC have a common catalytic triad of three amino acid, serine (nucleophile), aspartate (electrophile) and histidine base. Another interesting feature of serine proteases is the conservation of glycine residue in the vicinity of catalytic serine residue to form the motif Gly- Xaa- Ser- Yaa- Gly which is a characteristic feature of SE and SF (repressor Lex A) clans (Brenner, 1988). They have broad substrate specificities including esterolytic and amidase activity). These are generally of low molecular weight 15-35 KDa.

Serine proteases follow two-step reaction for hydrolysis in which a covalently linked enzyme–peptide intermediate is formed with a loss of the amino acid or peptide segment. Followed by acylation step, deacylation occurs by nucleophilic attack on the intermediate by water, resulting in hydrolysis of the peptide. Although the majority of serine proteases contain the catalytic triad Ser-His–Asp, a few use Ser-base catalytic dyad.

Serine proteinases can be classified into three groups based mainly on their primary substrate preference (a) trypsin-like, which cleave after positively charged residue; (b) chymotrypsin-like, which cleave after large hydrophobic residue; and (3) elastase-like, which cleave after small hydrophobic residues. A few of the serine proteases belonging to the subtilisin family show a catalytic triad composed of the same residue as in the chymotrypsin family; however, the residue occurs in a different order (Asp–His– Ser). Some important serine proteases and their applications are given in Table 2.

Cysteine/Thiol Proteases

The mechanism of action of cysteine proteases is very similar to that of serine proteases. Cysteine proteases catalyze the hydrolysis of carboxylic acid derivatives through a double–displacement pathway involving general acid–base formation and hydrolysis of an acyl-thiol intermediate. About 20 families of cysteine proteases have been recognized. The activity of all cysteine proteases depends on a catalytic dyad consisting of cysteine and histidine. The order of Cys and His residue differs among families. Generally, cysteine proteases are active only in the presence of reducing agents such as HCN or cysteine. Based on their side chain specificity, they are broadly divided into four groups: 1) papain-like, 2) trypsin-like with preference for cleavage at the arginine residue, 3) specific to glutamic acid, and 4) others. A few examples of important cysteine proteases are given in Table 3.

Aspartic Proteases

Aspartic acid proteases, commonly known as acidic proteases, are those that depend on aspartic acid residue for their catalytic activity. Barett (1995) grouped acidic proteases into three families, namely, pepsin (A1), retropepsin (A2) and enzymes from pararetroviruses (A3) and have been placed in clan AA. The members of families A1 and A2 are known to be related to each other, while those belonging to A3 family differ in some of the characteristics. The molecular masses of acid proteases are in the range of 30–45 kDa.

Microbial aspartic proteases can be broadly divided into two groups: (1) Pepsin-like enzymes produced by *Aspergillus, Penicillium, Rhizopus,* and *Neurospora*, and (2) Rennin-like enzymes produced by *Endothia* and *Mucor* spp. The application of a few aspartic proteases are given in Table 4. A general catalytic mechanism has been proposed for the hydrolysis of proteins by aspartic proteases such as penicillopepsin (James *et al.*, 1977) and endothia pepsin (Pearl, 1987). In most of the enzymes from the pepsin family, the catalytic aspartic residues are contained in an Asp-Thr-Gly-Xaa motive in both N and C terminal lobes of the enzyme where Xaa is Ser or Thr, whose side chains can have hydrogen bond with Asp. A marked conservation of cystine residue is also evident in aspartic proteases. The aspartic acid residue is situated within the motif Asp-Xaa-Gly, in which Xaa can be Ser or Thr.

Metalloproteases

Metalloproteases are the most diverse of the catalytic types of proteases and are widely produced

Table 2 : Some serine proteinases (EC 3.4.21) and their applications.

EC Number	Enzyme	Reaction Catalyzed	Application
3.4.21.1	Chymotrypsin	Preferential cleavage : Try↓, Trp↓, Phe↓, Leu↓.	Peptide mapping, peptide synthesis and peptide finger printing, anti-inflammatory agent, reduces soft tissue inflammation and edema.
3.4.21.4	Trypsin	Preferential cleavage : Arg↓, Lys↓	Disintegrates primary tissues in cultures, samples preparation for flow cytometric DNA analysis, preparing proteins and F_{ab} fragments from I_gM antibodies, reducing cell density in tissue culture. Tryptic mapping.
3.4.21.5	Thrombin	Selective cleavage of Arg↓ Gly bonds in fibrinogen to form fibrin and release fibrino-peptides A and B.	Topically reduces small vessel bleeding in surgery, clots blood in clinical diagnostic tests, determination of fibrinogen, stimulates cell proliferation in cell culture, defibrinates whole blood and plasma.
3.4.21.6	Coagulation Factor Xa	Selective cleavage of Arg↓Thr and then Arg↓Ile bonds in prothrombin to form thrombin.	Processing fusion proteins at definite cleavage sites, determination of prothrombin and heparin, coagulation studies.
3.4.21.7	Plasmin	Preferential cleavage : Lys ↓ > Arg ↓, converting fibrin into soluble products.	Coagulation research, degradation of extracellular matrix protein, protein structure and sequence analysis.
3.4.21.9	Endopeptidase	Activation of trypsinogen by selective cleavage of -$(Asp)_4$- Lys^6 ↓Ile- bonds.	Cleaves recombinant fusion proteins containing the recognition sequences, proteins modification and amino acid sequence determination.
3.4.21.19	Glutamyl Enteropeptidase	Preferential cleavage : Glu↓, Asp↓(phosphate buffer, pH 7.8), Glu↓(ammonium bicarbonate buffer, pH7.8; ammonium acetate buffer, pH 4.0)	Peptide mapping of basic proteins by proteolysis.
3.4.21.35	Tissue Kallikrein	Preferential cleavage of Arg↓ bonds in small molecule substrates. Highly selective hydrolysis of Met↓ or Leu↓ releases kallidin from kininogen. Rat enzymes cleaves two Arg ↓ bonds to liberate bradykinin directly from autologous kininogens.	Liberates vasoactive peptides from inactive kininogen precursors, protein structure and sequence studies.
3.4.21.36	Pancreatic Elastase	Preferential cleavage; Ala↓	Tissues dissociation, with collagenase, trypsin and chymotrypsin, protein sequence studies, solubilization of membrane proteins.

Table 2 : Contd.

EC Number	Enzyme	Reaction Catalyzed	Application
3.4.21.50	Lysyl Endopeptidase	Preferential cleavage : Lys↓, including -Lys↓Pro-	Sequence and structural studies, synthesis of Lys-X compounds.
3.4.21.62	Subtilisin	Hydrolysis of proteins with broad specificity for peptide bonds, and a preference for a large uncharged residue in P1.	Meat tenderizer, cheese flavor developer, improves dough texture, flavour and colour in cookies, cakes and pretzels, hydrolyzes fabric protein stains, recovery of fish oils, silver recovery from films, silk degumming.
3.4.21.64	Endopeptidase K	Hydrolysis of keratin and other proteins with subtilisin-like specificity.	Removes proteins in isolation of intact, native high molecular weight eukaryotic nucleic acids, specific modification of cell surface proteins and glycoproteins for analysis of membrane structure, solubilizes membrane-bound enzymes (acetyl cholinesterase).
3.4.21.66	Thermitase	Hydrolysis of proteins including collagen.	Preparing DNA samples from blood or in agarose plugs for pulsed field gel electrophoresis; preparing crude cell lysates suitable for RT-PCR.
3.4.21.70	Pancreatic Endopeptidase E	Preferential cleavage of Ala-↓-	Protein hydrolysis.
3.4.21.72	IgA-Specific Serine Endopeptidase	Cleavage of immunoglobulin A at Pro-↓- bonds in the hinge region.	Cleavage of fusion proteins if the amino acid sequence begins with X-Pro, where X =Thr, Ser or Ala.

Table 3 : Cysteine Proteinases (EC 3.4.22) and their applications.

EC Number	Enzyme	Reaction Catalyzed	Application
3.4.22.2	Papain	Hydrolysis of proteins with a preference for residues bearing a large hydrophobic side chain at P2 position.	Animal feed, brewing and distilling, wine and fruit juice, industrial cleaning of dairy products, leather and wool, cosmetics and bath powders, solubilization of integral membrane proteins, production of glycopeptides from purified proteoglycans, release and determination of synthetic colours in foods, peptide mapping by limited proteolysis, solubilization of oat proteins, prepration of Fab fragments of IgG antibodies, meat tenderizing and extracts, fish solubilization.
3.4.22.8	Clostripain	Preferential cleavage : Arg ↓ , including Arg ↓Pro but not Lys↓	Peptide mapping, fingerprinting and sequence analysis, hydrolysis/condensation of amide bonds.
3.4.22.14	Actinidain	Hydrolysis of proteins with a preference for residues bearing a large hydrophobic side chain at the P2 position.	Structure/function relationships between cysteine proteases and inhibitors.
3.4.22.17	Calpain	Preferential cleavage : Tyr ↓ , Met↓ or Arg↓, with Leu or Val as the P2 residue.	Used in connection with dystrophy, cataract, inflammation and accumulation of memory.
3.4.22.32	Stem Bromelain	Strong preference for Z-Arg –Arg↓ NHMec among small molecules along with broad protein cleavage specificity.	Protein hydrolysis; resolution of-β-hydroxy -α-amino acids by the action of protease on their N-acetyl methyl esters.

Table 4 : Aspartic proteinases (EC 3.4.23) and their applications.

EC Number	Enzyme	Reaction catalysed	Application
3.4.23.1	Pepsin A	Preferential cleavage : hydrophobic and preferably aromatic residues in P1 and P1′ position.	Nonsepecific hydrolysis of proteins and peptides in acidic media. Preparation of $F(ab')_2$ fragments of IgG antibodies. In growth media cleaves proteins to smaller molecules as nitrogen source. Peptide synthesis and milk clotting. Processing of collagen to remove telio-proteins.
3.4.23.4	Chymosin	Broad specificity similar to that of pepsin A. Clots milk by cleavage of a single Ser-Phe $^{105}\downarrow$ Met-Ala bonds in the κ-chain of casein.	Aids proper setting of cottage cheese and other cultured cream products, provides coagulation strength for cutting in pH 4.6-4.75 and increases viscosity.
3.4.23.5	Cathepsin D	Specificity similar to but narrower than pepsin A. Does not cleave the Gln $^{4}\downarrow$ His bond in the B chain of insulin.	Immunogen for antisera production, tracer for iodination.
3.4.23.15	Rennin	Cleavage of Leu↓ bond in angiotensinogen to form angiotensin I.	
3.4.23.18	Aspergillopepsin I	Generally favours hydrophobic residues in P1 and P1′, but also accepts Lys in P1 leading to activation of trypsinogen. Does not clot milk.	Manufacturing of proteins without bitterness and flavors from natural proteins, bio-polishing of wool fabrics. Protein hydrolysis, decomposition of soybean and rice protein, flour treatment and manufacture of baked goods, lactose reduction and flavor modification in diary applications.
3.4.23.20	Penicillopepsin	Hydrolysis of proteins with broad specificity similar to that of pepsin A. Prefers hydrophobic residue at P1 and P1′, but also cleaves $Gly^{20}\downarrow Glu$ in the β-chain of insulin.	Manufacture of peptides without bitterness.
3.4.23.24	Candidapepsin	Preferential cleavage at the carbcxyl of hydrohobic amino acids but fails to cleave Leu^{15} –Tyr, Tyr^{16}- Leu and Phe^{24} – Phe of nsulin B cha ns. Activates trypsinogen and degrades keratin.	Produce overlap peptides in sequencing studies.

Table 5 : Metalloproteinases (EC 3.4.24) and their application.

EC Number	Enzyme	Reaction catalysed	Application
3.4.24.3	Microbial Collagenase	Digestion of native collagen in the triple helical region at Gly bonds. Preference with synthetic peptides: Gly at P3 and P1′′, Pro and Ala at P2 and P2′ and hydroxyproline, Ala or Arg at P3′.	Determination of collagen ; debriding agent; Dissociation of tissue for establishment of primary cell culture (epithelial, lung, kidney, adrenal, bones, pancreas, thyroid, heart, salivary); Dissolves connective tissues and exposes embedded cells without destroying cell membrane and other structures. Isolation of rat pancreatic islets and hepatocytes.
3.4.24.27	Thermolysin	Preferential cleavage : ↓Leu > ↓ Phe	Probing the surface of intact pea chloroplasts. identify the location of disulfide bridges resistant to other proteases.
3.4.24.28	Bacillolysin	Similar bond cleavage to thermolysin (EC 3.4.24.27) ↓ Leu > ↓ Phe	Fortifies brewing malt proteases, softens wheat gluten in baking, lactose reduction and flavor modification in dairy products, upgrades feed vegetable and animal protein, rapidly and gently separates intact epidermis from dermis and intact epithelial sheets in culture from substratum; general use : textiles, animal feed, baking, silage, wine and fruit juice, industrial cleaning, leather and wool; prevent unwanted clumping of cells cultured in suspension ; silk degumming.
3.4.24.31	Mycolysin	Preferential cleavage : bonds with hydrophobic residues in P1′.	Degrading proteins completely to amino acids without decomposition, degrade proteins during isolation of DNA and RNA without phenol, isolation of chondrocytes.
3.4.24.32	β-Lytic Metalloendo-peptidase	Cleavage : the insulin β-chain at Gly ^{23}↓ Phe > Val ^{18}↓Cys (SO_3 H) and N-acetylmuramoyl↓Ala.	Preparation of protoplast and spheroplasts, bacterial enzymes ; isolation of bacterial proteins via cell wall lysis.
3.4.24.33	Peptidyl- Asp Metalloendo-peptidase	Cleavage : Xaa ↓Asp, Xaa↓ Glu and Xaa↓ cysteic acid.	Useful in sequencing because of limited specificity, cleavage of fusion proteins.

by microorganisms such as collaginase, thermolysin, mycolysin, etc. (Table 5). They are characterized by the requirement for a divalent metal ion for their activity They include enzymes from a variety of origins such as collagenases from higher organisms and actinomycetes, hemorrhagic toxins from snake venoms, and thermolysin from bacteria (Hibbs *et al.*, 1985; Okada *et al.*, 1986; Wilhelm *et al.*, 1987; Shannon *et al.*, 1989). About 30 families of metalloproteases have been recognized including endopeptidases and exopeptidases. Families of metalloproteases have been grouped into different clans based on the nature of the amino acid that completes the metal-binding site. Based on specificity of their action, metalloproteases can be divided into 4 groups—Neutral, Alkaline, Myxobacter I, and Myxobacter II. The mechanism of action of metalloproteases is slightly different from that of other proteases. These enzymes depend on the presence of bound divalent cations and can be inactivated by dialysis or by the addition of chelating agents. Most of the metalloproteases are enzymes containing the His-Glu-Xaa-His motive, which has been shown by X-ray crystallography to form a part of the site for binding of metal, usually zinc.

The neutral metalloproteases show specificity for hydrophobic amino acids, while the alkaline metalloproteases possess a very broad specificity. Myxobacter protease I is specific for small amino acid residues on either side of the cleavage bond, whereas protease II is specific for lysine residues on the amino side of the peptide bond.

Glutamic Proteases

Glutamic proteases represent a distinct, recently reclassified group of peptidases that are thought to be found only in fungi. Glutamic proteases are included in clan G with a single representative family. The type example of this family is Scytaliduglutamic peptidase (Sims *et al.*, 2004).

Threonine Proteases

These are characterized by a threonine nucleophile at the N terminus of the mature enzyme. This clan contains one family (T-5), while three other families (T1, T2 and T3) have been included in clan PB (MEROPS peptidase data base). The EC number assigned to threonine peptidases is 3.4.25.

Microbial proteases have also been classified in a number of other ways, e.g., (i) based on their ability to hydrolyze specific substrate (keratinases, collagenases, elastases, etc.), (ii) on the basis of similarities to well recognized proteolytic enzymes such as pepsin, trypsin, chymotrypsin, (iii) pH range over which they are active (acid, neutral and alkaline proteases), (iv) their sensitivity to various inhibitors (Keay, 1971). Based on their sensitivity to inhibitors, four types of proteinase can be distinguished (Tables 2–5). Classification of enzymes based on their substrate specificities is although arbitrary but indicates specificity of enzyme for practical purposes. While, the most satisfactory scheme for classification of microbial proteases is based on their pH optima.

SOURCE OF PROTEASES

Proteases occur naturally in all organisms including plants, animals and microorganisms, and are involved in a multitude of physiological and biochemical reactions. An account of proteases produced of each group is given in the following text.

(i) Plants

The major proteases produced by plants are papain, bromelain and ficin. Papain is obtained from the latex of *Carica papaya* fruits. The crude preparation of this enzyme has a broader specificity due to the presence of several proteinase and peptidase isozymes. The performance of this enzyme depends much on the plant source, habitat and on the method used for its extraction and purification. The bromelain is obtained from the stem and juice of pineapple. The major supplier of this enzyme is Thailand and Bangkok. Ficin is obtained from latex of Figs (*Ficus* spp.). The keratinases are also produced by some plant species, which degrade hair (Rao *et al.*, 1998).

(ii) Animals

The most familiar proteases of animal origin are pancreatic trypsin, chymotrypsin, papain and rennins (Boyer, 1971; Hoffman, 1974). The trypsin is a main intestinal enzyme responsible for hydrolysis of food proteins. This enzyme has received greater attention for biocontrol of insect pests and some specialized medical applications, but has limited application in food industries because its use generates hydrolysate with a bitter taste. Chymotrypsin is obtained from animal pancreatic extract. Its purified forms are used for diagnostic and analytical purposes. Pepsin is an acid protease found in the stomach of almost all the vertebrates. The active enzyme is released from its zymogen, i.e., pepsinogen by autocatalysis in the presence of hydrochloric acid. Similarly rennin is produced as an inactive precursor prorennin, in the stomach of all nursing animals. It is converted to active rennin by the activity of pepsin or by its autocatalysis.

The plant and animal proteases have contributed significantly to meet out current world demand to certain extent, but their production depends on the availability of related producers, plants and livestock for slaughter. The growing demand of enzymes in various sectors has led to an increased interest in microbial proteases.

(iii) Microorganisms

Microbial proteases replaced the traditional proteases from animal and plant sources and in addition have found various applications because of their broad biochemical diversity and their susceptibility to genetic manipulations. Proteases are widely distributed among the microorganisms including fungi, bacteria and actinomycetes. Fungal species of certain genera such as *Aspergillus*, *Penicillium*, *Paecilomyces*, *Rhizopus* and *Rhizomucor* are well-known producers of proteases. Among these *Aspergillus (A. flavus, A. oryzae* and *A. niger*) and *Rhizomucor pusillus* and *R. miehei* are important source of protease for industries. Fungi have been recognized as a major source of acid protease which are stable at acidic pH range between pH 2.5-5 with pH optima between 4.0 to 4.5. *Rhizopus* and *Mucor* species are used in the preparation of oriental foods such as Koji and Temph and in cheese making industries as a substitute of natural rennet (Sternberg, 1972; Nand *et al*., 1980). Other fungi such as *Rhizomucor pusillus, R. miehei, Penicillium roquefortii, P. camemberti* and *Endothia parasiticus* are important source of milk cloting enzymes (Sardinas, 1966; Arima *et al*., 1967; 1970; Tam and Whitakar, 1972).

Most commercial proteases, mainly neutral and alkaline proteases, are produced by species of genus *Bacillus*. Neutral proteases produced by *Bacillus* spp. are active in a narrow pH range (pH 5-8) and have relatively low thermotolerance. Because of their intermediate rate of reaction they generate less bitterness in hydrolysate as compared to animal proteases, and hence are valuable for use in food industries. The bacterial alkaline protease, particularly those produced by *Bacillus* species (e.g., *B. licheniformis, B. subtilis, B. alcalophilus, B. lentus* and other), are characterized by their high activity at alkaline pH (i.e., pH 10), broad substrate specificity and higher optimal temperature (approximately 60°C) and thus they are widely used as detergent supplements (Jasvir *et al*., 1999; Banik and Prakash, 2004; Tang *et al*., 2004; Gupta *et al*., 2005; Prakasham *et al*., 2005). Other bacteria belonging to genus *Pseudomonas, Flavobacterium, Halobacterium, Alcaligens, Arthrobactor, Clostridium, Staphylococcus, Serratia* and *Vibrio* have also been explored for production of alkaline proteases (Thangam and Rajkumar, 2002; Yogei *et al*., 2002; Gupta *et al*., 2005). Among actinomycetes, species of *Streptomyces* and *Streptoverticillium* have been reported to produce gelatinase and keratinases of industrial significance (Bockle and Muller, 1997; Chitte *et al*., 1999; Szabo *et al*., 2000 and Jain *et al*., 2004). Proteases such as collagenase and elastase are known to be synthesized and produced by almost all groups of microorganisms including fungi, bacteria and actinomycetes (Tsai *et al*., 1983; Ivanko *et al*., 2002; Ivanko and Varbanets, 2004; Gupta and Ramnani, 2006).

INDUSTRIAL SIGNIFICANCE AND APPLICATIONS

Proteases have a large variety of applications mainly in detergent and food industries. Proteases are required to replace hazardous chemicals used in processing of leather and degumming of silk. In view of the recent trend of developing eco-friendly technologies, they are finding major applications in bioremediation and solid waste management. They are extensively used in pharmaceutical industries, particularly for debridement of wounds. Depending on the process in which proteases are used, they may be required either in bulk, as crude preparations, e.g., in detergent industries or in small quantities and in finest pure form in medicine. Production of proteases and their applications in various industries are as follows:

(a) Detergent Industry

Enzymes based detergents popularly known as 'Green Chemicals' find a wide range of application in laundry, dishwashing, textile, dairy and food industries. The enzymes, i.e., proteases, lipases, cellulases and amylases are indispensable ingredients in these detergents. The use of proteases in laundry detergents accounts approximately 25% of the total worldwide sale of enzyme (Rao *et al*., 1998). Rohm and Hass of Germany used pancreatic proteases as a detergent additive as early as in 1913. The product was marketed as a presoak detergent under the trade name Burnus. The first detergent containing the bacterial enzyme was introduced in 1956 under the trade name B10-40. In 1960 an alkaline protease, subtilisin Carlsberg (Alcalase) from *Bacillus licheniformis* was developed by Novo Industry A/S Denmark and marketed under the trade name BIOTEX. While in 1989, two more detergent enzymes: a cloned Savinase type, Maxacal and Gistbrocades, Delf, Netherlands and another genetically engineered enzyme i.e., 'Subtilisin novo' from Genencor International of California. The detergent manufacturers explored new reveration enzymes for better wash performance and stability towards oxidative bleaching. Agents such as 'Maxapem', an alkaline protease and 'Durazyme', a protein engineered detergent enzyme were developed by Solvay enzymes, Germany and Novo Nordisk, Denmark, respectively (Anonymous).

The use of enzyme based detergents is preferred over the conventional synthetic detergent in view of their better cleaning properties, biodegradability, low toxicity, non-corrosiveness, increased efficiency and environment friendliness. Thus, these are also referred as green chemicals.

Alkaline serine proteases from *Bacillus licheniformis* is the predominant commercial detergent enzymes (Nihalar and Satyanarayana, 1992; Wolff *et al.*, 1996; Gupta *et al.*, 2002; Tang *et al.*, 2004), but there are also reports of alkaline proteases from other sources suitable for detergents (Borriss, 1987; Jasvir *et al.*, 1999; Mane and Bapat, 2001; Adinarayana and Ellaiah, 2002; Hui *et al.*, 2004). All of the proteases currently used in detergent formulations are serine proteases because thiol proteases like papain may be oxidized by bleaching agents and metalloprotease would lose their metal cofactor due to complexing with water softening agents.

Aelhe *et al.* (1993) carried out protein engineering of alkaline proteases from *Bacillus alcalophilus* to improve its washing performance. An alkaline protease from *Conidiobolus coronatus* was found to be compatible with commercial detergent use in India (Phadatare *et al.*, 1997). In 2002, Madan *et al.*, isolated alkaline proteases from a mutant of *Bacillus polymyxa* and reported that enzymes retained more than 50% activity after 30 minutes incubation at 35°C in the presence of detergents such as Avis and Vim ultra, indicating its suitability in detergent industry, while Saeki *et al.* (2002) reported that a novel strain of *Bacillus* produces an oxidatively stable serine protease that can be used in bleach based detergents. Mulimani *et al.* (2002) have also reported a bleach stable-alkali tolerant protease from *Aspergillus flavus.* An alkalophilic proteases from a new isolate of *Bacillus* sp. APR-4 was found to be stable at 65-70°C in presence of 5% detergent (Farishta, Fena). The protease also removed egg yolk stain within 10 min from cotton fabric when used at 100 m/mL with 1% detergent at pH 9.0.

Nowadays proteases are also incorporated in dishwashing detergents. Enzyme based diswashing detergents are less abrasive in function and are suitable for cleaning of delicate chinaware (Banerjee *et al.*, 1999).

(b) Cheese Production

One of the most important and challenging industrial applications of proteases is in the process of cheese manufacture. Microbial rennet obtained from *Mucor pusillus, M. miehei, Endothia parasitica* and *Bacillus subtilis* (Arima *et al.*, 1970; Sood and Kosikowski, 1979; Garg and Johri, 1994) are used as an effective substitute for calf, goat and swine derived rennet. Milk treated with microbial rennet causes casein to precipitate in the form of curd. Microbial rennets are highly specific in their ability to clot milk and are able to cleave the Phe-Met bond of k-casein, the bond cleaved by rennin. The split k-casein loses its stabilizing activity and the micelles of casein coagulate in presence of calcium (Sternberg, 1972; Rao *et al.*, 1998). Microbial proteases along with lipases when added to fresh milk or cheese curd produces products with enhanced dairy and cheese flavors. These enzyme blends also accelerate the ripening of cheese and produce intense flavor in snack food, soups and cheese sauces (Kinsella and Hwang, 1976).

Many other fungal acid proteases also have milk clotting activity, e.g., proteases from *Penicillium caseicolum*, *P. roquefortii*, etc. (Law and Wigmore, 1982). Wang *et al.* (1969) carried out detailed study of milk clotting activity of proteinases produced by *Rhizopus*. Although many more proteases are known but only a few are useful in cheese-making, because excessive proteolytic activity results in spoilage due to extensive breakdown of casein (North, 1982; Yamashita *et al.*, 1994).

Rennets of *M. pusilus* and *M. miehei* have molecular mass approximately 38.5 and 42 K Da and isoelectric points 3.9 and 4.1, respectively (Etoh *et al.*, 1979; Maheshwari *et al.*, 2000). The *M. pusillus* enzyme was reported to be stable from pH 3.0 to 6.0 and showed maximum activity at 55°C (Yu *et al.*, 1968). Gray *et al.* (1986) have reported that recombinantly produced *M. miehei* protease in *Aspergillus nidulans* had similar specific activity to that produced by *M. miehei*. Fernandez-Lahore *et al.* (1999) have worked out that fourth day extract of solid state culture of *Mucor* sp. M-105 strain showed a high milk clotting activity and its proteolytic activity ratio was similar to that of commercial preparations from microbial origins used in cheese manufacture.

(c) Food Industry

A number of traditional fermentation processes in food industry involve proteases. They have been used in preparations of several oriental foods, e.g., soya sauce (Yong and Wood, 1974; Impoolsup *et al.*, 1981). Acid, alkaline and neutral proteases have been used for the processing of proteinaceous food to increase their digestibility, dispersibility and solubility (Loffler, 1986; Frokjaer, 1994; Hashem, 1999). The use of proteases in food industries is rapidly increasing because of their higher specificity, finer product control, high efficiency and product formation under mild conditions (Anglo and Orillo, 1977; Ward, 1985). Variety of proteins, particularly those of soyabean, pea, yeast, maize, meat, fish, wheat flour, etc. are being processed by use of microbial proteases. Proteases in particular improve the flavor and texture as well as reduce the bitterness of proteins. They also improve the palatability of reformulated low carbohydrate/high protein foods (Kinsella and Hwang, 1976; North, 1982; Douglas and Parsons, 2000 and Kocher *et al.*, 2003).

Enzyme modified proteins are being used as active ingredients in dietary supplements, functional beverages, fermented foods, baked foods, etc. (Neklyudov *et al.*, 2000; Dhar and Sreenivasulu, 1986). Such proteins with altered fat and water binding properties have enhanced the ability to replace fats and carbohydrates in food products and also reduced allergenicity of food protein (Gupta *et al.*, 2005). However, type of proteases to be employed for processing depends upon the nature and type of protein, e.g., protease that cleave hydrophobic amino acid and proline are used for debittering (Rebecca *et al.*, 1991; Perea *et al.*, 1993). Acid proteases find application in processing of products of dairy industry, etc. (Palvannan, 1998).

Wheat flour used in bakery industry contains an insoluble protein gluten which determines the properties of dough. Proteases when applied for processing of dough weakens the gluten strands and thus reduce processing time of bakery products (Rao *et al.*, 1998). In bakeries, proteases are helpful in preparation of cookies, crackers and crisp bread dough, etc. They improve handling, reduces dough mixing time, provide anti-staling properties, and allow control over crumb, textures, colour, taste and volume (Barrett, 1975; Woods *et al.*, 1980). Proteases produced by *Aspergillus oryzae* have been used in baking industry to improve bread texture and also to gain dough uniformity (Barrett, 1975).

Soya bean is one of the richest protein sources of plant origin. Microbial proteases have been employed from ancient times to prepare soya sauce and other soya products. Fungal proteases are largely employed for processing of soya sauce. Treatment of soyproteins with alcalase at pH 8 yields soluble hydrolysates with high solubility and reduced bitterness (Nakadi *et al.*, 1972). They are also being employed for tenderization of meat and fish proteins, etc. The proteolytic enzymes from *Trichoderma reesei* MCG 80 have been reported to tenderize meat by acting upon myofibrillar protein producing uniform texture (Robbins *et al.*, 1986; Garg and Johri, 1999). In 1994, Dalev utilized proteases for conversion of waste feathers into protein concentrates.

(d) Leather Industry

Leather processing industries require a significant amount of enzymes. Proteases are employed mainly for the purpose of unhairing, bathing and liming (Mukhopadhya and Chandra, 1993). Traditional methods used for processing of leather makes use of harmful chemicals such as sodium sulfides and alkali that produce effluents and cause serious pollution problems. Enzymatic process in turn does not produce such effluents (Anderson, 1998; Anwar and Saleemuddin, 1998; Gupta *et al.*, 2002). Enzymic processes are more safer, efficient and ecofriendly than the traditional chemical methods.

Hides and skin received from slaughter houses and meat industry are composed of nearly 30% proteins, 10% fats and 60% water and are covered with hairs, dirt, blood and other connective tissues. To clean leather, a consortia/combination of various enzymes such as lipases, proteases, etc. are used. Proteolytic enzymes, especially alkaline proteases, remove hairs from hides and cause relaxation of collagen fibre without damaging them. This helps in retaining elastomeric value. They also gently remove non-fibrillar proteins such as albumin and globulin. Thus, relatively large amounts of enzymes are required (0.1-1.0% w/w) and the process is carried out in controlled conditions to avoid reducing the quality of leather (Rao *et al.*, 1998).

After dehairing, the hides are often defleshed and subsequently baited and delimed to produce soft leather clothing and goods. The use of alkaline proteases, having pH optima near 9-10, reduces the time required for dehairing from days to hours (Takami *et al.*, 1992; Riffel *et al.*, 2003; Gupta *et al.*, 2005). Traditionally, pancreatic proteases were used for baiting, which have been replaced by bacterial alkaline or neutral proteases, fungal acid proteases or a mixture of these depending upon the type of leather. *Bacillus* sp. are specific producers of extracellular proteases and find application in leather industry for unhairing and deliming. Protease obtained from different bacterial genus such as *Pseudomonas aeruginosa, Bacillus licheniformis, Bacillus subtilis* are used for baiting and leather processing (Ward, 1985; George *et al.*, 1995; Mane and Bapat, 2001). Novo Nordisk manufactures three different proteases, Aquaderm, NUE and Pyrase for use in soaking, dehairing and baiting of leather, respectively. Mukhopadhyay and Chandra (1993) have reported that alkaline protease produced by different bacterial strains can be used to inhair goat-skin, while in 2005, Najafi *et al.*, reported that protease isolated from *Pseudomonas aeruginosa* PD100 could act in the presence of SDS and Tween 80 and can be employed for dehairing of cow skin.

Bajza and Vrcek (2001) reported that the enzymatic process yields product of improved quality. The use of blench stable and alkali tolerant protease, produced by *Aspergillus flavus,* in leather processing is reported by Mulimani *et al.* (2002). Huang *et al.* (2003) demonstrated dehairing function of alkaline serine protease produced by *Bacillus pumilus*.

Biocatalystic leather processing involves the use of a mixture of enzymes, among which proteases, lipases and carbohydrases are well exploited for various pre-tanning stage (Saravanabhavan *et al.*, 2004; Thanikaivelan *et al.*, 2004). Keratinolytic proteases lacking collagenolytic and having mild elastolytic activities are increasingly being explored for the dehairing process. They would help in selective breakdown of keratin tissues in the follicle, thereby pulling out intact hairs without affecting the tensile strength of leather (Macedo *et al.*, 2005).

(e) Textile Industry

The biological treatment of wool and silk textile is a more ecofriendly finishing process because it is less abrasive to chemical treatment (Gupta *et al.*, 2005). Structurally, both wool and silk are protein fibers and consist of protein core and an outer layer. The outer scaly layer of wool comprises a fiber cuticle that is insoluble and hard to remove (Word *et al.*, 1993). In preparation of wool for textile use, the scaly layer of wool is peeled off partly or completely by digestion. Traditionally, digestion was achieved by using oxidizers such as NaClO and $KMnO_4$, but this often results in disproportionate treatment causing loss in strength of fiber.

The enzymatic treatment of wool textile using proteases results in shrink resistance, improved lusture, softness, plump, reduced facing up which significantly improves the pilling performance of garments along with reduced prickle (Bishop *et al.*, 1998; Breier *et al.*, 1999; Xiao-Wei *et al.*, 2005). In addition, treatment with improved protease selectively removes parts of wool cuticle (Heine *et al.*, 2001; Schroeder *et al.*, 2006). Effective modification of wool fabrics using enzymatic hydrolysis can be attained without causing serious loss in fiber strength (Sawada and Ueda, 2004; Queiroga *et al.*, 2007). The use of less aggressive chemical processes in combination with enzymatic treatments have also gained widespread acceptance (Jose, 1996; Cortez *et al.*, 2002). Several other enzymes like cellulase, xylanase along with proteases are being used today for processing of textiles. In 2001, Yongquan reported the use of *Aspergillus usamii* L86 enzymes including acid proteases and CMCase in degradation of inner and outer layer to peel off the scaly layer of wool.

In textile industries, proteases are also employed to remove the stiff and dull gum layer of sericine from the raw silk fiber to achieve improved luster and softness (Silva *et al.*, 2004; Najafi *et al.*, 2005). Earlier, degumming was performed in an alkaline solution containing soap

(Kanehisa, 2000). This is a harsh treatment, since fibrin is also attacked. Use of proteolytic enzymes provides better method as it protects fibre damage and that results in comparatively stronger silk threads than traditional treatments (Rissen and Antranikian, 2001; Freddi *et al.*, 2003).

Vankar *et al.* (2007) have reported enzymatic natural dyeing of cotton and silk fabrics with natural dyes. They used an enzyme blend containing proteases, amylase, diastases and lipase that showed rapid and higher dye absorption. However, the repeated use of enzymes containing detergents for soaking and washing of wool fabrics and garments may cause loss in strength of wool (Friedman, 1971). Recently, transglutaminase mediated grafting of silk and wool protein leads to improved physical and mechanical properties (Cortez *et al.*, 2004).

(f) Blood Decolorization

Blood, an under utilized source of animal protein, is a waste product of slaughter houses. In order to convert it into food grade protein, it is hygienically collected and treated with proteases. The red cell fraction with 75% protein is hydrolysed and decolourized using fungal proteases (Drepper *et al.*, 1981). Alcalase have also been employed for decolourization of slaughter house blood (Novo technical leaflets). The protein hydrolysate thus obtained is spray dried and agglomerated and can be incorporated in sausages and meat pies.

(g) Therapeutic Usage

The wide diversity and specificity of proteases of microbial origin are used to great extent in developing effective therapeutic agents. Proteases find application in dental care and medical therapy. In dental care, a mutanase-proteinase mixture has been used to reduce dental plaque and improve the clinical condition of palatal mucosa or denture stomatitis (Budtz-Joergensen *et al.,* 1977). Proteases are used as digestive aids in gastrointestinal disorders such as dyspepsia, hydrolysis of fibrin and fibrinogen, treatment of dermal ulcers and burns, and as potential bacteriocidal agent (North, 1982).

Oral administration of proteases from *Aspergillus oryzae* (Luizym and Nortase) has been used as an digestive aid to correct certain lytic enzyme deficiency syndrome (Rao *et al.*, 1998). Collagenase preparations incorporated into ointments were used for topical application in the treatment of burns (Mandl, 1961; Zimmermann, 1971) and the debridement of ulcera and decubiti (Mazurek, 1971). Gelatinases are specific in playing a prominent role in treatment of diseases such as cancer and arthritis (Browner *et al.*, 1995; Purva *et al.*, 1998). Keratinases are used in medicine and in pharmaceuticals for elimination of acne or psoriasis, elimination of human callus for preparation of vaccines for dermatophytosis and additives in skin lightening agents as they stimulate keratin degradation (Vignardet *et al.*, 2001).

(h) Peptide Synthesis

Peptides differ from amino acids and proteins in their nutritional and physiological properties. Some of them possess biological activity of pharmaceutical and medicinal value (Kimura *et al.*,

1999 a and b). Peptides can be synthesized chemically, but enzymatic peptide synthesis is possible using proteases, that catalyzes the reverse reaction under certain kinetically controlled conditions. Enzymatic methods offer several advantages over chemical synthesis such as shifting of thermodynamic equilibrium to favor synthesis, enantio-selectivity, low cost and no requirement for side chain protection. They do not require protecting groups, organic solvents and hazardous chemicals (Morihara, 1974). Peptide synthesis using proteases was first reported by Bergman and Frankel-Courat (1937). Since last three decades proteases are in frequent use for peptide synthesis using reverse enzymatic reaction of hydrolysis (Aso, 1989; Barbas and Wong, 1987; Clapes *et al.*, 1987; Blanco *et al.*, 1991; Chen *et al.*, 1998; Gill and Vulfson, 1993; Kasche, 1996; Kumar *et al.*, 2003; Obreza and Gobee, 2004). Kumar and Bhalla (2005) reviewed the role of microbial proteases in peptide synthesis and have given a comprehensive recount of site specificity of various proteases. Aspartate, serine and metalloprotease are being employed for the synthesis of peptides. The carboxypeptidases γ exhibited a remarkable esterase and amidase activity typical of serine protease, thus these are implicated in peptide synthesis (Gill *et al.*, 1996; Kasche; 1996; Morihara *et al.*, 1980, Morihara, 1987). Proteases have been used for the synthesis of several μ-α protected dipeptides with Z-Phe-Leu-NH_2 as a model (Clapes *et al.*, 1997). Kumar *et al.*, (2003) have evaluated the efficacy of proteases of *Bacillus* sp. APR – 4 in various dipeptide synthesis. Synthesis of tripeptides has also been carried out by several workers (Barros *et al.*, 2000; So *et al.*, 2000; Filippova and Lysogorskaia, 2003).

Immobilization is the most widely used strategy to achieve operational stability of enzymes and for product recovery without much loss of catalyst. It is the choicest technique in the application of enzyme catalysis to synthetic reactions in organic solvents, particularly in peptide synthesis due to their increased stability in organic solvents, easy recovery of enzyme catalyst and their reusability (Wilson *et al.*, 1994). Zhang *et al.* (1996) studied the effect of immobilizing support on the catalytic properties of the enzyme by studying the dependence of water, pH, ionic strength and temperature on the yield of model dipeptide Boc-Phe-Val-Ome in ethyl acetate under thermodynamic control. Several model dipeptide and tripeptide and oligopeptide also have been synthesized in water restricted media using free and immobilized, suspended and chemically modified enzymes (Barbas and Wong, 1987; Noritomi and Kise, 1987; Wong *et al.*, 1988; Kimura *et al.*, 1990 a and b; Gill *et al.*, 1996; Clapes *et al.*, 1997; Barros *et al.*, 1999; Filippova and Lysogorskaia, 2003). Peptide synthesis using immobilized protease has been widely preferred. Some immobilized proteases used in peptide synthesis are shown in Table 6.

A variety of peptides including aspartame, lysine sweet peptide, kyotorphin, angiotensin, the enkephalins and dynorphin have been successfully synthesized applying enzymic approach. The proteases used in the synthesis of small peptides of pharmaceutical and nutritional interest include enkephalin (Kullman, 1979; Kimura *et al.*, 1990a), aspartame precursor (Nakanishi *et al.*, 1990) and some nutritional dipeptides and tripeptides (Kimura *et al.*, 1990b and Monter *et al.*, 1991; Serralheiro *et al.*, 1994). Recently Kumar and Bhalla (2005) reviewed the work on microbial proteases in peptide synthesis. The readers can find a comprehensive account of the research approaches and applications of peptides in this review. Biologically active peptides synthesized using proteases are shown in Table 7. Peptide research is gaining importance because of their wider applicability, and hence to overcome problems in peptide synthesis, several technologies are being actively explored. Some advances in this direction include (i) Modification in biocatalyst by protein engineering (Abrahamsen *et al.*, 1991; Wang *et al.*, 1992; Gill *et al.*,

Table 6 : Protease inhibitors and their properties.

Serine Protease Inhibitors			
Inhibitors	Inhibition spectrum	Working concentration	Properties and Storability
Leupeptin	Inhibits trypsin-like serine proteaes such as trypsin, chymotrypsin, chymase, pepsin and thrombin. Inhibits selected cysteine processes such as calpain, cathepsin B, H & L and papain.	10–100 μM	Soluble in water, stable 6 months at –20°C
PMSF	Broad spectrum serine protease inhibits some cysteine proteases such as papain.	0.1–1.0 mM	Properties are lost during storage, prepared fresh in anhydrous ethanol or isopropanol (200 mM)
AEBSF	Broad spectrum serine protease inhibitor. Inhibits some cysteine proteases such as papain	0.1–1.0 mM	Solution in water (100 mM) stable 1 month at –20°C
Chymostatin	Inhibits chymotrypsin-like serine proteases such as chymase cathepsins A, B, D and G. Inhibits some cysteine proteases such as papain	10–100 μM	Solution in DMSO (10 mM) stable at –20°C
3,4–Dichloroisocoumarin	Broad spectrum serine protease inhibitor	5–100 μM	10 mM in DMSO stable at –20°C
TLCK	Inhibits chymotrypsin-like serine proteases	10–100 μM	Prepared fresh at 10 mM in 1 mM HCl
TPCK	Inhibits chymotrypsin-like serine proteases	10–100 μM	10 mM in ethanol stable at 4°C
DEP	Highly toxic, cholinesterase inhibitor. Broad spectrum serine protease inhibitor Hydrolyzes rapidly in aqueous solutions	0.1 mM	Prepare in anhydrous isopropanol at 200 mM

Table 7: Peptide synthesis by immobilized proteases

Protease	Support used for immobilization	Peptide formed
Papain	Celite, CM-cellulose, QAE-Sephadex, Egg white protein	Boc-Phe-Val-OMe
Papain	QAE-Sephadex A-25, Trisacryl MCM	Boc-Arg-Glu-OBu^t
Neutrase	Celite 545, Polyamide EP-700	Several N-α protected dipeptides with Z-Phe-Leu-NH_2 as a model
Subtilisin Carlsberg	Celite, porous glass	Z-Ser-Phe-NH_2
Subtilisin	Cryogel of Poly (vinyl alcohol)	Z-Ala-Ala-Xaa-Yaa-pNa
Thermolysin	Amberlite XAD-7	Z-L-Tyr-Gly-Gly-L-Phe-L-Leu-Oet
Thermolysin	Controlled pore glass derivatives, Polyamide, Celite support	α-L-Asp-Phe-OMe
α-Chymotrypsin	Porous chitosan beads	Ac-Tyr-GlyNH_2, Ac-Tyr-Ala-NH_2, Ac-Tyr-Leu-NH_2
Proteinase	Pore glass beads	Bz-Ala-Tyr-NH_2

1996; Takagi *et al.*, 2000; Wilking and Steward, 2004); (ii) Stereochemical modification involving recemization free condensation of oligopeptide fragments using proteases, accompanied by high level of racemization (Xans *et al.*, 1992; Barros *et al.*, 2000; Schug and Linder, 2005); (iii) Substrate mimetics, which are excellent tools for protease mediated peptide synthesis, that enable the coupling of peptides independently of the primary specificity of the enzyme without undesired cleavage of the newly formed peptide bonds. However, the approach may lead to unwanted cleavage due to native proteolytic activity of the biocatalyst (Grunberg *et al.*, 2000). Enzymatic peptide synthesis using eutectic mixture of substrates offers an interesting alternative to conventional solvent based media (Gill and Vulfson, 1993, 1994).

(i) Waste Treatment and Management of Waste

Proteinaceous waste including spoiled vegetables, eggs, meat, hairs, feathers, etc. are disposed in the environment everyday, while hair salons, poultry farms and certain industries such as leather industry, gelatin factories, pharmaceutical companies release a large amount of proteinaceous waste that poses a serious threat to the environment by polluting air, water and land. Extracellular proteinases from microbial origin have been found to play an important role in the management of proteinaceous waste and to obtain value added products from them.

In nature such waste materials are attacked by a variety of proteolytic microorganisms including fungi, bacteria and actinomycetes. Unattended growth of these organisms on the proteinaceous substrates causes problems of skin infections in the subjects exposed to them. Skin infections by fungi are specially important causing dermatomycosis in man and animals (Pier *et al.*, 1994; Gonzalez *et al.*, 1995; Jain and Jain, 2005). Thus, there is a strong need of a process to treat such waste in a controlled manner for its management in such a way so that certain products of economic value can be obtained from waste materials (Shin, 1993; Jain, 2000).

Protease produced by *Bacillus* sp. AH101was found to degrade human hair and useful for removal of hair clogs in drain pipes of bathrooms. This enzyme was found to be active at a wide range of temperature, 40-90°C (Takami *et al.*, 1992). The keratinolytic enzymes from *Streptomyces thermoviolacus* SD 8 was implicated for the treatment of keratinic waste in sewage (Chitte *et al.*, 1999). The enzyme showed its stability at pH 6.5–8.5 and at 65°C. An interesting application of alkaline protease was developed by Fujiwara and his co-workers in 1989. They reported the use of an alkaline protease to decompose gelatinous coating of X-ray films, while proteases of *Bacillus* sp. have been successfully used for bioprocessing of waste X-ray and photographic films for recovery of silver (Ishikawa *et al.*, 1993; Gajju *et al.*, 1996; Singh *et al.*, 1999). Nakiboglu *et al.* (2000) used enzyme extracts obtained from *Bacillus subtilis* to recover silver of high purity from waste photographic films.

A large amount of feather waste come out from poultry farms. This waste poses nuisance problem by contaminating the environment by pathogenic fungi. The feathers contain a large amount of keratin protein and hence are resistant for attack by most proteases. However, the organisms inhabiting feathers have been found to degrade them by virtue of keratinolysis followed by proteolytic attack. Novel and Nickerson (1959) carried out decomposition of native keratin by *Streptomyces fradiae*. Malviya *et al.* (1992) evaluated keratinolytic potential of fungal isolates from gelatin factory campus and claimed that the strain completely degraded keratinic substrates.

Sangal and Brandelli (2000) isolated a novel feather degrading *Vibrio sp.* with high keratinolytic activity, which completely degraded feather. Another report of a metalloprotease with keratinolytic activity was made by Allpress *et al.* (2002) that is employed for the processing of poultry waste. Gessesse *et al.* (2003) purified alkaline proteases produced by alkalophilic *Nesternkonia* and *Bacillus pseudofirmus* which can be employed for poultry waste management and the hydrolysates obtained can be used as an animal feed supplement.

Keratinases are also used for nutritional improvement of feathers as live stock feed resources (Onifade *et al.*, 1998 and Shih, 1993). Dalev (1994) used alkaline protease from *Bacillus subtilis* for the management of waste feathers from poultry slaughter houses. Proteases are also used to recover proteins from parts of animal and fish, which would otherwise go to waste after butchering (Watanabe, 2004).

Proteases along with other enzymes can be used for anaerobic digestion of poultry and other organic waste to generate fuels such as natural gas (Brutt and Ichida, 1999). They have also been applied for generating protein rich concentrated feather meal as poultry feed and for organic farming as a semi-slow-release nitrogen fertilizer (Hadas and Kautsky, 1994; Choi and Nelson, 1996).

FUTURE PROSPECTS

Besides the use of proteases in different sectors like detergent, medicine, cosmetics, leather processing, feed etc they find application in newer fields like prion degradation for treatment of the dread mad cow disease (Langeveld *et al.*, 2003). It would also prove useful for decontaminating medical instruments, lab equipment and interchangeable items like contact lenses and dentistry tools. They may also find application in production of biodegradable films, coatings and glues from proteinaceous and keratinic waste products like hairs, feathers, skin, fur, etc., which can be used for compostable packaging, agricultural films or edible film applications (Schrooyen *et al.*, 2001; Schrooyen *et al.*, 2004). For such applications, keratin structure can be chemically modified and hydrolysed to produce stable dispersions.

There is an urgent need today to develop biofuels to meet the demands of growing industrial sectors and population. Use of proteases along with other enzymes can speed up the biomethanation process from several days to hours. For this, search for proteases with wide substrate specificity and stability is required. Through protein engineering of proteases, the stability of enzyme at different pH, temperature, increase in enzyme activity (rate of reaction), robustness can be achieved (Kano *et al.*, 1997; Bryan, 2000; Tindbaek *et al.*, 2004). The engineered proteases can be further applied for various purposes such as generation of biofuel, bioremediation, improvement of wash performance in laundry detergents, therapeutics, treatment of cancer, vaccine development, curing of AIDS, etc.

Industrial applications of proteases have increased greatly during the past few years. Prospects of such applications are outstanding in particular with development of environment-friendly bioprocesses and in novel applications yet to be realized. Proteases offer distinct advantages for their use in industrial processes including great specificity of action, thermal and pH stability, etc. Due to such inherent benefits, many industries are keenly interested in switching to methods employing proteases over chemical methods as has been done in leather industries. Examples of some applications of proteases under thorough investigation include some novel paradigms.

Attempts are being made for conversion of feathers into mixtures of amino acids for poultry feed in our own laboratory and to develop process for feather waste management (unpublished data). Another recent application of proteases has been in clinical field. Proteolytic enzymes are used for debridement of wounds, and promising clinical results have been reported with streptokinase, trypsin, and chymotrypsin. However, clinical and therapeutic applications imply use of highly purified or crystalline enzymes and hence cost-effective and rapid processes for purification are desirable. Current research is devoted for finding novel and robust industrial proteases and methods of their application, improving yields of industrial microbial enzymes.

Several bacterial alkaline proteases including subtilisin Carlsberg, subtilisin BPN' and Savinase, are commercially available, with their major application as detergent enzymes. Induced mutagenesis has led to generation of robust protease preparations with improved catalysis, improved thermostability and high stability towards oxidizing agents and altered washing conditions. Newer preparations, such as Durazym, Maxapem are also a result of techniques of site-directed mutagenesis (SDM). The isolation and cloning of the gene encoding an alkaline protease represents an essential step in the engineering of most efficient producer microorganisms. Many bacterial genes encoding alkaline proteases have been cloned and expressed in hosts such as *E. coli* and *B. subtilis*. Feng *et al.* (2001) developed an engineered strain of *B. pumilis*, c172-14 by introducing the pBX 96 plasmid that produced 43% higher levels of protease than the parent strain has been developed by introducing the pBX 96 plasmid into the host strain.

Directed evolution and gene shuffling has also helped in generation of subtilisin variants with improved specificities and stability (Strausberg *et al.*, 1995). There still remains a huge possibility of modifying proteases through molecular approaches. Exploration of the microbial diversity including actinomycetes, fungi and marine microorganism are underway and have come up with some of the promising protease producers (Neklyudov *et al.*, 2000).

A number of proteases from microbial sources including bacteria, actinomycetes and yeasts have been crystallized for elucidation of their 3-dimensional structures. Three-dimensional structures of proteases from *B. amyloliquefaciens* (Bott *et al.*, 1988; Erwin *et al.*, 1990), Carlsberg subtilisin from *B. licheniformis* (Bode *et al.*, 1987; Niedhart and Petsko, 1988), subtilisin (Kuhn *et al.* 1998) and, opticlean from *B. alcalophilus* (Sobek *et al.*, 1990), proteinase K from *Tritirachium album* (Betzel *et al.* 1988), thermitase from *Thermoactinomyces vulgaris* (Teplyakov *et al.* 1990), and serine protease from *Penicillium cyclopium* (Day *et al.*, 1986; Koszelak *et al.* 1997) have been determined by X-ray crystallography. The crystallographic data revealed the mechanism and specificity of protease action. This has helped in understanding structure-function relationship of several proteases including subtilisin (Pantoliano *et al.*, 1988). Localization and functional relationship of Ca^{2+} ions in subtilisin Carlsberg (Bode *et al.*, 1987) and proteinase K (Betzel *et al.* 1988) has been elucidated on the basis of X-ray data.

Looking into the commercial success of proteases, workers have now started aiming at the discovery and engineering of novel enzymes that are more robust with respect to their pH and temperature kinetics using metagenomics (Lorenz *et al.*, 2001).

In future, protein engineering will offer possibilities of generating proteases with entirely new functions and newer applications. The pursuit of other, newer approaches, such as novel discovery strategies targeting new dimensions of molecular diversity and technologies to improve specificities and operational characteristics by induced evolutionary changes primary structures of proteases will certainly be the major field of development in coming years.

REFERENCES

Abe, K., Goto, M. and Nakashio, F. (1997). *J. Ferment. Bioeng.* **83:** 555-560.

Abrahamsen, L., Tom, J., Burnier, J., Butcher, K.A., Kossakoff, A. and Wells, J.A. (1991). *Biochemistry.* **30**: 4151-4159.

Adinarayana, K. and Ellaiah, P. (2002). *J. Pharm. Pharm. Sci.*, **5(3):** 272-278.

Aehle, W., Sobek, H., Amory, A., Vetter, R., Wilke, D. and Schomburg, D. (1993). *J. Biotechnol.* **28(1):** 31-40.

Allpress, J.D., Mountain, G. and Gowland, P.C. (2002). *Letters in Applied Mirobiology.* **34(5):** 337-342.

Andersen, L.P. (1998). Method for dehairng of hides or skins by means of enzymes. U.S. Patent No. 5, 834, 299.

Anglo, P.G. and Orillo, C.A. (1977). *Cohn. Philipp. J. Sci.* **106:** 1-10.

Anwar, A. and Saleemuddin, M. (1998). Alkaline proteases. A Review. *Bioresource Technology.* **6(3):** 175-183.

Arima, K., Iwasaki, S. and Tamura, G. (1967). *Appl. Microbiol.* **16:** 1727-1733.

Arima, K., Yu, J. and Iwasaki, S. (1970). *Methods in Enzymology,* **19:** 446-459.

Aso, K. (1989). *Agric. Biol. Chem.* **53:** 729-733.

Bajza, Z. and Vrcek, V. (2001). *Waste Management,* **21:** 79-84.

Banerjee, U.C., Sani, R.K., Azmi, W. and Soni, R. (1999). *Process Biochemistry.* **35(1):** 213-219.

Banik, R.M. and Prakash, M. (2004). *Microbiol. Res.* **159(2):** 135-140.

Barbas, C.F. and Wong, C.H. (1987). *J. Chem. Soc. Chem. Commun.* 533-534.

Barett, A.J. (1994). *Methods Enzymol.* **244:** 1-15.

Barett, A.J. (1995). *Methods Enzymol.* **248:** 183.

Barrett, F.F. (1975). *In:* Enzymes in food processing 2nd ed. (Ed. Reed, G.). Academic Press Inc., New York, pp 301-330.

Barros, R.J., Wehtje, E. and Aldercreutz, P. (1999). *Enzyme Microb. Technol.* **24:** 480-488.

Barros, R.J., Wehtje, E. and Aldercreutz, P. (2000). *Biotechnol Bioeng.* **67:** 319-326.

Betzel, C., Pal, G.P. and Saenger, W. (1988). *Eur. J. Biochem.* **178**: 155-171.

Bishop, D., Shen, J., Heine, E. and Hollfelder, B. (1998). *J. Text. Inst.* **89:** 546-553.

Blanco, R.M., Alvaro, G. and Guisan, J.M. (1991). *Enzyme Microb. Technol.* **13:** 573-583.

Bockle, B. and Muller, R. (1997). *Appl. Environmental Microbiology* **63(2):** 790-892.

Bode, W., Papamokos, E. and Musil, D. (1987). *Eur. J. Biochem.* **166:** 673-692.

Borriss, R. (1987). *In:* Biotechnology. (Eds. Rehm, H. and Reed. G.) Weinheim Verlag Chemie, pp. 35-62.

Bott, R., Ultsch, M., Kossiakoff, A., Graycar, T., Katz, B. and Power, S. (1988). *J. Biol. Chem.* **263:** 7895-906.

Boyer, P.D. (1971). The enzymes. 3rd ed. Academic Press Inc., New York.

Breier, R. and Eugen, P. (1999). Felt-free finishing of wool using proteolytic enzyme, useful for flocks, slubbing, Yarn and Fabric. DE19807456.

Brenner, S. (1988). *Nature,* **334:** 528-530.

Brockle, B., Gahunky, B. and Miiler, R. (1995). *Appl. Environ. Microbiol.* **61:** 3705-3710.

Browner, M.F., Smith, W.W. and Castelhano, A.L. (1995). *Biochemistry* **34:** 6601-6610.

Brutt, E.H. and Ichida, J.M. (1999). Keratinase produced by *Bacillus licheniformis. U.S. Patent,* 5,877,000.

Bryan, P.N. (2000). *Biochem Biophys.* **1543:** 223-238.

Budtz-Joergensen, E., Kelstrup, J. and Scand, J. (1977). *Dent. Res.*, **85:** 209-215.

Chen, Y.X., Zhang, X.Z., Zheng, K., Chen, S.M., Wang, Q.C. and Wu, X.X. (1998). *Enzyme Microb. Technol.* **23:** 243-248.

Chitte, R.R., Nalawade, V.K. and Dey, S. (1999). *Letters in Applied Microbiology.* **28(2):** 131-136.

Choi, J.M. and Nelson, P.V. (1996). *J. Am. Hort. Sci.* **121:** 634-638.

Clapes, P., Velencia, G., Reig, E. and Gracia-Anton, J.M. (1987). *Appl. Biochem. Biotechnol.* **15:** 89-96.

Cortez, J., Bonner, P.L.R. and Griffin, M. (2002). A method for enzymatic treatment of textiles. W00204739.

Cortez, J., Philip, B.L.R. and Griffin, M. (2004). *J. Biotechnology.* **116:** 379-386.

Dalev, P.G. (1994). *Biores. Tech.* **48:** 265-267.

Day, J., Koszelak, S., Cascio, D. and McPherson, A. (1986). *J. Biol. Chem.* **261:** 1957-1961.

Dhar, S.C. and Sreenivasulu, S. (1986). *Leather Sci.* **4:** 23.

Douglas, M.W. and Parsons, C.M. (2000). *J. Appl. Poult. Res.* **9(1):** 74-80.

Drepper, G., Drepper, K. and Ludwig-Bush, H. (1981). *Fleischwirtschaft.* **61:** 1393-1395.

Erwin, C.R., Barnett, B.L., Oliver, J.D. and Sullivan, J.F. (1990). *Protein. Eng.* **4:** 87-97.

Etoh, Y., Shoun, H., Beppu, T. and Arima, K. (1979). *Agric. Biol. Chem.* **43:** 209-215.

Feng, Y.Y., Yang, W.B., Ong, S.L., Hu, J.Y. and Ng, W.J. (2001). *Appl. Microbiol. Biotechnol.* **57:** 153-160.

Fernandez-Lahore, H.M., Auday, R.M., Fraile, E.R., Cascone, O., Biscoglio de Jimenez Bonino, M., Pirpignani, L. and Machalinski, C. (1999). *J. Peptide Research,* **53(6):** 599.

Filippova, I.Y. and Lysogorskaia, E.N. (2003). *Bioorg. Khim.* **29:** 544-550.
Friedman, M. (1971). *Textile Research J.* **41(4):** 315-318.
Frokjaer, S. (1994). *Food Technol.* **48(10):** 86-88.
Fujiwara, N. and Yamamota, K. (1987). *J. Ferment. Technol.* **65(3):** 345.
Gajju, H., Bhalla, T.C. and Agrawal, H.O. (1996). *Indian J. Microbiology,* **36:** 153-155.
Garg S.K. and Johri, B.N. (1999). *In:* Thermophilic moulds in Biotechnology (Ed. Johri, B.N., Satyanarayana, T. and Olsen, J.) Kluwer Academic Publishers, The Netherlands, pp. 191-218.
Garg, S.K. and Johri, B.N. (1994). *Food Rev. International,* **10:** 313-355.
George, S., Raju, V., Krishnan, M.R.V., Subramanian, T.V. and Jayaraman, K. (1995). *Proc. Biochem.* **30(5):** 457-462.
Gessesse, A., Hatti-Kaul, R., Gashe, A.B. and Mattiason, B. (2003). *Enz. Microbiol. Technology* **32:** 512-524.
Gill, I. and Vulfson, E.N. (1993). *J. Am. Chem. Soc.* **115:** 3348-3349.
Gill, I. and Vulfson, E.N. (1994). *Trends Biotechnol.* **12:** 118-122.
Gill, I., Fandino, R.L., Jobra, X. and Vulfson, E.N. (1996). *Enzyme Microb. Technol.* **18:** 162-183.
Godfrey, T. and West, S. (1996). Industrial Enzymology, 2nd ed., Macmillan Publishers Inc., New York. pp. 3.
Gonzalez, J.F., Barcena, M.C., Gomez, F. and Amigot, A. (1995). *Mycopathologia* **129:** 79-80.
Gray, G.L., Hayenga, K., Cullen, D., Wilson, L.J. and Norton, S. (1986). *Gene,* **48:** 41-53.
Grunberg, R., Domgall, I., Gunther, R., Rall, K., Hofman, H.J. and Bordusa, F. (2000). *Eur. J. Biochem.* **267:** 7024-7030.
Gupta, R. and Ramanani, P. (2006). *Appl. Microbiol. Biotechnol.* **70(1):** 21-33.
Gupta, R., Beg, Q.K., Khan, S. and Chauhan, B. (2002). *Appl. Microbiol. Biotechnol.* **60(4):** 381-395.
Gupta, R., Chauhan, B., Ramnani, P. and Singh Rajni (2005). *In:* Microbial Diversity: Current Perspectives and Potential Applications (Eds. Satyanarayana, T. and Johri, B.N.), I.K. International Pvt. Ltd., New Delhi, pp. 769-789.
Hadas, A. and Kautsky, L. (1994). *Fertil. Res.* **38:** 165-170.
Hartley, B.S. (1960). *Ann. Rev. Biochem.* **29:** 45-72.
Hashem, A.M. (1999). *Bioresource Technol.* **70:** 203-207.
Heine, E., Hartwig, H. and Lone, D. (2001). Method for enzymatic treatment of wool. US2001003220.
Hibbs, M.S., Hasty, K.A., Sayer, J.M., Kang, A.H. and Mainardi, C.L. (1985). *J. Biol. Chem.* **260:** 2493-2500.
Hoffman, T. (1974). *Adv. Chem. Ser.* **136:** 146-185.
Huang, Q., Peng, Y., Li, X., Wang, H. and Zhang, Y. (2003). *Curr. Microbiol.* **46(3):** 169-173.
Hui, Z., Doi, H., Kanouchi, H., Matsuura, Y., Mohri, S., Nonomura, Y. and Oka, T. (2004). *Biochem. Biophys. Res. Commum.* **321(1):** 45-50.
Impoolsup, A., Bhumlratana, A. and Flegel, T.W. (1981). *Appl. Environ. Microbiol.* **42:** 619-628.
Ishikawa, H., Ishimi, K., Sugiura, M., Sowa, A. and Fujiwara, N. (1993). *J. Fermentation and Bioengineering,* **76(4):** 300-305.
Ivanko, O.V. and Varbanets, L.D. (2004). *Mikrobiol. Z.* **66(2):**11-24.
Ivanko, O.V., Varbanets, L.D. and Valahurova, O.V. (2002). Collagenase activity of *Streptomyces* sp. 1349. *Mikrobiol Z.* **64(6):** 21-27.
Jain, P.C. (2000). *In:* Environmental Pollution and its Management. Eds. Shrivastava, P. APH Publishing Corporation, New Delhi, pp 88-107.
Jain, P.C. and Jain, P.K. (2005). *In:* Microbial diversity: current perspectives and potential applications. (Eds. Satyanarayana, T. and Johri, B.N.), I.K. International Pvt. Ltd., New Delhi, pp. 1005-1024.
Jain, P.K., Jain Richa, and Jain, P.C. (2004). *Hind. Antibiot. Bull.* **46:** 29-33.
James, M.N.G., Hsu, I.N. and Delbaere, T.J. (1977). *Nature,* **267:** 808-813.
Jasvir, S., Gill, N., Devasahayam, G. and Sahoo, D.K. (1999). *Appl. Biochem. Biotechnol.* **76(1):** 57-63.
Jose, C. (1996). *J. Soc. Dyers and Colorists* **112:** 326-329.
Kanehisa, K. (2000). Woven or knit fabrics manufactured using yan dyed raw silk. U.S. Pat. No. **6:** 080,689.
Kano, H., Taguchi, S. and Momose, H. (1997). *Appl. Microbiol. Biotech.* **47(1):** 46-51.
Kasche, V. (1996). *In:* Proteolytic enzymes: a practical approach. (Eds. Beynon, R.J. and Bond, J.S.), IRL Press, Oxford. pp 125-143.
Keay, L. (1971). *Process Biochem.* **6:** 17-21.
Kimura, Y., Nakanishi, K. and Matsuno, R. (1990a) *Enzyme Microb. Technol.* **12:** 273-280.
Kimura, Y., Muraya, K., Araki, Y., Matsuoka, H., Nakanishi, K. and Matsuno, R. (1990b). *Agric Biol. Chem.* **54:** 3331-3333.
Kinsella, J.E. and Hwang, D.H. (1976). *Crit. Rev. Food Sci. Nutr.* **8:** 191-228.
Kocher, A., Choct, M., Ross, G., Broz, J. and Chung, T.K. (2003). *J. Appl. Poult. Res.* **12(3):** 275-283.

Koszelak, S., Ng, J.D., Day, J., Ko, T.P., Greenwood, A. and McPherson, A. (1997).

Kuhn, P., Knapp, M., Soltis, M., Ganshaw, G., Thoene, M. and Bott, R. (1998). *Biochemistry* **37:** 13446-13452.

Kullman, W. (1979). *Biochem. Biophys. Res. Commun.* **91:** 693-698.

Kumar, D. and Bhalla, T.C. (2005). *Appl. Microbiol. Biotech.* .....1-23.

Kumar, D., Chand, D., Sankhian, U.D. and Bhalla, T.C. (2003). *Indian J. Microbiol.* **43:** 131-133.

Kuriyama, N., Kuriyama, H., Julin, C.M., Lamborn, K. and Israel, M.A. (2000). *Hum. Gene Ther.* **11:** 2219-2230.

Labbe, J.P., Rebegrotte, P. and Turpine, M. (1974). *C.R. Acad. Sci. Paris.* **278(D):** 2699.

Langeveld, J.P.M., Wang, J.J., van de Wiel, D.F.M., Shih, G.C., Garssen, G.J., Bossers, A. and Shih, J.C.H. (2003). *J. Infect. Dis.* **188:** 1782-1789.

Law, B.A. and Wigmore, A.S. (1982). *J. Society of Diary Technology.* **35(2):** 75-77.

Loffler, A. (1986). *Food Technol.* **40(1):** 63-70.

Lorenz, P., Liebeton, K., Niehaus, F., Eck, J. and Zinke, H. (2001). **5:** 379.

Macedo, A.J., de-Silva, W.O.B., Gava, R., Driemeier, D., Henriques, J.A.P. and Termignoni, C. (2005). *Appl. Environ. Microbiol.* **71:** 594-596.

Madan Meenu, Dhillon, S. and Singh, R. (2002). *Indian J. Microbiology* **42:** 155-159.

Malviya, H.K., Parwekar, S., Rajak, R.C. and Hasija, S.K. (1992). *Indian J. Experimental Biology.* **30:** 103-106.

Mandl. I. (1961). Advances in Enzymology. Vol. 23, Ed. Nord, F.F., Interscience, New York pp. 163.

Mane, R.R. and Bapat, M.M. (2001). *Indian J. Experimental Biology,* **39:** 578-583.

Mazurek, I. (1971). *In:* Collagenase: First Interdisciplinary Symposium. (Eds. Mandl, I., Gordon and Breach), New York.

Monter, B., Herzog, B., Stehle, P. and Furst, P. (1991). *Biotechnol. Appl. Biochem.* **14:** 183-191.

Morihara, K. (1974). *Adv. Enzymol.* **41:** 179-243.

Morihara, K. (1987). *Trends Biotechnol.* **5:** 164-170.

Morihara, K., Oka, T., Tsuzuki, H., Tochino, Y. and Kanaya, T. (1980). *Biochem. Biophy. Res. Commun.* **92:** 396-402.

Mukhopadhyay, R.P. and Chandra, A.L. (1993). *Indian J. Exp. Biol.* **31(6):** 557-558.

Mulimani, V.H., Patil, G.N. and Prashanth, S.J. (2002). *Indian J. Microbiol.* **42:** 55-58.

Najafi, M.F., Deobagkar, D. and Deobagkar, D. (2005). *Electronic J. Biotechnology,* **8(2):** 5 pages.

Nakadi, T., Nasuno, S. and Iguchi, N. (1972). *Agric. Biol. Chem.* **36**: 261-268.

Nakanishi, K., Takeuchi, A. and Matsuno, R. (1990). *Appl. Microbiol. Biotechnol.* **32:**633-636.

Nakiboglu, N., Toschali, D. and Yasa, I. (2001). *Turk. J. Chem.* **25:** 349-353.

Nand, K., Srikanta, S., Rao, K.S.N. and Murthy, V.S. (1980). *Nahrung,* **24:** 859-868.

Neidhart, D.J. and Petsko, G.A. (1988). *Protein Eng.* **2:** 271-276.

Neklyudov, A.D., Ivankin, A.N. and Berdutina, A.V. (2000). *Appl. Biochem. Microbiol.* **36:** 452-459.

Nihalar, D. and Satyanarayana, T. (1992). **32(4):** 443-449.

Noritomi, H. and Kise, H. (1987). *Biotechnol Lett.* **9:** 383-386.

North, M.J. (1982). *Microbiological Reviews* **46:** 308-340.

Noval, J.J. and Nickerson, W.J. (1959). *J. Bacteriol.* **77:** 251-263.

Novo technical leaflets: Decolourization of slaughterhouse blood by application of Alcalase 0.61 (1981). Novo Alle, DK-2880 Bagsvaerd, Denmark, Novo Industri A/s Enzyme Division.

Obreza, A. and Gobec, S. (2004). *Curr. Med. Chem.* **11:** 3263-3278.

Okada, Y., Nagase, H. and Harris, E.S. (1986). *J. Biol. Chem.* **261:** 14245-14255.

Onifade, A.A., Al-Sane, N.A., Al–Musallam, A.A. and Al-Zarban, S. (1998). *Bioresource Technoliogy,* **66:** 1-11.

Palvannan, T., Thangavelu, A.U., Jayapragasam M. and Boopathy, R. (1998). *J. Scientific and Industrial Research.* **57:** 625-628.

Pantoliano, M.W., Whitlow, M., Wood, J.F., Rollence, M.L., Finzel, B.C., Gilliland, G.L., Poulos, T.L. and Bryan, P.N. (1988). *Biochemistry,* **27:** 8311-8317.

Pearl, L.H. (1987). *FEBS Lett.* **214**: 8-12.

Perea, A., Ugalde, U., Rodriguez, I. and Serra, J.L. (1993). *Enzyme Microb. Technol.* **15:** 418-423.

Phadatare, S., Rao, M. and Deshpande V. (1997). *Arch. Microbiol.* **166:** 414-417.

Pier, A.C., Smith, J.M.B., Alexiou, H., Ellis, D.H., Lund, A. and Pritchard, R.C. (1994). *J. Med. Vet. Mycol.* **32(1):** 133-150.

Porto, A.L.F., Campos Takaki, G.M. and Limafilho, J.L. (2001). *Appl. Biochem. Biotech.* **60:** 115-122.

Prakasham, R.S., Rao, Ch.S., Rao, R.S., Rajesham, S. and Sarma, P.N. (2005). *Appl. Biochem. Biotechnol.* **120(2):** 133-144.

Purva, S.K. Soni L.K. and Gupta, J.K. (1998). *Indian J. Microbiol.* **38:** 149.

Queiroga, A.C., Pintado, M.M. and Malcata, F.X. (2006). *Enzyme and Microbial Technology* **40(6):** 1491-1495.

Rawling, N.D. and Barett, A.J. (1993). *Biochem. J.* **290:** 205-218.
Rao, M.B., Tanksale, A.M., Ghatge, M.S. and Deshpande, V.V. (1998). *Microbiol. Mol. Biol. Rev.* **62:** 597-635.
Rebeca, B.D., Pena-Vera, M.T. and Diaz-castaneda, M. (1991). *J. Food Sci.* **56:** 309-314.
Riffel, A., Ortolan, S. and Brandelli, A. (2003). *J. Chemical Technology and Biotechnology* **78**: 855-859.
Rissen, S. and Antranikian, G. (2001). *Extremophiles,* **5:** 399-408.
Robbin, F.M., Allens, A.L., Walker, J.E. and Cohen, S.H. (1986). Meat tenderization with a proteolytic enzyme from *Trichoderma reesei.* U.S. Pat. No. **4:** 689.589.
Saeki, K., Hitomi, J., Okuda, M., Hatada, Y., Kageyama, H., Takaiwa, M., Kubota, H., Hagihara, H., Kobayashi, T., Kawai, S. and Ito, S. (2002). *Extremophiles* **6(1):** 65-72.
Sangali, S. and Brandelli, A. (2000). *J. Appl. Microbiology,* **89(5):** 735-743.
Saravanbhavan, S., Aravindhan, R., Thanikaivelan, P., Rao, J.R., Nair, B.U. and Ramasami, T. (2004). *Clean Technol. Enviorn. Policy* **7:** 3-14.
Sardinas, J.L. (1966). Milk-curdling enzyme elaborated by *Endothia parasitica.* U.S. Pat. No. **3:** 275,453.
Sawada, K. and Ueda, M. (2004). *J. Chemical Technology and Biotechnology,* **79:** 376-380.
Schroeder, M., Lenting, H.B.M., Kandelbauer, A., Silva, C.J.S.M., Cavaco-Paulo, A. and Gubitz, G.M. (2006). Restricting detergent protease action to surface of protein by chemical modification. **72(4):** 738-744.
Schrooyen, P.M.M. and Radulf, B. (2004). Keratin based products and methods for their productions. U.S. Patent 20,040,210,039.
Schrooyen, P.M.M., Dijkstra, P.J., Oberthur, R.C., Bantjes, A. and Feijen, J. (2001). *J. Agric. Food Chem.* **49:** 221-230.
Schug, K.A. and Linder, W. (2005). *J. Am. Soc. Mass Spectrum.* **16:** 825-834.
Serralheiro, M.L.M., Parazeres, D.M.F. and Cabral, J.M.S. (1994). *Enzyme Microb. Technol.* **16:** 1064-1073.
Shannon, J.D., Baramova, E.N., Bjarnason, J.B. and Fox, J.W. (1989). *J. Biol. Chem.* **264:** 11575-11583.
Shih, J.C.H. (1993). *Poultry Science* **72:** 1617-1620.
Silva, C.J.S.M., Sous, F., Gubitz, G. and Cavaco-Paulo, A. (2004). *Food Technol. Biotechnol.* **42:** 51-56.
Sims, A.H., Dunn-coleman, N.S., Robson, G.D. and Oliver, S.G. (2004). *FEMS Microbiology Letters,* **239(1):** 95-101.
Singh, J., Vohra, R.M. and Sahoo, D.K. (1999). *Biotechnol. Lett.* **21:** 921-924.
So, J.E., Shih, J.S. and Kim, B.G. (2000). *Enz. Microb. Technol.* **26:** 108-114.
Sobek, H., Hecht, H.J., Hofmann, B., Aehle, W. and Schomburg, D. (1990). *FEBS Lett.,* **274:** 57-60. 91:755-760.
Sood, V.K. and Kosikowski, F.V. (1979). *J. Dairy Sci.* **62:** 1865-1872.
Sternberg, M. (1972). *Biochem. Biophys. Acta.* **285:** 383-392.
Strausberg, S.L., Alexander, P.A., Gallagher, D.T., Gilliland, G.L., Barnett, B.L. and Bryan, P.N. (1995). *Biotechnology,* **13:** 669-673.
Szabo, L., Benedek, A., Szabo, M.L. and Barabas, G. (2000). *World J. Microbiol. Biotechnol.* **16:** 252-255.
Takagi, H., Hirai, K., Maeda, Y., Matsuzawa, H. and Nakamori, S. (2000). *J. Biochem.* **127:** 617-625.
Takami, H., Nakamura, S., Aono, R. and Horikoshi, K. (1992). *Biosci. Biotech. Biochem.* **56(10):** 1667-1669.
Tam, J.J. and Whitaker, J.R. (1972). *J. Dairy Sci.* **55:** 1523-1531.
Tang, X.M., Sen, W., Lakay, F.M., Shao, W.L., Wang, Z.X., Prior, B.A. and Zhuge, J. (2004). *Biotechnol Lett.* **26(12):** 975-979.
Teplyakov, A.V., Kuranova, I.P., Harutyunyan, E.H., Vainshtein, B.K., Frommel, C., Hohne, W.E. and Wilson, K.S. (1990). *J. Mol. Biol.* **214:** 261-279.
Thangam, E.B. and Rajkumar, G.S. (2002). *Biotechnol. Appl. Biochem.* **35(2):** 149-154.
Thanikaivelan, P., Rao, J.R., Nair, B.U. and Ramasami, T. (2004). *Trends Biotechnol.* **22:** 181-188.
Tindbaek, N., Svendsen, A., Oestergaard, P.R. and Draborg, H. (2004). *Protein Engineering, Design and Selection.* **17:** 149-156.
Tsai, Y.C., Yamasaki, M., Yamamoto-Suzuki, Y. and Tamura, G. (1983). *Biochem. Int.* **7(5):** 577-83.
Vankar, P.S., Shanker, R. and Verma, A. (2007). *J. Cleaner Production,* **15(15):** 1441-1450.
Vignardet, C., Guillaume, Y.C., Michel, L., Friedrich, J. and Millet, J. (2001). *Int. J. Pharm.* **224**: 115-122.
Wang, H.L., Ruttle, D.L. and Hesseltine, C.M. (1969). *Can. J. Microbiol.* **15:** 99-104.
Wang, P., Hill, T.G., Warthow, C.A., Huston, M.E., Ohler, L.M., Smith, B.M., Bednarski, M.D. and Callstrom, M.R. (1992). *J. Am. Chem. Soc.* **114:** 378-380.
Ward, O.P. (1985). *In:* Comprehensive Biotechnology–The Principles, Applications and Regulation of biotechnology in Industy, Agriculture and Medicine. (Eds. Blanch H.W., Drew, S. and Wang, D.I.C.), Pergman Press. Oxford, New York **3:** 789-818.
Watanabe, K. (2004). *Appl. Microbiol. Biotechnol.* **63:** 520-526.
Wilhelm, S.M., Collier, I.E., Kronberger, A., Eisen, A.Z., Marmer, B.L., Grant, G.A., Bauer, E.A. and Goldberg, G.I. (1987). *Proc. Natl. Acad. Sci. USA* **84:** 6725-6729.

Wilking, S.D. and Steward, N. (2004) *J. Biotechnol.* **112:** 109-114.
Wilson, S.A., Daniel, R.M. and Peek, K. (1994) *Biotechnol Bioeng.* **44:** 337-346.
Wolff, A.M., Showell, M.S., Venegas, M.G., Barnett, B.L. and Wertz, W.C. (1996). In. Subtilisin enzymes: practical protein engineering. (Eds. Bott, R. and Betzel, C.), Plenum Press, New York, pp 113-120.
Wong, C.H., Matos, J.R., West, J.B. and Barbas, C.F. (1988). *J. Ind. Microbiol.* **29(3):** 171-177.
Woods, F.C., Bruinsma, B.L. and Kinsella, J.W. (1980). *Cereal Chem.* **57:** 290-293.
Word, R.J., Willis, H.A., George, G.A., Guise, G.B., Denning, R., Evans, D.J. and Short, R. (1993). *Textile Res. J.* **63:** 362-368.
Xans, N., Albericio, F., Jorba, X., Calvet, S., Clapes, P., Torres, J.L. and Valencia, G. (1992). *Int. J. Pept. Protein Res.* **39:** 528-532.
Xiao-Wei, Yu, Wen-Jun, G., Yong-Quan, L., Ting-Jing, G. and Ji-Dong, Z. (2005). *Brazilian Archives of Biology and Technology* **48(5):** 675-680.
Yogei, A.; Tuc, Y. and Soyletir, G. (2002). *New Microbiol.* **25(2):** 223-229.
Yamashita, T., Higashi, S., Higashi, T., Machida, H., Iwasaki, S., Nishiyama, M. & Beppu, T. (1994). *J. Biotechnol.* **32:** 17-28.
Yong, F.M. and Wood, B.J.B. (1974). *Adv. Appl. Microbiol.* **17:** 157-194.
Yongquan, Li. (2001). *Mycosystema.* **20:** 196-200.
Yu, J., Iwasaki, S., Tamura, G. and Arima, K. (1968). *Agric. Biol. Chem.* **32:** 1051-1052.
Zhang, X.Z., Wang, X., Chen, S., Fu, X., Wu, X. and Li, C. (1996). *Enzyme Microb. Technol.* **19:** 538-544.
Zimmermann, W.E. (1964). *In:* Physiology and treatment of burns. (Ed. Lorthioir, J.) Presses Acad. Europ. Brussels. pp. 413.

Chapter 12

Electrocatalytically Active Laccases: Application for Biosensor Development

SARIKA KUHAD AND RAMESH CHANDER KUHAD

Lignocellulose Biotechnology Laboratory, Department of Microbiology, University of Delhi South Campus, Benito Juarez Road, New Delhi–110 021, (India)
E-mail: kuhad@hotmail.com

The most promising approach for the development of electrochemical biosensors is to establish a direct electrical communication between the biomolecules and the electrode surface. In regard to this approach, laccase has a great potential, as it is capable of electrocatalyzing reactions (oxygen electroreduction) via direct electron transfer and simultaneously, also reduce the reaction over-voltage considerably. The construction of laccase-based biosensor has many advantages because of its simplicity (as it does not require any cofactor for its catalysis), low cost of fabrication and the reagentless detection of analyte. This review focuses on advances, directions and strategies in the development of third generation electrochemical biosensors utilizing laccase as biomolecule. Subjects covered include a brief description of the fundamentals of the electron transfer phenomenon and amperometric biosensor development. The analytical applications of laccase-based biosensor for insulin immunomonitoring, determination of phenols, reagentless detection of enzyme inhibitors, rapid detection of *Escherichia coli* density, monitoring of lignin and detection of plant flavonoids are also discussed.

INTRODUCTION

During the past two decades, bioelectrochemistry has received increased attention (Gorton *et al.,* 1999; Armstrong and Wilson, 2000; Gilardi and Fautuzzi, 2001). Progress of bioelectrochemistry has been integrated into analytical applications, e.g., in biosensor (Janata, 1992). The combination of oxidoreductases and amperometric electrodes is by far the most commonly studied biosensor concept and through various strategies the enzyme reaction can be followed easily (Habermüler *et al.,* 2000). These devices have many favorable analytical characteristics, such as selectivity, sensitivity, portability, speed, low cost and potential for miniaturization. Thus, biosensors offer exciting opportunities for numerous decentralized analytical applications and they are quickly becoming useful tools in medicine, food quality control, environmental monitoring and other practical fields. In principle, biosensors can be tailored to match individual analytical demands for almost any target molecule or compound that interacts selectively with a biological system (Vo-Dinh *et al.*, 2000; Rosatto, 2001; Scouten *et al.*, 1995).

Several enzymes are able to catalyze direct (mediatorless) electron transfer. This phenomenon is called bioelectrocatalysis. An important point characterizing bioelectrocatalysis is the catalytic removal of the reaction over-voltage. In terms of energy profile also, it reduces the energy of activation for the electrochemical transformation of the substrate into a product. The application of the phenomenon of bioelectrocatalysis for the development of biosensors is associated with the ability of biocatalysts to form molecular transducers. The molecular transducer is a molecular layer formed by electrocatalytically active enzyme in intimate contact with the electrode surface, and this is capable of selective transformation of a chemical signal of the substrate into an electrical one. Among electrocatalytically active oxidoreductases, laccase, the lignin-degrading enzyme is of special interest. This enzyme is able to catalyse important reaction for biosensor development, namely oxygen electroreduction. Notably, the reduction of reaction over-voltage caused by these enzymes is as high as 400 ± 500 mV. This factor results in a high potential for the use of this enzyme in the development of biosensors (Ghindilis, 2000).

BASIC CONCEPT

Biosensor is a device consisting of biologically active material used in close conjunction with a device that will convert a biochemical signal into a quantifiable electrical signal.

The main components of a biosensor (Fig. 1) include biocatalyst, which converts substrate to product. This reaction is determined by the transducer (key component), which converts it to an electrical signal. The analog signal produced at this stage may be output directly, but is usually converted to a digital signal and passed to a microprocessor stage where the data is processed, converted to concentration units and output to a display device or data store.

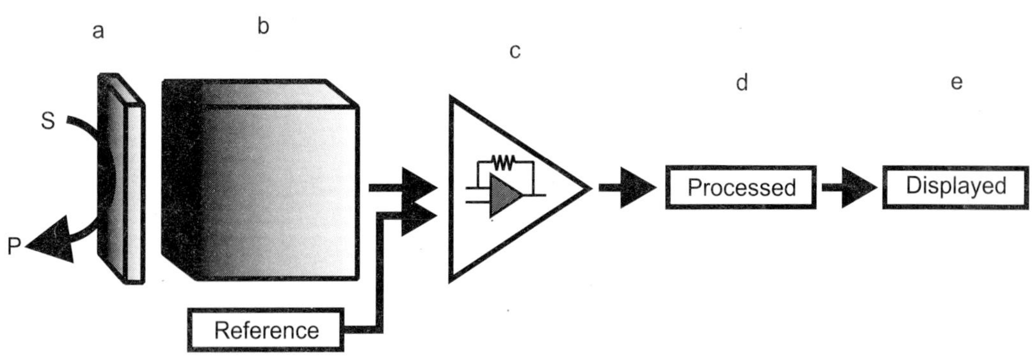

Fig. 1: Schematic diagram showing the main components of a biosensor (a: Biocatalyst; b: Transducer; c: Amplifier; d: Microprocessor and e: Display device)

The transducer makes use of a physical change accompanying the reaction, which are as follows:

(i) calorimetric biosensors: the heat output (or absorbed) by the reaction,
(ii) potentiometric biosensors: changes in the distribution of charges causing an electrical potential to be produced,
(iii) amperometric biosensors: movement of electrons produced in a redox reaction,
(iv) optical biosensors: light output during the reaction or a light absorbance difference between the reactants and products, or
(v) piezo-electric biosensors: effects due to the mass of the reactants or products.

A successful biosensor must possess at least some of the following beneficial features:

(i) The biocatalyst must be highly specific for the purpose of the analyses, be stable under normal storage conditions and should show good stability over a large number of assays.
(ii) The reaction should be independent of physical parameters such as stirring, pH and temperature. This would allow the analysis of samples with minimal pretreatment. If the reaction involves cofactors or coenzymes, these should, preferably, also be co-immobilised with the enzyme.
(iii) The response should be accurate, precise, reproducible and linear over the useful analytical range, without dilution or concentration. It should also be free from electrical noise.
(iv) If the biosensor is to be used for invasive monitoring in clinical situations, the probe must be tiny and biocompatible, having no toxic or antigenic effects. If it is to be used in fermenters, it should be sterilizable. This is preferably performed by autoclaving, but no biosensor enzymes can presently withstand such drastic wet-heat treatment. In either case, the biosensor should not be prone to fouling or proteolysis.
(v) The complete biosensor should be cheap, small, portable and capable of being used by semi-skilled operators.
(vi) There should be a market for the biosensor. There is clearly little purpose developing a biosensor if other factors (e.g. government subsidies, the continued employment of skilled analysts, or poor customer perception) encourage the use of traditional methods and discourage the decentralization of laboratory testing.

DIFFERENT APPROACHES FOR THE DEVELOPMENT OF BIOSENSORS

An amperometric biosensor may be more attractive due to its high sensitivity and wide linear range. Thus, amperometric enzymatic electrodes hold a leading position among the presently available biosensor systems. These devices combine the selectivity of the enzyme for the recognition of a given target analyte with the direct transduction of the rate of the biocatalytic reaction into a current signal, allowing a rapid, simple and direct determination of various compounds (Wang, 1999).

There are three so-called 'generations' of biosensors (Fig. 2). The electronic coupling between redox enzymes and electrodes for the construction of amperometric biosensors can be based on the electroactivity of the enzyme substrate or product (first generation biosensors); utilization of redox mediators, either free in solution or immobilized with the biomolecule (second generation

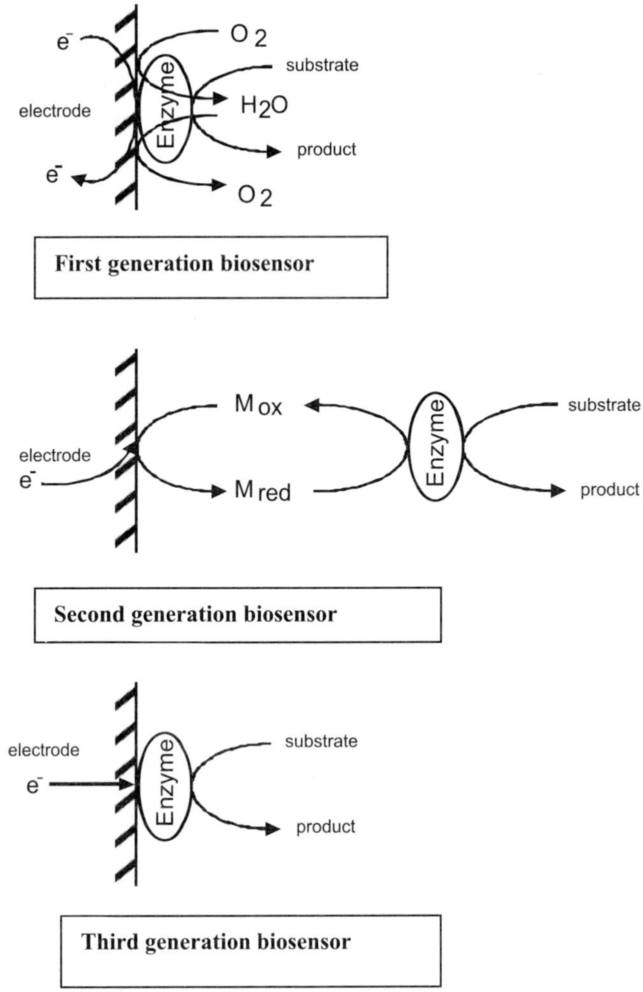

Fig. 2: Schematic representation of different biosensor generations (Freire *et al.*, 2003).

biosensors), or direct electron transfer (DET) between the redox-active biomolecule and the electrode surface (third generation biosensors).

The drawbacks with first generation biosensors, such as too high-applied potentials, put the focus on the use of mediators, which are small redox active molecules that could react with the active site of the enzyme and with the electrode surface, shuttling the electrons between the enzyme and the electrode. The use of mediators became possible in decreasing the applied potential for many redox enzyme-based biosensors. Unfortunately, the redox mediators used in conjunction with redox enzymes facilitate not only the electron transfer between electrode and enzyme but also various interfering reactions (Gorton *et al.,* 1999).

In the third generation biosensors, the electron transfer is associated with, or occurs during, the catalytic transformation of the substrate to the product. The redox enzyme acts as an electrocatalyst, facilitating the electron transfer between the electrode and the substrate molecule involving no mediator in this process (Ghindilis *et al.,* 1997). Thus, this kind of biosensor usually offers better selectivity, because they are able to operate in a potential range closer to the redox potential of the enzyme itself, becoming less exposed to interfering reactions. The higher integration between the biomolecule and the electrode surface can also improve the sensitivity of this kind of biosensor.

APPLICATIONS OF BIOSENSOR USING LACCASE

Although third generation biosensors present favorable characteristics, only a few groups of enzymes or proteins were found to be capable of interacting directly with an electrode while catalyzing the corresponding enzymatic reaction. Depending on the practical significance of the substrates of these enzymatic reactions, electroanalytical applications of bioelectrocatalysis began to appear in the late eighties (Ghindilis *et al.,* 1997). The phenomenon of direct electron transfer in enzymes was first described for a laccase (Tarasevich *et al.,* 1979). Laccases belong to the small group of enzymes called blue copper proteins or blue copper oxidases. The others members of this group are ascorbate oxidase and ceruloplasmin. More than 60 types of laccase have been reported from different sources, such as insects, plants, fungi and bacteria (Kuhad *et al.,* 1997). Although the copper center is similar for all laccase enzymes, differences in thermodynamic and kinetic properties are observed, depending on the source of the enzyme (Thurston, 1994; Yarapolov *et al.,* 1994). In general, the laccases exhibit four neighboring copper atoms, which are distributed among different binding sites. They are classified into three types: copper types 1, 2 and 3, which are differentiated by specific characteristics that allow them to play an important role in the catalytic mechanism of the enzyme. According to Call and Mucke (1997), copper types 1 and 2 are involved in electron capture and transfer, while copper types 2 and 3 are involved in binding with oxygen.

A comparative study of the redox transformations between laccase from different sources (*Coriolus hirsitus* and *Rhus vernicifera*) and other copper containing oxidases (ascorbate oxidase and ceruloplasmin), which have very close substrate specificities and high similarities of the prosthetic group structure, was performed (Ghindilis *et al.,* 1987 and Yaropolov *et al.,* 1996). It was observed that both ceruloplasmin and tyrosinase showed electrochemical redox reactions in anaerobic conditions. However, they did not have bioelectrocatalytic properties, while laccase from both sources possesses a significant ability to catalyze oxygen electroreduction. Therefore, it was proposed that redox transformation of the prosthetic groups of the enzymes is not sufficient to obtain electrocatalysis. An important parameter, which defines the possibility of bioelectrocatalysis with redox enzymes, is the kinetic mechanism of their catalytic action in homogeneous reactions. The major difference between these enzymes is in their mechanism of action. Laccase catalyzes according to a "ping-pong" or sequencial mechanism (Petersen, 1978), while ceruloplasmin and ascorbate oxidase act by formation of a ternary donor-enzyme-acceptor complex (Ghindilis *et al.,* 1987 and Yaropolov *et al.,* 1996). In the case of a mechanism of catalysis with the formation of a ternary complex, rigorous conditions for the structure of this complex are required and the limiting step of oxygen reduction is different from that observed for the laccase enzyme.

Immunosensors

The application of enzyme-catalysed direct electron transfer for the development of immunosensors has several advantages. For immunosensors, the detection principle is based not on detection of a substrate by a molecular transducer (as in the case of enzyme sensors), but on detection of formation/disintegration of the molecular transducer itself. The enzyme laccase has shown potentials. Its substrates are atmospheric oxygen and an electron directly transferred from the electrode. This permits design of a virtually reagentless immunosensor. The basic principle of the detection of the immunointeraction is presented in Fig. 3.

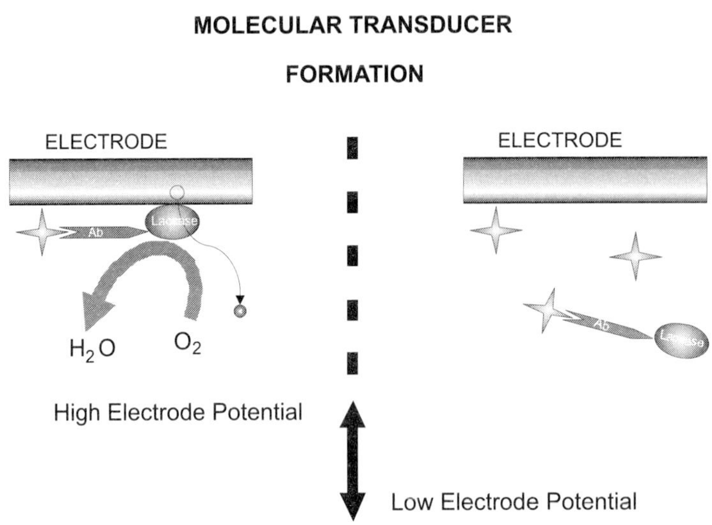

Fig. 3: Molecular transducer formed by laccase based bioelectrocatalysis during immunointeraction (Ghindilis, 2000).

The electrode is modified with antigen or hapten, and when laccase conjugate interacts with the antigen it forms a molecular transducer. The electrode surface is coated with a layer of laccase-antibody conjugate to carry out immunointeraction. There is electrode potential shift between the electrode before immunointeraction, and after the interaction (approx. 400 mV). The potential shift itself is a characteristic of the area of electrode surface covered by laccase immunoconjugate. Therefore, the electrode potential shift can be used for quantification of the immunointeraction occurring on the surface of the electrode. It is notable that the conjugate in the bulk solution not attached to the electrode surface is unable to catalyse the electron transfer and does not affect the electrode potential. Therefore, phase separation is not needed. A sensing element on the basis of either a sandwich-assay or a displacement-assay format does not need components to start the measurement other than the analyte. This allows for the development of a reagentless immunoprobe.

The sensing element based on a sandwich assay format requires a separation membrane impermeable to the immunoconjugate, but permeable to the antigen (analyte). The primary

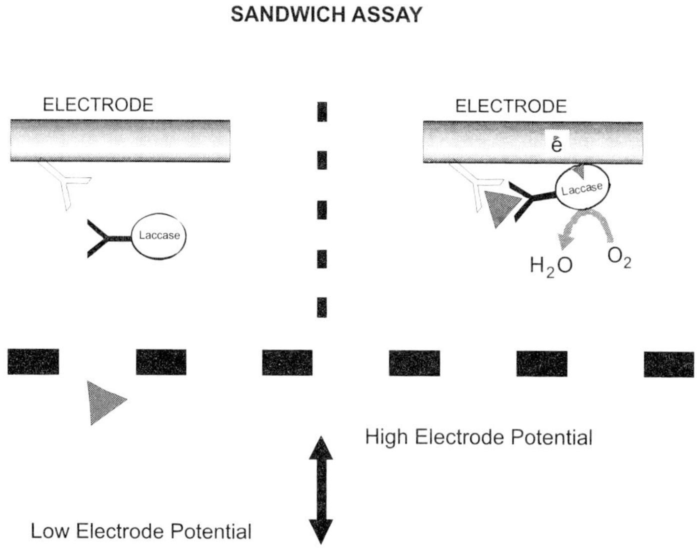

Fig. 4: The sensing element based on a sandwich assay format (Ghindilis, 2000).

antibody is immobilized on the electrode surface, whereas the secondary antibody coupled with a laccase label is non-immobilized. The diffusion of the analyte through the separating membrane towards the electrode results in the formation of the complex antibody$_1$-antigen-antibody$_2$-laccase (Fig. 4). This leads to an increase in the electrode potential caused by the electrocatalytic removal of reaction over-voltage. Such a sensor can be designed as a reagentless assay probe.

On the basis of a displacement scheme, free antigen (analyte) causes the displacement of the laccase-antigen conjugate from the electrode surface. This results in a decrease in the electrode potential that is proportional to the concentration of free antigen (Ghindilis, 2000).

An alternative version of the displacement based sensing element employs an immuno-interaction separated from the electrode surface. In this case, an antibody-laccase conjugate is modifed additionally by biotin. This conjugate is bifunctional, having a site for immunointeraction with the analyte, as well as a site for specific interaction with an avidin-modified electrode. The electrode surface modified with avidin is able to capture specifically this conjugate due to the avidin-biotin interaction. This permits reagentless quantification of the immunoconjugate on the electrode when immunointeraction occurs elsewhere (Fig. 5). The rate of increase of avidin electrode potential is proportional to the concentration of the conjugate. This detection principle allows for the development of a reagentless detection with immunochromatography or immuno-filtration.

The application of biotin-modified laccase immunoconjugates for the development of all solid immunoprobes based on the displacement scheme has the advantage of high technological flexibility. The probe employs a separating membrane and an immunomembrane situated at some distance from the avidin-modified electrode (Fig. 6). Diffusion of the analyte through the separating membrane onto the immunomembrane results in displacement of the laccase-antibody-biotin conjugate from the immunosorbent. The conjugate then diffuses towards the

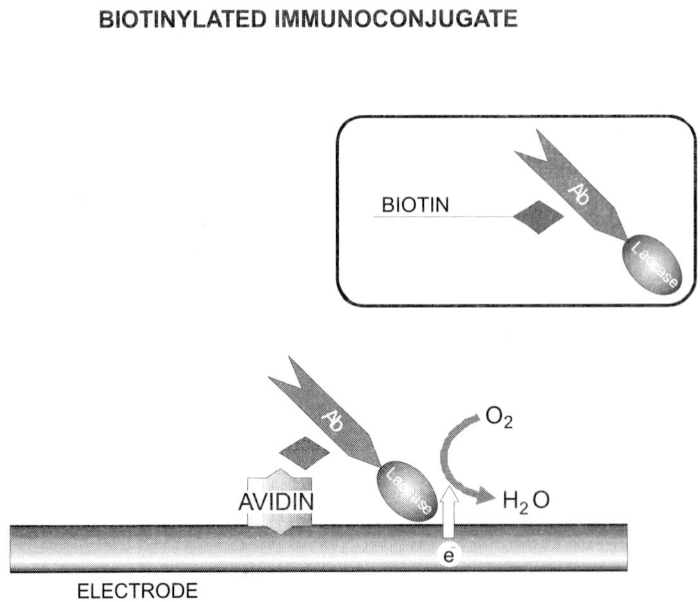

Fig. 5: Formation of biotin-labelled immunoconjugate on the surface of an avidin-modified electrode resulting in signal transduction (Ghindilis, 2000).

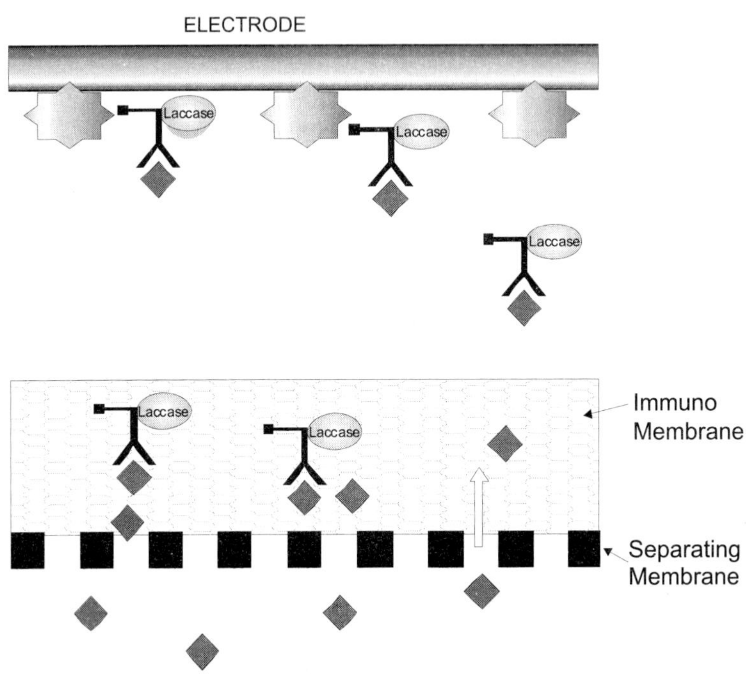

Fig. 6: Schematic representation of a displacement-based sensing element employing a bifunctional laccase conjugate (Ghindilis, 2000).

electrode and attaches to its surface by means of the biotin-avidin interaction. This results in an increase of the electrode potential, the rate of which is proportional to the concentration of the analyte. A low detection limit for insulin immunomonitoring below micromolar levels has been achieved (Ghindilis, 2000).

Hence, the advantages of the application of laccase catalysed direct electron transfer for development of immunosensors are associated with: (i) the reagentless nature of the assay procedure; (ii) all solid format of the immunoprobes; and (iii) simplicity and low cost of fabrication, which allows for manufacture of disposable sensing elements.

A potential for the practical application of immunoprobes on the basis of laccase-catalysed direct electron transfer is associated with high throughput screening for new drug development, individual quantitative immunoassay and application in combination with non-invasive or minimal invasive sampling systems for scheduled routine immuno-monitoring. Two new enzymatic methods have been developed to quantify morphine and codeine simultaneously in a flow injection system (FIA). The enzyme sensor for morphine is based on laccase (LACC) and PQQ-dependent glucose dehydrogenase (GDH) immobilized at a Clark oxygen electrode. Morphine is oxidized by laccase under consumption of oxygen and regenerated by glucose dehydrogenase. Since laccase cannot oxidize codeine, this sensor is selective for morphine. Morphine is detected between 32 nM and 100 μM. The method has been applied for the quantitative analysis of codeine or morphine in drugs (Bauer *et al.*, 1999).

Determination of Phenols

Phenolic compounds constitute a large group of organic compounds, which are widely distributed throughout the environment. These are used in many industrial processes, e.g., in the manufacture of plastics, paper, dyes, pesticides and antioxidants (Canofeni *et al.*, 1994, Nistor *et al.*, 1999). Certain phenols and related aromatic compounds are highly toxic, carcinogenic and allergenic (Deichmann and Keplinger, 1981; Vianello *et al.*, 2006), and due to their toxic effects, their determination and removal in the environment are of great importance. In general, phenolic compounds are subjected to chromatographic separation before detection. However, it utilizes expensive and nonportable equipment and the separation is time-consuming and often requires pre-concentration. Many biosensors have been developed, which permit the detection of phenols in aqueous solutions at concentrations in the low micromolar range with minimal sample preparation (Vianello *et al.*, 2004). These have been developed with enzymes such as tyrosinase, peroxidase, laccase etc. Among them, laccase have wide substrate specificity, and also, it does not require H_2O_2 as co-substrate and any cofactors for its catalysis and hence the construction of biosensor becomes simple. Therefore, biosensor-utilizing laccase has a great potential for the determination of phenolic compounds (Duran and Esposito, 2000; Freire *et al.*, 2002).

Freire and coworkers (2002) combined laccase biocatalysis with the screen-printed technique, which resulted in effective single-use biosensors for phenolic compounds. Laccase was immobilized over the screen-printed electrode surface by simultaneous cross-linking and covalent binding methods. The sensitivity of this biosensor was in the nanomolar range, with detection limits of 50 and 30 nmol L^{-1} for guaiacol and chloroguaiacol, respectively. The new device displayed an excellent stability allowing measurements for over 1 month. The experiments described above illustrated the ability to employ this biosensor for monitoring phenol compounds in environmental samples. The field monitoring capability is coupled to a selective, sensitive, reversible response and low cost of production.

Haghighi and coworkers (2003) reported the use of laccase modified graphite electrode as the working electrode in an amperometric flow-through cell for monitoring phenolic compounds in a single line flow injection system. Moreover, the sensor characteristics were calculated and compared with those known for biosensors based on laccase from *Coriolus hirsutus*, cellobiose dehydrogenase (CDH) from *Phanerochaete chrysosporium* and horseradish peroxidase (HRP). This comparison clearly illustrated the different selectivity, which justified the idea to combine different enzymes on electrode arrays for increasing the recognition power of detection systems directed towards analysis of complex samples such as wastewaters. Quan and Shin (2004) utilized DeniLite laccase covalently immobilized on Platinum (Pt) electrode. It showed very stable and fast response and outstanding long-term stability for the detection of catechol and catecholamines with submicromolar detection limit.

Cross-linked enzyme crystals (CLEC) of laccase from *Trametes versicolor* were embedded in 30% polyvinylpropylidone (PVP) gel and mounted into an electrode to make the sensor. The biosensor was used to detect the phenols in 50-1000 μmol concentration levels. Phenols with lower molecular weight (2-amino phenol, catechol) gave a short reponse time, whereas the higher molecular weight substrates like catechin and ABTS had comparatively a long response time (Roy *et al.*, 2005).

An amperometric biosensor, using laccase from *Rigidoporus lignosus* as bioelement was reported. A bioreactor connected to an amperometric flow cell by a FIA system was filled with the immobilized enzyme and the operational conditions of this biosensor were optimized as regards pH. This biosensor permited the detection of phenolics in aqueous solutions at concentrations in the nanomolar range and was successfully used to detect phenolics in wastewaters from olive oil mill without sample preparation (Vianello *et al.*, 2006).

Reagentless Detection of the Enzyme Inhibitors

The optimization of a reagentless enzyme sensor for the detection of azide (inhibitor) based on the mediated reduction of O_2 by a laccase enzyme co-immobilized in a redox hydrogel on electrode surfaces was reported. The sensor response was shown to be influenced by the enzyme loading, the electrolyte pH and ionic strength. The resulting enzyme inhibition biosensor could detect levels of azide as low as 2.5 μM under these conditions. Constant potential amperometric detection at the laccase enzyme electrode in a flow injection set-up yields a peak current for inhibition of the mediated reduction of O_2. Reproducible peak currents and areas are obtained for repeated injections of 100 ìM azide. Reproducible response curves can be obtained by injection of a 25 mM azide sample and assuming that the peak height and peak area obtained represent 100% inhibition of the enzyme (Leech and Daigle, 1998).

Rapid Detection of *Escherichia coli* (*E. coli*) Density

Tang and coworkers (2006) developed a new amperometric method for rapid detection of

Escherichia coli (*E. coli*) density using a bienzyme biosensor. The bienzyme biosensor was fabricated based on the covalent immobilization of laccase and horseradish peroxidase (HRP) at indium tin oxide (ITO) electrode by (3-aminopropyl) triethoxysilane (APTES) monolayer. The bienzyme biosensor showed a high sensitivity in determination of the polyphenolic compounds, which was microbially generated from the salicylic acid (SA) added into the culture medium during the course of *E. coli* metabolism. Since the amount of polyphenolic compounds depends on *E. coli* density, the bienzyme biosensor was applied for the rapid and high sensitive detection of *E. coli* density after the *E. coli* solution was incubated in culture medium with salicylic acid for 2.5 h at 37°C. The amplified response current at the biosensor was linear with the *E. coli* density ranging from 1.6×10^3 to 1.0×10^7 cells/mL. The detection limit was 9.7×10^2 cells/mL.

Laccase-based Biosensors for Monitoring Lignin

A fast and easy method for analytical determination of lignin and its degradation products is of great importance for various industries and for monitoring environmental pollution. Four different ways to detect Kraft and soluble pine lignin using either soluble or immobilised *Trametes hirsutus* laccase have been investigated by Shleev and coworkers (2006). First, the concentration of lignin was measured using laccase in homogeneous medium with a Clark electrode. Then, laccase immobilized onto aminated porous glass beads was used in a reactor submerged into the lignin containing solution and the reaction was followed with a Clark electrode. The third and fourth techniques involved lignin detection using laccase modified spectrographic graphite electrodes, either used in flow injection mode or in batch mode, where lignin solutions were injected into an electrochemical cell. It was shown that the use of laccase modified graphite electrodes in conjunction with flow injection for amperometric lignin detection is an attractive perspective, and data form the basis for designing laccase-based biosensors to be applied for measuring lignin and its model compounds in waste water, e.g., from the pulp and paper industry.

Amperometric Detection of Plant Flavonoids

The selective and sensitive detection of flavonoids and phenols from various sources is very important and their concentration should be carefully controlled both in medicinal and natural vegetable products. For example, naturally occurring polyphenols in beer are members of the proanthocyanidin series. These are monomers or polymers of three basic units: epicatechin, catechin, and gallocatechin, and after polymerization may form unfavorable hazes (Asano *et al.*, 1984).

The performed investigations by Jarosz-Wilkolazka and coworkers (2004) demonstrated that laccase from white rot fungus *Cerrena unicolor* could be applied for the construction of biosensors useful for determination of flavonoids and related compounds in flow-injection system. In this study, laccase from *C. unicolor* as a highly active enzyme was adsorptively immobilized on the surface of graphite electrode. The electrode inserted into flow-injection system served as a biosensor for detection of polyphenols, including flavonoids. Catechin hydrate, epicatechin, epicatechin gallate, prodelphinidin, and caffeic acid were used as a target compounds in determination. Electrodes modified with laccase yield responses for both simple compounds (caffeic acid) and compounds with three or more phenolic and non-phenolic rings, although with different sensitivity.

CONCLUSION

Biosensor technology is an open field for innovative approaches to analysis. Research to elucidate and modulate electron transfer mechanisms, direct or mediated, is indispensable for the success of biosensors as a future technology. Moreover, they are also important for understanding the fundamental principles of biological recognition and communication between enzyme and transducer in amperometric biosensors.

The mechanism of direct electron transfer in biosensors is not well understood as yet. There are a wide range of questions to be clarified: i) which parameters (structural or kinetic) determine the possibility that a redox enzyme catalyzes a given electrode reaction; ii) what is the role of the enzyme prosthetic group in the electrocatalysis and in the electron transfer process; iii) what is the influence of the nature of the electrode material and the structure of the electrode surface; and iv) what is the relationship between the catalytic mechanism of the enzyme action and the manifestation of direct electron transfer ability.

Despite of these challenges, laccase-based biosensors are taken into great consideration for the development of new approaches and devices in order to solve many analytical problems. The main advantage obtained with third generation biosensors using laccase as biomolecule is the development of simple analytical devices (reagentless sensors) with high sensitivity (due to a closer integration between the recognizer element, which is the signal generator, and the transducer) and, especially, high selectivity (due to the specific interactions between the biomolecules and their substrates). Having all these characteristics in a single sensing element is the goal of most analytical chemists.

REFERENCES

Armstrong, F.A. and Wilson, G.S. (2000). *Electrochimica Acta* **45(15-16):** 2623-2645.

Asano, K., Ohtsu, K., Shinagawa, K. and Hashimoto, N. (1984). *Agric. Biol. Chem. Jpn.* **48:** 1139-1146.

Bauer, C.G., Kühn, A., Gajovic, N., Skorobogatko, O., Holt, P.J., Bruce N.C., Makower A. Lowe C.R. and Scheller F.W. (1999). *Fresenius J. Anal. Chem.* **364** : 179-183.

Call, H.P. and Mucke, I. (1997). *J. Biotechnol.* **53:** 63.

Canofeni, S., Di Sario, S., Mela, J. and Pilloton, R. (1994). *Analytical Letters.* **27(9):** 1659-1669.

Deichmann, W.B. and Keplinger, M.L. (1981). *In Patty's Industrial Hygiene and Toxicity*, (Eds. Clyton, G.G. and Clyton, F.E.), Wiley, New York, pp. 2567.

Durain, N. and Esposito, E. (2000). *Appl. Catalysis B: Environmental,* **28(2):** 83-99.

Freire, R.S., Duran, N., Wang, J. and Kubota, L.T. (2002). *Analytical Letters,* **35(1):** 29-38.

Ghindilis, A. (2000). *Biochemical Soc. Transactions,* **28:** 84-89.

Ghindilis, A.L., Atanasov, P. and Wilkins, E. (1997). *Electroanalysis,* **9:** 661.

Ghindilis, A.L., Yaropolov, A.I. and Berezin, I.V. (1987). *Dokl. Phys. Chem.* **293:** 273.

Gilardi, G. and Fantuzzi, A. (2001). *Trends in Biotechnology,* **19(11):** 468-476.

Gorton, L., Lindgren, A., Larsson, T., Munteanu, F.D., Ruzgas, T. and Gararyan, I. (1999). *Anal. Chim. Acta,* **400:** 91-108.

Habermüller, K., Mosbach, M. and Schuhmann, W. (2000). *Fresenius J. Analytical Chemistry,* **366(6-7):** 560-568.

Haghighi, B., Gorton, L., Ruzgas, T. and Jonsson, L.J. (2003). *Analytica Chimica Acta* **487:** 3-14.

Janata, J. (1992). *Analytical Chemistry,* **64(12):** 196R-219R.

Jarosz-Wilkolazka, A., Ruzgas, T. and Gorton, L. (2004). *Enzyme and Microbial Technology,* **35:** 238-241.

Kuhad, R.C., Singh, A. and Eriksson, K.E. (1997). *In:* Advances in Biochemical Engineering Biotechnology, (ed. K.E. Eriksson), Springer-Verlag, Germany, pp. 46-125.

Leech, D. and Daigle, F. (1998). *Analyst,* **123:** 1971-1974.

Nistor, C., Emneìus, J., Gorton, L. and Ciucu, A. (1999). *Analytica Chimica Acta,* **387(3):** 309-326.

Petersen, L. and Degh, H. (1978). *Biochim. Biophys. Acta,* **526:** 85.

Quan, D. and Shin, W. (2004). *Electroanalysis,* **16:** 576-1582.

Rosatto, S.S., Freire, R.S., Duran, N. and Kubota, L.T. (2001). *Quim. Nova* **24:** 77.

Roy, J.J., Abraham, T.E., Abhijith, K.S., Kumar, P.V.S. and Thakur M.S. (2005). *Biosensors and Bioelectronics,* **21:** 206-211.

Scouten, W.H., Luong, J.H.T. and Brown, R.S. (1995). *Trends Biotechnol.* **13:** 178.

Shleev, S., Persson, P., Shumakovich, G., Mazhugob, Y., Yaropolov, A., Ruzgas, T. and Gorton, L. (2006). *Enzyme and Microbial Technology,* **39:** 835-840.

Tang, H., Zhang, W., Geng, P., Wang, Q., Jin, L., Wu, Z. and Lou, M. (2006). *Analytica Chimica Acta,* **562:** 190-196.

Tarasevich, M.R., Yaropolov, A.I., Bogdanovskaya, V.A. and Varfolomeev, S.D. (1979). *Bioelectrochem. Bioenerg.* **6:** 393.

Thurston, C.F. (1994). *Microbiology,* **140:** 19.

Vianello, F., Cambria, A., Ragusa, S., Cambria, M.T. and Rigo, A. (2004). *Biosens. Bioelectron.* **20:** 315-321.

Vianello, F., Ragusa, S., Cambria, M.T. and Rigo, A. (2006). *Biosensors and Bioelectronics,* **21:** 2155-2160.

Vo-Dinh, T. and Cullum, B. (2000). *Fresenius J. Anal. Chem.* **366:** 540.

Wang, J. (1999). *J. Pharm. Biomed. Anal.* **19:** 47.

Yaropolov, A.I., Kharybin, A.N., Emneus, J., Marko Varga, G. and Gorton, L. (1996). *Bioelectrochem. Bioenerg.* **40:** 49.

Yaropolov, A.I., Skorobogatko, O.V., Vartanov, S.S. and Varfolomeyev, S.D. (1994). *Appl. Biochem. Biotechnol.* **40:** 257.

Chapter **13**

Potential Applications of Microbial Phytases in Aquaculture

VENKATRAO KONASANI AND T. SATYANARAYANA*

Department of Microbiology, University of Delhi South Campus, Benito Juarez Road, New Delhi–110 021, (India)
E-mail : tsnarayana@gmail.com

Phytate phosphorus accounts for 60-80 % of the total phosphorus in plant organic matter, which is practically unavailable to the monogastrics such as fishes due to the lack of adequate levels of phytases in their digestive tracts. Phytases are a class of phosphatases, which catalyse the stepwise hydrolysis of phytic acid into lower inositol phosphates and inorganic phosphate. Various microbial phytases have been used in fresh water and marine aquaculture. The supplementation of phytases in the aquaculture feed has been shown to be effective in making phosphorus, minerals and proteins available, and thus, improving the growth performance of fishes, besides mitigating phosphorus pollution of water bodies.

INTRODUCTION

Aquaculture, one of the major food producing industries in the world, is growing rapidly to serve the world's increasing food demand. Several issues have been raised in this regard in the past few years, which need to be critically debated. One of these issues is the development of high quality fish feed from inexpensive sources and making it free from anti-nutritional factors (ANFs). Feed accounts for 30-60% of the total production cost of the farmed fishes due to extensive reliance on fishmeal. Fishmeal is the major constituent of the fish feed. It serves as a good source of amino acids, fatty acids and phospholipids. It is mainly prepared from the fishes caught from the wild. The wild sources have already reached their maximum biological limits. Even though fishmeal is highly palatable to fishes, there are some problems associated with extensive reliance on it. Fishmeal varies in quality, availability and price. It is predicted that global fishmeal production will remain at about 6 million tons (Higgs *et al.,* 1995). Moreover, it is possible that fishmeal production may decline from the present level because of increase in direct use of fishes in human diets to support population growth. Due to the increased demand, fishmeal prices are likely to be higher in the future. The cost and availability of the fishmeal forced the manufacturers to search for alternative cheap sources. Several plant proteins, especially some oil seed cakes appear to offer a great potential in this area (Rumsey *et al.,* 1993, Mwachireya 1995). Plant proteins are more advantageous than fishmeal due to their availability and low cost. Several anti-nutritional factors like phytic acid, lectins, saponins, phyto-estrogens, anti-vitamins, tannins, alkaloids, glucosinolates and others are present in plant based feeds. The presence of these anti-nutritional factors, however, limits the usage of plant based feed in aquaculture, since these affect growth and physiological characteristics of the fishes.

NUTRITIONAL IMPACT OF PHYTIC ACID

Utilization of oil cakes and other plant materials in aquaculture feed is limited by the presence of phytic acid and other anti-nutritional factors. Phytic acid (myo-inositol 1, 2, 3, 4, 5, 6, hexa kis phosphate) is commonly present in plant seeds and is a principal phosphorus source comprising 0·5 – 5% (w/w) of total dry weight (Reddy *et al.,* 1989). Phytic acid was discovered in 1855 by Hartig (1855), who isolated it from small non-starch grains from the seeds of several plant species. Pfeffer (1872) differentiated three different types of compounds in the grains: crystals of calcium oxalate, a protein substance and 'globoids'. Globoids are nitrogen free organic substances containing calcium, magnesium and phosphorus (P). These 'globoids' were later named 'inosite-phosphoric acid' as they could be hydrolyzed into inosite and phosphorus (Schulze *et al.,* 1896). Phytic acid is commonly found in different organs of plants (seeds and grains, roots, tubers, fruit, nuts, vegetables and pollens). According to Reddy *et al.* (2002), phytic acid mainly exists as salts of calcium, potassium and magnesium (phytate) and accumulates in protein bodies (globoids). In seeds and grains, phytate may account for over 60-80% of the total stored phosphorus. The distribution of phytate in seeds and grains varies in monocot and dicot plants. In monocots, such as rice, phytate is concentrated in the germ or aleuronic layer, and only traces are found in the endosperm. In dicot seeds (e.g., peas, tobacco), phytate is mainly found in the cotyledons. The amount of phytate in other organs of a plant is relatively low when compared with that in seeds and grains. In some species, phytate may account for 20 - 34% of the total phosphorus in tubers and fruits (Reddy *et al.,* 1989). Due to the presence of six phosphate groups, it forms chelate complexes with divalent cations like Zn^{2+}, Fe^{2+}, Cu^{2+}, and Mg^{2+}. Since monogastric animals like fishes have a little or no phytase activity in their gastrointestinal tract, they are not able to utilize the phosphorus present in the form of phytic acid, and therefore, it is excreted out undigested. Accumulation of phytic acid and its chelates in the ponds causes eutrophication.

METHODS OF PHYTATE REDUCTION IN THE FEEDS

A number of physical, chemical and enzymatic methods have been used for reducing the phytate content. Phytates can be partly degraded by extrusion processing (Sandberg *et al.,* 1987), and by soaking to activate endogenous phytase in wheat bran (Morris and Elli, 1980). Toma and

Tabekhia (1979) reported that cooking milled rice led to a large reduction in phytate content, while Satoh *et al.* (1998) recorded 30% reduction in phytate due to extrusion cooking of canola meal. A reduction in phytate content of soybean meal by treatment with alkaline solution (Hartmann 1979) and ion exchange resin (Niiyama, 1992) had also been shown to be possible. The soaking of phytase-supplemented diet in water at room temperature for 8–15 h improved P digestibility (Kemme and Jongbloed, 1993). According to a study of Näsi *et al.* (1995), soaking of a diet with whey for 3 h at 40°C also ameliorated the apparent P absorption. The steeping of a diet for 9 h at room temperature reduced the phytate content by 45% (Skoglund *et al.,* 1997). A similar observation was also recorded by Larsen *et al.* (1999). Microwave heating had, however, little effect on the phytate reduction in soybean meal (Hafez *et al.,* 1989).

Disadvantages of Physical and Chemical Methods of Phytate Reduction

Feed pre-treatment methods like dry/wet heating and solvent extraction for removing anti-nutrients has been successfully used, but sometimes they cause adverse effects on nutritional quality of the feed. Heat treatment alters the chemical nature of the feed and decreases the nutritional quality of proteins and carbohydrates. Excessive heating of oilseed meal during processing led to loss in the content and digestibility of amino acids (Rackis, 1974). Mwachireya *et al.* (1999) compared the effect of physical, chemical and enzymatic processing on the digestibility of commercial canola meal. Aqueous-methanol washing of sieved canola meal increased the phytate levels and neutral detergent fibre (insoluble), whereas further treatment with phytase in acidified aqueous media followed by filtration increased the levels of crude protein and concurrently decreased the levels of phytate (33%) as compared to the commercial canola meal.

PHYTASE: AN ECO-FRIENDLY FEED ENZYME

Phytases are a class of phosphatases, which catalyse sequential hydrolysis of phytic acid (*myo*-Inositol hexa kis phosphate) to inorganic phosphate and myo-inositol. Phytases are widely distributed among plants and microorganisms including fungi, bacteria and yeasts. Phytases are generally classified as acid, neutral and alkaline phytases based on the optimum pH for their activity. Based on their initial point of hydrolysis, phytases are classified as 3-phytases and 6-phytases. Recently Mullaney and Ullah (2003) classified phytases based on structural properties into histidine acid phosphatases, purple acid phosphatases and β-propeller phytases.

Microbial Sources

Phytases are widespread in nature including plants, animals and microorganisms. Among these sources, phytase activity has been frequently found in microorganisms.

Fungal Sources

Fungal phytases have generally received more attention due to high yields and acid tolerance (Kim *et al.,* 1998). More than 200 fungal species have been isolated out of which *Aspergillus* spp. are the most common producers of phytase. *Mucor, Rhizopus,* and *Penicillium* species are the other phytase-producing fungi. Howson and Davis (1983) examined eighty-four species from twenty-five fungal genera for the production of extracellular enzymes capable of hydrolyzing phytate. Among these, *Aspergillus ficuum* NRRL 3135 had the greatest phytase activity. *Aspergillus terreus*, *A. ficuum*, *A. oryzae* (Shimizu, 1992), *A. niger* (Volfova *et al.,* 1994), *A. fumigatus* (Rodriguez *et al.,* 2000), and *A. carbonarius* (Al-Asheh and Duvnjak, 1994) are some of the phytase-producing species of *Aspergillus*. *Rhizopus spp.* such as *R. oryzae, R. oligosporus,* and *R. stolonifer* are known to produce phytase (Howson and Davis, 1983). Thermophilic fungi such as *Thermomyces lanuginosus* and *Sporotrichum thermophile* have been reported to produce phytase at 65°C and 45°C, respectively (Berka *et al.,* 1998, Singh and Satyanarayana, 2006 a, b).

Bacterial Sources

Most bacteria produce low yields of phytase (Wodzinski and Ullah, 1996), which function at neutral pH (Liu *et al.,* 1998). Natural or genetically modified strains of bacteria such as *Escherichia coli*, *Bacillus subtilis*, *B. amyloliquefaciens* and *Klebsiella* sp. are some of the potent phytase producers. Two periplasmic *E. coli* phytases, called P1 and P2, were purified to about 16,500-fold with a molecular mass of 42 kDa (Greiner *et al.,* 1993). A bacterial strain capable of producing a thermo-acid-tolerant phytase was isolated from soil around haystacks and designated as strain PH01 (Popanich *et al.,* 2003). Another thermostable extracellular phytase from *Bacillus* sp. DS11 isolated from the Korean cattle shed soil was purified to homogeneity and had a molecular mass of 44 kDa (Kim *et al.,* 1998). When localization of phytase activity of *Enterobacter* sp. 4 was studied, 81.7% of the enzyme activity was found in the extracellular fraction, 4.4% in the periplasmic space, and the remainder was in the intracellular and cell-bound fractions (Yoon *et al.,* 1996). A cytoplasmic phytase from *Klebsiella terrigena* was purified about 410-fold to homogeneity, which had a molecular mass of 40 kDa (Greiner *et al.,* 1997). Phytase activity of 12 species of sour dough lactic acid bacteria was screened, wherein *Lactobacillus sanfranciscensis* possessed the highest levels of activity at pH 4.5 and 40°C (De Angelis *et al.,* 2003). Sreeramulu *et al.* (1996) screened different lactic acid bacteria for extracellular phytase production and found *Lactobacillus amylovorus* as the best producer.

Yeast Sources

Thermostable and acid stable phytases have been reported in *Pichia spartinae* and *P. rhodanensis*, which were active in the temperature range between 70 and 80°C at pH 4.5 (Nakamura 2000). When the localization of phytase in *P. anomala* was investigated, 78% of it was confined to cell wall, 16 % intracellular, 5% cell membrane associated and the remainder was extracellular (Vohra 2002). Cell-bound phytase was optimally active at pH 4.0 and 60°C (Vohra and Satyanarayana, 2001, 2002). *Schwanniomyces castellii* produced extracellular phytase, which was stable at 74°C for 1 h (Segueilha *et al.,* 1992). While Sano *et al.* (1999) characterized the extracellular phytase of *Arxula adenenivorans.*

PRODUCTION OF MICROBIAL PHYTASES

Phytase Production in Submerged Fermentation (SmF)

Several process parameters like agitation, aeration, pH, composition and viscosity of the culture medium, inoculum size as well as age and morphology are some of the parameters that influence the process productivity in submerged fermentation. The optimum temperature for the production of phytases from most of the microorganisms lies in the range 25 to 37°C. Generally, phytase producing microorganisms are mesophiles with some exceptions like *Thermomyces lanuginosus* (Berka *et al.,* 1998), *Talaromyces thermophilus* (Pasamontes *et al.,* 1997), *Rhizomucor pusillus* (Chadha 2004) and *Sporotrichum thermophile* (Singh and Satyanarayana 2006 a, b). Papagianni *et al.* (2001) studied the effect of inoculum age and morphology on phytase production in both SSF and SmF. While Gargova *et al.* (1997) screened 203 fungal strains for the phytase production and selected *A. niger sp. 307* due to its best extracellular phytase production. The phytase production by *Aspergillus ficuum NRRL* 3135 had been studied by several investigators (Shieh and Ware, 1968, Howson and Davis, 1983, Gibson 1987, Ullah and Gibson, 1987, Chelius and Wodzinski, 1994). Other fungi that are known to produce phytases include *Rhizopus oligosporus* (Howson and Davis, 1983), *A. oryzae* (Shimizu 1993), *A. carbonarius* (Al-Asheh and Duvnjak 1994), *A. niger* (Volfova *et al.,* 1994; Vats *et al.,* 2004). Kaur and Satyanarayana (2005) optimized medium components, cane molasses and urea using statistical tools for the cost-effective production of cell-bound phytase of *Pichia anomala*, and they recently reviewed the production of yeast phytases (Kaur and Satyanarayana, 2006).

Phytase Production in Solid State Fermentation (SSF)

SSF is defined as the fermentation involving solid substrates in the absence or near absence of free water. The substrate must possess enough moisture to support the growth and metabolism of microorganisms (Pandey, 2001). Solid state fermentation has received much attention in the recent years due to its advantages over submerged fermentation (SmF), which include high yield, simple equipment and process control, less expensive, lower waste water output, and simpler fermentation media (Pandey 2000). Roopesh *et al.* (2005) compared the production of phytase by *Mucor recemosus* on different oil cakes and wheat bran and their combinations (1:1). The mixture of wheat bran (WB) and sesame oil cake supported a high phytase production. Bogar *et al.* (2003) compared the phytase production by three *Mucor* and eight *Rhizopus* strains by solid state fermentation using three common feed ingredients: canola meal, coconut oil cake and wheat bran. The phytase production by *Mucor recemosus* was higher in SSF than that in SmF based on the initial substrate. Chadha *et al.* (2004) studied the phytase production using thermophilic fungus, *Rhizomucor pusillus* in solid-state fermentation. *Mucor racemosus* secreted a high phytase (26 IU g^{-1} DM) when cultivated on coconut oil cake supplemented with a solution containing glucose, ammonium sulphate, and casein in SSF (Ramachandran *et al.,* 2005). Gautam *et al.* (2002) used polystyrene as an inert solid support in SSF for the production of phytase from *Rhizopus oligosporus*. Mandviwala and Khire (2000) used thermotolerant *Aspergillus niger* for the production of extracellular phytase on agricultural residues such as wheat bran, mustard oil cake, cowpea meal, groundnut oil cake, coconut oil cake, cotton oil cake and black bean flour in SSF. Maximum enzyme production [108 U g^{-1} dry mouldy bran (DMB)] was attained on cowpea meal.

Fermentation of foods involves the action of microorganisms or enzymes that cause desirable biochemical changes and significant modifications in favour and texture. Several attempts have been made to reduce the phytic acid content in feed by using fermentation technology. Recently, Singh and Satyanarayana (2006a) have shown the possibility of dephytinization of sesame oil cake by cultivating thermophilic mould *Sporotrichum thermophile* in SSF. Batal *et al.* (2001) reported reduction of phytic acid in rapeseed meal by cultivating *Aspergillus niger*; 70% reduction in phytic acid content was observed at 70% moisture content. Nair and Duvnjak (1990) optimized the process parameters in SSF for the reduction of phytic acid using *Aspergillus ficuum* NRRL 3135 in canola meal. A complete reduction of phytic acid content in canola meal was recorded after 48 h of fermentation. Furthermore, the substitution of water in the semi-solid medium with acetate buffer resulted in a faster reduction of the phytic acid content and 15% increase in the amount of protein after 120 h. Solid-state fermentation using *Aspergillus carbonarius* with canola meal as a substrate showed that the production of phytase was growth-associated. Maximum enzyme production was achieved after 72 h. Twenty five percent and 10% increase in the protein contents of the canola meal was attained after 48 and 72 h, respectively. Total carbohydrate concentration, however, declined by 25% by the end of fermentation. The phytic acid reduction was optimum with moisture content between 53% and 60% (Al-Asheh and Duvnjak, 1995). The SSF process appears to be beneficial for the upgradation of canola meal by reducing its phytic acid content and increasing the amount of protein.

COMMERCIAL PHYTASES

Recent market trends have clearly shown that the enzymes have emerged as big feed supplements. Feed enzymes (protease, xylanase, phytase and amylase) are the newest segment of animal nutrition market, which is growing fast. Several major animal nutrition companies are getting involved in the area of making dietary enzymes. The first commercial phytase, which entered into the market in 1991, was manufactured by Gist Brocades (now DSM Corporation) and it was marketed under the trade name Natuphos® (Table 1). It is available in the form of granules, powder and liquid formulations. Natuphos® is a recombinant enzyme from *Aspergillus niger* that is classified as a 3-phytase, with hydrolysis of the phosphate moiety being initiated at the 3-position on the phytate molecule. Later many phytases appeared in the market but only a limited number of phytases are currently available. Most of the commercial phytases have been expressed in filamentous fungi. DSM corporation marketed its phytase under the trade name Ronozyme®, which was derived from *Peniophora lycii* and expressed in a yeast expression system. Biofeed phytase is the phytase product of Novozymes, which is derived from *Peniophora lycii* and expressed in *Trichoderma reesei*. Diversa corporation markets two phytases Phyzyme® and Quantum® phytase in collaboration with Danisco and Syngenta, respectively. AB Enzymes also

Table 1: Commercial phytases

Product	Source	Expressed in	Producer
Natuphos®	*Aspergillus niger*	NA*	BASF Corporation
Ronozyme®	*Peniophora lycii*	NA	Roche Ltd.
Biofeed phytase	*Peniophora lycii*	*Trichoderma reesei*	Novozymes
Phyzyme®	*Schizosaccharomyces pombe*	NA	Diversa corporation/Danisco
Quantum™ phytases	*E. coli*-derived phytase	NA	Diversa corporation/Syngenta
Cenzyme	Fungal	NA	Cenzone
Finase®	*E. coli*	*Trichoderma reesei*	AB Enzymes GmbH
EcoPhos®	appA2 gene of an *E. coli*	*E. coli*	ECP, Phytex LLC, Portland, ME

NA*= Not Available

entered into phytase market with the trade name Finase®, which is derived from *Aspergillus awamori* and expressed in *Trichoderma reesei.*

CHARACTERISTICS OF MICROBIAL PHYTASES

pH and Temperature Optima

Phytases can be classified into two major types acidic phytases and alkaline phytases, based on their pH optima. Most of the isolated phytases are acidic type, having pH optima around 5.0 (Table 2). The phytases isolated from *Bacillus* spp. showed optima at neutral or alkaline pH (Choi *et al.,* 2001, Kim *et al.,* 1998). *A. niger* phytase (phyA) exhibited two different pH optima, 5 and 5.5, respectively (Han and Lei, 1999). Mullaney *et al.* (2002) employed site-directed mutagenesis to remove the drop of activity in the pH range between 3 and 5 for maintaining a high specific activity at the stomach pH.

Table 2: Optimum pH and temperature for the activity of microbial phytases

S.No.	Organism	Optimum pH	Optimum Temp	Reference
1	*Aspergillus carneus*	5.6	40	Ghareib (1990)
2	*Pichia anomala**	4	60	Vohra and Satyanarayana (2002)
3	*Aspergillus carbonarius*	4.7	53°C	Al-Asheh and Duvnjak (1994)
4	*Pantoea agglomerans*	4.5	60	Greiner R.(2004)
5	*Cladosporium* sp. FP	3.5	40	Quan *et al.* (2004)
6	*Aspergillus niger* van Teighem	2.5	52-55	Vats and Banerjee (2005)
7	*Candida krusei* WZ-001	4.6	40	Quan *et al.* (2002)
8	*Citrobacter freundii*	4-4.5	40	Leo *et al.* (2006)
9	*Aspergillus caespitosus*	5.5	80	Guimaraes *et al.* (2004)
10	*Rhizopus oligosporus*	5.0	65	Casey and Walsh (2004)
11	*Klebsiella* sp.	—	—	Sajidan *et al.* (2004)
12	*Pseudomonas syringae*	5.5	40	Cho *et al.* (2003)
13	*Lactobacillus sanfranciscensis*	4	45	De Angelis *et al.* (2003)
14	*Citrobacter braakii*	4	50	Kim *et al.* (2003)
15	*Aspergillus niger*	5	65	Casey and Walsh (2003)
16	*Bacillus subtilis*	7.5	55	Wang *et al.* (2001)
17	*Bacillus* sp	6.5-8.5	40	Choi *et al.* (2001)
18	*Penicillium simplicissimum*	4	55	Tseng *et al.* (2000)

*Cell-bound

The temperature optima of most microbial phytases range between 45 and 60°C. These relatively high optimal temperatures preclude a full activity of phytases at the stomach temperatures of fishes and result in even poorer performance of phytases in the stomach of fish. The phytase to be applicable in aquaculture should have the optimum pH at an acidic range, which is prevailing in the stomach of the most fish species. The Arrhenius activation energy for the hydrolysis of sodium phytate by the phytate degrading enzymes from different sources ranges between 25 and 53 kJ mol^{-1} (Konietzny and Greiner, 2002).

Phytases: Specificity and Affinity for Substrate

Phytases commonly show broad substrate specificity. They hydrolyse the phosphate ester bond in a number of molecules like adenosine monophosphate (AMP), adenosine diphosphate (ADP), adenosine triphosphate (ATP), guanosine monophosphate (GMP), guanosine triphosphate (GTP), nicotinamide–adenine dinucleotide phosphate (NADP), p-nitrophenyl phosphate, phenyl phosphate, 1-naphthyl phosphate, 2-naphthylphosphate, fructose 1,6-diphosphate, fructose 6-phosphate, glucose 1-phosphate, glucose 6-phosphate, galactose 1-phosphate, α-glycerophosphate, β-glycerophosphate, pyridoxalphosphate, o-phospho-l-serine, pyrophosphate, etc. While the phytases isolated from *A. niger, A. terreus* CBS *and E. coli* were specific to phytic acid, phytases with broad substrate specificity are preferable for applications in food and feed industries (Wyss *et al.,* 1999a).

Most of the phytate-degrading enzymes characterized so far showed the highest affnity to phytate. The K_m values of the phytate-degrading enzymes studied so far lie within the range from <10 to 650 μM. Relatively low K_m values have been reported for the phytate-degrading enzymes from *A. niger* (10–40 μM), *A. terreus* (11–23 μM), *A. fumigatus* (< 10 μM), *Schwanniomyces castellii* (38 μM), *K. aerogenes* (62 μM), while phytases isolated from *Pichia anomala* (200 μM), *P. spartninae* (300 μM), *P. rhodanensis* (250 μM) showed high k_m values (Vohra and Satyanarayana, 2003).

BENEFITS OF PHYTASE SUPPLEMENTATION

Effects of Phytase Supplementation on P Availability

Phosphorus is a critical element and essential for every form of the life including aquatic organisms. Fishes and other aquatic organisms can obtain substantial amount of the required elements directly from their rearing water, but phosphorus is one essential element that must be supplemented in the diet. Phosphorus is a component of the teeth, bone, scales and is a precursor of phospholipids, nucleic acids and energy currency of life i.e. ATP. Daily requirements of phosphorus are in the range of 0.3-0.9% of diet for various fish species (NRC 1993). The fish with scales need more phosphorus in diet than the scale-less. In order to fulfill the P requirements of fishes, manufacturers add inorganic phosphorus supplements like mono- and di-calcium phosphates to the feed. Addition of high levels of phosphorus to the feeds causes several problems such as eutrophication of ponds.

Phytase supplementation proved to be effective in increasing P digestibility in poultry chicks. Improvement in phytate P availability was observed in Rainbow trout fed with soybean based feed supplemented with phytase (Table 3) (Cain and Garling 1995, Rodehutscord and Pfeffer 1995, Vielma *et al.,* 1998, Lanari *et al.,* 1998, and Storebakken *et al.,* 1998). Weerd *et al.,* (1999) reported a positive effect of phytase treatment on phosphorus digestibility and retention in African cat fish (*Clarius gariepinus*) fed with soybean based diets. Similar results were observed by Jackson *et al.* (1996) in channel catfish (*Ictalurus punctatus*). Up to the addition of 1000U of microbial phytase kg^{-1} diet, there is a clear dose response relationship between the inclusion level of microbial phytase and the digestibility of phosphorus in maize and soybean meal with a low activity of phytase from plant origin (Weremko *et al.,* 1997). For diets based mainly on maize and soybean meal, 500U phytase was found to be equivalent to 1 g phosphorus from mono-calcium phosphate. This is the equivalency recommended presently by the manufacturers (Hoppe and Schwarz 1993). Satyanarayana and Vohra (2003) developed a synergistic feed composition to enhance phosphorus availability in non-ruminants.

Effects of Phytase Supplementation on Protein and Carbohydrate Digestibility

At acidic pH in the gut, phytate forms binary complexes with proteins through electrostatic interaction of its charged phosphate groups with the free amino group on arginine or lysine residues present within the proteins. Formation of these binary phytate and protein complexes may affect the digestibility of proteins (Selle *et al.,* 2000). Phytate can also form ternary complexes with proteins in the presence of multivalent cations. These associations of phytic acid with cations result in the formation of either soluble or insoluble chelate complexes. There is a possibility to form four types of complexes which can alter protein digestibility. These include complexes already present in feed stuffs, *de novo* formation of protein-phytate, and phytate-free amino acid complexes during intestinal transit in the animal. Formation of these protein and phytate complexes makes the proteins less accessible to proteolytic enzymes, and some times, there is also a possibility for the *de novo* formation of phytate-protease complexes. Formation of phytate/protein complexes and non-availability of metal ions (cofactors), which are required for optimal activity of some proteolytic enzymes, leads to less digestibility of proteins. Sajjadi and Carter (2004) concluded that there was a significant negative effect of phytic acid on protein digestibility in Atlantic salmon, and *in vitro* trypsin activity was decreased by 40% in the presence of phytic acid. Carmovale (1988) tested the *in vitro* digestibility of protein by using the following food sources: faba bean, pea's whole flour, protein concentrate, protein isolate, zein, casein and others. In all cases, there was a negative effect of phytic acid up to the addition of 10 mg in any protein containing 10 ng of nitrogen after incubation for an hour at room temperature. This is evident strongly that the phytic acid-protein interactions negatively affect protein digestibility *in vitro*. Rainbow trout (*Oncorhynchus mykiss)* fed with the purified diets containing 0.5 or 1% phytic acid had a significant reduction in growth and feed efficiency (Spinnelli 1983), in contrast to the purified diets containing up to 1.5% of phytic acid, which had no adverse affect on growth or feed efficiency of catfish (*Ictalorus punctatus*) (Gatlin and Philips 1989). Improvement in protein digestibility in phytase pretreated and phytase added soybean meal based diet was

Table 3: Effects of phytase supplementation in aquaculture

S. No.	Fish species	Plant protein source and Inclusion rate	Phytase source	Effect on fish growth and nutrient availability	References
1.	*Oncorhynchus mykiss*	Diets were mainly based on soybean products and contained P only from plant materials	BASF Natuphos® 5000	At 10°C, P utilization increased from 6 to 25%, but intake of the diet as well as growth rate of trout were very low, irrespective of whether phytase was supplemented or not.	Rodehuttscord and Pfeffer, (1995)
2.	*Oncorhynchus mykiss*	—	BASF Natuphos®	Phytase supplementation did not significantly increase bone ash of fish fed with soy diets.	Vielma (2000)
3.	*Sebastes schlegeli*	30% Soybean meal and 1000U kg^{-1} phytase	BASF Natuphos® 5000	Improvement in apparent phosphorous digestibility and growth	Yeol Yoo *et al.* (2005)
4.	*Morone saxatilis*	Practical diets	BASF Natuphos® 5000	2400 PU kg^{-1} of enzyme resulted in bone mineralization and serum phosphorus concentrations equal to that observed with 13 g kg^{-1} dietary potassium mono-phosphate (PMP) addition (9 g kg^{-1} total P).	Hughes *et al.* (1998)
5.	*Oncorhynchus mykiss*	Canola meal	Finases 2x, Alko ltd, Finland	Dephytinized canola meal was found to be a potential substitute for fishmeal for rainbow trout and salmonoids	Mwachireya *et al.* (1999)
6.	*Oncorhynchus mykiss*	Full fat soybean	BASF Natuphos® 5000	Phytase supplementation (400 FTU/kg diet) in expelled soybeans increased apparent digestibility coefficients (ADCs) of crude protein, sulfur, total-phosphorus, and phytate-phosphorus significantly.	Cheng *et al.* (2003)
7.	*Oncorhynchus mykiss*	Soybean meal	Ronozyme® P and Experimental phytase	Top-sprayed phytase could increase the utilization of soybean meal-derived protein in rainbow trout	Vielma *et al.* (2004)
8.	*Oreochromis nyloticus*	24.5% SBM 11.5% Wheat gluten	Ronozyme®	Phytase supplementation significantly increased energy, protein and phosphorus utilization.	Liebert *et al.* (2005)
9.	*Salmo salar* L.	Feed with phytic acid+ 2000 U kg^{-1} phytase Feed without phytic acid+ phytase (2000U kg^{-1})	BASF Natuphos® 5000	The addition of phytic acid to practical diets significantly reduced the protein digestibility, but the phytase supplementation (2000 U phytase kg^{-1} of feed) neutralized the effect of phytic acid	Sajjadi *et al.* (2004a)

Table 3: Contd.

S. No.	Fish species	Plant protein source and Inclusion rate	Phytase source	Effect on fish growth and nutrient availability	References
10.	*Salmo salar* L.	Canola meal based feed + phytase (2000U kg^{-1})	BASF Natuphos® 5000	Increased phosphorus digestibility in Atlantic salmon	Sajjadi *et al.* (2004b)
11.	*Oncorhynchus mykiss*	Soy proteins	BASF Natuphos® 5000	Weight gain was better in fish fed with dephytinized soy proteins than in the fish fed with untreated regular soy fortified with P.	Vielma *et al.* (2002)
12.	*Paralichtys olivaceous*	Soybean based meal	Novo CT, Novo Nordisk	Two phytase dietary applications, pretreatment and supplementation, improved P availability of soybean meal in Japanese flounder. Pretreatment resulted in higher digestibility values than did supplementation, but the latter is convenient for practical use because it needs a shorter preparation.	Masumoto *et al.* (2001)
13.	*Oncorhynchus mykiss*	Soybean meal	BASF Natuphos® 5000	A clear positive dose-response of phytase on dietary phytate digestibility and phosphorus availability that improved significantly by the high level of phytase. (1500 FTU/kg)	Forster *et al.* (1999)
14.	*Oncorhynchus mykiss*	Herring meal 25%, Soy bean meal (Defatted meal, Dehulled meal)	BASF Natuphos 5000	The addition of phytase in diet B increased significantly the apparent phosphorus availability from 58.6 to 68.1%.	Lanari *et al.* (1998)
15.	*Dicentrarchus labrax* (juvenile)	Soybean meal	BASF Natuphos® 5000	The inclusion of 1000 and 2000 IU·kg^{-1} of microbial phytase in the soybean meal based diet significantly increased the ADC of phosphorus to 71.5 % and 79.8 %, respectively.	Olivateles *et al.* (1998)
16.	*Oncorhynchus mykiss*	Soy protein concentrate	BASF Natuphos® 5000	The dietary phytase supplementation was effective in reducing P load of rainbow trout fed soybean protein concentrate-based diet.	Vielma *et al.* 1998
17.	*Labeo rohita juveniles*	Soybean meal-25%, wheat flour-15%, Mustard oil cake-4%, Rice polish-28.55% and deoiled rice bran-19.8%+Phytase (500 U kg^{-1})	Natuphos® 5000G, BASF India	Dietary supplementation of microbial phytase (500 Ukg^{-1}) significantly increased bone Na, Ca, K, P and Fe (40.7%) contents.	Baruah *et al.* 2005

Table 3: Contd.

S. No.	Fish species	Plant protein source and Inclusion rate	Phytase source	Effect on fish growth and nutrient availability	References
18.	*Clarias gariepinus*	58% fishmeal+18% SBM 6% fishmeal + 69% SBM 29% fishmeal + 41% SBM Phytase (1000Ukg^{-1})	BASF Natuphos® 5000	Phytase supplementation resulted in a positive effect on phosphorus digestibility and retention, and consequently the phosphorus conversion efficiency and the phosphorus budget.	Weerd *et al.* 1999
19.	*Pangasius pangasius*	Soybean meal-45%, wheat flour-15%, corn flour-15%, cotton oil seed cake-3% + different phytase levels (0, 150, 250, 350, 500, 1000, 2000 FTUkg^{-1})	Natuphos® 5000G, BASF India	A phytase level of 500 FTU kg^{-1} or above was found to be effective in improving the apparent absorption of Mg and Mn. Iron and cobalt contents were maximum at a supplemental level of 350 FTU kg^{-1} diet. Apparent absorption of P from diet was improved by supplementing phytase.	Debnath *et al.* (2004)
20.	*Morone saxatilis*	4 different diets (soybean meal, corn gluten meal, isolated soy protein, wheat middlings) with phytase (1000Ukg^{-1})	BASF Natuphos® 5000	Significant increases in apparent digestibility of P after phytase supplementation were evident in all four ingredients tested, but there was no phytase effect on protein digestibility.	Papatryphon *et al.* (2001)
21.	*Oncorhynchus mykiss*	Soybean meal based diets with phytase (5000 UL^{-1})	RONOXYMETM® Phytase, F. Hoffmann-La Roche Ltd, Basel,	95-98% reduction in phosphorus excretion, 90-93% increase in apparent absorption of phosphorus with phytase supplementation when compared to fish fed with commercial trout feed	Sugiura *et al.* (2001)

observed (Spinnelli 1983, Suiguira, 2001). Phytase supplementation significantly increased the apparent digestibility coefficients (ADC) of dry matter, crude protein, amino acids in rainbow trout (Cheng *et al.*, 2005).

Phytate may influence starch digestibility through interaction with amylase, proteins associated with starch, calcium ions (which are required for amylase activity), or with starch itself (Sharma *et al.*,1978, Deshpande and Cheryan 1984, Thompson and Yoon 1984, Thompson *et al.*, 1987, and Johnston *et al.*, 2004).

Phytase Supplementation and Mineral Availability

As phytate forms complexes with minerals, hydrolysis of phytate increases the mineral bioavailability. Phytase supplementation increased the retention of calcium, zinc and magnesium in bone and elevated the concentration of ash (Young *et al.*, 1993, Harper *et al.*, 1997, O'Quinn *et al.*, 1997, Liu *et al.*, 1998, Radcliffe *et al.*, 1998). Phytase supplementation also improved the retention of magnesium in bone (Eardman 1979). The effect of phytate on zinc availability has been widely studied, and there have been several reports demonstrating that phytates can reduce zinc bioavailability. Mineral availabilities clearly depend on three factors: (a) the % of phytate content, (b) chemical form in which phytate is present in that feed (sodium/calcium/magnesium phytates), and (c) the composition of the feed used.

Phytase Supplementation and Feed Intake Rate and Growth Performance in Fish

Phytase supplementation may also stimulate appetite, and therefore, increase in growth directly through feed intake rate (Jackson *et al.*, 1996, Li and Robinson 1997, Hauler and Carter 1997). Rodehuttscord and Pfeffer (1995) recorded increased feed intake in rainbow trout fed with a diet containing phytase at a suboptimal phosphorus level. The increased availability of P due to phytase was considered to be the reason for increased feed intake and higher growth performance. Higher feed intake rate was observed in catfish fed with diets containing at least 250U phytase kg^{-1} (Li and Robinson, 1997) and 500U phytase kg^{-1} (Jackson *et al.*, 1996). Hauler and Carter (1997) showed that phytase in a plant based diet with a slightly deficient level of available phosphorus (5g kg^{-1}) increased feed intake in Atlantic salmon. In all the above cases, the phytase supplemented feeds containing suboptimal levels of phosphorus showed increase in feed intake. In contrast, Sajjadi and Carter (2004), who used phytase-supplemented feed containing optimum levels of phosphorus, did not find any difference in feed intake. The use of dephytinized rapeseed protein concentrate in rainbow trout diet had no significant effect on growth, protein utilization, health and survival (Tuskerzic *et al.*, 1995).

Effect of Phytase Supplementation on Pollution in Aquaculture

Fish production through intensive cultivation has been doubled in the last few years. This increase in production leads to an increase in the release of metabolic wastes into aquatic systems (Singh *et al.*, 2006). Environmental phosphorus pollution from animal waste caused serious problems in the areas of intensive fish production, where more than 70 percent of manure phosphorus remains in the farm. Run-off can lead to pollution of surface waters. Waste disposal back into the river, estuary or coastal ecosystem is the most prominent problem in the aquaculture. Effluents from farms that do not use fertilizer or inorganic phosphorus supplements have much less impact on the environment compared to those that do. Strategies should be made to control or reduce the pollution in aquatic systems. Fish feed is a critical factor which plays a role in aquatic pollution. In intensive fish farming systems, a poor degradation of phytate leads to detrimental effects on the aquatic environment like eutrophication effects on fresh water. Phytate is the major storage form of phosphorus in cereal grains, legumes and others, which can be used to replace fishmeal in aqua feed. Phytate passes undigested through digestive tracts of monogastric animals like fishes due to the lack or less phytase activity in their stomach and excreted out into surrounding water. The microorganisms present in water degrade this excreted phytic acid to orthophosphate and cause the eutrophication of pond water leading to off flavor in fishes. As phytate phosphorus is unavailable to fishes, in order to meet the phosphorus requirements, the feed should be supplemented with inorganic phosphate supplements. The amount of phosphorus excretion in aquaculture effluents can be minimized by nutritional methods that enhance the phosphorus availability and reduce the phosphorus in metabolic wastes (Coloso *et al.*, 2003). Soluble P excretion by fish is independent of dietary phosphorus until a certain level, but then increases sharply above this dietary level. Phytate-degrading enzymes were originally proposed as an animal feed additive to enhance the value of plant material in animal feed by liberating orthophosphate (Mitchell *et al.*, 1997). More recently, addition of phytate-degrading enzymes has also been seen as a way to reduce the level of phosphate pollution in areas of intensive livestock management. A number of studies have already shown that the addition of phytate-degrading enzymes enhances phosphate utilization from phytates and drastically reduces orthophosphate excretion (Cromwell *et al.*, 1995, Simons *et al.*, 1990).

Proteolytic Resistance

An effective phytase should have the resistance against the hydrolytic activity of proteases present in digestive tracts. Fungal and bacterial phytases show different sensitivities to pepsin and trypsin (Kerovuo *et al.*, 1998, Singh and Satyanarayana 2006). The protease resistance of the phytase can be improved by modifying or by blocking the protease sensitive sites which are present in the exposed loops at the surfaces of the molecules (Wyss *et al.*, 1999b). Phytase from *Citrobacter braakii* is resistant to proteases such as trypsin, pepsin, papain, pancreatin, and elastase (Kim *et al.*, 2003).

Thermostable Phytases and Feed Applications

Since feeds are often heated to high temperature (60–80°C) in order to make them powder or in the form of pellets, all feed enzymes need to be heat stable to avoid substantial activity loss during this process. Thermostability and optimal activity at high temperatures are inherent properties of thermophilic enzymes. Several strategies have been made to improve thermostability of phytases. Chemical coating of phytases has been used to improve their heat stability, but the coating may somewhat compromise the release and function of the enzymes in stomach. Some researchers tried application of stabilizers to protect phytase enzyme from the high temperatures that are used for feed pelleting process. Thermophilic enzymes are more preferable for the feed applications. Spraying liquid phytase preparations to feed after pelleting can bypass the activity loss by pelleting, but increases labor and equipment costs. Several biotechnological attempts have been made for improving the thermostability of phytases. Application of cell bound phytase in aquaculture has more advantages than that of the extracellular phytases, due to over dilution of the later. Yeasts containing cell-bound phytases are more advantageous due to the fact that they act as Single Cell Protein (SCP). Salts and polyols have been used to increase the thermostability of enzymes (Obon *et al.,* 1996, Lamosa *et al.,* 2000), but the enzyme activity will be influenced by high concentrations of salts. Chen *et al.* (2001) used sorghum liquor supplemented with starch for improving the thermostability of phytase. When sorghum liquor wastes were mixed at the ratio of 1:1:10 (v/w/w), the residual phytase activities after 30 min of treatment at 70 and 80°C were about 90% and 18%, respectively of that at 37°C. After 10 min treatment, the residual activity was 67% at 80°C and 10% at 90°C.

FUTURE PERSPECTIVES AND CONCLUSIONS

For supplementation in aquaculture feed, the phytase of interest should be optimally active in the conditions prevalent in digestive tract of the fishes. An ideal phytase for feed supplementation must show good activity at acidic pH of gastrointestinal tract, have resistance to proteases, and show activity in a broad range of temperatures. Unfortunately, so far, there is no single phytase suitable for feed supplementation for all fish species. Further, investigators are, therefore, needed to develop an ideal phytase for aquaculture feed supplementation. The investigations are also needed on phytase application in fish feed and phytate hydrolysis in the gut environment. The use of phytase in fish feed will expand along with the need of cost-effective feed and environmental protection concern. The development of an engineered phytase with these features would be a potential candidate for application in aquaculture.

Due to the high cost and low availability of fish meal, plant-based feeds are preferable. Plant-based feeds, however, have their own limitations such as the presence of anti-nutritional factors like phytic acid. The reduction of phytate content by supplementing the fish feed with phytase results in the improvement of P utilization and decreases P excretion into the effluent water. Phytase supplementation not only increases P utilization, but also ameliorates mineral availability, and protein and carbohydrate digestibility. Phytase supplementation, thus, results in improved growth performance of fishes.

ACKNOWLEDGEMENTS

The financial assistance from the Department of Biotechnology and Indian Council of Agricultural Research, Government of India, during the course of writing this article is gratefully acknowledged.

REFERENCES

Al-Asheh, S. and Duvnjak, Z. (1995), *Appl. Microbiol. Biotechnol.* **43**: 25-30.

Baruah, K., Pal, A.K., Sahu, N.P., Jain, K.K, Mukherjee, S.C. and Debnath, D. (2005). *Aquatic Res.* **36:** 803-812.

Batal, E. and Karem, H.A. (2001). *Food Res. International,* **34**: 715-720.

Berka, R.M., Rey, M.W., Brown, K.M., Byun, T. and Klotz, A.V. (1998). *Appl. Environ. Microbiol.* **64:** 4423-4427.

Bogar, B., Szakacs, G., Pandey, A., Abdulhameed, S., Linden, J.C. and Tengerdy, R.P. (2003). *Biotechnol. Progr.* **19:** 312-319.

Carmovale, E., Lugaro, E. and Boccia, G.L. (1988). *Cereal Chem.* **65:** 114-117.

Chadha, B.S., Harmeet, G., Mandeep, M., Saini, H.S. and Singh, N. (2004). *World J. Microbiol. Biotechnol* **20:** 105-109.

Chelius, K.M. and Wodzinski, R.J. (1994). *Appl. Microbiol. Biotechnol.* **41:** 79-83.

Chen, C., Hunag, C. and Cheng, K. (2001). *Biotechnol. Lett.* **23:** 331-333.

Cheng, Z.J., Hardy, R.W. and Usry, J.L. (2003). *Aquaculture,* **218:** 553-565.

Cheng, Z.J.J. and co-workers. (2005). *J. World Aquaculture Soc.* **35:** 1-15.

Choi, Y.M., Suh, H.J. and Kim, J.M. (2001). *J. Protein Chem.* **20:** 287-292.

Coloso, R.M., King, K., Fletcher, J.W., Hendrix, M.A. , Subramanyam, M., Weis, P. and Ferraris, R.P. (2003). *Aquaculture* **220:** 801-820.

Cromwell, G.L., Coffey, R.D., Parker, G.R., Monegue, H.J. and Randolph, J.H. (1995). *J. Animal Sci.* **73:** 2000-2008.

De Angelis, M., Gallo, G., Corbo, M.R., McSweeney, P.L.H., Faccia, M., Giovine, M. and Gobbetti, M. (2003). *International J. Food Microbiol.* **87:** 259-270.

Debnath, D., Sahu, N.P., Pal, A.K., Jain, K.K., Yengkokpam, S. and Mukherjee, S.C. (2004). *Aquatic Res.* **36:** 326-335.

Deshpande, S.S. and Cheryan, M. (1984). *J. Food Sci.* **49:** 516-519.

Eardman, J.W. (1979). *J. Am. Oil Chemists Soc.* **56:** 736-741.

Forster, I., Higgs D.A., Dosanjh B.S., Rowshandeli M. and Parr, J. (1999). *Aquaculture* **179:** 109-125.

Gargova, S., Roshkova, Z. and Vancheva, G. (1997). *Biotechnol. Techniques* **11:** 221-224.

Gatlin, D.M. and Phillips, H.F. (1989). *Aquaculture,* **79:** 259-266.

Gautam, P., Sabu, A., Pandey, A., Szakacs, G. and Soccol, C.R. (2002). *Bioresour. Technol.* **83:** 229-233.

Gibson, D.M. (1987), *Biotechnol. Lett.* **5**: 305-310.

Greiner, R., Haller, E., Konietzny, U. and Jany, K.D. (1997). *Arch. Biochem. Biophys.* **341:** 201-206.

Greiner, R., Konietzny, U. and Jany, K.D. (1993). *Arch. Biochem. Biophys.* **303:** 107-113.

Hafez, Y.S., Mohammed, A.I., Perera, P.A., Singh, G. and Hussein A.S. (1989). *J. Food Sci.* **54:** 958-962.

Han, Y.M. and Lei, X.G. (1999). *Arch. Biochem. Biophys.* **364:** 83-90.

Hartig, T. (1855). *Über das Kleberme. Bot. Ztg.* **13:** 881.

Hartman, G.H. (1979). *J. Am. Oil Chem. Soc.* **59:** 731.

Hauler, R.C. and Carter, C.G. (1997). *Proc. Nutr. Soc. Aust.* **21:** 139.

Higgs, D.A., Dosanjh, B.S., Prendergast, A.F., Beames, R.M., Hardy, R.W., Riley, W. and Decon, G. (1995). *In:* Nutrition and Utilization Technology in Aquaculture (Eds. Lim, C.E. and Sessa, D.J.), AOCS Press, Champaign, IL. pp. 130-156.

Hoppe, P.P. and Schwarz, G. (1993). *In:* Enzymes in Animal Nutrition (Eds. Wenk, C. and Boessinger, M.), Proc. 1st Symp. Kartause Ittingen, Switzerland. Schriftenreihe aus dem Institut fur Nutztierwissenschaften, Gruppe Erna¨hrung, ETH-Zurich. p. 187.

Howson, S.J. and Davis, R.P. (1983). *Enzyme Microb. Technol.* **5:** 377-382.

Hughes, K.P. and Soares Jr. J.H. (1998). *Aquaculture Nutr.* **4:** 133-140.

Igbasan, F.A., Männer, K., Miksch, G., Borriss, R., Farouk, A., and Simon, O. (2000). *Arch. Animal Nutr.* **53:** 353-373.

Jackson, L.S., Li, M.H. and Robinson E.H. (1996). *J. World Aquaculture Soc.* **27:** 309-313.

Johnston, S.L., Williams, S.B., Southern, L.L., Bidner, T.D., Bunting, L.D., Matthews, J.O. and Olcott, B.M. (2004). *J. Animal Sci.* **82:** 705-714.

Kaur, P. and Satyanarayana, T. (2005). *Process Biochem.* **40:** 3095-3102.

Kaur, P. and Satyanarayana, T. (2006). *In:* Current Concepts in Botany (Eds. Mukherji, K. and Manoharachary, C.), I.K. International Publishing House, New Delhi, India. pp. 377-394.

Kemme, P.A. and Jongbloed, A.W. (1993). *J. Animal Sci.* **71:** 79-86.

Kies, A.K., Van Hemert, K.H.F. and Sauer, W.C. (2001). *World Poult. Sci. J.* **57:** 109-125.

Kim, Y., Kim, H.K., Bae, K.S., Yu, J.H. and Oh, T. (1998). *Enzyme Microb. Technol.* **22:** 2-7.

Konietzny, U. and Greiner, R. (2002). *Int. J. Food Sci. Technol.* **37:** 791-812.

Lamosa, P., Burke, A., Peist, R., Huber, R., Liu, M.Y., Silva, G., Rodrigues- Pousada, C., LeGall, J., Maycock, C. and Santos, H. (2000). *Appl. Environ. Microbiol.* **66:** 1974-1979.

Lanari, D., Agaro, E.D. and Turri, C. (1998). *Aquaculture,* **161:** 345-356.

Larsen, T., Skoglund, E., Sandberg, A.S. and Engberg, R.M. (1999). *Can. J. Animal Sci.* **79:** 477-483.

Li, M.H. and Robinson, E.H. (1997). *J. World Aquaculture Soc.* **28:** 402-406.

Liebert, F. and Portz, L. (2005). *Aquaculture,* **248:** 111-119.

Liu, B-L., Rafiq, A., Tzeng, Y.M. and Rob, A. (1998). *Enzyme. Microb. Technol.* **22:** 415-424.

Mandviwala, T.N. and Khire, J.M. (2000). *J. Indust. Microbiol. Biotechnol.* **24:** 237-243.

Masumoto, T., Tamura, B. and Shimeno, S. (2001). *Fisheries Sci.* **67:** 1075-1080.

Morris, E.R. and Ellis, R. (1980). *J. Nutr.* **110:** 2000-10.

Morz, Z., Krasuki, W. and Grela, E. (1995). *In Manipulating Pig Production* (Eds. Hennessy, D.P. and Cranwell, P.D.), Werribee, Australasian Pig Science Association. **5:** 185.

Mullaney, E.J. and Ullah, A.H.J. (2003). *Biochem. Biophys. Res. Commun* **312:** 179-184.

Mullaney, E.J., Daly, C.B., Kim, T., Porres, J.M., Lei, X.G., Sethumadhavan, K. and Ullah, A.H.J. (2002). *Biochem. Biophys. Res. Commun.* **297:** 1016-1020.

Mwachireya, S.A., Beames, R.M., Higgs, D.A. and Dosanjh, B.S. (1999). *Aquaculture Nut.* **5:** 73- 82.

Mwachireya, S.A.K. (1995). *M. Sc. Thesis*, University of British Columbia, Vancouver, Canada.

Nair, V.C. and Duvnjak, Z. (1990). *Appl. Microbiol. Biotechnol.* **34:** 183-188.

Nakamura, Y., Fukuhara, H. and Sano, K. (2000). *Biosci. Biotechnol. Biochem.* **64:** 841-844.

Nasi, J.M., Helander, E.H. and Partanen, K.H. (1995). *Animal Feed Sci. Tech.* **56:** 83-98.

Niiyama, Y., Sakamoto, S., Okada, K., Matsuo, T. and Kimoto, K. (1992). *Nutr. Sci. Soy Prot. Japan,* **13:** 80-85.

NRC. (1993). National Academy Press, Washington, D.C.

O'Quinn, P.R. (1997). *J. Animal Sci.* **75:** 714-719.

Obon, J.M, Manjon, A. and Iborra, J.L. (1996). *Enzyme Microb. Technol.* **19**: 352-360.

Oliva-Teles, A., Pereira, J.P., Gouveia, A. and Gomes E. (1998). *Aquatic Living Resources,* **11:** 255-259.

Pandey, A., Szakacs, G., Soccol, C.R., Rodriguez, J.A. and Soccol, V.T. (2001). *Bioresour. Technol.* **77:** 203-214.

Papagianni, M., Nokes, S.E. and Filer, K. (1999). *Process Biochem.* **35:** 397-402.

Papatryphon, E. and Soares, J.H. (2001). *Aquaculture Nutr.* **7:** 161-167.

Pasamontes, L., Haiker, M., Wyss, M., Tessier, M. and van Loon, A.P.G.M. (1997). *Appl. Environ. Microbiol.* **63:** 1696-1700.

Pfeffer. (1872). *Jahrb. Wiss. Bot.* **8:** 465.

Popanich, S., Klomsiri, C. and Dharmsthiti, S. (2003). *Bioresour. Technol.* **87:** 295-298.

Quan, C.S., Tian, W.J., Fan, S.D. and Kikuchi, Y.I. (2004). *J. Biosci. Bioeng.* **97:** 260-266.

Quan, C.S., Zhang, L.H., Wang, Y.J. and Ohta, Y.Y. (2001). *J. Biosci. Bioeng.* **92:** 154-160.

Rackis J.J. (1974). *J. Am. Oil Chem. Soc.* **51:** 161-174.

Radcliffe, J.S. , Zhang, Z. and Kornegay, E.T. (1998). *J. Animal Sci.* **76:** 1880-1886.

Ramachandran, S., Roopesh, K., Nampoothiri, K.M., Szakacs, G. and Pandey, A. (2005). *Process Biochem.* **40:** 1749-1754.

Reddy, N.R., Pierson, M.D., Sathe, S.K. and Salunkhe, D.K. (1989). In *Phytates in Cereals & Legumes.* CRC Press, Boca Raton, FL.

Reddy, N.R. and Sathe, S.K. (2002). In. *Food Phytates.* CRC Press, London.

Rodehutscord, M. and Pfeffer, E. (1995). *Water Sci. Technol.,* **31:** 143-147.

Rodriguez, E., Mullaney, E.J. and Lei, X.G. (2000). *Biochem. Biophys. Res. Commun.* **268:** 373-378.

Rodriguez, E., Han, Y. and. Lei, X.G. (1999). *Biochem. Biophys. Res. Commun.* **257:** 117-123.

Roopesh, K., Ramachandran, S., Nampoothiri, K.M., Szakacs, G. and Pandey, A. (2006). *Bioresour. Technol.* **97:** 506-511.

Rumsey G.L., Hughes S.G. and Winfree R.A. (1993). *Animal Feed. Sci. Technol.* **40:** 135-151.

Sajjadi, M. and Carter, C.G. (2004a). *Aqua. Nutr.* **10:** 135-142.

Sajjadi, M. and Carter, C.G. (2004b). *Aquaculture,* **240:** 417-431.

Sandberg A.S., Andersson H., Carlsson N.G. and Sandström B. (1987). *J. Nutr.* **117:** 2061-2065.

Sano, K., Fukuhara, H. and Nakamura, Y. (1999). *Biotechnol. Lett.* **21:** 33-38.

Satoh S., Higgs D.A., Dosanjh B.S., Hardy R.W., Eales J.G. and Deacon G. (1998). *Aquaculture Nutr.* **4:** 115-122.

Satyanarayana, T. and Vohra, A. (2003). Indian Patent no. 197593.

Schulze and Winterstein (1896). *Zeitsch. Physiol. Chem.* **22:** 90.

Segueilha, L., Lambrechts, C., Boze, H., Moulin, G. and Galzy, P. (1992). *J. Ferment. Bioeng.* **74:** 7-11.

Selle, P.H., Ravindran, V., Caldwell, R.A. and Bryden, W.L. (2000). *Nutr. Res. Rev.* **13:** 255-278.

Sharma, C.B., Goel, M. and Irshad, M. (1978). *Biochem.* **17:** 201-204.

Shieh, T.R. and Ware, J.H. (1968). *Appl. Microbiol.* **169:** 1348-1351.

Shimizu, M. (1992). *Biosci. Biotechnol. Biochem.* **56:** 1266-1269.

Shimizu, M. (1993) *Biosci. Biotechnol. Biochem.* **57:** 1364-1365.

Simons, P.C.M., Versteeg, H.A.J., Jongbloed, A.W., Kemme, P.A., Slump, P., Bos, K.D., Wolters, M.G.E., Beudeker, R.F. and Verschoor, G.J. (1990). *Br. J. Nutr.* **64:** 525-540.

Singh, B. and Satyanarayana, T. (2006a). *Appl. Biochem. Biotechnol.* **133:** 239-50.

Singh, B. and Satyanarayana, T. (2006b). *J. Appl. Microbiol.* **101:** 344-352.

Singh, B., Kaur, P. and Satyanarayana, T. (2006). *In:* Microbes: Health and Environment (Eds. Chauhan, A.K. and Verma, A.) I.K. International Publishing House, New Delhi, India. pp. 289-326.

Skoglund, E., Larsen, T. and Sandberg, A.S. (1997). *Can. J. Animal Sci.* **77:** 471-477.

Spinelli, J., Houle, C.R. and Wekell, J.C. (1983). *Aquaculture,* **30:** 71-83.

Sreeramulu, G., Srinivasa, D.S., Nand, K. and Joseph, R. (1996). *Lett. Appl. Microbiol.* **23:** 385-388.

Storebakken, T., Shearer, K.D. and Roem, A.J. (1998). *Aquaculture,* **161:** 365-379.

Sugiura, S.H., Gabaudan, J., Dong, F.M. and Hardy, R.W. (2001). *Aquaculture Res.* **32:** 583-592.

Teskeredzic, Z., Higgs, D.A., Dosanjh, B.S., McBride, J.R., Hardy, R.W., Beames, R.M., Simell, M., Vaara T. and Bridges, R.B. (1995). *Aquaculture,* **131:** 261-277.

Thompson, L.U. and Yoon, J.H. (1984). *J. Food Sci.* **49:** 1228-1229.

Thompson, L.U., Button, C.L. and Jenkins, D.J.A. (1987). *J. Clin. Nutr.* **46:** 467-473.

Toma R.B. and Tabekhia, M.M. (1979). *J. Food Sci.* **44:** 629-632.

Ullah, A.H.J. and Gibson, D.M. (1987). *Prep. Biochem.* **17:** 63-91.

Vielma J., Lall S.P., Koskela J., Schner F.J. and Mattila P. (1998). *Aquaculture,* **163:** 307-321.

Vielma J., Ruohonen, K., Gabaudan, J. and Vogel, K. (2004). *Aquaculture Res.* **35:** 955-964.

Vielma, J., Ruohonen, K. and Peisker, M. (2002). *Aquaculture* **204:** 145-156.

Vielma, J., Makinen, T., Ekholm, P. and Koskela, J. (2000). *Aquaculture,* **183:** 349-362.

Vohra, A. and Satyanarayana, T. (2001). *Biotechnol. Lett.* **23:** 551-554.

Vohra, A. and Satyanarayana, T. (2002). *World J. Microbiol. Biotechnol.* **18:** 687-691.

Vohra, A. (2002). *Ph.D. Thesis:* University of Delhi South Campus, New Delhi, India

Vohra, A. and Satyanarayana, T. (2003). *Crit. Rev. Biotechnol.* **23:** 29-60.

Volfova, O., Dvorakova, J., Hanzlikova, A. and Jandera, A. (1994). *Folia Microbiologica,* **39:** 481-484.

Weerd, V.H., Khalaf, Aartsen and Tijssen, (1999). *Aquaculture Nutr.* **5:** 135-142.

Weremko, D., Fandrejewski, H. and Zebrowska, T. (1997). *Asian Australas J. Animal Sci.* **10:** 551.

Wodzinski, R.J. and Ullah, A.H.J. (1996). *Adv. Appl. Microbiol.* **42:** 263-302.

Wyss, M., Brugger, R. and Kronenberger, A. (1999a). *App. Environ. Microbiol.* **65:** 367-373.

Wyss, M., Pasamontes, L. and Friedlein, A. (1999b). *App. Environ. Microbiol.* **65:** 359-366.

Yoo, G.Y., Wang, X., Choi, S., Han, K., Kang, J.C. and Bai S.C. (2005). *Aquaculture* **243:** 315-322.

Yoon, S.J., Choi, Y.J., and Min, H.K. (1996). *Enzyme Microb. Technol.* **18:** 449-454.

Young, L.G., Leunissen, M. and Atkinson, J.L. (1993). *J. Animal Sci.* **71:** 2147.

Zamudio, M., González, A. and Medina, J.A. (2001). *Lett. Appl. Microbiol.* **32:** 181-184.

Chapter **14**

Laccase Regulation and Laccase-Dependent Bioremediation

R.P. Singh*, Vivekanand and Pallavi Dwivedi

Department of Biotechnology, Indian Institute of Technology, Roorkee
Roorkee–247 667, (India)
E-mail: rpsbsfbs@iitr.ernet.in

Increasing concentration of various recalcitrant organopollutants through industrial, military and farming activities is a matter of concern for environmentalists. Degradation of all xenobiotic compounds in an eco-friendly manner is not possible by conventional techniques. Bioremediation serves as the best alternative in order to get rid of from these non-biodegradable wastes. White rot fungi are the most efficient group among all other microbes for degradation of organopollutants in economical and eco-friendly manner. Due to their low oxido-reductive potential and non-specificity of substrate, laccases are the most potent enzyme among all other lignin modifying enzymes (LiP and MnP) for degradation of phenolic as well as non-phenolic compounds along with aromatic amines and toxic dyes, etc. Several isoforms of laccase are known in white rot fungi and all these isoforms are supposed to be encoded by a multigene family. In order to understand its transcriptional and translational regulation laccase gene has been isolated, cloned, characterized and expressed from several microbes into different homologous as well as heterologous hosts. Induction of laccase biosynthesis in presence of several low molecular weight compounds which are known as mediators and various metal ions shows inducible nature of enzyme. The active site of enzyme consists of four copper ions which are involved in a coordinated oxygen reduction. The three-dimensional structure of laccases was determined from several white rot fungi and it shows that protein consists of three sequentially arranged domains, each of them with a β-barrel type topology. This review accounts into the state of art in the research related to laccase mediated bioremediation, its molecular genetics, regulation of expression and structural features. It is followed by its prospective use for dye decolorization, effluent treatment along with degradation of recalcitrant pollutants.

INTRODUCTION

The problem of contaminated land is continuously increasing nowadays due to ever-increasing industrial, military and farming activities. Many organopollutants like PAH (polyaromatic hydrocarbons), PCBs (poly-chlorinated biphenyls), TNT (trinitrotoluene) and DDT which are well known for their persistent, mutagenic and bioaccumulating nature, are continuously released into the environment. The annual production of hazardous organopollutants is about eighty billion pounds in the United States and only 10% of these are disposed of safely (Reddy and Mathew, 2001). Removal of these organopollutants from environment in an ecologically responsible, safe and cost-effective way is a major concern for environmentalists. Bioremediation, which can be defined as the degradation of organopollutants into its non-toxic form due to metabolic activities of living organisms, primarily microorganisms, offers a best solution to this problem. Table 1 shows the major contaminants suitable for bioremediation. It uses naturally occurring bacteria, fungi or plants to degrade or detoxify substances hazardous to human health and/or environment. The microbes may be indigenous to a contaminated area or they may be isolated from else where and brought to the contaminated site. Bioremediation is typically more economical than conventional methods such as incineration and land filling. Since, bioremediation is based on natural degradative capability of microbes, it has got more public acceptance as compared to other technologies. Laccase being a major tool involved in bioremediation, the present article emphasizes the status of the efforts made so far in understanding the molecular architecture and regulation of expression of laccase.

Table 1. Major contaminants which can be degraded by biological agents

Class of Contaminant	Specific Example	Source	Microbial Group	Examples
Pesticides, Hydrocarbons Polyaromatic compounds	Atrazine, 2,4-D, propham, carbofuran, Naphthalene,	Farming, coal mining, oil drilling wood burning	Anaerobic bacteria	*Pseudomonas*, *Alcaligenes*, *Sphingomonas*, *Rhodococcus*, *Mycobacterium*
Polychlorin-Biphenyls, Chlorinated solvents	Trichloroet-hylene, per-chloroethylene, 4-chlorobiph--enyl, 4,4-di--chlorobiphenyl	Farming, wood preser-vatives	Anaerobic	*Dehalospirillum multivorans*
Polyaromatic Compounds, dyes, PCBs	Naphthelene, anthracene, pyrene, poly R-478, reactive orange, RBBR	Plastic and textile Indus-try, hydraulic fluids	White rot fungi	*Phanaerochaete chrysosporium*, *Trametes*, *Phlebia*, *Pleurotus ostreatus*
Chlorinated aliphatic compounds	Trichloroe--thylene, 1,2-dichloroethane	Military com--plexes		Methylotrophs

BIOREMEDIATION CAPABILITIES OF WHITE ROT FUNGI

Within past two decades most research within the field of bioremediation has focused on bacterial and fungal based bioremediation and has attracted interest. White rot fungi can withstand toxic levels of most organopollutants, whereas the natural degradation of many organopollutants is limited by bacteria because of the toxic effects (Aust *et al.*, 2003). White rot fungi is a major physiological group of fungi that can degrade lignin or lignin like compounds. The branching, filamentous mode of fungal growth allows more efficient colonization and thus exploration of contaminated soil as compared to the other microbes. *Phanerochaete*, *Trametes*, *Bjerkandera* and *Pleurotus* are four main genera of white rot fungi that have potential for bioremediation (Hestbjerg *et al.*, 2003).

All lignin modifying enzymes (LMEs) that are produced by white rot fungi are known for very low substrate specificity. Due to their low substrate specificity, they can degrade or mineralize a wide range of highly recalcitrant organopollutants that are structurally similar to lignin (Pointing, 2001; Cajthaml *et al.*, 2002; Mansur *et al.*, 2003; Veignie, 2004). The three main LMEs are lignin peroxidase, manganese peroxidase and laccase, and all the three enzymes are stimulated by nutrient limitation (Mansur *et al.*, 2003; Aust *et al.*, 2003).

The biodegradation mechanism in part depends on the compound targeted for degradation but involves some consistent steps in the process regardless of the substrate. The degradation of pollutants is catalyzed by lignin modifying enzymes of white rot fungi using a non-specific free radical mechanism (Pointing, 2001; Law *et al.*, 2003). When an electron is added or removed from the ground state of a chemical, it becomes highly reactive, allowing it to give or take electrons from other chemicals. This provides the basis for the non-specificity of the enzymes and the ability of the enzymes to degrade xenobiotic and other compounds (Pointing, 2001; Reddy and Mathew, 2001). The main reactions that are catalyzed by the ligninolytic enzymes include depolymerization, demethoxylation, decarboxylation, hydroxylation and aromatic ring opening, etc. Many of these reactions result in oxygen activation, creating free radicals that perpetuate oxidation of the organopollutants (Reddy and Mathew, 2001). Once the peroxides have opened the aromatic ring structures by introducing oxygen, other more commonly occurring species of fungi and bacteria can mineralize the products intracellularly into end products such as CO_2 and other benign compounds.

Several laboratory scale experiments all around the world prove the organopollutant degrading capability of white rot fungi. Cajthaml *et al.* (2002) looked at the ability of purified enzymes to degrade several PAHs (anthracene, phenanthrene, pyrene, fluoranthrene). They found in *in vitro* studies that the lignolytic fungi extensively degrade these compounds through both intra and extracellular processes. After 50 days of incubation of these pollutants with purified enzymes, there remained less than 1% of anthracene, 5% of phenanthrene, 32% of fluoranthrene and 7% pyrene. Levin *et al.* (2003) analysed the biodegradation of two PAHs, nitrobenzene and anthracene by white rot fungi, *Trametes trogii*. They found that the fungus was tolerant to extremely high levels of these compounds (250-500 ppm), and it was able to metabolize 90-97% of high concentration of anthracene and nitrobenzene respectively. The removal of TNT from TNT contaminated soil by white rot fungi was studied by Boopathy (2000). The TNT levels in the soil sample were 4000-12000 mg kg^{-1} and they followed two approaches, soil slurry reactor and *in situ* bioremediation. The soil slurry reactor removed all the TNT (below detection limit) within three months, while the bioremediation method took nearly 12 months to decrease TNT level to below detection limit. However, more research needs to be carried out to determine the applicability of utilizing these organisms in contaminated field sites.

LIGNOLYTIC ENZYME SYSTEM OF WHITE ROT FUNGI

The three main lignin modifying enzymes produced by white rot fungi are lignin peroxidase (LiP), (E.C. 1.11.1.14); manganese peroxidase (MnP), (E.C. 1.11.1.13) and laccase (Lac), (E.C. 1.10.3.2). Degradation of lignin in their natural lignocellulosic substrates (Becker and Sinitsyn, 1993; Hatakka, 1994) as well as the degradation of various xenobiotic compounds (Barr and Aust, 1994; Scheibner *et al.*, 1997; Pointing, 2001;) including dyes (Glenn and Gold, 1983; Pasti-Grigsby *et al.*, 1992; Paszczynski *et al.*, 1992; Spadaro *et al.*, 1992) takes place by direct involvement of lignin modifying enzymes produced by white rot fungi.

Manganese peroxidase (MnP) is the most common lignolytic peroxidase produced by almost all white rot basidiomycetes and other litter-decomposing fungi. MnP preferentially oxidizes Mn^{2+} to Mn^{3+} (Glenn *et al.*, 1986), which is stabilized by chelaters such as oxalic acid, which is also excreted by the fungi. Chelated Mn^{3+} acts as a highly reactive low molecular weight, diffusible redox-mediator. Thus, MnP is able to oxidize and depolymerize their natural substrate.

Lignin peroxidases (LiP) catalyze the oxidation of non-phenolic aromatic lignin moieties and similar compounds. LiP catalyzes several oxidations in the side chains of lignin and related compounds (Tien and Kirk, 1983) by one-electron abstraction to form reactive radicals (Kersten *et al.*, 1985). The cleavage of aromatic ring structures has also been reported (Umezawa and Higuchi, 1987). The role of lignin peroxidase in lignolysis could be the further transformations of lignin fragments, which are initially released by MnP. Table 2 summarizes the properties of all lignin modifying enzymes of white rot fungi.

Table 2. Lignolytic system of white rot fungi.

Enzyme	Molecular Weight (kDa)	Prosthetic group	Mediators	Reaction mechanism	Substrate
Lignin peroxidase (LiP) EC 1.11.1.14	38-47	Glycoprotein Heme prosthetic group	No	H_2O_2 dependent oxidation	Lignin and lignin related compounds phenolic as well as non-phenolic
Manganese peroxidase (MnP) EC 1.11.1.13	32-62.5	Glycoprotein Heme prosthetic group	Thiols unsaturated fatty acids	H_2O_2 dependent oxidation Mn^{+2} to Mn^{+3}	Oxidation of phenolic compounds
Laccase (Lac) EC 1.10.3.2	60-390	Four Cu^{+2} ions	ABTS, HBT, syring-aldigine	O_2 as oxidant, release H_2O molecule	Degradation of dihydroxy and diamino aromatic compounds

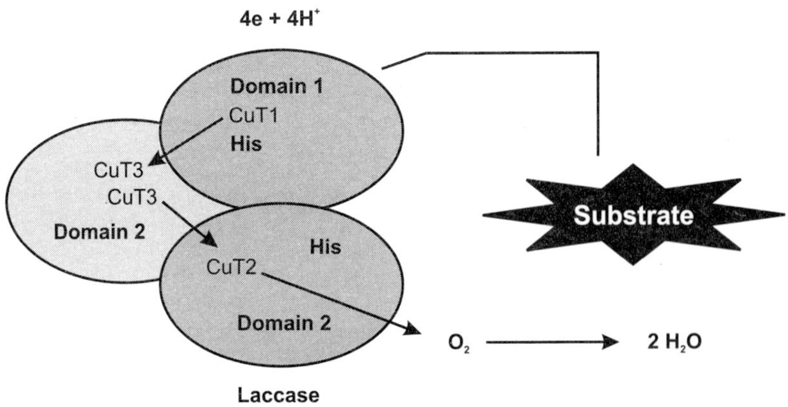

Fig. 1. Proposed reaction mechanism of laccase catalyzed oxidation of substrates.

The last but the most common lignin modifying enzyme, which is produced by almost all wood- and litter-transforming basidiomycetes, is laccase. Laccases catalyze the oxidation of a variety of aromatic hydrogen donors with the concomitant reduction of oxygen to water (Fig. 1). Moreover, laccases not only oxidize phenolic and methoxyphenolic acids, but also decarboxylate them and attack their methoxy groups (demethylation). Laccases have been reported to oxidize many recalcitrant substances, such as chlorophenols (Ricotta *et al.*, 1996; Grey *et al.*, 1998; Fahr *et al.*, 1999), PAHs (Majcherczyk *et al.*, 1998), lignin-related structures (Boyle *et al.*, 1992; Bourbonnais *et al.*, 1996), organophosphorus compounds (Amitai *et al.*, 1998), non-phenolic lignin model compounds (Majcherczyk *et al.*, 1999), phenols (Bollag *et al.*, 1988; Xu, 1996) and last but not least, aromatic dyes (Rodrý´guez *et al.*, 1999; Abadulla *et al.*, 2000).

Direct and specific interactions between lignin (or recalcitrant structural analogs) and lignin modifying enzymes are highly improbable due to the random polymeric nature of lignin and the bulk of lignin modifying enzymes (Evans and Hedger, 2001). Rather low-molecular weight, diffusible redox mediators provide high redox potentials (>900 mV) to attack lignin and are able to migrate into the lignocellulose complex. They could be involved in the lignin modifying enzymes-catalyzed generation of reactive radical moieties from a variety of lignin-like substrates, and also in the formation of reactive oxygen species (ROS), which either directly or indirectly could attack lignin or xenobiotic molecules (Hammel, 1996; Van Aken and Agathos, 2001; 2002).

MOLECULAR GENETICS OF LACCASE

The oxidation of phenolic compounds and aromatic amines into molecular oxygen and water is catalyzed by laccases, which are multi copper oxidative enzymes (Thurston, 1994). Despite their low oxido-reductive potential as compared to lignin peroxidase and manganese peroxidases, it can also catalyze the degradation of non phenolic compounds in presence of mediators like 2, 2’-azinobis-(3-ethylbenzethiazoline-6-sulphonate (ABTS) and 1-hydroxybenzotriazole (HBT) (Bourbonnais *et al.*, 1995; 1998). Thus laccase resulted into complete lignin degradation and it plays a pivotal role in carbon cycling. Fungal laccases are known to play various physiological roles such as conidia pigment production in *Aspergillus nidulans* (Clutterbuck, 1972), fruiting body formation in *Lentinus edodes* (Leatham and Stahmann, 1981), *Schizophyllum commune* (Leonard, 1971) and in *C.congregatus* (Ross, 1982), pathogenicity in several fungi such as *Cryptococcus neoformans* (Salas *et al.*, 1996) and *Cryphonectria parasitica* (Rigling and Van Alfen, 1991), and degradation of lignin and other recalcitrants in white rot fungi (Kim *et al.*, 1995; Cullen, 1997). Besides this, it also catalyzes the degradation of several organopollutants, industrial effluent, toxic dyes. It also has application in pulping and bleaching in paper industry. Therefore, it is necessary to gain knowledge about laccase gene structure, its regulation at molecular level, and high level expression of engineered protein in suitable host in order to achieve scaled up production. Rapidly developing experimental techniques such as transcript analysis in complex substrates, gene replacement and heterologous expression will help in establishing the fundamental knowledge about laccase gene.

Genome Organization

Laccases appeared to be encoded by complex gene families of structurally releated genes. Multiplicity is very common among laccases. Almost all basidiomycetes fungi produce more than one type of laccase. Genetic linkage based on restriction fragment length polymorphism (RFLP) has been useful for identifying gene clusters. Several isoforms of laccases have been reported in single species. (*lcc1*-5). Five isozymes of laccase are produced by *Trametes sanguinea* (Hoshida *et al.*, 2001), *T. villosa* (Yaver and Golightly, 1996) and *T. versicolor* (Jonsson *et al.*, 1995), while four isoforms of laccase are reported in *Trametes* sp. Strain C30 by Klonowska *et al.* (2005). Mikuni and Morohoshi (1997) had isolated three isozymes of laccase in *Coriolus versicolor*. The most interesting example of laccase isozymes are seen in *Coprinus congregatus*. It produces several laccase isozymes during its development: a hyphal tip isozyme (Ross, 1982), primordial isozyme (Choi *et al.*, 1987) and sclerotial isozyme (Choi *et al.*, 1987). These isozymes result into different patterns when analyzed by native polyacrylamide gel electrophoresis (PAGE). The saprophytic fungi are more diverse in their laccase gene than the mycorrhizal fungi. All of these isoforms are supposed to be encoded by a multigene family.

Structure and Cloning of Laccase Gene

Laccase gene resemble the feature of a typical eukaryotic gene structure. On the basis of intron/exon sequences, laccases can be divided into subfamilies. Cloning and sequencing of at least 17 fungal laccase genes have been completed. These include four from *Rhizoctonia solani* (Wahleithmer *et al.*, 1995), five from *T. villosa* (Yaver and Golightly, 1996; Yaver *et al.*, 1996), two from *Agaricus bisporus* (Perry *et al.*, 1993) and single genes from *Coriolus hirsutus* (Kojima *et al.*, 1990), *Cryphonectria parasitica* (Choi *et al.*, 1992) and *Pleurotus ostreatus* (Giardina *et al.*, 1995). Table 3 shows the diversity among laccase genes from various organisms. Diversity *amA* gene coding for the multi-copper phenol oxidase laccase has been isolated and characterized from the white rot basidiomycetes *T. versicolor* (Jonsson *et al.*, 1995). The gene is preceded by a putative TATA box, which is approximately 130 bp upstream of the start codon. The gene is predicted to have 10 introns, ranging from 51-60 bp. The mature translation product consists of 498 amino acids and is preceded by a 22-residue signal peptide. Within the promoter of *lcc3*, there are putative TATAAA and CAAT motifs, which are 77 and 390 nt upstream of the initiator ATG respectively, while in the promoter of *lac4*, these motifs are 128 and 480 nt upstream of the initiator ATG respectively. The *lcc5* promoter contains TATAAA and CAAT motifs at 107 and 267 nt, respectively, upstream of the ATG. Two laccase genes (*clvg1* and *cvl3*) have been cloned and sequenced from *Coriolus versicolor* (Mikuni and Morohoshi, 1997; Iimura *et al.*, 1992).

The cDNAs encoding laccase LccI and a putative Lcc IV and the gene for *Lcc I* from the white rot basidiomycetes *Trametes versicolor* were cloned, sequenced and characterized (Ong *et al.*, 1997). Five laccase genes have been isolated, cloned and characterized from *T. Sanguinea* M 85-2 by PCR screening with degenerate primers that were designed based on the conserved copper-binding sequences of laccase (Hoshida *et al.*, 2001).

Table 3. Diversity profile of laccase genes

Organism/Gene	Size of gene	Signal sequence	Mature protein	Introns	References
	(bp)	(aa)	(aa)		
Trametes villosa					
lcc 3	3016	21	491	12	Yaver and Golightly,
lcc 4	2906	22	498	10	(1996)
lcc 5	2812	23	504	11	
T. versicolor					
Lcc I	1932	20	519	10	Ong *et al.* (1997)
Lcc II	1560	21	499	-	Cassland and
Lcc III	1563	-	-	-	Jonsson,1999
Lcc IV	2561	23	527	-	Ong *et al.* (1997)
Coriolus versicolor					
CVLG1	3099	23	503	11	Mikuni and Morohoshi, (1997)
CVL3	2684	21	520	08	Iimura *et al.* (1997)
Pycnoporus cinnabarinus					
lcc3-1	2629	21	497	10	Eggert *et al.* (1998)
lcc3-2	2840	23	498	10	Temp *et al.* (1999)
Coprinus congregatus					
clac2	1815	16	362	-	Kim *et al.* (2001)

A successful attempt of cloning of a laccase gene (*clac2*) from *Coprinus congregatus* was made by Kim *et al.* (2001). The cloned cDNA consisted of 1817 bp and its open reading frame consisted of 1086 bp corresponding to 362 amino acids with a signal sequence of 16 amino acids. A unique feature of this gene is that it contains only two copper binding regions, while other fungal laccases usually have four such regions. Eggert *et al.* (1998) studied the cloning and expression of laccase encoding gene from *Pycnoporus cinnabarinus* under xylidine induced conditions. The *lcc 3-1* cDNA was 1828 bp long and contained a open reading frame of 1554 bp. It was predicted that the mature laccase polypeptide contains 497 amino acids preceded with 21 amino acids signal peptide. Characteristic fungal promoter sequences CAAT and TATA boxes are present 68 bp upstream of the translation start site. A consensus polyadinylation signal, AATAA, with slight modification (AATAAA) was found 167 bp downstream of the laccase stop codon. Initially, it was thought that only single isoform of laccase was produced by this fungus but results from southern analysis had indicated the presence of a small laccase gene family in *P. cinnabarinus* (Eggert *et al.*, 1998). Three laccase genes (*LAC1, LAC2* and *LAC3*) in the plant pathogenic ascomycetes fungus *Gaeumannomyces graminis* var. tritici (Litvintseva and Henson, 2002) were identified. *LAC1* contained a 1977 bp open reading frame (orf) that was interrupted by two introns of 87 and 80 bp. *LAC2* contained a 1731 bp orf with no introns and *LAC3* contained a 2212 bp orf that was interrupted by six introns ranging from 63 to 87 bp. *LAC1*, *LAC2* and *LAC3* encoded polypeptides of 609, 577 and 581 amino acids respectively. All the three genes revealed typical promoter sequences along with four putative metal response elements (MRE) consensus sequences situated around the TATA box of the *LAC2* gene promoter and two consecutive putative MRE sequences located 751 and 690 bp upstream of the TATA box of the *LAC1* gene promoter. Cloning of a laccase encoding cDNA from the edible straw mushroom, *Volvariella volvacea* was performed using reverse transcription polymerase chain reaction (RT-PCR) and rapid amplification of c DNA ends (Chen *et al.*, 2004). The cDNA of the putative laccase gene (*lac 4*) consisted of 1689 bp, including an open reading frame encoding a 23 amino acid signal peptide of the protein containing 540 amino acids. A gene coding for laccase in *C. subvermispora* (*Cs-lcs 1*) has also been cloned and sequenced (Lobos *et al.*, 2001). Southern blot analysis showed that *Cs-lcs* 1 is the only gene coding for laccase in this fungus. It contained an open reading frame of 2215 base pairs, encoding a mature protein of 499 amino acids with a signal peptide of 21 residues. The gene possesses 11 introns with splicing junctions and internal lariat forming sites adhering to GT-AG and CTRAY rules respectively. The upstream region of *Cs-lcs* 1 contained a TATA box, two CAAT sites, five putative MREs and an ACE1 element.

Regulation and Expression of Laccase Gene

Cloning and expression of laccase gene in heterologous host is a suitable method to understand its regulation at molecular level. The heterologous expression of fungal laccase has been straightforward, but yields of the purified enzyme achieved in these systems are still far from what is expected for a pilot scale-up. An *A. oryzae* TAKA amylase system has been successfully used for the production of *T. vilosa* and *R. solani* laccases (Wahleithmer *et al.*, 1995; Yaver *et al.*, 1996). The *Phlebia radiate* laccase has been efficiently expressed in *T. reesei* under the control of the *T. reesei cbh1* promoter (Saloheimo and Nikku-Paavola, 1991). The level of *lcc1* RNA was induced by about 17 fold by the addition of the 2, 5-xylidine to a culture, while the levels of *lcc2* RNA was not significantly affected (Yaver *et al.*, 1996). Similarly, the transcription of two laccase genes (*lcc3-1* and *lcc3-2*) from *P. cinnabarinus* was studied using competitive RT-PCR (Temp *et al.*, 1999). Transcription analysis showed that the transcription of *lcc3-1* was induced in presence of 2,5-xylidine, while the transcription of another isoform was unaffected by the presence of 2,5-xylidine.

A new laccase isozyme *lac3* from *Trametes* sp. strain C30 has been expressed in *S. cerevisiae* (Klolowska *et al.*, 2005). Hoshida *et al.* (2001) reported the heterolous expression of *lcc1* cDNA from the white rot fungus *T. sanguinea* into *S. cerevisiae* and *Pichia pastoris*. pRCUP2 was the vector used for the expression in *S. cerevisiae* and the expression was controlled by copper concentration. pPIC3.5 and pPIC9 were used for the expression in *Pichia pastoris* and the expression was induced by methanol. Transcription of laccase genes in *G. graminis* var. tritici was induced in different conditions. *LAC1* was transcribed constituitively in all conditions tested. In contrast, transcription of *LAC2* was copper inducible (Litvintseva and Henson, 2002). Copper-mediated induction was detected for the *P. anserine* (Fernadez-Larrea and Stahl, 1996), *T. versicolor* (Jonsson *et al.*, 1995), and *P. ostreatus* (Palmieri *et al.*, 2000) laccase genes. Laccase activity was also affected by pH and temperature. Optimal laccase production at pH 6.5 indicated that low pH is detrimental for active laccases due to susceptibility of proteases at acidic pH. In addition to this, lower cultivation temperature improves the production of heterologous protein (Guo *et al.*, 2006). The similar observations were also detected by Cassland and Jonsson (1999) in *S. cerevisiae* system. A 16-fold higher laccase activity was found when a transformant was kept at 19°C instead of 28°C. Similar experiments with *P. pastoris* expressing the *T. versicolor* laccase gene *lcc1* also showed that the expression level was favoured considerably by lower cultivation temperature. In order to improve the yield of heterologous proteins by fungi, different approaches have been reported such as increase of gene number, use of strong promoters, appropriate signal sequences and protein fusions. An interesting alternative could be the use of culture conditions such as solid-state fermentation which has been shown to improve enzyme titers in a wide range of fungal strains (Pandey, 2003; Viniegra-Gonzalez *et al.*, 2003). Laccase IV gene from *Trametes versicolor* has been cloned and expressed in *A. niger* C28B25. Production of recombinant protein was studied under different fermentation conditions like submerged and solid-state fermentation (Tellez-Jurado *et al.*, 2006) systems.

STRUCTURE OF LACCASE

In order to modify enzymes for their suitability in industrial applications, determination and analysis of structure is an essential prerequisite. X-ray crystallography, NMR and Mass spectrophotometry are potential techniques which have been used extensively for structural determination of the proteins. Laccase is a blue copper polyphenol oxidase, which catalyzes oxidation of phenols, organic amines and various inorganic substrates. The active site of laccase enzyme consists of four copper ions involved in a coordinated oxygen reduction. The enzyme is a member of the BMCO (Blue Multicopper Oxidase) family, which is characterized by presence of at least one type-1 (T1) copper together with a minimum of at least three additional copper ions: one type-2 (T2) and two type-3 (T3) copper ions arranged in a trinuclear cluster. The different copper centres can be described and identified on the basis of their spectroscopic properties. The T1 copper is characterized by a strong absorption around 600 nm, whilst the T2 copper exhibits only weak absorption in the visible region. The substrates are oxidized by the T1 copper and the extracted electrons are transferred, likely through a strongly conserved His-Cys-His tripeptide motif, to the T2/T3 site where reduction of molecular oxygen to water takes place (Fig. 1). The three-dimensional structure of fungal laccases was reported from *Coprinus cinereus* (Ducros *et al.*, 1998), *Trametes versicolor* (Piontek *et al.*, 2002), *P. cinnabarinus* (Antorini *et al.*, 2002), *Melanocarpus albomyces* (Hakulinen *et al.*, 2002) and *Rigidoporus lignosus* (Garavaglia *et al.*, 2004). In *R. lignosus*, the crystal structure was determined at 1.7 Å resolution from perfectly twinned crystals by molecular replacement technique. The protein folds into three sequentially arranged domains, each of them with a β-barrel type topology. Domain one encompasses residues 1-157, domain two, residues 158-341 and domain three, residues 342-500. As observed in all laccase, the type-1 copper is located in domain 1, whereas the trinuclear copper cluster (T2/T3) is embedded between domains 1 and 3. The structure is stabilized by two disulphide bridges: Cys85-Cys487 located between domain 1 and 3; and Cys117-Cys210 connecting domains 1 and 2 (Garavaglia *et al.*, 2004). The laccase of *T. villosa* contains five Cys residues and 10 conserved His residues. Cys-453 of *lcc2* carries the sulphydryl group, which is a ligand to the type-1 copper. Cys-85/Cys-488 form one of the disulphide bridge, and Cys-117/Cys-205 form the other disulphide bridge (Cassland and Jonsson, 1999). All the ten His copper ligands identified in the *lcc1* gene product (Jonsson *et al.*, 1995) are present in the translation product of *lcc2*. In addition, three other His residues, which seem to be highly conserved among laccases, are present in the *lcc2* sequence, namely His-153, His-216 and His-402. Five potential N-glycosylation sites were identified in the *lcc2* translation product, namely Asn-51, Asn-208, Asn-217, Asn-333 and Asn-436. The redox potential of different laccases might be very important for their practical utilization. An interesting amino acid residue in this respect is Met/Phe/Leu. All so far characterized, *T. versicolor* sequence has a Phe residue in this position. Some plant laccases and ascomycetes have a Met residue in the corresponding position. Alternatively, a Leu residue in the corresponding position is found in some basidiomycetes, ascomycetes and plant laccases. The results of site-directed mutagenesis and the studies performed by Xu (1996) supported the hypothesis that laccases harboring Phe at this position should have type one copper centres with high redox potentials, whereas laccases with Met at this position should have low redox potentials. On the basis of three known residues at this position, laccases can be classified into three classes, 1, 2 and 3: Lac1 (Met), Lac2 (Leu) and Lac3 (Phe) (Eggert *et al.*, 1998).

APPLICATIONS

Almost every natural product, irrespective of its molecular weight or structural complexity, is degraded by one or another microbial species in some particular environment. Accumulation and specific substitution pattern of some groups such as halogen, sulpho-, azo- or nitro- confer xenobiotic character to a synthetic compound. Moreover, the electron-withdrawing characters of these substituents generate electron deficiency and thus makes the compounds less susceptible to oxidative catabolism. As a consequence, many of these chemicals tend to persist under aerobic environmental conditions. Biodegradation of a xenobiotic substance can be accomplished when the catabolic activities, present in mixed microbial communities, complement each other. Thus, synergistic interactions can lead to complete mineralization of even complex xenobiotic compounds. Mineralization of xenobiotic by a single organism can be achieved by taking advantage of natural or induced gene transfer to construct hybrid degradative pathways. Mobilization of blocks of genes, encoding catabolic bottleneck reactions, may speed up the evolutionary potential of natural communities, so that under appropriate selective conditions new multifunctional pathways are generated.

Organopollutants Degradation

The fate of environmental pollutants is largely determined by abiotic processes such as photooxidation the metabolic activities of microorganisms. Since catabolic enzymes are more or less specific, they can act on compounds other than their natural substrates also. This explains why the majority of xenobiotics are subject to fortuitous metabolism. Co-metabolic transformation of a xenobiotic compound often generates reactive metabolites such as epoxides, dihydrodiols, aromatic diols, aromatic amines or at least products that are more easily oxidized than the original chemical. It is followed by its further chemical or biological transformation to completely eliminate the xenophor. Lignolytic fungi can be used for remediation of pollutants in water and soil. White rot fungi applied for soil bioremediation have to compete with indigenous soil microorganisms. The effect of competition on both indigenous soil microflora and white rot fungi was evaluated with regard to degradation of polycyclic aromatic hydrocarbons (PAH) with different persistence in soil. Many soils on former industrial sites, particularly gas plant sites, have become contaminated with polycyclic aromatic hydrocarbons (PAH). Polyaromatic hydrocarbons (PAH) are known for their mutagenic or carcinogenic effects, therefore clean-up of such sites is desirable (Dipple *et al.*, 1990). The lignolytic enzymes are assumed to be responsible for the ability of white rot fungi to degrade a range of aromatic pollutants, including PAH (Field *et al.*, 1993; Sack *et al.*, 1997; Mester and Tien, 2000). The ability of *P. ostreatus* and *D. squalens* to mineralize ^{14}C-PAH in axenic culture was determined using sterile soil.

Although the activities of lignolytic enzymes produced by both fungi were basically same, a much greater portion of ^{14}C-PAH was mineralized by *P. ostreatus* as compared to *D. squalens* (inder Wiesche *et al.*, 2000). The degradation of wide range of organopollutants with diverse ring structure (dyes, PAHs, PCBs) using lignolytic enzymes of *Phanerochaete chrysosporium*, *Trametes versicolor*, *Coriolopsis polyzona*, *Pleurotus ostreatus* and *Irpex lacteus* was studied by Novotny *et al.* (2004). A strong correlation was established between the levels of extracellular lignolytic enzymes involved and degradation rates. There was a strong correlation between the rate of decolorization of RO 16 and Acid Black in the liquid medium cultures and the concentrations of lignolytic enzymes. The transformation of chlorinated hydroxybiphenyls by laccase produced by *Pycnoporus cinnabarinus* has also been investigated. Because of their low water solubility and low chemical reactivity, polychlorinated biphenyls (PCBs) are used mainly as solvent extenders, heat transfer fluids, plasticizers, organic solvents, flame retardants and hydraulic fluids. Due to their hydrophobicity and stability, PCBs persist for a long time in the environment.

Dye Decolorization

Major consumption of synthetic dyes is by textile industry and it requires an eco-efficient solution for treatment of the coloured effluent released from it. Approximately, 600 thousand tons of dyes per annum are currently consumed by India, Russia, Eastern Europe, China, South Korea and Taiwan (Ishikawa *et al.*, 2000), and up to 15% of the used dyestuffs are released into the environment during processing (Vaidya and Datye, 1982). Conventional wastewater treatment techniques are inefficient in order to decolorize dye containing effluents (Willmott *et al.*, 1998; Shaul *et al.*, 1991). In addition to their visual effect and adverse impact in terms of chemical oxygen demand, many synthetic dyes are toxic, mutagenic and carcinogenic (Michaels and Lewis, 1985; Chung *et al.*, 1992). The effluent faces increasingly stringent policies of environmental protection. Action of purified laccases and its role in colour removal remains poorly understood. Unfortunately, conventional waste water treatment plants are unavailable to perform a complete dye removal. About 90% of reactive textile dye persists after activated sludge treatment and is discharged into the receiving water bodies (Pierce, 1994). Several biotechnological processes have been suggested for decolorization and removal of dyes from industrial effluents (McMullan *et al.*, 2001; Stolz, 2001). Although bacteria could decolorize textile wastewater, under anaerobic environments, azo dyes potentially form carcinogenic aromatic amines (McMullan *et al.*, 2001) while aerobic bacteria usually tend to be specific towards a particular dye (Reddy, 1995). However, the most efficient lignin-degrading organisms, the white rot basidiomycetes (Kirk and Farrell, 1987; Breen and Singleton, 1999), have been reported to degrade a wide range of xenobiotic compounds and other pollutants (Reddy, 1995; Pointing, 2001) under aerobic conditions. Cripps *et al.* (1990) had reported the decolorization of azo dyes by a white rot fungus. Presently, fungal bioremediation is regarded as a promising and environment-friendly alternative for textile effluent treatment. Thus, new or little studied basidiomycetous fungi have been investigated for their potential use in the treatment of coloured effluents (Pointing *et al.*, 2000). Major class of microorganisms most effective in catabolizing synthetic dyes is the white rot fungi. The lignolytic system of white rot fungi has been shown directly involved in the degradation of various xenobiotic compounds and dyes. Primarily, the decolorization of sulfonated polymeric dyes was used to assay lignolytic activities (Glenn and Gold, 1983; Freitag and Morrell, 1992) and to assess the biodegradation capabilities of white rot fungi (Field *et al.*, 1993; Cripps *et al.*, 1990). Later on, numerous white rot fungi strains were used for the decolorization of distinct synthetic (textile) dyes and synthetic effluents, i.e., dye mixtures (Jarosz-Wilkolazka *et al.*, 2002; McMullan *et al.*, 2001). Several groups had attempted to correlate the production of lignin modifying enzymes in white rot fungi and the rates of decolorization. A direct correlation between lignin modifying enzyme production and industrial effluent decolorization was given by Wesenberg *et al.*, 2002, suggesting a differential inducing effect of the effluent on the lignin modifying enzyme production pattern. Dias *et al.* (2003), compared laccase catalyzed decolourization of xenobiotic azo dyes by a basidiomycetous fungus Euc-1 (On the basis of phylogenetic analysis, strain Euc-1 resembles the genera *Trametes* and *Pycnoporus*). Solid-plate decolorization studies showed that basidiomycetes Euc-1 decolourized 11 azo dyes, whereas *P. chrysosporium* decolorized only two. The decomposition of azo dyes by oxidative methods such as laccase and ultrasound treatments was investigated by Tauber *et al.* (2005). For the first time it was shown that a combination of laccase and ultrasound treatments can have synergistic effects, which was shown by higher degradation rates. This finding shows the possibility of saving time and energy by applying a simultaneous combination of laccase and ultrasound treatments for decolorization of azo dyes. The laccase cDNA of *Panus rudis* was isolated by RT-PCR and the highest identity was found to be 67% with the sequence of laccase from *Lentinula edodes*.This laccase had excellent ability to decolorize anthraquinone dye (Acid Green 27) without any redox mediators, as well as azo and indigo dyes (Acid Violet 7 and Indigo Carmine) mediated by ABTS (Zhang *et al.*, 2006).

A white rot fungus *Thelephora* sp. was used for decolorization of azo dyes such as orange G, congo red and amido black. Fungal decolorization was 33.3%, 97.1% and 98.8% for orange G, congo red and amido black, respectively. A maximum decolorization of 12.0% and 15.0% for congo red and amido black, respectively, was reported by using laccase. A dye industry effluent was treated by the fungus in batch and continuous modes. A maximum decolorization of 61% was achieved on the third day in the batch mode and a maximum decolorization of 50% (Selvam *et al.*, 2003) was obtained by the seventh day in the continuous mode.

Effluent Treatment

Inks, lacs and explosives industries are major consumer of nitrocellulose polymer and toxic effluents are generated from these industries during nitrocellulose manufacturing. Effluent delignification results from the alkaline treatment of cotton fibres. This effluent has dark colour, a large amount of organic matter and a high degree of toxicity. The conventional biological treatment of this kind of effluent by activated sludge was largely detected to be ineffective (Mittar *et al.*, 1992). Tremendous potential of fungal strains for treatment of effluents from cellulose industry had been demonstrated by several studies, since, they not only metabolize lignin and its derivatives, but also reduce the effluent, colour, organic matter and toxicity (Eaton and Chang, 1980; Sundman *et al.*, 1981; Mehna *et al.*, 1995). In order to have an effective biological treatment system for cellulose industry effluents, the microorganism must be able to grow rapidly, predominate in the environment with competition, and degrade low and high molar weight lignin derivatives. These requirements are fulfilled by the white rot fungi. However, soft-rot fungi, Deuteromycetes and Ascomycetes have also been found to be useful in the biological treatment of effluents (Malarczyk *et al.*, 1998; Nagarathnamma and Bajpai, 1999). During the screening for the effluent treatment, the fungi *Aspergillus* 2BNL1, *Aspergillus* 2AAL1 and *L. edodes* had resulted into the highest rate of effluent colour reduction (Souza *et al.*, 2005). The molar mass distribution analysis showed that the three fungi were able to change the pattern of the untreated chromatogram, degrading high molecular mass compounds in few hours of treatment.

CONCLUSION AND FUTURE PERSPECTIVES

Fungal degradation of xenobiotic compounds appears to be a promising alternative to replace or supplement conventional treatment methods. Laccase-mediated bioremediation possess ecological and economic benefits over conventional processes. Most of the pollutants, which are generally considered as non-degradable or recalcitrant, are suitable for laccase-mediated bioremediation due to low substrate specificity of the enzyme. Information is available regarding gene structure, regulation, and expression into suitable host, but more efforts are needed to determine the applicability and practicability of utilizing the organism in contaminated field sites. Practical applicability of laccase-mediated degradation of contaminants requires scaled up production of enzymes with desirable features at economic prices. Modification of critical fermentation parameters and evaluation of suitable fermentation strategy may help in the scaled up production of enzyme for degradation of xenobiotics. Genetic manipulation of laccase gene could be a promising approach for simultaneous degradation of various pollutants and effective decolorization of industrial effluents, which generally contain mixture of hazardous dyes. Further, applicability of laccase for bioremediation could be extended by achieving wider substrate range, catalytic activity and stability under desired condition through genetic manipulation.

Development of suitable enzyme preparation for bioremediation is followed by its application under field conditions. Transfer of developed technology from *in vitro* conditions to the field condition for scaled up production of laccase for bioremediation is a major challenge. Laboratory experiments for laccase-mediated degradation of toxic compounds show promising results, but the results may not be significant under natural environment. Several natural environmental variables such as presence of mixture of xenobiotics, native soil fauna, temperature, pH and moisture, etc. affect the laccase-mediated bioremediation and decrease its efficiency. The base of knowledge on this process is rapidly coming out from laboratory studies, so the next step is to utilize this pool of information in an exploratory way in the field.

REFERENCES

Abadulla, E., Tzanov, T., Costa, S., Robra, K-H., Cavaco-Paulo, A. and Gu¨bitz, G.M. (2000). *Appl. Environ. Microbiol.* **66:** 3357-62.

Amitai, G., Adani, R., Sod-Moriah, G., Rabinovitz, I., Vincze, A. and Leader, H. (1998). *FEBS Lett.* **438:** 195-200.

Antorini, M., Herpoel-Gimbert, I., Choinowski, T., Sigoillot, C., Aster, M., Winterhalter, K. and Piontek, K. (2002). *Biochim. Biophys. Acta.* **1594:** 109-114.

Aust, S.D.P.R. and Swaner J.D. Stahl. (2003). *In. Pesticide decontamination and detoxification.* (Zhu, J.J.P.C., Aust, S.D. and Lemley Gan, A.T.), Oxford University Press, Washington, D.C., 3-14.

Barr, D.P. and Aust, S.D. (1994). *Environ. Sci. Technol.* **28:** A78-87.

Becker, H.G. and Sinitsyn, A.P. (1993). *Biotechnol. Lett.* **15:** 289-94.

Bollag, J.M., Shuttleworth, K.L. and Anderson, D.H. (1988). *Appl. Environ. Microbiol.* **54:** 3086-91.

Boopathy, R. (2000). *International Biodeterioration & Biodegradation,* **46:** 29-36.

Bourbonnais, R., Paice, M.G., Freiermuth, B., Bodie, E. and Bornemann, S. (1996). *Appl. Environ. Microbiol.* **63:** 4627-32.

Bourbonnais, R., Leech, D. and Paice, M.G. (1998). *Biochimica Biophysica Acta,* **1379:** 381-390.

Bourbonnais, R., Paice, M.G., Reid, I.D., Lanthier, P. and Yaguchi, M. (1995). *Appl. Environ. Microbiol.* **61:** 1876-1876.

Boyle, C.D., Kropp, B.R. and Reid, I.D. (1992). *Appl. Environ. Microbiol.* **58:** 3217- 24.

Breen, A. and Singleton, F.L. (1999). *Current Opinion in Biotechnology* **10:** 252-258.

Cajthaml, T.M., Möder, P., Kacer, V. and Šašek P. Popp. (2002). *J. Chromatography **A*** **974:** 213-22.

Cassland, P. and Jonsson, L.J. (1999). *Appl. Microbiol. Biotechnol.* **52:** 393-400.

Chen, S., Ge, W. and Buswell, J.A. (2004). *FEMS Microbiol. Lett.* **230:** 171-176.

Choi, G.H., Larson, T.G. and Nuss, D.L. (1992). *Molecular Plant Microbe Interaction* **5:** 119-128.

Choi, H.T., Wilks, R.W. and Ross, I.K. (1987). *Mycology,* **79:** 166-172.

Chung, K.T., Stevens, S.E. and Cerniglia, C.E. (1992). *Crit. Rev. Microbiol.* **18:** 175- 90.

Clutterbuck, A.J. (1972). *J. Gen. Microbiol.* **70:** 423-435.

Cripps, C., Bumpus, J.A. and Aust, S.D. (1990). *Appl. Environ. Microbiol.* **56:** 1114-8.

Cullen, D. (1997). *J. Biotechnol.* **53:** 273-289.

Dias, A.A., Bezerra, R.M., Lemos, P.M. and Pereira, A.N. (2003). *World J. Microbiology and Biotechnology,* **19:** 969-975.

Dipple, A., Cheng, S.C. and Bigger, C.A.H. (1990). *In: Mutagens and Carcinogens in the Die*t (Eds. Pariza, M.W., Aeschbacher, H.U., Felton, J.S. and Sato, S.), Wiley-Liss, New York, pp. 109-127.

Ducros, V., Brzozowski, A.M., Wilson, K.S., Brown, S.H., Østergaard, P. and Schneider, P. (1998). *Nature Struct. Biol.* **5:** 310-316.

Eaton, D. and Chang, H. (1980). *Tappi,* **63:** 103-108.
Eggert, C., Lafayette, P.R., Temp, U., Eriksson, K.L. and Dean, J.F.D. (1998). *Appl. Environ. Microbiol.* 1766-1772.
Evans, C.S. and Hedger, J.N. (2001). *In:* Fungi in bioremediation. (Eds. Gadd, G.M.) British Mycological Society. Cambridge Univ. Press. pp. 1-20.
Fahr, K., Wetzstein, H.G., Grey, R. and Schlosser, D. (1999). *FEMS Microbiol Lett.* **175:** 127-32.
Fernadez-Larrea, J. and Stahl, U. (1996). *Mol. Gen. Genet.* **252:** 539-551.
Field, J.A., de Jong, E., Feijoo-Costa, G. and de Bont, J.A.M. (1993). *Trends Biotechnol* **11:** 44-9.
Freitag, M. and Morrell, J.J. (1992). *Can. J. Microbiol.* **38:** 811-22.
Garavaglia1, S., Cambria, M.T., Miglio, M., Ragusa, S., Iacobazzi, V., Palmieri, F. Ambrosio, C.D., Scaloni, A. and Rizzi1, M. (2004). *J. Mol. Biol.* **342:** 1519-1531.
Giardina, P., Cannio, R., Martirani, L., Marzullo, L., Palmieri, G. and Sannia, G. (1995). *Appl. Environ. Microbiol.* **61:** 2408-2413.
Glenn, J.K., Akileswaran, L. and Gold, M.H. (1986). *Arch. Biochem. Biophys.* **251:** 688-96.
Glenn, J.K. and Gold, M.H. (1983). *Appl. Environ. Microbiol.* **45:** 1741-7.
Grey, R., Hofer, C. and Schlosser, D. (1998). *J. Basic Microbiol.* **38:** 371-82.
Guo, M., Lu, F., Du, L., Pu, J. And Bai, D. (2006). *Appl. Microbiol. Biotechnol.* **71:** 848-52.
Hakulinen, N., Kiiskinen, L.L., Kruus, K., Saloheimo, M., Paananen, A., Koivula, A. and Rouvinen, J. (2002). *Nature Struct. Biol.* **9:** 601-605.
Hammel, K.E. (1996). *New J. Chem.* **20:** 195-8.
Hatakka, A. (1994). *FEMS Microbiol. Rev.* **13:** 125-35.
Hestbjerg, H.P.A., Willumsen, M., Christensen, O., Andersen, C.S. and Jacobsen. (2003). *Environmental Toxicology and Chemistry,* **22:** 692-98.
Hoshida, H., Nakao, M., Kanazawa, H., Kubo, K., Hakukawa, T., Morimasa, K., Akada, R. and Nishizawa, Y. (2001). *J. Bioscience and Bioengineering,* **92:** 372-380.
Iimura, Y., Takenouchi, K., Nakamura, M., Kawai, S., Katayama, Y. and Morohoshi, N. (1992). *In: Biotechnology in the pulp and paper manufacture* (Eds. Kuwahara, M. and Shimada, M.), Uni, Tokyo. pp. 427-431.
Inder Wiesche, C., Martens, R. and Zaarazil, F. (2000). *Water, air, and soil pollution* **3:** 73-79.
Ishikawa, Y., Esker, T. and Leder, A. (2000). Chemical economics handbook: dyes. Menlo Park (CA): SRI Chemical and Health Business Services;
Jarosz-Wilkolazka, A., Kochmanska-Rdest, J., Malarcyk, E., Wardas, W. and Leonowicz, A. (2002). *Enzyme Microb. Technol.* **30:** 566-72.
Jonsson, L., Sjo¨stro¨m, K., Ha¨ggstro¨m, I. and Nyman, P.O. (1995). *Biochim. Biophys. Acta,* **1251:** 210-215.
Kersten, P.J., Tien, M., Kalyanaraman, B. and Kirk, T.K. (1985). *J. Biol. Chem.* **260:** 2609-12.
Kim, S., Leem, Y., Kim, K. And Choi, H.T. (2001). *FEMS Microbiol. Lett.* **195:** 151-156.
Kim, H.Y., Leem, Y.E., Choi, H.T. and Song, H.G. (1995). *Korean J. Mycol.* **23:** 298-304.
Kirk, T.K. and Farrell, R.L. (1987). *Ann. Rev. Microbiology,* **41:** 465-505.
Klonowskaa, A., Gaudina, C., Assob, M., Fournelb, A., R´egliera, M. and Trona, T. (2005). *Enzyme Microbial Technol.* **36:** 34-41.
Kojima, Y., Tsukuda, Y., Kawai, Y., Tsukamoto, A., Sugiura, J., Sakaino, M. and Kita, Y. (1990). *J. Biol. Chem.* **265:** 15224-15230.
Law, W.M.W.N., Lau, K.L., Lo, L.M. and Wai, S.W. (2003). *Chemosphere,* **52:** 1531-37.
Leatham, G.F. and Stahmann, M.A. (1981). *J. Gen. Microbiol.* **125:** 147-157.
Leonard, T.J. (1971). *J. Gen. Microbiol.* **106:** 162-167.
Levin, L.A. and Viale, A. Forchiassin. (2003). *International Biodeterioration & Biodegradation,* **52:** 1-5.
Litvintseva, A.P. and Henson, J.M. (2002). *Appl. Environ. Microbiol.* 1305-1311.
Lobos, S., Tello, M., Polanco, R., Larrondo, L.F., Manubens, A., Salas, L. and Vicuna, R. (2001). *Curr. Sci.,* **81:** 992-997.
Majcherczyk, A., Johannes, C. and Huttermann, A. (1998). *Enzyme Microb. Technol.* **22:** 335-41.
Malarczyk, E., Kornittowicz, T. and Leonowicz, A. (1999). *In:* Proc. Seventh International Conference on Biotechnology in the Pulp and Paper Industry.Vancouver, Canada, C219.
Mansur, M.M.E., Arias, J.L., Copa-Patino, M., Flärdh and González, A.E. (2003). *Mycologia,* **95:** 1013-20.
McMullan, G., Meehan, C., Conneely, A., Kirby, N., Robinson, T. and Nigam, P. (2001). *Appl. Microbiol. Biotechnol.* **56:** 81-7.
Mehna, A., Bajpai, P. and Bajpai, P.K. (1995). *Enzyme Microbial.Technol.* **17:** 18-22.
Mester, T. and Tien, M. (2000). *Int. Biodeterior. Biodegrad.* **46:** 51-59.
Michaels, G.B. and Lewis, D.L. (1985). *Environ. Toxicol. Chem.* **4:** 45-50.
Mikuni, J. and Morohoshi, N. (1997). *FEMS Microbiol. Lett.* **155:** 79-84.

Mittar, D., Khanna, P.K., Marwaha, S.S. and Kennedy, J.F. (1992). *J. Chem. Tech. Biotechnol.* **53:** 81-92.
Nagarathnamma, R. and Bajpai, P. (1999). *Appl. Environ. Microbiol.* 1078-1082.
Novotny, C., Svobodova, K., Erbanova, P., Cajthaml, T., Kasinath, A., Lang, E. and Sasek, V. (2004). *Soil Biology and Biochemistry,* **36:** 1545-1551.
Ong, E., Brent, W., Pollock, R. and Smith, M. (1997). *Gene,* **196:** 113-119.
Palmieri, G., Giardina, P., Bianeo, C., Fontanella, B. and Sannia, G. (2000). *Appl. Environ. Microbiol.* **66:** 920-924.
Pandey, A. (2003). *Biochem, Eng. J.* **13:** 81-4.
Pasti-Grigsby, M.B., Paszczynski, A., Gosczynski, S., Crawford, D.L. and Crawford, R.L. (1992). *Appl. Environ. Microbiol.* **58:** 3605-13.
Paszczynski, A., Pasti-Grisgby, M.B., Gosczynski, S., Crawford, R.L. and Crawford, D.L. (1992). *Appl. Environ. Microbiol.* **58:** 3598-604.
Perry, C., Smith, M., Britnell, C., Wood, D. and Thurston, C. (1993). *J. Gen. Microbiol.* **139:** 1209-1218.
Pierce, J. (1994). *J. Soc. Dyers and Colourists,* **110:** 131-134.
Piontek, K., Antorini, M. and Choinowski, T. (2002) *J. Biol. Chem.* **277:** 37663-37669.
Pointing, S.B. (2001). *Appl. Microbiol. Biotechnol* **57:** 20-33.
Pointing, S.B., Bucher,V.V.C. and Vrijmoed, L.L.P. (2000). *World J. Microbiol. and Biotechnol.* **16:** 199-205.
Reddy, C.A. and Mathew, Z. (2001). *In Fungi in bioremediation.* (Ed. Gadd, G.M.) Cambridge University Press. Cambridge, U.K.
Reddy, C.A. (1995), *Current Opinion in Biotechnol.* **6:** 320-328.
Ricotta, A., Unz, R.F. and Bollag, J. (1996). *Bull. Environ. Contam. Toxicol.* **57:** 560-7.
Rigling, D. and Van Alfen, N.K. (1991). *J. Bacteriol.* **173:** 8000-8003.
Rodrý´guez, E., Pickard, M.A. and Vazquez-Duhalt, R. (1999). *Curr. Microbiol.* **38:** 27-32.
Ross, I.K. (1982). *J. Gen. Microbiol.* **128:** 2763-2770.
Sack, U., Hofrichter, M. and Fritsche, W. (1997). *FEMS Microbiol. Lett.* **152:** 227-234.
Salas, S.D., Bennett, J.E., Kwon-Chung, K.J., Perfect, J.R. and Williamson, P.R. (1996). *J. Exp. Med.* **184:** 377-386.
Saloheimo, M. and Niku-Paavola, M-L. (1991). *Biotechnology,* **9:** 987-990.
Scheibner, K., Hofrichter, M. and Fritsche, W. (1997). *Biotechnol. Lett.* **19**: 835-9.
Selvam, K., Swaminathan, K. and Chae, K. (2003). *Biores. Technol.* **88:** 115-119.
Shaul, G.M., Holdsworth, T.J., Dempsey, C.R. and Dostal, K.A. (1991). *Chemosphere,* **22:** 107-19.
Souza, J.V.B., Silva, E.S. and Paiva, T.C.B. (2005). *In:* Semina ´rio de Hidro ´lise Enzima ´tica de Biomassa,2 .6 December, Brazil, pp. 166.
Spadaro, J.T., Gold, M.H. and Renganathan, V. (1992). *Appl. Environ. Microbiol.* **58:** 2397-401.
Stolz, A. (2001). *Appl. Microbiol. Biotechnol.* **56:** 69-80.
Sundman, G., Kirk, T.K. and Chang, H.M. (1981). *Tappi J.* **64:** 145-148.
T´ellez-Jurado, A., Arana-Cuenca, A., Gonz´alez Becerra, A.E., Viniegra-Gonz´alez, G. and Loera, O. (2006). *Enzy. Microbial. Technol.* **38:** 665-669.
Tauber, M.M., Guebitz, G.M. and Rehorek, A. (2005). *Appl. Envirn. Microbiol.* **71:** 2600-2607.
Temp, U., Zierold, U. and Eggert, C. (1999). *Gene* **236:** 169-177.
Thurston, C.F. (1994). *Microbiology,* **140:** 19-26.
Tien, M. and Kirk, T.K. (1983). *Science,* **221:** 661-3.
Umezawa, T. and Higuchi, T. (1987). *FEBS Lett.* **218:** 255-60.
Vaidya, A.A. and Datye, K.V. (1982). *Colourage,* **14:** 3-10.
Van Aken, B. and Agathos, S.N. (2001). *Adv. Appl. Microbiol.* **48:** 1-77.
Van Aken, B. and Agathos, S.N. (2002). *Appl. Microbiol. Biotechnol.* **58:** 345-51.
Veignie, E.C., Rafin, P., Woisel, F. and Cazier. (2004). *Environmental Pollution* **129:** 1-4.
Viniegra-Gonzalez, G., Favela-Torres, E., Aguilar, C.N., Romero-Gomez, S.U., Diaz Godinez, G. and Augur, C. (2003). *Biochem. Eng. J.* **13:** 157-67.
Wahleithner, J.A., Xu, F., Brown, K.M., Brown, S.H., Golightly, E.J., Halkier, T., Kauppinen, S., Pederson, A. and Schneider, P. (1995). *Current Genetics,* **29:** 395-403.
Wesenberg, D., Buchon, F. and Agathos, S.N. (2002). *Biotechnol. Lett.* **24:** 989-93.
Willmott, N., Guthrie, J. and Nelson, G. (1998). *J. Soc. Dyers. Colour.* **114:** 38-41.
Xu, F. (1996). *Biochemistry,* **35:** 7608-14.
Yaver, D.S. and Golightly, E.J. (1996). *Gene,* **181:** 95-102.
Yaver, D.S., Xu, F., Golightly E.J., Brown, K.M., Brown, S.H., Rey, M.W., Schneider, P., Halkier, T., Mondorf, K. and Dalboge, H. (1996). *Appl. Environ. Microbiol.* **62:** 834-841.
Zhang, M., Wu, F., Wei, Z., Xiao, Y. and Gong, W. (2006). *Enz. Microbial. Technol.* **39:** 92-97.

Chapter **15**

Biotechnological, Genetic Engineering and Nanotechnological Potential of Actinomycetes

R. Balagurunathan and M. Radhakrishnan

Department of Microbiology, Sri Sankara Arts and Science College
Kanchipuram–631 561, Tamil Nadu, (India)
E-mail: rbalaguru@yahoo.co.in

Actinomycetes have been looked upon as potential sources of bioactive compounds and they are the richest sources of secondary metabolites. Actinomycetes synthesized about two-thirds of currently reported bioactive substances belonging to a great variety of chemical groups. Diversity and bioactivity of actinomycetes are well documented in various normal ecosystems all over the world. At present bioprospecting of actinomycetes from extreme ecosystems by adopting selective methodologies are undertaken by the global scientists. Actinomycetes are unambiguously important in commercial side as cell factories or bioreactors for applications in industrial, agricultural, environmental and pharmaceutical fields. Five major groups of bioproducts produced by actinomycetes include primary metabolites, secondary metabolites, bioconversion products, microbial cell products and recombinant products. In recent years many works have been carried out on actinomycetes in the genetic engineering field with respect to genetic structure, genetic manipulation, antibiotic production and resistance, plasmids, vectors and regulation of gene expression. Recently, actinomycetes such as *Rhodococcus* and *Thermomonospora* sp have been recognized as a potential organism for biosynthesis of nanoparticles like gold and silver. It is unquestionable that the actinomycetes will represent the biggest chance to obtain unique compounds that will be important in medical, agricultural and industrial sectors including nanomedicine in future. Therefore, it will play a major role in production of efficient, more productive and cost-effective compounds for the disease control in both agricultural and human diseases.

INTRODUCTION

Microorganisms being the pioneer colonizers of this earth planet, as come to stay has cosmopolitan conglomerates of highly compatible organisms. Bestowed with inherent physiological and functional diversity, microorganisms have found applications in agriculture, medicine, industry and environment. Microorganisms with its 3.8 billion years of biosynthetic experience remain nature's best chemists and treasure house for a variety of novel biologically active metabolites. Microorganisms act as mini-biological factories to produce various high value metabolites. Since civilization, the use of microorganisms to produce natural products and processes that benefit and improve our socio-economic lifestyles has been a part of our human history. They are the easiest manipulated sources of value added products like drugs, therapeutic proteins, vaccines, diagnostics and others.

Among the various industrially important microorganisms, actinomycetes are of prime importance and are primarily recognized as organisms of academic curiosity and potential antibiotic producers.

Actinomycetes are a group of Gram-positive filamentous bacteria with high guanine + cytosine (G+C) content. These come under the order actinomycetales, whose members have an unparalled ability to produce diverse secondary metabolites. Members of this group form substrate and/or aerial mycelium and asexual spores.

In addition to its widespread distribution in terrestrial and aquatic habitats, they are also abound in habitats with extremes of temperature, pH and salt stress. Actinomycetes are also present in association with healthy plants (endophytes), animals including marine organisms. These bacteria are primarily saprophytic and are best known from soils where they contribute significantly to the turn over of complex biopolymers such as lignocellulose, hemicellulose, pectin, keratin and chitin (Mincer *et al.*, 2002).

BIOPROSPECTING OF ACTINOMYCETES

Actinomycetes are the most industrially important group of microbes to be exploited more in terms of biotechnological applications. The important reasons for the global bioprospecting of actinomycetes include:

- excellent tract record – previous reports about actinomycetes and their products and about 16,500 bioactive compounds;
- genetic diversity – existence of various genetically diverse groups of actinomycetes strains;
- biosynthetic/chemical diversity – synthesis of structurally different compounds – *S. hygroscopicus* is reported to produce 286 different compounds;
- inherent and novel biological activity – diverse activities exhibited by same or different compounds produced by actinomycetes. For e.g., Munumbicin is reported to have activity against the diseases of TB, anthrax, malaria, etc.
- excellent biodiversity – existence of actinomycetes even in diverse ecosystems including extreme conditions leads to about 4000 species.

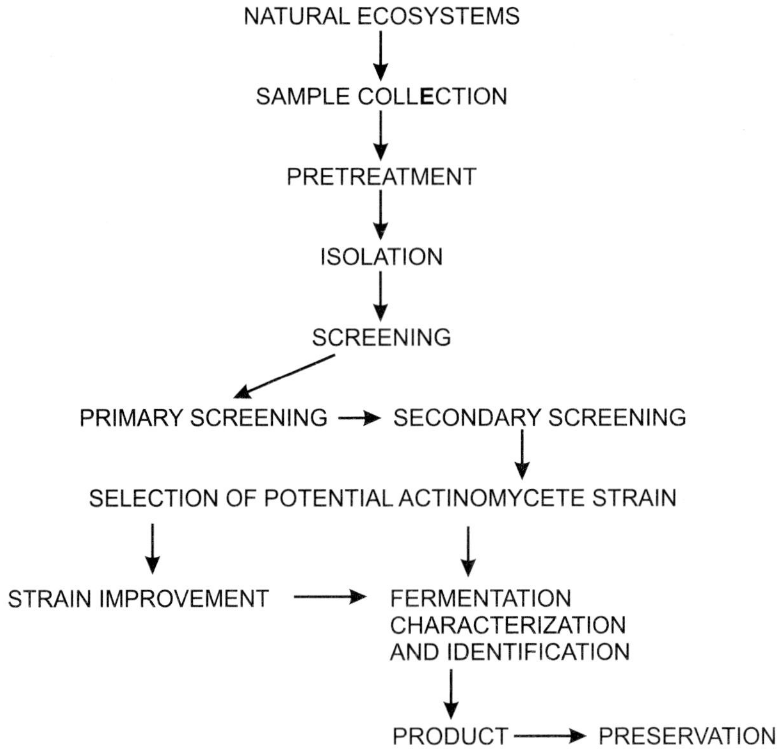

Fig. 1. Bioprospecting of actinomycetes for their products.

ACTINOMYCETES IN DIVERSE ECOSYSTEMS

Actinomycetes are the most successful group of bacteria that occur in wide range of natural and man made environments. They also exist in extreme environments like mountain soils at higher altitude, desert, deep sea ecosystem and salt pans.

For several decades, microbiologists have been routinely practicing and isolating actinomycetes from soil and water samples collected from routine environments. For e.g., soil from cultivable lands, gardens and water samples from pond, lake and river. The existing isolated actinomycetes are mainly from normal habitats, and various useful products are produced from these organisms. Bioprospecting actinomycetes from routine habitats leads to re-isolation of same actinomycetes and re-discovery of known metabolites (Jenson *et al.,* 2003). More importantly, each and every genera that comes under actinomycetes prefers specific habitat. For example, streptomycetes prefer soil rich in compost; *Micromonospora* sp. are dominant in aquatic habitats and *Dactylosporangium* is present in leaf litter. A single gram of rich agricultural soil can contain 10^6 Streptomycete colony forming units and 10^4 to 10^5 *Micromonospora* and other genera. The percentage of distribution of certain actinomycetes is given in Table 1.

Table 1: Distribution of certain actinomycetes

Actinomycetes genera	Percentage of distribution
Streptomycetes	95.34
Nocardia	1.98
Micromonospora	1.40
Thermomonospora	0.22
Actinoplanes	0.20
Microbispora	0.18
Streptosporangium	0.10
Actinomadura	0.10
Actinosynnema	0.10
Dactylosporangium	0.06

Recently, it was found that unusual environments like marine ecosystems, salt pan and mountains were reported as source for novel actinomycetes, which may be a potential source for novel bio products.

For example, *Salinispora, Verrucosispora* are some novel actinomycete genera reported from deep sea that produce antibiotics like salinosporamide and abyssomycin, respectively (Baltz, 2007). In addition, *Gordonia rhizosphera* sp. nov from mangrove rhizosphere (Tkechi AND Hatano, 1998), *Nocardiopsis aegyptia* sp nov from marine sediment (Sabry *et al.*, 2004), *Prauserella halophila* sp nov from saline soil (Li *et al.*, 2003) and Actinomyces *Marimammalium* sp nov (Hoyles *et al.*, 2001) was also reported. Actinomycetes like *Streptomyces, Micromonospora* and *Nocardia* have also been reported from healthy plants as endophytes from which various bioproducts are isolated (Balagurunathan and Radhakrishnan, 2007).

ISOLATION OF ACTINOMYCETES

Since the discovery of *Actinomyces* by Ferdinand Cohn in 1875, so far more than 145 actinomycete genera and more than four thousand species (Berdy, 2005) have been reported. Isolation of actinomycetes from different samples collected from various ecosystems is the first step for its bioprospecting. Before isolation, pretreatment of samples, especially soil and sediment, is the prerequisite for decreasing bacterial and fungal population and increasing actinomycetes recovery from the collected samples. For example, dry heating soil samples at 55°C for 10 minutes and supplementing the isolation media with antibacterial agents (Nalidixic acid – 100 mg/ml) and antifungal agent (20 mg/ml) facilitates the isolation of actinomycetes (Balargurunathan and Radhakrishnan, 2007). In general, media supplemented with complex substrates was used for the isolation of actinomycetes. Media such as chitin agar, humic acid vitamin (HV) agar, starch casein agar, arginine glycerol salt agar, and tap water agar are well suited for the isolation of actinomycetes. Majority of the reported actinomycetes are mesophiles, and incubation should be at 28°C for 1 month since they are slow growing organisms. The incubation period may be up to 3 months for the initial isolation of rare actinomycetes.

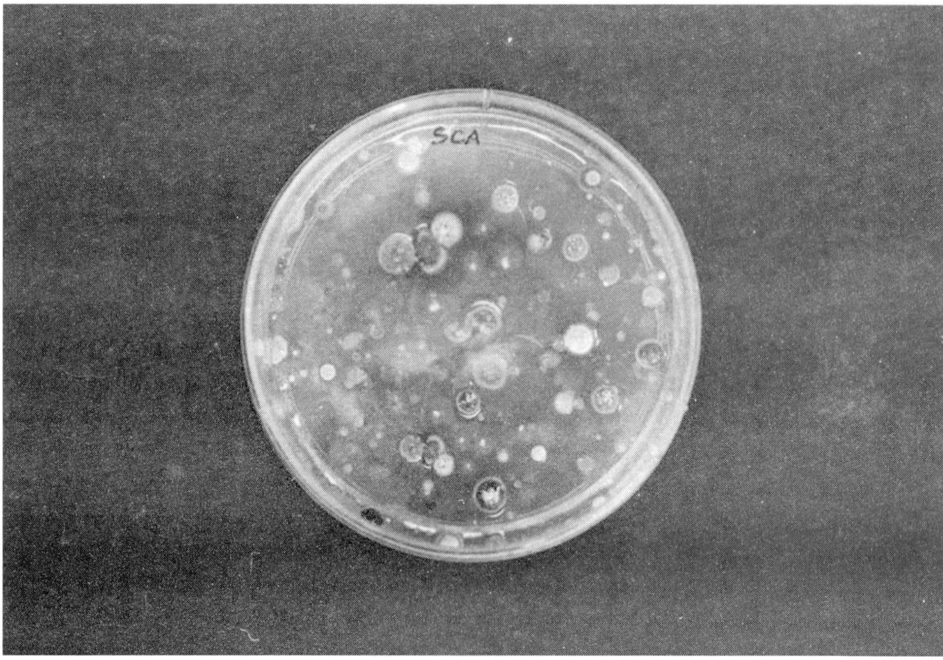

Fig. 2. Isolation of actinomycetes.

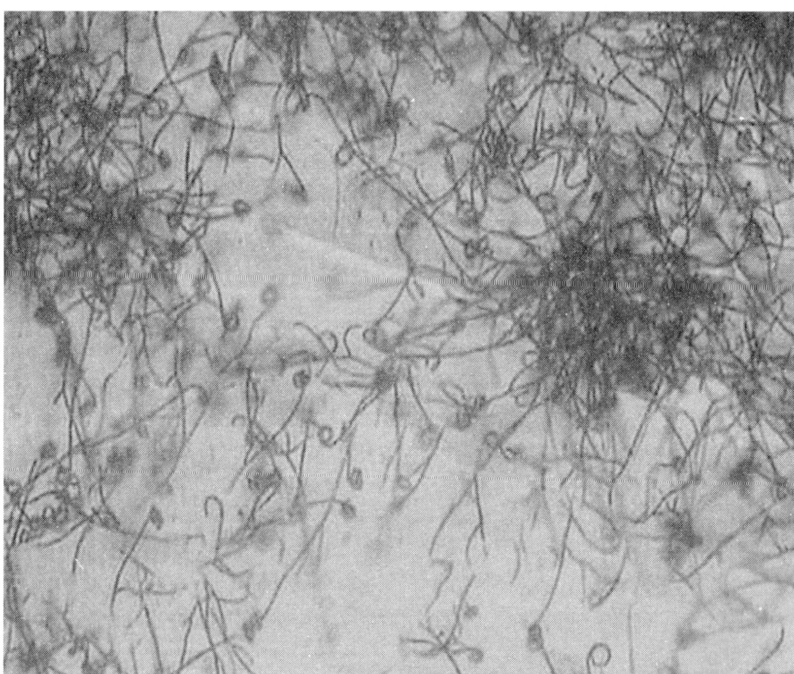

Fig. 3. Microscopic appearance of actinomycetes.

SCREENING OF ACTINOMYCETES

Screening may be defined as the use of highly selective procedures to allow the detection and isolation of microorganisms of interest from a large microbial population. The usual screening procedure includes primary screening by cross streak method and secondary screening by disc diffusion tests on the fermentation extracts. The secondary screening method selects the potential strain for further fermentation process and product identification. Further the potential strains will be subjected to strain improvement for product yield by modern methods like mutation—UV treatment, chemical mutagens—protoplast fusion and inducing antibiotic resistance in addition to subculture on agar slants, glycerol stock, soil stock and lyophilization.

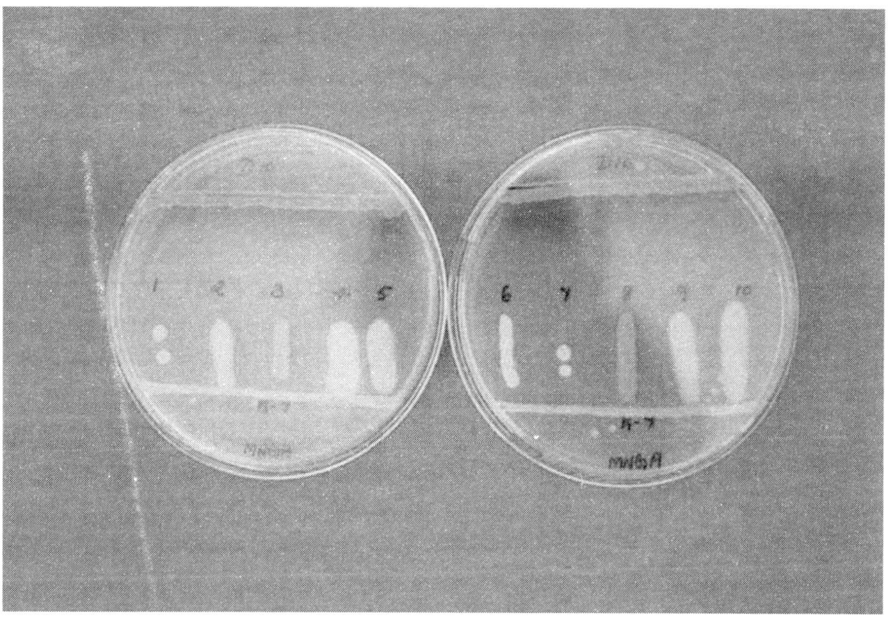

Fig. 4. Primary screening of actinomycetes for antagonistic activity by cross streak method.

RANGE OF BIOPRODUCTS FROM ACTINOMYCETES

Actinomycetes are important due to many reasons, but the principle one is that they produce valuable products. These may be very large proteins, nucleic acids, carbohydrate polymers or even cells, or they can be smaller molecules that include primary and secondary metabolites. The power of actinomycetes in the competitive world of commercial synthesis can be appreciated by the fact that even simple molecules are made by fermentation rather than chemical synthesis. Most of the actinomycete natural products are so complex and contain so many centers of asymmetry that they will probably never be made commercially by normal synthesis.

The main reason for the use of microorganisms including actinomycetes to produce compounds that can otherwise be isolated from plants and animals or synthesized by chemist is the ease of increasing production by environmental and genetic manipulation.

Five major groups of bioproducts are produced by actinomycetes:

- Primary metabolites
- Secondary metabolites
- Bioconversion products
- Actinomycete cells as products
- Recombinant products

Primary Metabolites from Actinomycetes

After inoculation of microbial strain in to suitable nutrient medium when microbial growth is in exponential/trophophase many intermediate metabolic products are produced. Therefore, the metabolites produced during trophophase are known as "primary metabolites". These primary metabolites include enzymes, amino acids, proteins, carbohydrates, lipids and vitamins which are needed for growth and alcohol and organic acids which are needed in energy yielding catabolism. Various primary metabolites are reported to produce by certain actinomycete genera particularly by *Streptomyces*. Some important primary metabolites produced by actinomycetes are given in Table 2.

Table 2: Primary metabolites produced by actinomycetes

Primary metabolites	Producing actinomycetes
Enzymes	
Lipase	*Streptomyces levendulae*
Amylase	*Streptomyces sp, Thermomonospora sp*
Protease	*S. hygroscopicus*
Cellulase	*S. viridosporus*
Chitinase	*S. antibioticus, Amycolatopsis*
Xylanase	*Streptomyces, Nocardia sp*
L-asparaginase	*S. plicatus*
L-glutaminase	*Streptomyces sp.*
Pectate lyase	*S. nitrosporeus*
Invertase	*S. antibioticus*
Nitrilase	*Dactylosporangium, Rhodococcus sp*
Nitrile hydratase	*Rhodococcus rhodochrous*
Vitamins	
Vitamin B12	*S. olivaceus, Micromonospora sp, & Nocardia gardneri*
Amino acids	
L-phenylalanine	*Rhodococcus sp*
Lysine	*Nocardia alkaloglutinosa*
Nucleotide	
Guanosine	*S. griseus*
Carotenoids	
Lycopene	*S. chrestomyceticus*
Siderophores	
Madurastatin	*Actinomadura madurae*
Nocobactin	*Nocobactin*
Desferrioxamine B	*S. griseus*

Secondary Metabolites from Actinomycetes

For more than a century, chemists have been concerned with the discovery, isolation and characterization of chemical substances in the earth biota. New substances continue to be discovered in substantial numbers. By and large these do not fit into the well established systems for intermediary metabolism. Hence, diverse collection of compounds is collectively called "secondary metabolites"

The secondary metabolites are compounds with varied sophisticated chemical structures, produced by strains of certain microbial species, especially filamentous forms and certain plants. The important secondary metabolites include antibiotics, anticancer agents, antivirals, antiparasitic, insecticides, immunosuppressives, toxins, growth promoters, insecticides and pigments. Although antibiotics are the important secondary metabolites, there are other such metabolites with an enormous range of biological activities, acquiring actual/potential industrial importance that are also produced by actinomycetes. These compounds do not play a physiological role during exponential phase of growth.

A characteristic of secondary metabolism is that the metabolites are usually not produced during the phase of rapid growth (trophophase), but are produced during a subsequent production stage (idiophase). So, secondary metabolites are also termed as idiolites. Production of secondary metabolites starts when the growth is limited by the exhaustion of one key nutrient source such as carbon, nitrogen/phosphate. Some important secondary metabolites produced by actinomycetes are given in Table 3.

Table 3: Some important secondary metabolites produced by actinomycetes

Biological activity	Secondary metabolite	Producer
Antibacterial	Streptomycin	*Streptomyces griseus*
	Chloramphenicol	*Streptomyces venezulae*
	Kanamycin	*Streptomyces kanamyceticus*
	Tetracycline	*Streptomyces rimosus*
	Erythromycin	*Saccharopolyspora erythraea*
	Rifampicin	*Amycolatopsis meditteranei*
	Gentamycin	*Micromonospora sp*
	Abyssomycin	*Verrucosispora sp*
Antifungal	Amphotericin B	*Streptomyces nodosus*
	Candicidin	*Streptomyces griseus*
	Nystatin	*Streptomyces noursei*
Antivirals	Fattiviracin	*Streptomyces sp*
Insecticide and antiparasitic	Avermectin	*Streptomyces avermetilis*
	Milbemycin	*Streptomyces hygroscopicus*
Anticancer	Actinomycin D	*Streptomyces antibioticus*
	Bleomycin	*Streptomyces verticillus*
	Mitomycin D	*Streptomyces levendulae*
	Salinosporamide	*Salinospora sp.*
Anticholestrolemic	Pravastatin	*Streptomyces carbophilus*
Growth promoter	Monensin	*Streptomyces cinnamomiensis*
	Tylosin	*Streptomyces fradiae*
Herbicide	Bialophus	*Streptomyces fradiae*
Immunosuppressives	Rapamycin	*Streptomyces hygroscopicus*
	Tacrolimus (FK 506)	*Streptomyces sp*
Pigments	Actinorhodin	*Streptomyces*
	Prodigiosin	*Streptoverticillium sp*

Bioconversion Products from Actinomycetes

Biological conversion or biotransformation has received increasing attention instead of conventional chemical methods. Microbial conversion comprises the conversion of one substance (substrate) to another (product) by a microorganism. In microbial transformation, enzymes simply act as catalysts (biocatalysts) for chemical reactions. In addition to their natural substrates, many of these enzymes can utilize other structurally related compounds as substrates and therefore occasionally catalyze unnatural reactions when foreign substrates are added to the reaction mixture. Thus, microbial transformation constitutes a specific chemical synthesis category. The combined use of enzymes with chemical synthesis can lead to the production of useful substances with higher flexibility (Zhou *et al.,* 2005). Some important type of bioconversion reactions includes oxidations, reductions, hydrolytic reactions and condensations, although a vast array of biotransformations have found industrial application. Microbial conversion of steroids, pesticides, antibiotics and xenobiotics are some of the economically important bioconversions mediated by certain microorganisms. Among the countable number of microbes involved in bioconversions, actinomycetes are important one. Actinomycetes reported to be involved in bioconversion reactions include species of *Streptomyces, Nocardia, Rhodococcus* and *Dactylosporangium.* Some important bioconversion products produced by actinomycetes are given in Table 4.

Table 4: Bioconversion products produced by actinomycetes

Actinomycetes	Substrate	Product
Nocardia sp	N – dedecyl benzene	Phenyl acetic acid
Streptomyces sp	Chitin	Chitosan
Streptomyces aureofaciens	Nitropentachlorobenzal	Pentachloraniline
Streptomyces aureofaciens (ATCC10762)	Anhydrotetracycline	Tetracycline
Streptomyces griseus	Streptomycin	Streptomycin – P
Streptomyces roseochromogenes	9 alpha fluorocartisol	9 alpha fluoro – 16 alpha hydroxycartisol
Streptomyces aureus	Progesterone	15 alpha hydroxyl 4 pregene
Streptomyces sp	2 amino 4 alkyl imidazole	3, zodione 2 nitro 4 alkyl imidazole

The chemical industry has recently begun to focus on enzymes as novel catalysts for organic synthesis. The advantages of bioconversions are that the pH and temperature conditions are less severe than for chemical processes, and very pure products can also be obtained without secondary by-products. Many aliphatic nitriles, for example, acetonitrile and acrylonitrile are produced as solvents, precursors of plastics, synthetic fibers and pharmaceuticals. Nitrile converting enzymes such as nitrilase and nitrile hydratase (NHase) have potential as biocatalysts for the synthesis of high value acids/amides like acrylamide and these enzymes are now commercially produced from the actinomycete genera *Rhodococcus species* (Zhou *et al.,* 2005)

Actinomycete Cell as Products

Microbial cells are used as products like biofertilizers, Single Cell Protein (SCP), probiotics, biopesticides and biocontrol agents. To utilize microbial cell as products, they are cultivated in large scale by using natural substrates and harvested by simple bioseparation methods. In addition to primary and secondary metabolites, certain actinomycetes used as all the above described product preparation. First of all, *Frankia*, a symbiotic nitrogen fixing actinomycetes used as nitrogen fixing biofertilizers. *Streptomyces fradiae*, *Streptomyces lydicus*, *S albocinerescens and S.aurantiacus* (Balagurunathan *et al.*, 2007) are used as biocontrol agents to control some microbial plants diseases. The dried mycelial preparation of *Thermomonospora fusca*, a thermophilic actinomycete, used as single cell protein feed especially for animals and poultry. Recently it was proved that the aquaculture feed supplemented with marine *Streptomyces* biomass as probiotic prevent microbial diseases in fishes and prawns (Devendiran *et al.*, 1997, Balagurunathan *et al.*, 2007). Surprisingly, when applying aquaculture feed supplemented with live *Streptomyces* biomass reduce White Spot Syndrome Virus (WSSV) infection (11-83%) in prawns.

Recombinant Products from Actinomycetes

Recombinant DNA (rDNA) technology is the basis for the development of tools to identify, isolate and manipulate the individual genes that govern the specific character in plants, animals and microorganisms. Recombinant DNA technology can have a positive impact on antibiotic production from actinomycetes in two ways. First, this technology can be used to develop new, structurally unique antibiotics with increased activities against selected targets and decreased side effects. Second, genetic manipulation can be used to enhance yields and hence lower the cost of production.

Some new (hybrid) antibiotics produced from *Streptomyces* by rDNA technology include dihydrogranaticin, dihydrogranatirhodine and mederrhodine A. These antibiotics represent minor structural variants of the preexisting antibiotics—actinorhodin, midermycin, granaticin and hydrogranaticin. These hybrid antibiotics probably arise when an intermediate compound in one biosynthetic pathway acts as a substrate for an enzyme from other pathways.

ACTINOMYCETES AND GENETIC ENGINEERING

In recent years the literature on the genetics and molecular biology of *Streptomyces* has increased considerably. Many excellent reviews have appeared, presenting updated summaries on genetic structure, genetic manipulation, antibiotic production and resistance, plasmids and vectors and regulation of gene expression.

The *Streptomyces* Genome

Actinomycetes, as typical prokaryotes, contain one major genetic lineage group, presumably in the form of a circular chromosome and a variety of mostly extrachromosomal genetic elements, plasmids. The DNA content of the haploid genetic information in *Streptomyces* has been determined by various research groups. Genome size estimates vary from 3.69 to 7.3×10^9 daltons (Table 5). It can now be assumed that *Streptomyces* strains contain 3.7 to 5×10^9 daltons of DNA per haploid genome, i.e., 1.5 to 2 times as much as *Escherichia coli.*

Table 5: Genome size of *Streptomyces* DNA

Organisms	Genome size (md x 10^3)[a]
Streptomyces coelicolor A3(2)	7.09 + 0.73 (mycelium) 7.23 + 0.51 (spores)
Streptomyces rimosus ATCC 10970	6.77 + 0.26 (mycelium) 6.33 + 0.30 (spores)

(1 megadalton (md) corresponds to 1500 base pairs (bp) i.e. 1 kilobase pairs (kb) equals to 0.66 md.)

Extra-Chromosomal Genetic Material in Actinomycetes–Plasmids

Actinomycetes in general, *Streptomyces* in particular, carry a wealth of autonomously replicating extra chromosomal genetic elements. They vary in size from 1.8 to > 200 kb and in copy number from 0.2 up to 600. In extreme cases, extra chromosomal DNA may amount to 10-20 % of the total cellular DNA. In most cases the plasmid size is around 10-40 kb and the copy number is < 30. Actinomycete plasmids have much attention, primarily due to their wide spread occurrence; their involvement in fertility; the assumption that they carry structural or regulatory genes involved in antibiotic biosynthesis and their possible involvement in genome rearrangement.

With the evolution of genetic engineering, plasmids have become indispensable tools for cloning, analysis and transfer of genes. Many Streptomycetes isolated from nature carry detectable extra chromosomal elements and in several cases the simultaneous occurrence of more than one type of plasmid in a host has been reported. In most cases plasmids were detected in the form of covalently closed circular (-CCC) DNA but, occasionally, linear elements were also found. In few cases plasmids were found to be prophages. Plasmids have also been characterized from representative strains of other actinomycete genera including *Streptosporangium, Micromonospora* sp, *Streptoverticillium, Nocardia, Amycolatopsis* and *Frankia* sp.

Genetic Aspects of Antibiotic Biosynthesis in Actinomycetes

Actinomycetes synthesize about two third of currently known antibiotically active substances belonging to a great variety of chemical groups. It is therefore understandable that genetic aspects of antibiotic biosynthesis have attracted great interest in order to gain basic information on the organization and regulation of antibiotic biosynthesis and possibly to exploit such knowledge for influencing antibiotic production, quantitatively/qualitatively. In an early period of research on genetics of antibiotic biosynthesis, considerable emphasis was placed on whether antibiotic biosynthetic structural genes were chromosomally located or whether they encoded extra chromosomally. More detailed analysis has shown that antibiotic biosynthetic structural genes lie mostly on the chromosome (Table 6). Extra chromosomal determination was suggested by (a) the inherent instability of antibiotic formation (b) the observation, that antibiotic – negative mutant frequency increased after curing procedure; (c) some segregational evidence and (d) observation on infectious transfer.

Table 6: Chromosomal and extra chromosomal location of antibiotic biosynthetic structural genes in actinomycetes

Antibiotics	Organism
Chromosomal location	
Actinorhodin	*Streptomyces coelicolor A(3)2*
Chloramphenicol	*Streptomyces venezulae*
Erythromycin	*Saccharopolyspora erythraea*
Oxytetracycline	*Streptomyces rimosus*
Rifamycin	*Amycolatopsis mediterranei*
Extrachromosomal location	
Methylenomycin	*Streptomyces coelicolor A(3)2* *Streptomyces violaceoruber*
Location on self-transmissible element	
Tylosin	*Streptomyces fradiae*

Gene Transfer Techniques in Actinomycetes

In general, conjugation and transformation are the major mechanisms for the genetic exchange in *Streptomyces*. Although transducing phages have also been characterized, transduction is of minor importance for the introduction of foreign DNA in actinomycetes. Recently, generalized transduction was reported for some *Streptomyces* strains.

Conjugation

As in other bacteria, self-transmissible plasmids are responsible for the conjugal gene transfer in *Streptomyces*. Following the discovery of plasmids SCP1 and SCP2 involved in the genetic exchange in *Streptomyces coelicolor*, a great variety of plasmids have been detected in other *Streptomyces* strains. Conjugative plasmids transfer takes place only on agar medium and does not occur in liquid culture. Conjugation in Streptomycetes is very efficient. Up to 100 % of the progenies of a mating receive the conjugative plasmid and upto 0.1–1 % are recombinants. Although the mechanism of conjugative gene transfer in *Streptomyces* remains unclear, it seems to be different from the transfer systems of other Gram-positive and Gram-negative bacteria. The transfer event is macroscopically visible on agar plates by the formation of pocks. These pocks resemble characteristic inhibition zones, surrounding the donor colonies, where the growth of the aerial mycelium is retarded and indicate spreading of a plasmid within the recipient mycelium. Although conjugation is not a widespread technique for the introduction of foreign DNA into Streptomycetes, it is the only route to transfer very large DNA fragments, which have the capacity to encode even complex pathways, between *Streptomyces* strains.

Polyethylene Glycol (PEG) Induced Protoplast Transformation

PEG mediated introduction of DNA into protoplasts is a standard procedure for transformation. Streptomyces mycelium grown in the presence of 0.5 % glycine can be converted into protoplasts by the action of hens egg white lysozyme. The membrane active substance, PEG, mediates the efficient introduction of DNA in these protoplasts. Following regeneration of the protoplasts on isotonic media, transformants are selected by overlaying with the desired antibiotic. The transformation efficiency dramatically depends on the host strain. Using *Streptomyces lividans,* transformation rates up to $10^7 - 10^8$ per microgram of CCC DNA can be obtained. The main barrier for introducing foreign DNA into *Streptomyces* protoplasts seems to be the potent restriction system of Streptomycetes.

Electrophoration

Electrophoration protocols for the efficient introduction of DNA have also been developed for *Streptomyces*. Mycelium can be electrophorated directly and does not necessarily require protoplasting and protoplast regeneration step. Cultivation conditions, pretreatment of mycelia, buffer composition and electrical field strength are the important factors that affect the electrophoration process. Application of electrical field strength of 10 kV/cm at 400 w/25 mF to a 24 hours culture pretreated with lysozyme (100 μg/ml) and resuspended in sucrose-glycerol-PEG buffer yielded transformation rates of upto 10^5-10^6/ μg of DNA for *Streptomyces rimosus* R6. These frequencies are $10^2 - 10^3$ times higher than the efficiency of PEG-mediated protoplast transformation for that strain.

Electroduction

Electroduction is a simple and fast method that allows the direct transfer of bifunctional plasmids from Streptomycetes to *E. coli*. This procedure depends on the formation of transient pores in cell membranes under conditions of Electrophoration. Plasmid DNA is released in to the medium through this pores but can also be taken up by the pores. When *Streptomyces* mycelium carrying a shuttle plasmid is mixed with Electrophoration-competent *E. coli* cells, the plasmids are transferred to *E. coli* after an electrical pulse is applied. This method is of particular interest for the rapid analysis of recombinant *Streptomyces* plasmids in *E.coli* to overcome the difficulties associated with plasmid preparations obtained directly from *Streptomyces*.

Host Vector Systems

Host strains

In contrast with other eubacterial species, such as *E. coli K12* or *Bacillus subtilis*, many species of the genus *Streptomyces* have been analyzed to utilize as host strains because of their industrial importance. However, two strains gained importance by molecular biologists and genetic engineers: *Streptomyces coelicolor* A(3)2 and the closely related *Streptomyces lividans* 66. *Streptomyces coelicolor* is the best studied strain and has therefore developed into a model strain. However, *S. coelicolor* shows a strong restriction barrier against DNA from *E. coli*. Therefore, largely non-restricting and plasmid free *S. lividans* (no plasmid free *S. coelicolor* strain is available) is used in general as an intermediate host for cloning experiments.

Vector Systems for Streptomyces

Replicative vectors

The important replicative vector for *Streptomyces* includes the derivatives for pIJ101, pSG5 and SCP2*.

pIJ101 derivatives

pIJ101 derivatives are the most commonly used vectors for *Streptomyces*. These are based on the 8.83 kb high-copy-number. *S.lividans* broad host range plasmid pIJ101 based vectors have been used to construct genomic libraries, to analyze promoter activity, to create a positive selection vector and as inducible high copy number expression vectors.

pSG5 derivatives

pSG5 is a naturally temperature sensitive, high copy number rolling circle replicating plasmid of *Streptomyces ghanaensis*. It is distinguished from pIJ101 by its moderate copy number (approximately 50 copies per chromosome instead of 300 copies in the case of pIJ101). Many useful vectors have been constructed on the basis of the *ts* minimal replicon, such as pGM9 and pGM160, and are used particularly in gene disruption and gene replacement experiments. Derivatives of pSG5 have also been used as cloning vectors to identify genes of interest in *Streptomyces*.

SCP2 Derivatives*

The 31.4 conjugative plasmid SCP2* is an intensively studied low copy number plasmid from *S. coelicolor A(3)2*. The advantage of SCP2* derived vectors is their high stability and ability to carry large fragments of DNA. They have been used to clone the entire biosynthetic gene cluster encoding the antibiotic Actinorhodin and to produce the first hybrid antibiotics by genetic engineering.

Integrative Plasmid and Phage Vectors

Several integrating vectors have been constructed to allow stable integration into the host chromosome. They are mostly based on naturally integrative genetic elements of *Streptomyces*, some of which have a broad host range. Examples are IS117-derived vectors and pSAM2-derived vectors. Pai C31 is a temperate phage with a relatively wide host range. The wild type paiC31 phage lysogenizes nearly all its hosts by integration into the chromosomal *att* site.

Bi-functional cosmids

pKC505 is the most frequently used bifunctional cosmid vector in *Streptomyces* genetics. This cosmid was constructed by fusing the SCP2* transfer and replication region with a modified *E.coli* cosmid vector carrying the *aac(3)IV* gene for conferring apramycin (Amr) both in *Streptomyces* and *E. coli.*

Marker genes

Many resistance genes originating from antibiotic biosynthetic pathways have been cloned and used as marker genes for the construction of cloning vectors. The neomycin resistance gene *aphII* of *Tn5* and the gentamycin resistance gene *aacC1* of Tn1696 are expressed in Streptomycetes by their own promoters. Furthermore, several marker genes from Gram-negative bacteria have been tested in Streptomycetes. Both the *xylE* gene from *Pseudomonas putida* and *luxAB* gene from *Vibrio fisheri* have been used for construction of promoter probe plasmids. Similarly, expression of the green fluorescent protein from *Aequora Victoria* demonstrated in *S. lividans*.

Actinomycetes as host cell for the production of recombinant proteins

Actinomycetes are highly attractive as cell factories or bioreactors for applications in industrial, agricultural, environmental and pharmaceutical fields. Genome sequencing of several species of actinomycetes paved the way for biochemical and structural analysis of important proteins and the production of such proteins as recombinants on a commercial scale. In this regard there is a need for improved expression vectors that will be applicable to actinomycetes.

Since long, *E. coli* is the most commonly used host bacterium for the large scale production of recombinant proteins. However, the expression and isolation of all the proteins in *E. coli* is difficult on account of the problems of insolubility, cytotoxic, post-translational modifications, or inefficient translation. Eukaryotic cells such as yeast cells, insect cells or immobilized cell lines are particularly useful for the expression of proteins, but undergo post-translational modifications (Brooks, 2004). To overcome these problems, alternate host systems with suitable expression vector system, the actinomycetes, are considered to be as the best host systems.

There are two striking characteristics for looking actinomycetes as host cells in genetic engineering. First, they exhibit unique metabolic diversity and enzymatic capabilities. The compounds they produce as secondary metabolites are valuable for industrial and pharmaceutical purposes, particularly antibiotics (Weber *et al.*, 2003). Second, actinomycetes are expected to have different intracellular milieu as compared to conventional host cells such as *E. coli.* Therefore, it is important to provide a variety of host-vector systems (expression systems) in order to increase the opportunities to screen for the most suitable expression conditions or host cell (Komeda *et al.*, 1996). Already several *Streptomyces* secretion systems have successfully produced eukaryotic proteins (Mellaert *et al.*, 1988, Taguchi *et al.*, 1989; Fornwald *et al.*, 1993), the soluble form of human CD4 was efficiently produced using *Streptomyces lividans* as host cell. Until recently, no host cell, from which, all the proteins can be universally expressed in large quantities, has been found. Recent advancement in gene expression systems, knowledge regarding the intracellular environment, characterization and identification of plasmids has made it possible to develop practicable recombinant expression systems in actinomycetes (Nakashima *et al.*, 2005). So genetic engineers are now focusing on actinomycetes to exploit it as a best suitable host for production of numerous recombinant proteins.

ACTINOMYCETES IN NANOTECHNOLOGY

Over the last few years, the scientific and engineering communities have been witnessing an impressive progress in the field of nanotechnology. Nanotechnology deals with small structures and small-sized materials of dimensions in the range few nanometers to less than 100 nanometers. The term nanotechnology is also referred to as designing, characterization, production and application of structures, devices and systems by controlling shape and size at nanometer scale. The area of research in the field of nanotechnology is as diverse as physics, chemistry, material science, microbiology, biochemistry and molecular biology.

The first so-called scientific study of nano particles took place way back in 1831, when Michael Faraday investigated the ruby red colloids of gold and made public that the color was due to the small size of the metal particles. Gold and silver have found their way into glasses for over 2000 years, usually as nanoparticles.

Nanotechnology offers an extremely broad range of potential applications from electronics, optical communications and biological systems to new smart materials. In the last one decade, the field of nanotechnology gained tremendous importance and is expected to grow enormously in the years to come. The applications of nanotechnology in the field of agriculture, live stock, drug delivery systems of animals and humans, biocompatible implantable medical devices, preparation of innovative health care products of pharmaceutical industry, fish forming and aquaculture is becoming popular and is expected to grow efficiently. Global nanotechnology market is US \$ 225 billion in 2005, US \$ 7000 billion by 2008 and US \$ 1 trillion by 2015.

SYNTHESIS OF NANO PARTICLES—FACTS AND PROBLEMS

Nanoparticles are being viewed as fundamental building blocks of nanotechnology. They are the starting point for preparing many nano structured materials and devices. An important area in research in nanotechnology deals with the synthesis of nano particles of different chemical compositions, sizes and controlled monodispersity.

The nano particles of wide range of materials can be prepared by a number of methods by using precursors from liquids, solids or gas phase. They employ both physical and chemical deposition approach. They rely on their chemical reactivity/physical compaction to integrate nanostructure building blocks. Generally, the manufacturing techniques fall under two categories: bottom up and top down approach. The bottom approach to the build up of a material from the bottom that is atom-by-atom, molecule-by-molecule or cluster-by-cluster. Top down approach involves starting with a block bulk material and designing/milling it down to the desired shape. Both approaches play very important role in modern industry and most likely in nanotechnology as well. There are some merits and demerits in both the approaches. Some important physical and chemical methods for nanoparticle synthesis includes sol process, sol-gel process, synthesis using micelle, chemical precipitation, hydrothermal method, pyrolysis and chemical vapor deposition. Each and every method has its own merits and demerits.

Microbial Protocols for Nanoparticle Biosynthesis

The preparation of nano structured materials possesses several unique challenges. Several manufacturing techniques are in use that usually employ atomistic, molecular and particulate processing in a vacuum or a liquid medium. But most of these techniques are environmental unfriendly and expensive.

Currently there is a growing need to develop environmentally benign nanoparticle synthesis processes that do not use toxic chemicals in the synthesis protocol. As a result, researchers in the field of nanoparticles synthesis and assembly have turned to biological systems for inspiration. This is not surprising that many unicellular and multicellular organisms are known. to produce inorganic materials either intracellularly/extracellularly (Simkis and Wilbur, 1989; Mann, 1996).

Recently the utilization of biological systems has emerged as a novel method for the synthesis of nanoparticles. Some of the microorganisms including bacteria, actinomycetes, fungi and algae are reported to synthasize various nanoparticles in different size and shapes. Even though many biotechnological applications such as remediation of toxic metals employ microbes such as bacteria and yeast, it is only recently that material scientists have been viewing with interest such microbes as possible ecofriendly nanofactories.

Actinomycetes in Nanoparticle Biosynthesis

The process devised by the nature for the synthesis of inorganic materials on nanoscales has contributed to the development of a relatively new and largely unexplored area of research based on the use of microorganisms in the biosynthesis of nano materials. Various nanoparticles were reported to be synthesized by countable number of microorganisms (Table 7). Among the microorganisms, prokaryotic bacteria have received the most attention in the area of biosynthesis of nanoparticles.

Table 7: Synthesis of nanoparticles by different microorganisms

Microorganisms	Type of nanoparticles
Bacteria	
Bacillus subtilis	Gold
Shewanella algae	Gold
Pseudomonas stutzeri	Silver
Lactobacillus	Gold, silver
Clostridium thermoaceticum	Cadmium sulfide
Klebsiella aerogenes	Cadmium sulfide
Escherichia coli	Cadmium sulfide
Desulfobacteriaceae	Zinc sulfide
Thermoanaerobacter ethanolicus	Magnetite
Magnetospirillum	Magnetite
Yeast	
Candida glabrata	Cadmium sulfide
Torulopsis sp	Lead sulfide
Schizosaccharomyces pombe	Cadmium sulfide
Molds	
Verticillium	Gold, silver
Fusarium oxysporum	Gold, silver, Au-AG alloy, cadmium sulfide
Colletotrichum sp	Gold
Phoma sp	Silver
Algae	
Chlorella vulgaris	Gold
Phaeodactylum tricornutum	Cadmium sulfide

Recently actinomycetes, in general, and extremophilic actinomycetes, in particular, were recognized as potential synthesizers of gold and silver nanoparticles. More by the chance than design, an extremophilic actinomycete, *Thermomonospora* sp, when exposed to gold ions reduced the metal ions extracellularly, yielding gold nanoparticles with the much improved poly dispersity. This novel alkalothermophilic actinomycete, *Thermomonospora* sp having an optimum growth at pH 9 and temperature 50°C was isolated from self-heating compost (Ahmad *et al.*, 2003a). A complete reduction of the 10^{-3} M aqueous tetrachloro auric acid ($HAuCl_4$) solution at pH 9 and 50°C results in spherical and reasonably monodisperse nanoparticles (8 mm).

In contrast to extracellular synthesis of gold by *Thermomonospora* sp, alkalotolerant actinomycete, *Rhodococcus* sp (Ahmad *et al.*, 2003b) synthesized gold particles more concentrated form on the cytoplasmic membrane than on the cell wall. An another evidence for biosynthesis of nanoparticles from actinomycetes of extreme ecosystem is certain strains of *Streptomyces* sp. isolated from Thar Desert soil, Rajasthan and Andaman marine sediments were able to synthesize gold and silver nanoparticles intracellularly (Balagurunathan 2007, unpublished data).

If biosynthesis of metal nanoparticles using microbes is to be a viable alternative to chemical methods, currently, then greater control over particles size and polydispersity would need to be established. This was one of the goals in screening different species of fungi, bacteria and now actinomycetes. Thus a greater improvement in the monodispersity has been achieved using actinomycetes. This size and dispersity control may be the consequence of large amounts of proteins/enzyme secreted by actinomycetes in comparison with fungi. It is also to be noted that the synthesis conditions are completely different. In the case of actinomycete *Thermomonospora* sp, the reaction is carried out under alkaline conditions at slightly elevated temperatures. Under these extreme conditions, fungi such as *F.oxysporum* would not survive. The use of extreme biological conditions in the synthesis of nanoparticles could also be a contributory factor in the size and monodispersity control observed using actinomycetes. It has also been noted that the same actinomycete, *Thermomonospora* sp cannot synthesize different metal nanoparticles with similar monodispersity. This may be due to the difference in mode of interactions between different metal nanoparticles and the capping proteins (Mandal *et al.*, 2006).

In general, when compared to research on actinomycetes for their bio-products, especially secondary metabolites, nanoparticle research is still in the stage of infancy. So bioprospecting of actinomycetes, especially from extreme ecosystem, was needed to prove its potential in the field of nanotechnology.

FUTURE PROSPECTS

Actinomycetes have been the largest percentage of cultured members, which probably reflects both their relative easy of culturing and their importance in antibiotic discovery efforts. Taken at face value, these estimates suggest that the number of antibiotics from uncultured sources is 2^{-3} orders of magnitude greater than cultured sources. Although only 1-3% of all Streptomycete antibiotics have been discovered, to find the remaining 97-99% will require a combination of high throughput screening by modern technologies, selection against the most common antibiotics, methods to enrich rare and slow growing actinomycetes, a prodigious microbial collecting and culturing effort and combinatorial biosynthesis in Streptomycetes (Clardy *et al.*, 2006). Baltz (2006) estimated that less than one part in 10^{12} of the earth's soil surface has been screened for actinomycetes. The discovery of actinomycete secondary metabolites is unrivaled and unmatched in medical significance. Structurally and functionally diverse bioactive compounds have been isolated from this group of prokaryotes as antibiotics with antibacterial, antifungal, antiparasitic, antiviral and antitumor activity. Recently, the rate of discovery of new compounds from existing actinomycete genera obtained from terrestrial sources has decreased, while the rate of re-isolation of known compounds is increased. Thus, it is critical that novel actinomycetes from unexplored habitats be pursued as sources of novel antibiotics and other high value metabolites.

Finally it is unquestionable that the actinomycetes represent the biggest chance to obtain further medically, agriculturally and industrially useful compounds, which serve not only as direct drugs or other forms but also as lead compounds for structural modifications and templates for the rationale drug design and other derivatives. It is important note that in respect of the whole natural product researchers the actinomycetes, over other microorganisms, higher plants and animals, give better chance for the satisfactory scale up. Judicious and gainful utilization of this nature treasure, however, depends on the efforts of mankind.

REFERENCES

Ahmad, A., Senapathi, S., Kahn, M.I., Ramani, R., Srinivas, V. and Sastry, M. (2003b). *Nanotechnology*, **14:** 824-828.

Balagurunathan, R. (2007). Biosynthesis of gold and silver nanoparticles by actinomycetes isolated from less explored ecosystems (unpublished data).

Balagurunathan, *et al.,* (2007). Effect of Actinomycetes on the growth of fish (personal communication).

Balagurunathan, R. and Radhakrishnan, M. (2007). *In:* Microbiology – Applications and Current Trends. (Ed. Trivedi, P.C.) Pointer Publishers, Rajasthan, India, pp: 297-329.

Baltz, R.H. (2007). *Microbe,* **2(3):** 125-131.

Berdy, J. (2005). *J. Antibiotics*, **58(1):** 1-26.

Brooks, S.A. (2004). *Mol Biotechnology,* **28:** 241-256.

Fornwald, J.A., Donovan, M.J., Geber, R., Keller, J., Taylor, D.P., Arcuri, E.J. and Browner, M.E. (1993). *BioTechnology,* **11:** 1031-1036.

Hoyles, L., Pascual, C., Falsen, E., Foster, G., Grainger, J.M. and Collins, M.D. (2001). *Int. J. Sys. Evol. Microbiol.,* **51:** 151-156.

Jenson, P.R., Mincer, T.J. and Fenical, W. (2003). www.currentdrugdiscovery.com, 17-19.

Komeda, H., Hori, T., Kobayashi, M. and Shimizu, S. (1996). *Proc. Nat. Acad. Sci.* USA. **93:** 10572-10577.

Li, W.J., Xu, P., Tang, S.K., Xu, L.H., Kroppenstedt, R.M., Stackebrandt, E. and Jiang, C.L. (2003). *Int. J. Syst. Evol. Microbiol.,* **53:** 1545-1549.

Mandal, D., Bolander, M.E., Mukhopadhyay, D., Sarkar, G. and Mukherjee, P. (2006). *Appl. Microbiol. Biotechnol.,* **69:** 485-492.

Manju, K.G. and Devendran, K. (1997). *Ind. J. Exp. Biol.* **35:** 53-55.

Mann, S. (1996). Biomimetic material chemistry. VCH, New York. pp. 1-40.

Mellaert, L.V., Dillen, C., Proost, P., Sablon, E., DeLeys, R., Van Broekhoven, A., Heremans, H., Van Damme, J., Eyssen, H. and Annie, J. (1994). *Gene,* **150:** 153-158.

Mincer, T.J., Jenson, P.R., Kauffman, C.A. and Fenical, W. (2002). *Appl. Environ. Microbio.,* **68(10):** 5005-5011.

Nakashima, N., Mitani, Y. and Tamura, T. (2005). *Microbial Cell Factories,* **4:** 7.

Sabry, S.A., Ghanem, N.B., Abu-Ella, G.A., Schumann, P., Stackebrandt, E. and Kroppenstedt, R.M. (2004). *Int. J. Syst. Evol. Microbiol.,* **54:** 453-456.

Simkiss, K. and Wilbur, K.M. (1989). Biomineralization. Academic Press, New York.

Taguchi, S., Kumagai, I., Nakayama, J., Suzuki, A. and Miura, K. (1989). *Biotechnology,* **7:** 1063-1066.

Takeuchi, M. and Hatano, K. (1998). *Int. J. Sys. Bact.,* **48(3):** 907-912.

Werber, T., Welzel, K., Pelzer, S., Vente, A. and Wohlleben, W. (2003). *J. Biotechnology,* **106:** 221-232.

Zhou, Z., Hashimoto, Y. and Kobayashi, M. (2005). *Actinomycetologica,* **19(1):** 18-26.

Chapter **16**

Strategic Synthesis of Nanoparticles by Mycetes

M.K. RAI AND A.K. GADE

Department of Biotechnology, SGB Amravati University, Amravati–444 602 Maharashtra, (India)

Nanobiotechnology is a new branch of sciences dedicated to the improvement and utilization of devices and structures ranging from 1 to 100 nm in size, in which new chemical, physical and biological properties, not seen in bulk materials can be observed. In the field of nanobiotechnology the synthesis of nanoparticles is the greater challenge. While a number of physical and chemical methods are available and extensively used but are often energy intensive and employ toxic chemicals. On the other hand, the synthesis of nanoparticles by biological system occur close to ambient temperature and pressure, and at neutral pH. In biological systems, fungal systems scores over the bacterial system with respect to their extracellular synthesis, ease of recovery, high tolerance towards metals and a high wall-binding capacity, as well as intracellular metal uptake capabilities and easy to culture on a large scale. The process of reduction of silver ions by fungus being extracellular and fast may lead to development of an easy process for biosynthesis of silver nanoparticles.

INTRODUCTION

The term "nanotechnology" derive from the Greek word 'nano', meaning 'dwarf' and applies to the principles of engineering and manufacturing at a molecular level. The common definition of nanotechnology is manipulation, observation, measurement and synthesis at a scale of 1 to 100 nanometres. According to another definition, nanobiotechnology is a new branch of science dedicated to the improvement and utilization of devices and structures ranging from 1 to 100 nm in size, in which new chemical, physical, and biological properties, not seen in bulk materials, can be observed. There is a tremendous excitement in this field with respect to their fundamental properties, organization of superstructure, and applications.

Nanotechnology is usually represented by two different approaches: 'top-down' and 'bottom-up'. 'top-down' refers to making nanoscale structures by machining, templating and lithographic techniques, whereas 'bottom-up', or molecular nanotechnology applies to building organic and inorganic materials into defined structures, atom-by-atom or molecule-by-molecule, often by-self-assembly or self-organisation. Biologists and chemists are involved in the synthesis of inorganic, organic, hybrid and metal nanomaterials for use in nanodevices, development of novel nanoanalytical techniques, and manipulation of biological system for the synthesis of nanomaterials.

The unusual physico-chemical and optoelectronic properties of nanoparticles arise primarily due to confinement of electrons within particles of dimensions smaller than the bulk electron delocalization length, the process being termed quantum confinement (Brus, 1984; Fendler, 1992; Weller, 1993; Henglein, 1993 and Alivisatos, 1996). Recognizing the importance of nanomaterials in key future technologies, many countries have launched major initiatives into the development of a strong fundamental and applied knowledge base in the area of nanotechnology (www.nano.gov.). It appears that "There's plenty of room at the bottom" (The title of classic talk given by Richard Feynman delivered at the annual meeting of the American Physical Society at the California Institute of Technology in 1959 is possibly the first serious exposition on the problem of manipulating nanoscale objects) in this intriguing area.

Various strategies are employed to synthesize semiconductor and transition metal nanoparticles. Foremost among these are chemical methods because of their inherent advantage in producing large quantities of nanoparticles in relatively short periods of time with a fairly good control on the size distribution. Moreover, in chemical synthesis a variety of shapes of particles can be realized by adjusting the concentration of reacting chemicals and controlling the reaction environment (Matijevic and Cimas, 1987; Weller, 1993; Kundu *et al.,* 1997; Murray *et al.,* 2000). Current chemical methods for the synthesis of nanoparticles are energy intensive, employ toxic chemicals and often yield particles in non-polar organic solution, thereby precluding biomedical applications. On the other hand, in physical methods of nanoparticles synthesis (Takagi, 1986 and Ayyub *et al.,* 2001), such as sputter deposition, laser ablation or cluster beam deposition, thin films are directly grown. However, narrow size distribution of the particles or clusters is often difficult to achieve. Moreover, these methods are time consuming and still under development. Microorganisms typically live under comfortable conditions of temperature, pressure and acidity. These are candidates for development of manufacturing techniques that are more ecofriendly than today's often hot, high-pressure and caustic processes. Recognizing the importance of developing eco-friendly nanoparticle synthesis methods, more and more researchers have turned to synthesis by microorganisms. This comes from the fact that microorganisms while interacting with metal ions have shown to reduce the ions into metallic particles. Both bacteria and fungi have shown ability to reduce metal ions to form metallic nanoparticles.

Undoubtedly, nanomaterials will play a key role in many technologies of the future. One key aspect of nanotechnology concerns the development of reliable experimental protocols for the synthesis of nanomaterials over a range of chemical compositions, sizes and high monodispersity. In the context of the current drive to develop green technologies in materials synthesis, this aspect of nanotechnology assumes considerable importance. An attractive possibility is to use microorganisms in the synthesis of nanoparticles.

Many organisms, both uni- and multicellular, are known to produce inorganic materials either intra- or extracellularly (Simkiss and Wilbur, 1989 and Mann, 1996). Some well-known examples of bio-organisms synthesizing inorganic materials include magnetotactic bacteria, which synthesize magnetite nanoparticles (Frankel and Blakmore, 1991), diatoms, which synthesize siliceous materials (Mann, 1993) and S-layer bacteria, which produce gypsum and calcium carbonate layers (Pum and Sleytr, 1999). The secrets gleaned from nature have led to the development of biomimetic approaches for the growth of advanced nanomaterials.

The gold particles of nanoscale dimensions can be readily precipitated within the bacterial cells by incubation of the cells with Au^{3+} ions. This was demonstrated by Beveridge and co-workers (Beveridge *et al.,* 1980 and Fortin *et al.,* 2000). Silver nanoparticles of well-defined size and distinct morphology within the periplasmic space of the bacteria may be formed when bacterial culture isolated from a silver mine (*Pseudomonas stutzeri* AG259) was placed in a concentrated aqueous solution of $AgNO_3$, which resulted in the reduction of the Ag^+ ions and formation of silver nanoparticles (Klaus *et al.,* 1999; Joerger *et al.,* 2000 and Klaus-Joerger *et al.,* 2001). Nair and Pradeep (2002) have shown that nanocrystals of gold, silver and their alloys can be synthesized by reaction of the corresponding metal ions within the cells of lactic acid bacteria present in buttermilk. More recent and detailed investigations into the use of microbes in the deliberate synthesis of nanoparticles of different chemical compositions include bacteria for gold (Beveridge and Murray, 1980; Southam and Beveridge, 1996 and Fortin and Beveridge, 2000), silver (Joerger *et al.*, 2000; Klaus-Joerger *et al.*, 2001; Nair and Pradeep, 2002) CdS (Smith *et al.*, 1998), ZnS (Labrenz *et al.*, 2000), magnetite (Lovley *et al.*, 1987; Roh *et al.*, 2001; Philipse and Maas, 2002), iron sulfide (Watson *et al.*, 1999, 2000), yeast for PbS (Kowshik *et al.*, 2002a) and CdS (Kowshik *et al.*, 2002b), and algae for gold (Robinson *et al.*, 1997).

Ahmad and co-workers (Mukherjee *et al.*, 2001 and Ahmad *et al.*, 2002) have shown in their laboratory that fungi, when challenged with aqueous metal ions, lead to the formation of nanoparticles both intra- and extracellularly. A number of different genera of fungi have been investigated in this effort and it has been shown that fungi are extremely good candidates in the synthesis of metal and metal sulphide nanoparticles. They were also successful in the extracellular synthesis of gold nanoparticles of about 8 nm diameter by using the alkalothermophilic actinomycete *Thermomonospora* sp. (Ahmad *et al.*, 2003b).

Sastry *et al.* (2003) have reported that fungus *Verticillium* sp. and *Fusarium oxysporum*, when exposed to gold and silver ions, reduced the metal fairly rapidly and formed respective metallic nanoparticles. As can be seen from the above, the use of microorganisms in the deliberate and controlled synthesis of nanoparticles is a relatively new and exciting area of research with considerable potential for development. Fungi in general and yeast in particular play an important role in remediation of toxic metals through reduction of the metal ions. This was considered interesting as nanofactories. The biosynthesis of inorganic nanomaterials can be processed by using these dissimilatory properties of fungi.

Fungi Possess Some Advantages Over Bacteria

- Most filamentous fungi have a high tolerance towards metals, and a high wall-binding capacity, as well as intracellular metal uptake capabilities (Volesky and Holan, 1995; Dias *et al.,* 2002).

- Fungi are easy to culture on large scale, especially by the thin solid fermentation method, thus making it possible to easily obtain enough biomass for processing.
- The fungus could grow on the surface of an inorganic vector during culture, which could distribute silver nanoparticles in a more efficient way as a catalyst.
- Extracellular secretion of enzymes offers the added advantage that downstream processing and handling of the biomass would be much simpler.
- The process can be extended to the synthesis of nanoparticles of different chemical compositions and indeed, different shapes and sizes by suitable identification of enzymes secreted by the fungi.
- The fungal system provides economic viability, possibility of easily covering large surface areas by suitable growth of the mycelia.
- Eukaryotic nature of fungi that has made them favourite hosts for heterologous expression of high-value mammalian protein for manufacturing by fermentation.

BIOSYNTHESIS OF NANOPARTICLES USING FUNGI AND ACTINOMYCETES

A. Intracellular Synthesis of Silver Nanoparticle using *Verticillium sp.* (Sastri *et al.*, 2003)

Isolation of *Verticillium sp.* on PDA (potato dextrose agar)
Incubation at 25°C

↓

Grow in 500 ml Erlenmeyer flasks each containing 100 ml MGYP medium,
at 25°C 28°C, shaking at 200 rpm for 96 hrs

↓

Separation of mycelia by centrifugation (5000 rpm) at 10°C for 20 min

↓

Washing with distilled water

↓

Resuspension of 10 g mycelia in 100 ml 2×10^{-4} aqueous $AgNO_3$ solution
in 500 ml Erlenmeyer flasks at pH 5.5 – 6.0

↓

Keep mixture in shaker at 28°C (200 rpm) for 72 hrs

↓

Visual observation of biotransformation and biomass and
measurement of the UV-vis spectra from the fungal cells.

↓

Appearance of dark brown color in fungal biomass indicates
the formation of silver nanoparticles

↓

Gold and silver nanoparticles exibit striking colors
(Pink to blue for gold and light yellow to brown for silver

B. Extracellular synthesis of gold nanoparticles by using actinomycetes *Thermomonospora sp.* (Sastri *et al.,* 2003)

Isolation of *Thermomonospora* sp. on MGYP medium

↓

Grow in 250 ml Erlenmeyer flasks containing 50 ml MGYP medium at 50°C, shaking at 200 rpm for 96 hrs

↓

Separate mycelia by centrifugation (5000 rpm) at 20°C for 20 min

↓

Wash mycelia thrice with sterile distilled water

↓

Resuspend harvested mycelia (10 g of wet mycelia) in 50 ml of 10^{-3} M aqueous $AuCl_4$ solution in 250 ml Erlenmeyer flasks at pH 9

↓

Keep whole mixture into a shaker at 50°C (200 rpm) and maintain in dark

↓

Periodic sampling of aliquots (2 ml) of the aqueous component to monitor bioreduction of the $AuCl_4$ – ions and measuring the UV-vis spectra of the solution

↓

Gold surface plasmon band occurs at 520 nm and steadily increase in the intensity as function of time of reaction

C. Synthesis of biogenic Silica and Titania by *Fusarium oxysporum* (Bansal *et al.,* 2005)

Inoculation of mycelia of *Fusarium oxysporum* in a 500 ml Erlenmeyer flask containing 100 ml of MGYP medium and incubated for 72 hours under shaking conditions (200 rpm) at 27°C

↓

Harvest mycelia and wash thoroughly under sterile conditions

↓

Resuspend 20 g (wet weight) of the fungal biomass in 100 ml aqueous solutions of 10^{-3} M K2SiF6 (pH 3.1) and K2TiF6 (pH 3.5) separately in 500 ml Erlenmeyer flasks and keep in a shaker (200 rpm) at 27°C

↓

Reaction between the fungal biomass and the $SiF6^{2-}$ and $TiF6^{2-}$ ions was carried out for a period of 24 hours

↓

Collect bio-transformed products by filtration

↓

Characterize filtrate by TEM and FTIR

Table 1: List of organisms producing nanoparticles

Name of organism	Nanoparticle synthesized	Reference
A. Intracellular		
Haliotis rufescens	Calcitic crystals	Sarikaya and Aksay (1991)
Amphora ceffeaeformis	Gold	Robinson *et al.* (1997)
Diatoms	Silaceous material	Mann (1993)
Aquaspirillum magnetotacticum	Magnetite	Frankel and Blakmore (1991)
Bacillus subtilis	Metal nanoparticles	Beveridge *et al.* (1980)
S layer bacteria	Gypsum and calcium carbonate	Pum and Sleytr (1999)
Sulfate-Reducing Bacteria	Sphalerite (ZnS)	Labrenz *et al.* (2000)
	Iron sulphide	Watson *et al.* (2000)
Pseudomonas stutzeri AG259	Silver	Tanja Klaus *et al.* (1999)
Lactobacillus strains	Silver and gold	Nair and Pradeep (2002)
Klebsiella aerogenes	CdS and PbS	Holmes *et al.* (1995)
Torulopsis sp.	Pbs	Kowshik *et al.* (2002)
Schizosaccharomyces pombe	Cds	Kowshik *et al.*(2002)
B. Extracellular		
Saccharomyces cerevisae MKY3	Silver	Meenal Kowshik *et al.* (2003)
Fusarium oxysporum	Silver, gold, Silica & Titania and CdS, Magnetite	Mukherjee *et al.* (2001) Ahmad *et al.* (2003) Vipul Bansal *et al.* (2005) Absar Ahmed *et al.* (2002) Atul Bharde *et al.* (2006)
Thermomonospora sp.	Gold	Ahmad *et al.* (2003)
Phoma sp.32883	Silver	Chan *et al.* (2003)
Aspergillus fumigatus	Silver	Bhainsa *et al.*(2006)
Colletotrichum sp.	Gold	Shiv Shankar *et al.* (2003)
Aspergillus sp.	Silver	Unpublished Data of Gade and Rai, (2006)
Fusarium sp.	Silver	Unpublished Data of Gade and Rai, (2006)
Phoma sp.	Silver	Unpublished data of Gade and Rai, (2006)
C. Intra and extracellular Both		
Verticillium sp.	Silver and gold, Magnetite	Mukherjee *et al.* (2001) Mukherjee *et al.* (2001) Atul Bharde *et al.* (2006)

Commercialization of Nanoparticles

The unique properties of common materials observed only at the nano-scale dimensions have already found their way into commercial applications (Salata, 2004). For example, nanomaterials are present in some toothpastes, sanitary ware coatings, sunscreens and even food products. Manmade nanoparticles range from the well-established multi ton production of carbon black and fumed silica for applications in plastic fillers and car tyres to microgram quantities of fluorescent quantum dots used as markers in biological imaging. As nanosciences are experiencing massive investment worldwide (Paull, *et al.,* 2003 and Mazzola, 2004), there will be a further rise in consumer products relying on nanotechnology (http://www.nanobusiness.org).

Companies are developing pharmaceutical applications, mainly for drug delivery. Several companies exploit quantum size effects in semiconductor nanocrystals for tagging biomolecules, or use bio-conjugated gold nanoparticles for labelling various cellular parts. A number of companies are applying nano-ceramic materials to tissue engineering and orthopaedics. Most major pharmaceuticals exploit colloidal silver for antimicrobial formulations and dressings. The high reactivity of titania nanoparticles, either on their own or then illuminated with UV light, is also used for bactericidal purposes in filters. Enhanced catalytic properties of surfaces of nano-ceramics or those of noble metals like platinum are used to destruct dangerous toxins and other hazardous organic materials.

As nano-sciences are experiencing massive investment worldwide, there will be a further rise in consumer products relying on nanotechnology. While benefits of nanotechnology are widely publicised, the discussion of the potential effects of their widespread use in the consumer and industrial products are just beginning to emerge.

Are Nanoparticles Health Hazardous?

There is a strong likelihood that biological activity of nanoparticles will depend on physico-chemical parameters not routinely considered in toxicity screening studies. Physico-chemical properties that may be important in understanding the toxic effects of test materials include particle size and size distribution, agglomeration state, shape, crystal structure, chemical composition, surface area, surface chemistry, surface charge, and porosity.

Particles in the nano-size range can certainly enter the human body via the lungs and the intestines; penetration via the skin is less evident. It is possible that some particles can penetrate deep into the dermis layer of the skin. The chances of penetration depend on the size and surface properties of the particles and also on the point of contact in the lung, intestines or skin. After the penetration, the distribution of the particles in the body is a strong function of the surface characteristics of the particles. A critical size might exist beyond which the movement of the nanoparticles in parts of the body is restricted. The pharmaco-kinetic behavior of different types of nanoparticles requires detailed investigation, and a database of health risks associated with different nanoparticles (e.g., target organs, tissue or cells) should be created. The presence of the contaminates, such as metal catalysts present in nanotubes, and their effect along with effect of nanomaterials on the health should be considered.

The increased risk of cardiopulmonary diseases requires specific measures to be taken for every newly produced nanoparticle. There is no universal "nanoparticle" to fit all the cases, so each nanomaterial should be treated individually when health risks are expected. The tests currently used to measure the safety of materials should be applicable to identify hazardous nanoparticles. Proven otherwise, it would be a challenge for industry, legislators and risk assessors to construct a set of high throughput and low cost tests for nanoparticles without reducing the efficiency and reliability of the risk assessment. Nanoparticles designed for drug delivery or as food components need special attention.

APPLICATION IN AIDS AND CANCER

To avoid the side effects of photodynamic cancer therapy, use of hydrophobic version of the dye molecule was enclosed inside a porous nanoparticle (Roy *et al.,* 2003). The dye stayed trapped inside the Ormosil nanoparticle and did not spread to the other parts of the body. At the same time, its oxygen generating ability has not been affected and the pore size of about 1 nm freely allowed for the oxygen to diffuse out.

Elechiguerra and co-workers (2005) demonstrated that silver nanoparticles undergo a size dependent interaction with HIV-1, with nanoparticles exclusively in the range of 1–10 nm attached to the virus. The regular spatial arrangement of the attached nanoparticles, the center-to-center distance between nanoparticles, and the fact that the exposed sulfur-bearing residues of the glycoprotein knobs would be attractive sites for nanoparticle interaction suggest that silver nanoparticles interact with the HIV-1 virus via preferential binding to the gp120 glycoprotein knobs. Due to this interaction, silver nanoparticles inhibit the virus from binding to host cells, as demonstrated *in vitro*.

FUTURE TRENDS

Nanoscale structure materials and devices that can be fabricated today hold great promise for advanced diagnostics and biosensors, targeted drug delivery, smart drugs, and immunoisolation therapies. There are various uses of nanoparticles (Table 2). In the longer term, molecular machine systems and medical nanorobots will allow instant pathogen diagnosis and extermination, chromosome replacement and individual cell surgery *in vivo*, and the efficient augmentation and improvement of natural physiological function.

In the next couple of years, it is widely anticipated that nanotechnology will continue to evolve and expand in many areas of life and science, and the achievements of nanotechnology will be applied in medical sciences, including diagnostics, drug delivery systems and patient treatment. Kubik *et al.* (2005) reported that "Nanotechnology provides the potential for significant advances over the next 50 years" with potential applications of:

- biological nanosensors for diagnostics in the next 1-5 years,
- generation of artificial muscles, development of lab-on-a chip technology for more efficient drug discovery and targeted drug and gene delivery within the next 6-10 year, and
- later on (after 10-50 years) introduction of nanomachines for *in vivo* treatment and nanopumps/valves for tissue engineering and generation of artificial organs, in health care and medicine (Kubik *et al.,* 2005)

The major trend in further development of nanomaterials is to make them multifunctional and controllable by external signals or by local environment, thus essentially turning them into nanodevices.

Table 2: Applications of nanoparticles.

Applications	Reference
Fluorescent biological labels	Bruchez *et al.* (1998), Chan and Nie (1998) , Wang *et al.* (2002)
Drug and gene delivery	Mah *et al.* (2000), Panatarotto *et al.* (2003)
Biodetection of pathogens	Edelstein *et al.* (2000)
Detection of proteins	Nam *et al.* (2003)
Probing of DNA structure	Mahtab *et al.* (1995)
Tissue engineering	Ma *et al.* (2003), de la Isla *et al.* (2003)
Tumour destruction via heating (hyperthermia)	Yoshida and Kobayashi (1999)
Separation and purification of biological molecules and cells	Molday and MacKenzie (1982)
Magnetic Resonance Imaging (MRI) contrast enhancement	Weissleder *et al.* (1999)
Phagokinetic studies	Parak *et al.* (2002)
Interaction with HIV–1	Elechiguerra *et al.* (2005)
Nanobiosensor for the immunoassay of antibiotin	Riboh *et al.* (2003)

CONCLUSION

The use of fungi for catalyzing specific reactions leading to inorganic nanoparticles is a new and rational biosynthesis strategy. However, the exact reaction mechanism of synthesizing naoparticles by biological system is yet to be elucidated. Ahmed *et al.* (2003) have reported the involvement of certain NADH dependent reductase in the reduction of silver ions in case of *Fusarium oxysporum*. Extracellular synthesis of nanoparticles by filamentous fungi offers an advantage of obtaining large quantities in a relatively pure state and at a rapid rate. Furthermore, the extracellular synthesis of nanoparticles by filametous fungi would make the process simpler and easier for downstream processing. Compared to bacterial fermentations, in which the process technology involves the use of sophisticated equipment for getting clear filtrates from the colloidal broths, fungal broths can be easily filtered by filter press of similar simple equipment, thus saving considerable investment costs for the equipment. Use of specific enzymes secreted by organisms such as fungi in the synthesis of nanoparticles is exciting because of the possibility of genetically engineered microbes to overexpress specific reducing molecules and capping agents and thereby, control the size and shape of the biogenic nanoparticles. In addition, fungi and actinomycetes are known to secrete much higher amounts of proteins as compared to bacteria, thereby significantly increasing the productivity of this biosynthetic approach.

In future, it will be important to understand the biochemical and molecular mechanism of the synthesis of the nanoparticles by the fungal system in order to achieve better control over size and polydispersity.

REFERENCES

Ahmad, A., Mukherjee, P., Senapati, S., Mandal, D., Khan, M.I., Kumar, R. and Sastry, M. (2003). *Colloid Surf. B.,* **28:** 313-318.

Ahmad, A., Senapati, S., Khan, M.I., Kumar, R. and Sastry, M. (2003). *Langmuir,* **19:** 3550-3553.

Alivisatos, A.P. (1996). *Science,* **271:** 933-937.

Ayyub P., Chandra R., Taneja P., Sharma A.K. and Pinto R. (2001). *Appl. Phys.,* A **73:** 67

Beveridge, T.J. and Murray, R.G.E. (1980). *J. Bacteriol.* **141:** 876-887.
Bruchez, M., Moronne, M., Gin, P., Weiss, S., and Alivisatos, A.P. (1998). *Science,* **281:** 2013-2016.
Brus, L. E. (1984). *J. Chem. Phys.* **80**: 4403–4409.
Chan, W.C.W. and Nie, S.M. (1998). *Science*, **281:** 2016-2018.
Chen, J.C., Lin, Z.H. and Ma, X.X. (2003). *Letters in Applied Microbiology* **37:** 105-111.
de la Isla A., Brostow, W., Bujard, B., Estevez, M., Rodriguez, J.R., Vargas, S. and Castano, V.M. (2003). *Mat. Resr. Innovat.*, **7:** 110-114.
Dias, M.A., Lacerda, I.C.A., Pimentel, P.F., de Castro, H.F. and Rosa, C.A. (2002). *Letters in Applied Microbiology,* **34:** 46-50.
Edelstein, R.L., Tamanaha, C.R., Sheehan, P.E., Miller, M.M., Baselt, D.R., Whitman, L.J. and Colton, R.J. (2000). *Biosensors Bioelectron* **14:** 805-813.
Fendler, J.H. (1992). *Membrane Mimetic Chemistry Approach to Advanced Materials*, Springer-Verlag, Berlin.
Fortin, D. and Beveridge, T.J. (2000). *In: Biomineralization. From Biology to Biotechnology and Medical Applications* (Ed. Baeuerien, E.), Wiley-VCH, Weinheim, p. 7.
Frankel, R.B. and Blakemore, R.P. (1991). *Iron Biominerals* (Plenum, New York).
Henglein, A. (1993). *J. Phys. Chem.* **97:** 5457-5464.
Holmes, J.D., Smith, P.R., Evans-Gowing, R., Richardson, D.J., Russell, D.A. and Sodeau, J.R. (1995). *Arch. Microbiol.,* **163:** 143-147.
Riboh, J.C., Haes, A.J., Mcfarland, A.D., Yonzon, C.R. and Van Duyne, R.P. (2003). *J. Phys. Chem. B,* **107:** 1772.
Joerger, R., Klaus, T. and Granqvist, C.G. (2000). *Adv. Mater.* **12:** 407-409.
Elechiguerra, J.L., Justin L.B., Morones, J.R., Camacho-Bragado, A., Gao, X., Lara, H.H. and Yacaman, M.J. (2005). *J. Nanobiotechnology,* **3:** 6.
Klaus, T., Joerger, R., Olsson, E. and Granqvist, C.G. (1999). *Proc. Natl. Acad. Sci. USA.* **96:** 13611-13614.
Klaus-Joerger, T., Joerger, R., Olsson, E. and Granqvist, C.G. (2001). *Trends Biotechnol.* **19:** 15-20.
Kowshik, M., Deshmukh, N., Vogel, W., Urban, J., Kulkarni, S.K. and Paknikar, K.M. (2002b). *Biotechnol. Bioeng.* **78:** 583-588.
Kowshik, M., Vogel, W., Urban, J., Kulkarni, S.K. and Paknikar, K.M. (2002a). *Adv. Mater.* **14:** 815-818.
Kubik, T., Bogunia-Kubik, K. and Sugisaka, M. (2005). *Current Pharmaceutical Biotechnology.* **6:** 17-33.
Kundu, M., Khosravi, A.A., Singh, P. and Kulkarni, S.K. (1997). *J. Mater. Sci.* **32:** 245.
Labrenz, M., Druschel, G.K., Thomsen-Ebert, T., Gilbert, B., Welch, S.A., Kemnet, K.M., Logan, G.A., Summons, R.E., Stasio, G.D., Bond, P.L., Lai, B., Kelly, S.D. and Banfield, J.F. (2000). *Science,* **290:** 1744-1745.
Ma, J., Wong, H., Kong, L.B. and Peng, K.W. (2003). *Nanotechnology,* **14:** 619-623.
Mah, C., Zolotukhin, I., Fraites, T.J., Dobson, J., Batich, C. and Byrne, B.J. (2000). *Mol. Therapy,* **1:** S239.
Mahtab, R., Rogers, J.P. and Murphy, C.J. (1995). *J. Am. Chem. Soc.* **117:** 9099-9100.
Mann, S. (1996). *Biomimetic Materials Chemistry,* VCH Publishers.
Mann, S. (1993). *Nature,* **365:** 499-505.
Matijevic, E. and Cimas, S. (1987). *Colloid Polym. Sci.* **265:** 155.
Mazzola, L. (2003). *Nature Biotechnology,* **21:** 1137-1143.
Molday, R.S. and MacKenzie, D. (1982). *J. Immunol. Methods,* **52:** 353-367.
Mukherjee, P., Senapati, S., Mandal, D., Ahmad, A., Khan, M.I., Kumar, R. and try, M. (2001a). *Angew. Chem., Int. Ed. Engl.* **40:** 3585-3588.
Mukherjee, P., Senapati, S., Mandal, D., Ahmad, A., Khan, M.I., Kumar, R. and Sastry, M. (2001b). *Nano Lett.* **1:** 515-519.
Mukherjee, P., Senapati, S., Mandal, D., Ahmad, A., Khan, M.I., Kumar, R. and Sastry, M. (2002). *Chem. Biochem.* **3:** 461-463.
Murray, C.B., Kagan, C.R. and Bawendi, M.G. (2000). *Ann. Rev. Mater.Sci.* **30:** 545.
Nair, B. and Pradeep, T. (2002). *Cryst. Growth Des.* **2:** 293-298.
Nam, J.M., Thaxton, C.C. and Mirkin, C.A. (2003). *Science*, **301:** 1884-1886.
NanoBusiness Alliance. (2003). [http://www.nanobusiness.org].
Panatarotto, D., Prtidos, C.D., Hoebeke, J., Brown, F., Kramer, E., Briand, J.P., Muller, S., Prato, M. and Bianco, A. (2003). *Chemistry & Biology,* **10:** 961-966.
Parak, W.J., Boudreau, R., Gros, M.L., Gerion, D., Zanchet, D., Micheel, C.M., Williams, S.C., Alivisatos, A.P. and Larabell, C.A. (2002). *Adv. Mater,* **14:** 882-885.
Paull, R., Wolfe, J., Hebert, P. and Sinkula, M. (2003). *Nature Biotechnology,* **21:** 1134-1147.
Pum, D. and Sleytr, U.B. (1999). *Trends Biotechnol.* **17:** 8-12.
Robinson, M., Brown, L.N. and Beverley, D. (1997). *Biofouling.*, **11:** 59.

Roy, I., Ohulchanskyy, T.Y., Pudavar, H.E., Bergey, E.J., Oseroff, A.R., Morgan, J., Dougherty, T.J., Prasad, P.N. (2003). *J. Am. Chem. Soc.* **125:** 7860-7865.

Salata, O.V. (2004). *J. Nanobiotechnology,* **2:** 3.

Sarikaya, M. and Aksay, I.A. (1991). *In: Structure, Cellular Synthesis, and Assembly of Biopolymers*, (Ed. Case, S.T.) (Springer, Berlin), pp. 1-25.

Simkiss, K. and Wilbur, K.M. (1989). *Biomineralization*, Academic Press, New York.

Takagi, T. (1986). *Physics* D. **3:** 272.

Volesky, B. and Holan, Z.R. (1995). Biosorption of heavy metals. *Biotechnology Progress,* **11:** 235-250.

Wang, S., Mamedova, N., Kotov, N.A., Chen, W. and Studer, (2002). *J. Nano Letters* **2:** 817-822.

Weissleder, R., Elizondo, G., Wittenburg, J., Rabito, C.A., Bengele, H.H. and Josephson, L. (1990). *Radiology,* **175:** 489-493.

Weller, H. (1993). *Angew. Chem.* **32:** 42.

Weller, H. (1993). *Angew. Chem., Int. Ed. Engl.* **32:** 41-53.

Yoshida, J. and Kobayashi, T. (1999). *J. Magn. Magn. Mater.* **194:** 176-184.

Chapter 17

Discovery of Bioactive Molecules from Plants and Microorganisms

K.G. RAMAWAT, SHAILY GOYAL AND MEETA MATHUR

Laboratory of Biomolecular Technology, Department of Botany, M.L. Sukhadia University, Udaipur–313 001 (India)
E-mail: kg_ramawat@yahoo.com

Plants and microorganisms are sources of natural products constituting important drugs. The discovery of penicillin from *Penicillium flavum* has revolutionized the medical world and leads to search of new drugs from microorganisms, which resulted in several new antibiotics, antitumour agents, enzyme inhibitors including powerful hypocholesterolemic agents, immunosuppression agents and antimigraine agents, etc. Traditional system of medicine relies mainly on herb-based formulations and their bioactive molecules, e.g., guggulsterones from *Commiphora wightii*, morphinan alkaloids from the opium poppy and glycirrhetic acid from Liquorice. Besides the traditional use of such molecules, new molecules such as stilbenes, phytoestrogens and new generation of antibiotics are being discovered and added to the ever increasing list of natural products to combat against dreaded diseases. Various approaches to discover new molecules are presented in the light of High Throughput Screening (HTS) system for rapidly screening a large number of plant samples.

INTRODUCTION

Natural products obtained from plants and microorganisms are used directly or after modification to combat diseases. The discovery of penicillin from *Penicillium flavum* made a great impact on treatment strategy and microorganisms as a source of life saving drugs. During the last century, microbial and plant secondary metabolites helped to double our lifespan, reducing pain and suffering and revolutionized medicine. The persistent problem of the increased development of resistance by microorganisms against existing drugs has been overcome by (1) newly discovered antibiotics (e.g., candins, epithilones), (2) new semi-synthetic versions of old antibiotics (e.g., ketolides), (3) older underutilized antibiotics (e.g., teicoplanin); and (4) new derivatives of previously undeveloped narrow-spectrum antibiotics (e.g., streptogramins). Besides this, many antibiotics are used commercially or as medicine (for purposes other than their role as antimicrobial) as antitumour agents, enzyme inhibitors including powerful hypocholesterolemic agents, immunosuppressive agents, antimigraine agents, etc. (Demain and Zhang, 2005).

About 12.5% of the 422,000 plant species of higher plants are known as medicinal plants. As compared to this, only a very minor proportion (0.1-5%) of microorganisms have been explored for the production of secondary metabolites (Demain and Zhang, 2005). The proportion of medicinal plants to the total documented species in different countries varies from 4.4 to 20% (Schippmann *et al.*, 2002). About 25% of drugs in modern pharmacopoeia are derived from plants and many others are synthetic analogues built on prototype compounds isolated from plants. Up to 60% of prescribed drugs in Western world contain plant products or their derivatives (Lancet, 1994). In India, 1100 species are recognized as source of raw materials for Ayurvedic and Unani formulations (Gupta, 1986).

TRADITIONAL SYSTEM OF MEDICINE

Medicinal and aromatic plants play an important role in the healthcare (and also in rituals) of people around the world, especially in developing countries. In countries like India, these plants are involved in worshipping of God in all activities of life, from childbirth to cremation. The Indian system of medicine is as old as Indian civilization and written records are available dating 3000 years. Therefore, until the allopathic system came in force, man was dependent on plants for treating human and livestock diseases. About 80% of the population of most developing countries still use traditional medicine derived from plants for their routine illness. China, Cuba, India, Sri Lanka, Thailand and a few other countries have endorsed the official use of traditional systems of medicine in their healthcare programmes. For example, Government of India has officially recognized "Ayurveda", "Sidha" and "Unani" systems of medicines and these systems depend on plant materials for obtaining drugs to cure human ailments (Ramawat *et al.*, 2004). Therefore, there is a long history of use of plants as source of drug and this information was useful in obtaining and establishing modern medicines (Fig. 1) from these plants, e.g., isolation of guggulsterones from *Commiphora wightii*, (syn: *C. mukul*), morphinan alkaloids from dried latex of the opium poppy (*Papaver somniferum*) and glycirrhetic acid from Liquorice (*Glycyrrhiza glabra*). The medicinal usage of these plants was referred to by the first known written record on clay tablets from Mesopotamia in 2600 B.C. and these plants are still in use today for the treatment of various diseases as ingredients of herbal preparations used in traditional system of medicine (Chin *et al.*, 2006).

Morphine

Morphine was the first alkaloid isolated in pure form. Morphine, codeine, noscapine and papaverine isolated from *P. somniferum* were developed as single chemical drugs and are still used clinically. Traditional medicine and ethnobotanical exploration were helpful in identification of plant materials used for various diseases and isolation of pure compounds. The other approach is bioactivity-guided isolation methods, which has been successful in isolating anticancer agents

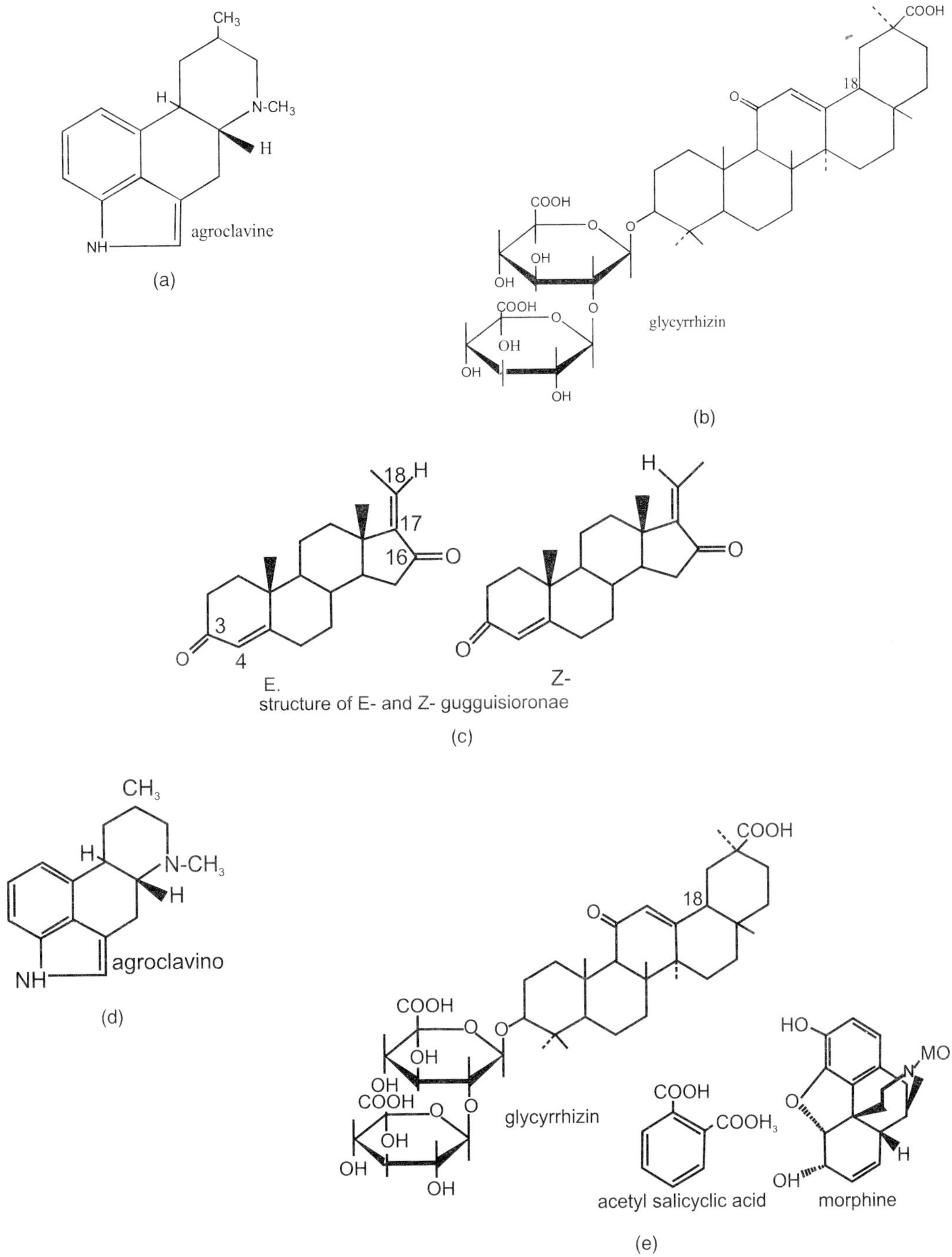

Fig. 1. Various Drugs obtained from plants

like paclitaxel from *Taxus brevifolia* (Wani *et al.*, 1971) and camptothecin from *Camptotheca accuminata* (Wall *et al.*, 1966).

Phytoestrogens

Kudzu (*Pueraria lobata*) is believed to have originated in China and has been part of culture for over two millennia. The Chinese value kudzu primarily for healing power. The tea of the root of kudzu was first described in the Chinese material medica (The Pharmacopoeia of Shen Nong, Anonymous ~200 B.C.) as sweet and acrid in taste, cool in nature and was used as an antipyretic, antidiarrhetic, diphoretic and anti-emetic agent. About a thousand years later, the drug was recommended for anti-alcohol therapy (Keung and Valee, 1998). It is during the past three decades that the serious attempts have been made to evaluate the drug on scientific parameters to know the active principles and their mechanism of action. Now it is known that the plant contains several isoflavones. Interest in polyphenols, including isoflavonoids has increased considerably because of beneficial effects proposed by epidemiologists, nutritionists and food manufacturers. Isoflavones are broadly defined as phytoestrogens, and can interfere with mechanisms controlled by the hormone through competition for its receptors. Phytoestrogens are defined as estrogenic compounds in plants. They have common chemical characteristics as phenolics substances, with some structural similarities to the natural estrogen 17 β-estradiol. The presence of a phenolic ring allows binding to the estrogen receptor (ER), which is followed by induction of specific estrogen-responsive gene products that lead to biological effects such as stimulation of cell growth. They also perturb the action of DNA topoisomerase II and ribosomal S6 kinase, which could explain their observed effects on cell cycle, differentiation, proliferation and apoptosis (Kurzer and Xu, 1997). The binding affinity of the plant estrogens is, however, much weaker than that of human estrogen produced. In addition, phytoestrogen can have non-estrogen receptor mediated effects. Plant estrogens are found in *Pueraria tuberosa, P. lobata, P. phaseolides, Genista tinctoria* and many other plant species. Of all plant estrogens soy isoflavones i.e., daidzein and genistein have been studied the most (Dixon and Ferreira, 2002).

Isoflavones are protective antioxidants. They reduce the formation of radicals and reactive oxygen by decomposition of hydrogen peroxide without generating radicals, by quenching active singlet oxygen, and by trapping and quenching radicals before they reach a cellular target. The antioxidant activity of phenolics is based on their ability to donate hydrogen atom to free radicals. Due to these properties, isoflavones are believed to have potential in the prevention of cancer, atherosclerosis and diabetes (Yousef *et al.*, 2004, Torres *et al.*, 2006).

There are many epidemiological observations revealing a protective role of isoflavone in modulating numerous markers of cardiovascular risks. Isoflavone intake lowers the LDL-cholesterol level among those with elevated LDL-C concentration. The structural similarity of isoflavones to the endogenous estrogens and the relation of reproductive hormones to the breast cancer have provided a rationale for evaluating the hormonal and anti-hormonal effect of phytoestrogen as one of the possibilities in reducing the cancer risk. Phytoestrogens also play an important role in protection against prostrate cancer in males other than breast cancer in females. Genistein strongly inhibits the growth of leukaemia cells when targeted to them by linkage to a monoclonal antibody, and a prenyl isoflavone derivative has been developed as an oral treatment for acute leukaemia (Jeon *et al.*, 2005, Lacey *et al.*, 2002).

Risk of osteoporosis is mostly concerned with the decline in estrogen production in post-menopausal women. Although hormone replacement therapy (HRT) is an efficient method that can compensate the low level of hormone, but there remains a substantial interest in finding dietary alternatives to HRT. In this regard, phytoestrogens (soya bean products) have received greater attention as an alternative improving bone health. Numerous studies relating the isoflavone intake and bone mineral density of pre- and post-menopausal women have shown a positive association. According to a study, genistein and daidzein significantly prevented bone loss in ovariectomized mice without exhibiting estrogenic action in the uterus (Kritz-Silverstein *et al.*, 2002).

Stilbenes

Vitis vinifera is one of the oldest plants known to man and grows in virtually every country in the world. Grapes were first cultivated in Caucasia in about 6000 B.C. Cultivation progressed into Egypt and Phoenicia in about 3000 B.C. and by 2000 B.C., viticulture reached Greece. By 1000 B.C., grape cultivation had reached Italy, Sicily and North Africa. Residents of Spain, Portugal and France began viticulture in 500 B.C. and finally the practice spread into eastern and Northern Europe and even the British Isles. Grapes may be eaten fresh, but many Old World varieties are dried first into raisins or currants. Grapes are also used to make grape juice and wine. Grapes grow well where the winter is not too cold and where there is a long, warm summer that enables the plants to ripen their fruit slowly. Grape has been used for good health since ancient times from China to Greece. Grape was believed to have curative and preventive powers.

Now it is well known that grapes are a rich source of polyphenolics. Polyphenolics have important effects in plant biochemistry and physiology, acting as antioxidants, enzyme inhibitors, pigments and light screens, and are involved in plant defence against pathogens. These natural products have been shown to posses multiple biological activities in mammalian systems such as effects on various enzymes systems *in vitro*, modulation of inflammatory cell functions, hepatoprotective effects, inhibition of lipid peroxidation and oxyradical production, which are implicated in several pathologic conditions (including aging, cancer, atherosclerosis, etc.), and anticarcinogenic and proliferative specific effects. There is a clear evidence that natural polyphenols have the potential to act in three general areas, i.e., transition metal ion complexation, as antioxidants in cellular pro-oxidant states, and by association with proteins and peptides (Haslam 1996; Aumont *et al.*, 2004; Piver *et al.*, 2004).

It is red wine, prepared from blackberries (*Vitis vinifera*) that contains high amounts of stilbenes including cis- and trans-resveratrol and their glucosides, which is studied most extensively after the epidemiological evidence showing that in France the consumption of red wine may counteract the effects of saturated fats and reduce the risk of coronary heart disease (Renaud and de Lorgeril, 1992), popularly known as "**French Paradox**".

Trans-resveratrol (trans-3,5,4'-trihydroxystilbene) is synthesized by grapes in response to fungal infection and is found in red wine at a concentration ranging from 1 to 10 mM (Soleas *et al.*, 1997). *Trans*-resveratrol has been reported to have both anticarcinogenic and cardioprotective activities, which could be attributed to its antioxidant and anticoagulative properties. This compound is not only a modulator of hepatic apolipoprotein and lipid biosynthesis (Cojocel and

Thompson, 2004) but also an inhibitor of platelet aggregation and eicosanoid production in human platelets and neutrophils (Bertelli *et al.*, 1996). Moreover, *trans*-resveratrol causes vasorelaxation *in vitro* by an endothelium-dependent process involving the NO second messenger system (Chen and Pace-Asciak,1996).

A number of recently published reports have explored the effects of stilbenes, including resveratrol, as cancer-chemopreventive agent in assay representing three major stages of carcinogenesis. Resveratrol was found to act as an antioxidant and antitumourigen and to induce phase II drug metabolizing enzymes (anti-initiation activity). It mediated anti-inflammatory effects and inhibited cyclooxygenase and hydroperoxidasc functions (antipromotion activity). It also induced human promyelocytic leukemia cell differentiation (anti-progressive activity) and nephroprotective effect. Considerable evidence suggests that COX-2 is important for tumor genesis. Though the mechanism of action of resveratrol is far from clear, it is evident that it has some role in the inhibition of cancerous growth. According to recent findings, *trans*-resveratrol inhibits some CYPs, which have been described as being crucial in tumour progression and activity (Ciolino *et al.*, 1998; Piver *et al.*, 2004). *Trans*-resveratrol induces MAPK- and p53-dependent apoptopsis in several human cancer cell lines by binding at extracelluar domain of integrin (Lin *et al.,* 2006). The other species extensively investigated for stilbene production in cell suspension cultures is *Vitis vinifera* (Vitrac *et al.*, 2004) and *V. trifolia* (Ramawat *et al.*, 2006). Resveratrol is under phase-I clinical trials for colon cancer (www.clinicaltrials.gov/ct/show/nct00256334) and against AIDS (www.ihv.org/clinical_trials/thera_vacc2.html).

APPROACHES FOR DRUG DISCOVERY

Isolation of compounds in pure form and evaluating them for their pharmacological properties leading to drug discovery is a long, tedious, time-consuming and expensive path. Drug discovery from natural sources has become very expensive (currently estimated at $500 million) and time consuming (5-6 years in 1980s to 15-22 years in 21st century). Two-third of the cost is involved on leads that fail during the clinical trials (Wilson, 2002). Fifty per cent of all potential drugs fail because of adsorption, distribution, metabolism, excretion or toxicity problems (Demain and Zhang, 2005). Due to long period of discovery, pre-clinical and clinical trials, it also requires large number of volunteers (1000-5000 for long-term effects). So, many companies have abandoned search for natural products and are giving attention to combinatorial chemistry and modification/analogue synthesis of existing drugs. Therefore, the new approaches like metabolomics and high-throughput screening are used to screen and evaluate several metabolites in short time. In post-genomic era, pharmaceutical researchers are evaluating vast number of protein sequences to formulate novel strategies for identifying valid targets and discovering lead against them (Betz *et al.*, 2002). Modern drug discovery often involves screening small molecules for their ability to bind to a pre-selected protein target. Drug discovery can also involve screening small molecules for their ability to modulate biological pathway in cells or organism, without regard for any particular protein target. Thus, with the establishment of various techniques of genomic sciences, like rapid DNA sequencing in combination with combinatorial chemistry, cell based assays and automated high throughput screening (HTS) have led to a new concept of drug discovery (Drew, 2000). In this concept, an interaction between biologist and chemists as well as scientific reasoning have been replaced by a very high number of samples processed. With the rapid industrialization, HTS system has been developed to screen not only a few hundreds but also hundreds of thousands of chemical compounds in short time. The term HTS was created in early 1990s to screen such large number extracts/compounds in short time. This requires identification of disease specific target by basic research or genomic approach, which is used to design/develop a bioassay used in HTS system (Kessmann and Schnurr, 2002). Under HTS setup, large numbers of hypothetical targets are incorporated into cell-based assays and exposed to large numbers of compounds repressing numerous variations on a few chemical themes. It was assumed that this experimental design would be suitable to identify many substances, which can modify the target in question. Such molecules are then isolated in greater quantity and evaluated on more complex models (cells, animals) to a certain efficacy. Such a model is depicted in Fig. 2. It is used for evaluation of antimitotic compounds. About 50 million screening tests have been conducted so far using different molecules, concentrations and bioassays (Drew, 2000). These new approaches towards drug discovery are presented in Fig. 3. These technologies generated vast information on natural products obtained from plants and microorganisms and made a historical mark in the modern medicine.

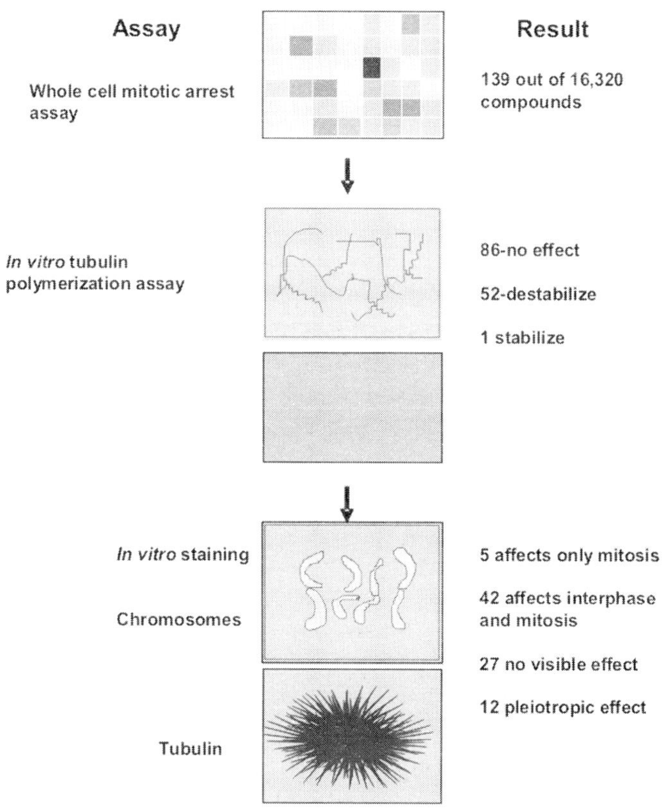

Fig. 2. Schematic presentation of the example of identification of 'hits' in a HRS assay that identified five small molecules affecting mitosis. Twelve antimitotic compounds tested on cells had pleiotropic effects and were not evaluated further (developed from Drew, 2000).

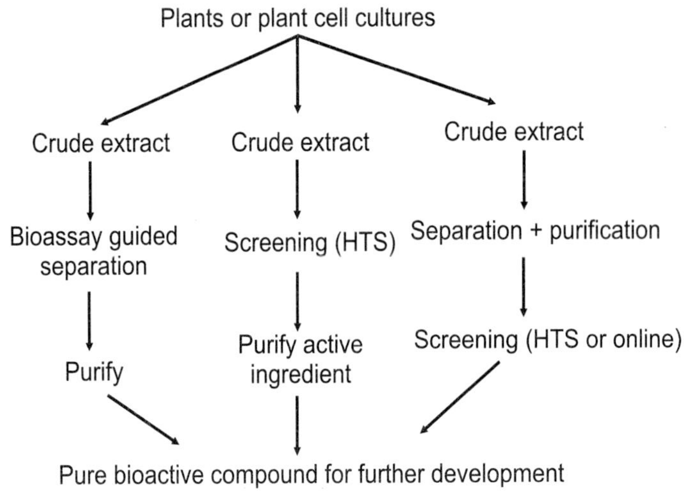

Fig. 3. Schematic presentation of various approaches towards isolation and identification of lead molecules

New bioactive molecules are desirable for pharmaceutical companies for their growth and economic viability. Search for new bioactive molecules was carried out either by activity based separation and purification or pure compounds obtained through slow and tedious chemical methods. These methods were time consuming, yielded a few number of products and resultant product process would get failed on critical evaluation parameters at various stages of clinical/ toxicological tests. The history of gene expression analysis began when laboratory methods were developed to examine expression of individual known genes. The northern blot technique developed during late 1970s hybridizes labelled DNA or RNA probes of known genes to RNA blots. The resulting expression patterns of mRNA transcripts can then be read (Lennon, 2000). Now technology has reached to its pinnacle where large number genes can be sequenced or gene products can be analyzed. The ability to rapidly survey and compare gene expression levels between reference and test samples using new technology such as differential display, READS™ (improved differential display), Expressed sequence tag (EST), Serial analysis of gene expression (SAGE) and DNA microarray made the drug discovery genomic oriented. By comparing gene products of two samples, it is possible to identify gene products or disease target. For finding new drugs in post-genomic era, pharmaceutical researchers must evaluate vast numbers of protein sequences and formulate novel intelligent strategies for identifying valid targets and discovering lead molecules against them. This is helpful in identification of small molecules that selectively target proteins. By identifying protein function first, efficiencies are gained that allow subsequent focus of resources on particular protein families of interest (Betz *et al.*, 2002). Although sequencing of human and other genomes is a result of the synergy between innovations in automation and bioinformatics, it is in principal a breakthrough in synchronous use and process management. New technologies offer rapid and simultaneous analysis of sequence and function of all genes in the genome. Mass sequencing is done today in factory like setups with many to several hundred sequencers run in a 24-hour, 7- day a week manner (Bertels *et al.*, 2000). The companies involved in such work are Incyte Pharmaceuticals, Human Genome Sciences, Millennium and PE Celara. These known sequences will be useful in rapid drug discovery for target diseases and identifying lead molecules (Kessmann and Schnurr, 2002).

Evolution of Drugs and Molecules

Secondary metabolism has evolved in nature in response to needs and challenges of natural environment. Nature has been carrying out its own version of combinatorial chemistry (Verdine, 1996) for over 3 billion years during which bacteria have inhabited the earth (Holland, 1998). During this long period, microorganisms adapted themselves to changing environment and modified secondary metabolites to their needs. The genes for non-required metabolites were switched off or altered. Combinatorial chemistry practiced by nature is much more sophisticated than that in the laboratory, yielding exotic structures rich in stereochemistry, concatenated rings and reactive functional groups (Verdine, 1996). As a result of this, a wide variety of natural products, from simple amines to complex molecules like azadirachtin, taxol, vincristine and vinblastine are produced. The total number of natural products produced by plants has been estimated to be over 500,000. One hundred sixty thousand natural products have been identified, a figure increasing by 10,000 per year (Anonynous, 2001). About 100,000 secondary metabolites of molecular weight less then 2500 Dalton have been characterized, about 50% from microorganisms and the other half from plants (Demain and Zhang, 2005; Roessner and Scott, 1996).

More than 2400 years ago, the Greeks used extracts of *Salix alba* and bark from several other species of willow to treat pain, gout and other illness. During the same period, Native Americans were using willow to relieve pain and fight fever. The effect of the willow was largely due to activity of a glycoside, salicin, which was first isolated in the nineteenth century from numerous species of *Salix* (e.g., *S. alba, S. helix, S. pentandra, S. praecox*) and wintergreen. The main commercial sources today are *S. fragilis* and *S. purpurea*, which are native to Europe and/or Asia. The Bayer Company in Germany produced synthetic salicylates on a large scale in 1874. Later on, it was discovered that its acetyl derivative was able to reduce its acidity, bad taste, and stomach irritation (Shapiro, 2003), leading to the genesis of "Asprin". There are several medicines whose genes can be traced to natural products, e.g., Acyclovir (used against herpes virus), cytarabine (for non Hodgkin's lymphoma) were isolated from a sponge (Shapiro, 2003) and Angiotensin-converting enzyme inhibitors (used for hypertension) are derivatives based on snake venom (Ondetti *et al.*, 1971).

Ergot is the name given to the sclerotium of the fungus *Claviceps purpurea* and some other *Claviceps* species that infect many wild grasses and cereals. Ergot as source of drug was known by the beginning of nineteenth century. The principal ergot alkaloids isolated are ergotamine, ergocarmine, ergocristine, ergocryptine, ergometrine and agroclavine. Ergometrine is responsible for uterine contractions, while ergotamine has a constricting effect on blood vessels. This was one of the compounds responsible for cutting of the blood supply, leading to the gangrene in the victims of St. Anthony's fire. The effects of all the ergot alkaloids appear to result from their actions as partial agonist or antagonist at adrenergic, dopaminergic and triptaminergic receptors (Kumar and Bhansali, 2006).

Lichens are associations of two or more components consisting of a fungus in association with a green algae or a cyanobacterium. Approximately 700 secondary metabolites of diverse chemical nature have been identified in Lichens (Huneck and Yoshimura, 1976). Majority of these compounds are derived from acetyl polymalonyl pathway. Polyaromatic products unique to lichens are depsides, depsidones, dibenzofurans, usnic acid and depsones. Other aromatic compounds obtained from this pathway are chromones, xanthones and anthraquinones. Products synthesized by Shikimic Pathway are yellow pigments, the pulvinic acid derivatives including vulpinic and rhizocarpic acids. Polyols and carbohydrates are synthesized by algal counterparts (Quilhot *et al.*, 2006). These products show wide ranging biological activities like antibiotic, anticancer, enzyme inhibitor, anti-analgesic, anti-inflammatory and anti-protozoa activity.

Antibiotics

Antibiotics are defined as low molecular weight organic natural products (secondary metabolites or idiolites) produced by microorganisms, which are active at low concentration against other microorganisms. In 1995, about 12000 antibiotics were known, out of which 55% were produced by streptromyces (actinomycetes), 11% from other actinomycetes, 12% from non-filamentous bacteria and 22% from filamentous fungi. Many more new products are still being discovered from microorganisms. About 350 antimicrobials are available in the market. The antibiotics include cephalosporin (45%), penicillins (15%), tetracyclines (6%) macrolides (5%), aminoglycosides, ansamycins, glycopeptides and polyenes (Strohl, 1997). Of the 25 top selling drugs in 1997, 42% were natural products or derived from natural products (Bull *et al.*, 2000), of which antibiotics contributed 67% of sales (US $ 55 billion world market).

Besides antimicrobial activity, other properties of antibiotics have been explored. An extremely important concept for further drug development of natural products is that a compound possesses more then one biological activities. In 2000, 57% of all drugs in clinical trials for cancer were either natural products or their derivatives (Cragg and Newman, 2000). The antibiotics used for cancer chemotherapy are cytarabine, actinomycin D, mitomycin, bleomycins and the anthracyclines – daurorubicin and doxorubicin. Cyclosporin A was originally discovered as a narrow spectrum antifungal peptide produced by the mould *Tolypocladium niverum* (Borel *et al.*, 1976). Discovery of its immunosuppressive activity led to its use in organ transplants to the overwhelming success of the organ transplant field. Later on, two new products obtained from actinomycetes, FK 506 (tacrolimus) and rapamycin were introduced to break the monopoly of cyclosporin. These new products were narrow spectrum polypeptide antifungal agents having hundred-fold more potency than cyclosporin as immunosuppressent.

Besides anticancer agents (e.g., pentostatin, peptomycin and epirubicin), there are several metabolites of microorganism origin currently in therapeutic use. These are immunosuppressive agents (e.g., cyclosporin), cholesterol lowering agents (e.g., lovastatin), anti-helmintic agents (e.g., ivermectin) and an anti-diabetic agent (acarbose) (Fig. 4, Sneader, 2005; Butler, 2005; Neuman *et al.*, 2000).

Everolimus is a 40-O- (2 hydroxyl ethyl) derivative of rapamycin obtained from *Streptomyces hygroscopicus*. This drug exhibits its immunosuppressive effect by blocking the growth factor mediated proliferation of hematopoietic (T-cells and B-cells) and non-

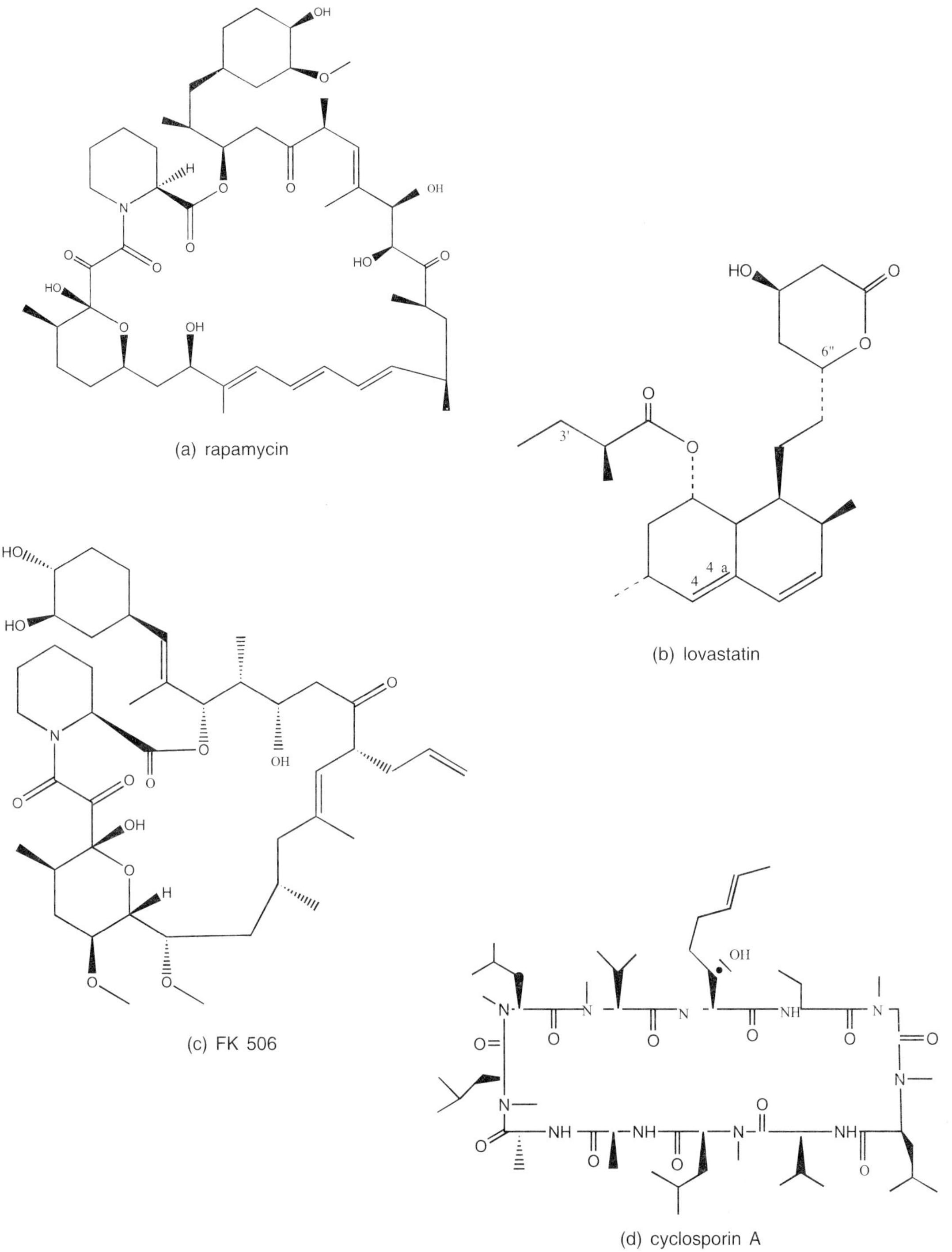

(a) rapamycin

(b) lovastatin

(c) FK 506

(d) cyclosporin A

Fig. 4. Immunosuppressive antimicrobial agents.

hematopoietic (vascular smooth muscle cells) cells through inhibiting P70S6 kinase leading to arrest of the cell cycle at the G1/S phase (Chapmann, 2004). Mycophenolic acid, isolated from *Penicillium brevicompactum*, effectively and selectively inhibits the proliferation of lymphocytes. The drug, mycophenolate sodium is used for the prophylaxis of organ rejection in patients receiving allogeneic renal transplants treated with cyclosporin and corticosteroids (Curran and Keating, 2005).

Cardiovascular

Before the isolation of ephedrine from *Ephedra sinica*, commonly known as Ma Hung in Chinese, epinephrine (adrenaline) was known. Epinephrine was isolated from sheep adrenal glands and used as bronchiodialator. *E. sinica* and *E. gerardiana* contains ephedrine, pseudoephedrine and several other derivatives (Arya and Ramawat, 1988). Ephedrine mimics the effect of epinephrine, causing elevation of blood pressure with inotropic and chronotropic actions on the heart. *Ephedra* is used in Chinese herbal medicine past 2000 years. However, FDA issued caution against its frequent use in 2004 looking to its harmful effect on heart.

The increased level of LDL and low level of HDL leads to the threat of arterioscleroses. It is estimated that ~30% of the cholesterol comes from daily diet and liver synthesizes 70% of the cholesterol. Due to deranged physiology, many people cannot maintain their cholesterol level and require cholesterolemic drugs. The statins inhibit *de novo* production of cholesterol in the liver. Statins are inhibitor of 3-hydroxy-3-methyl-glutaryl co-enzyme-A reductase, the regulatory and rate limiting enzyme of cholesterol biosynthesis. These extremely successful cholesterol-lowering agents are substituted hexahydronaphthalene lactones. Statins are obtained from *Penicillium citrinum*, *Monascus ruber* and *Aspergillus tareous*. The methylated form of compactin known as lovastatin reduces total plasma cholesterol by 20 to 40% (Knowles and Gromo, 2003).

CNS Activity

Pain is very common and use of medicinal plants to get relieved of pain is as old as civilization. Sumerians used the opium poppy extract 6000 years ago for pain relief. However, the earliest reference to the use of the juice of the poppy is from Theophrastus (~2300 years ago). Morphine was isolated by Serturner in 1806 followed by codeine in 1832 by Robiquet and then non-morphine alkaloid papaverine by Merck in 1848.

Apomorphine, a derivative of morphine is used as a dopamine receptor and antagonist to treat Parkinson's disease. This neuro-degenerating disease is caused by the loss of pigmented mesostriatal dopaminergic neurons linking the substantia nigra to the neostriatum. Subcutaneous apomorphine is currently used for management of fluctuating conditions of Parkinson disease (Deleu *et al.*, 2004). Morphine is obtained from the dried latex of *Papver somniferum*. Similarly, galantamine, an alkaloid obtained from *Galanthus nivalis* (Amyrillydaceae) has been launched in the market as a selective acetylcholine esterase inhibitor for the treatment of Alzheimer's disease. Slowing the process of neurological degeneration by inhibiting acetylcholine esterase as well as binding to and modulating the nicotinic acetylcoline receptor (Chin *et al.*, 2006).

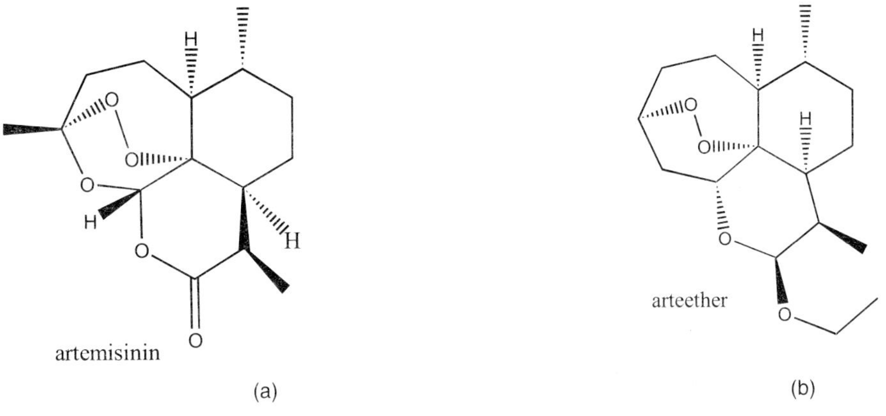

Fig. 5. Antimalarial compound obtained from *Artemisia annua*.

Antimalarial

Artemisia annua (Compositae) is known as Quinghao in Chinese traditional medicine where it has been used for centuries for the treatment for fevers and malaria. Artemisinin is a sesquiterpene lactone containing a rare peroxide linkage, which appears essential for activity. Arteether, an antimalarial agent, has been developed from artemisinin (Fig. 5). Other derivatives of artemesinin are in various stages of clinical treatment as antimalarial drug in Europe (Van Agtmael *et al.*, 1999).

Anticancer

More than 90 drugs are commercially available for cancer therapy, out of which 62% can be related to a natural product origin, excluding compounds like interferon (Newmann *et al.*, 2000). Cancer or neoplastic conditions are poorly defined in ancient traditional or folklore medicine, hence, leads are not available from such literature. A number of plant based compounds (Fig. 6a,b) are currently used as an anticancer agent such as taxol, derivatives of podophyllotoxins, vincristine, vinblastine and camptothecins (Butler, 2005; Ramawat *et al.*, 2006). Several derivatives of these established drugs as well as other compounds such as camptothecins and combretastatin are under clinical trials against various types of neoplasm affecting tubulin or topoisomerase I activity. Other compounds under clinical trials are homoharringtonine (Kantarjian *et al.*, 2001) from *Ceplalotaxus harringtonia,* ingenol 3-O-angelate from *Euphorbia peplus* which are being evaluated against hematologic malignancies and skin cancer, respectively (Kedei *et al.*, 2004).

Podophyllotoxins

The rhizome of *Podophyllum* species is known to contain several lignans. Lignans are dimerisation

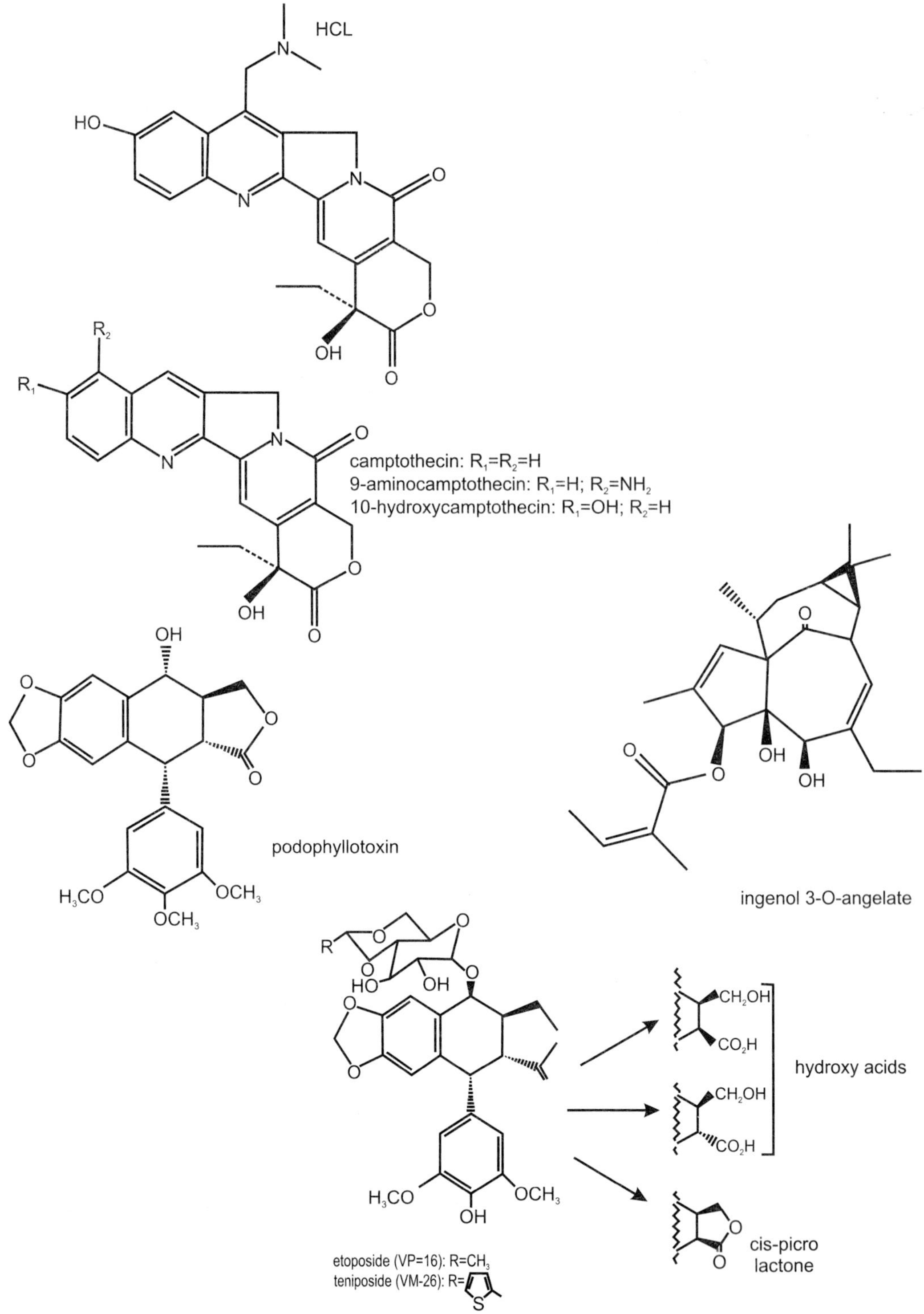
HCL
camptothecin: R1=R2=H
9-aminocamptothecin: R1=H; R2=NH2
10-hydroxycamptothecin: R1=OH; R2=H
podophyllotoxin
ingenol 3-O-angelate
etoposide (VP=16): R=CH3
teniposide (VM-26): R=
hydroxy acids
cis-picro lactone

Fig. 6a.

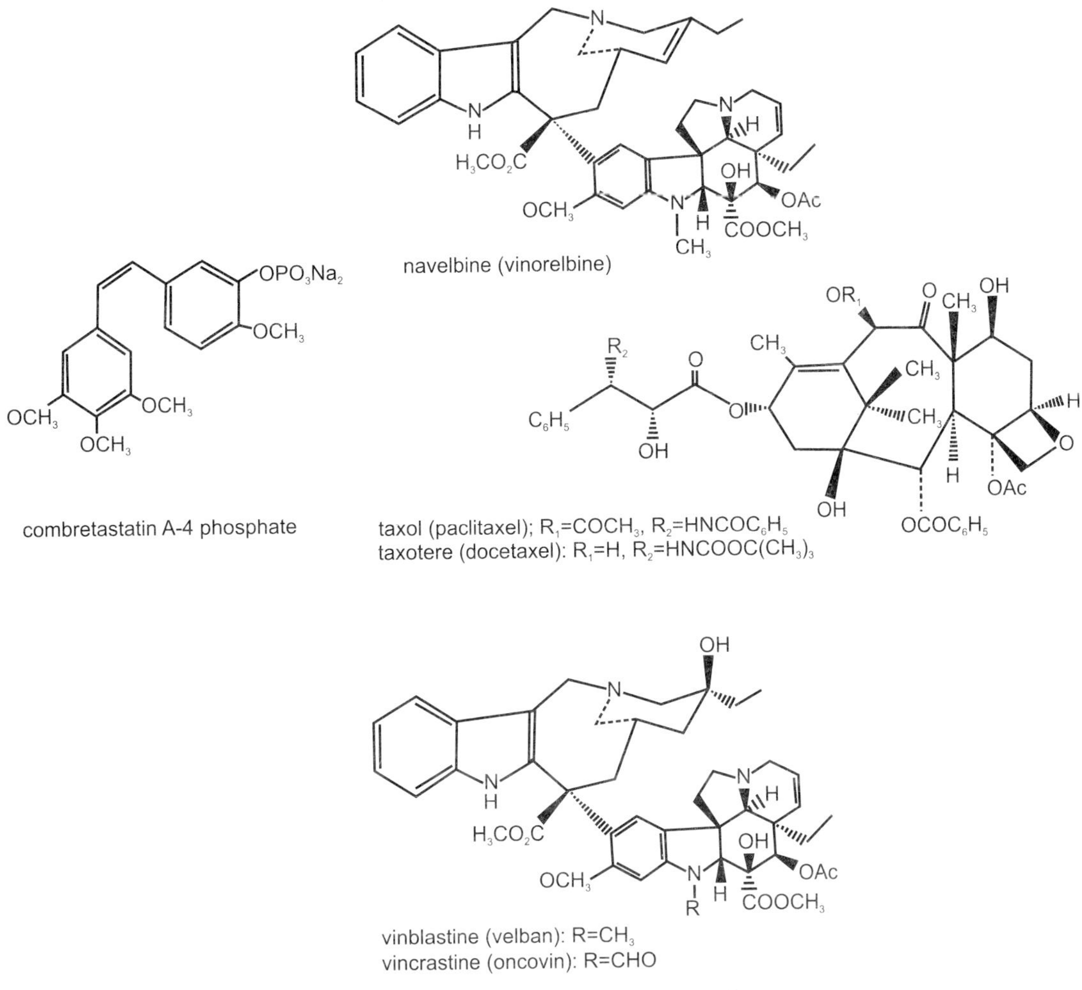

Fig. 6b. Different anticancerous plant products

products of phenylpropanoid pathway intermediates linked by the central carbons of their side chain. The lignans possess antitumour properties, podophyllotoxin being the most active cytotoxic compound. The lignans are too toxic for the treatment of neoplastic diseases in humans. Therefore, two semisynthetic glycosides of isomeric natural product, epipodophyllotoxin, called etoposide and teniposide have been developed by Sandoz (now part of Novartis) that show significant therapeutic activity in several human neoplasm, including small cell carcinomas of the lung, testicular tumours, Hodgkins disease and diffuse histocytic lymphoma (Lee, 2004; Newman and Cragg, 2005).

Although podophyllotoxin binds to tubulin at a site different from that for *Catharanthus* alkaloids, the drug has no effect on microtubular structure or function, but inhibits topoisomerase II (Lee, 2004).

Vinblastine

Medicinal properties of *Catharanthus roseus* have been described in traditional and folk medicine of several countries. Beneficial effect of its extract in diabetes mellitus was known, but later on active principles suppressing the neoplasm were isolated. The extracts yielded four active dimeric alkaloids—vinblastine, vincristine, vinleurosine and vinrosidine. Navelbine (Vinorelbine), vinblastine (velban) and vincristine (oncovin) are currently used clinically (Lee, 1999; Ramawat, 2006).

The catharanthus alkaloids are cell-cycle specific agents, similar to colchicine and podophyllotoxin, block mitosis and cause metaphase arrest. These alkaloids bind with tubulin protein and prevent tubulin formation from the protein. Though vincristine and vinblastine have anti-proliferative properties, both have different patterns of cytotoxic effect and are used in combination.

Taxol

Taxol, a diterpene, is the latest among anti-neoplastic drug of natural origin. Work on *Taxus* took a momentum after establishment of antitumour properties of taxol obtained from *T. brevifolia* during 1960s. Taxol is obtained from bark of Yew, *Taxus brevifolia, T. baccata* and other *Taxus* species and binds to tubulin (at site different from that used by *Catharanthus* alkaloids) and promotes the assembly of microtubules. Cells resistant to *Catharanthus* alkaloids are affected by taxol, which is effective against refracting metastatic ovarian cancer. Taxanes acetylated at C-13 were effective in inhibiting DNA and protein synthesis.

Bristol-Mayers Squibb, USA, Rhone-Paulence, France and Dabur, India are producing taxol (drug) at commercial level. Most of the derivatives are obtained through partial synthesis, e.g., taxotere. It is estimated that about 50 kg of taxol will be required every year for the treatment of approximately 12500 women in USA alone. The present world demand may touch 250 kg per year. The extraction of taxol from the bark leads to the death of the plant. Though taxol is obtained from the bark of the plant, the needles contain equivalent or even higher amounts of paclitaxel in four species of *Taxus, T. brevifolia, T. baccata, T. canadensis, T. cuspidata*, which is a renewable source as compared to removal of the bark and killing the plant. Semi-synthesis of paclitaxel from 10-deacetyl baccatin III (DAB) and its total synthesis have also been achieved (Ramawat, 2006; Newman and Cragg, 2005; Kingston, 2001).

Ellipticinium acetate, a derivative of ellipticine isolated from *Ochrosia elliptica* (Apocyanaceae), was approved for breast cancer in France (Chenieux *et al*., 1988). This molecule is very little in use because of preference for taxanes (Newman and Cragg, 2005).

Camptothecin

Camptothecin (CPT) was isolated by Wall *et al*. (1966) from the Chinese tree *Camptotheca acuminata* and is used to treat gastric, rectal, colon and bladder cancers. Govindachari and Viswanathan (1972) isolated camptothecin and 9-methoxy camptothecin from an unrelated plant, *Mappia foetida.* Camptothecins and its derivatives (9-amino CPT, 10-hydroxy CPT, camptothecin, topotecan and irinotecan) are potent antitumour and DNA topoisomerase I inhibiting agents. The basic molecule has poor solubility in aqueous phase and is found to be too toxic, therefore, attempts are being made to develop new effective soluble derivatives (Lee, 1999). Topotecan was approved for the use in the USA in 1996 and others are under clinical trials (Newman *et al.*, 2000).

OTHER COMPOUNDS

Cephalotaxus harringtonia (Cephalotaxaceae, Gymnosperm) contains several alkaloids, deoxyharringtonine, harringtonine, homoharringtonine, and isoharringtonine, that have shown anticancer activity against leukemia in mice. The drug is still under clinical trial (Newman and Cragg, 2005).

Tripterygium wilfordii, a lianas in the family Celastraceae and a native of East Asia is a source of antineoplastic diterpene triepoxides, tripdiolide and triptolide (Kupchan *et al.*, 1972).

The combretastatins obtained from *Combretum caffrum* are from a class of polyphenolics known as stilbenes, which act as anti-angiogenic agents, causing vascular shutdown in tumours and resulting in tumour necrosis (Holwell *et al.*, 2002). Its water-soluble derivative, combretastatin A4 phosphate is under phase II clinical trial (Newman and Cragg, 2005).

CONCLUSION

Drug development is a long and expensive process, and due to this reason, several pharmaceutical companies reduced their expenses on research and development for drug discovery from natural sources. From the above discussion it is very much clear that national resources, particularly plants, are an excellent source of life-saving drugs. New compounds are being discovered from lichens, microorganisms and marine organisms. Therefore, efforts are required to collaborate internationally to isolate, identify and establish biological/pharmacological properties of molecules (Cordell, 2003). New tools like high-resolution NMR and mass spectrophotometers and 2-D HPLC are helpful in identifying a plethora of molecules from plants and microorganisms. HTS systems are capable of evaluating a large number of bioactive molecules on various bioassays. At the same time, it is necessary to conserve traditional and folk knowledge about use of medicinal plants and develop appropriate technology to conserve the plants. In developing countries, still there are several species not evaluated for their phytochemical properties and the entire gamut of molecules in the established medicinal plants is yet to be known. It is urgently required to investigate plants used in traditional medicinal for the complete primary and secondary metabolites, and where a cocktail of plants is used, the effect of this cocktail on molecules reorientation may be explored. What we know is still a tip of iceberg and we have yet to explore the entire biological world for novel leads.

REFERENCES

Anonynous. (2001). Dictionary of natural products, Chapman and Hall/CRC Press, London.

Arya, H.C. and Ramawat, K.G. (1988). *In:* Phytochemicals in plant cell cultures, (Eds. Constabel, F. and Vasil, I.K.) Vol. 5. Academic Press NY. 237-243.

Aumont, V., Larronde, F. and Richard, T. (2004) *J. Biotechnol.* **109(3):** 287-94.

Bertels S., rormann, S., Jas, G. and Bindseil, K.U. (2000). *In:* Drug discovery from nature (eds. Grabley, S. and Thierike, R.) Springer Berlin 72-105.

Bertelli, A.A.E., Givanni, L., Stradi, R., Bertelli, A. and Tillement, J.P. (1996). *Int. J. Clin. Pharmocol. Res.* **26:** 77-81.

Betz, S.F., Baxter, S.M. and Fetrow, J.S. (2002). *Drug Discovery Today* **7:** 865-869.

Borel, J.F., Feurer, C., Gabler, H.U. and Stahelin, H. (1976). *Agents Actions* **6:** 468-475.

Bull, A.T., Ward, A.C. and Goodfellow, M. (2000). *Microbiol. Mol. Bio. Rev.* **64:** 573-606.

Butler, M.S. (2005). *Nat. Prod. Rep.* **22:** 162-195.

Chappmann, T.M. and Perry, C.M. (2004). *Drugs,* **64:** 861-872.

Chenieux, J.C., Ramawat, K.G. and Rideau, M. (1988). *In:* Medicinal and Aromatic plants I. (Ed. Bajaj, Y.P.S.) Springer Verlag Heidelberg pp: 448-463.

Chen, C.K. and Pace-Asciak, C.K. (1996). *Gen. Pharmac.* **27:** 363-366.

Chin, Y.W., Balunas, M.J., Chay, H.B. and Kinhorn, A.D. (2006). *Drug The AAPS J.* **8(2):** E239-253.

Cojocel, C. and Thompson, M.S. (2004). *Arc. Toxicol.* **78:** 525-532.

Cordell, G.A. (2003). *Phytochem. Rev.* **1:** 261-273.

Cragg, G.M. and Newman D. *Exp. Opin. Invest Drugs,* **9:** 2783-2797.

Curran, M.P. and Keating, G.M. (2005). *Drugs,* **65:** 799-805.

Lin, H.Y., Lansing, L., Merillon, J.M., Davis, F.B. and Tang, H.Y. *et al.*, (2006). FASEB *J.* **20(10):** 1742-44.

Deleu, D.Y., Hanssens, M.G. and Northway. (2004). *Drugs Aging,* **21:** 687-709.

Demain, A.L and Zhang, L. *In:* Natural Products: Drug discovery and Therapeutic Medicine, (Eds. Zhang, L. and Demain, A.L.) Humana Press, Totowa New Jersey 3-29.

Dixon, R.A. and Ferreira, D. (2002). Genistein *Phytochem.* **60:** 205-211.

Drews, J. (2000). *Science,* **287:** 1960-64.

Govindachari, T.R. and Viswanathan N. (1972). *Indian J. Chem.* **10:** 453-454.

Gupta, R. (1986). *In:* Agroforestry syatems: A New Challenge. Indian Society of Tree Scientists, Solan India 59-67.

Haslam, E. (1996). *J. Nat. Prod.* **59:** 205-215.

Holland, H.D. (1998). **275:** 38-39.

Holwell, S.E., Cooper, P.A. and Grosios, K, (2002). *Anticancer Research* **22:** 707-712.

Huneck, S. and Yoshimura, I. (1996). Identification of lichen substances Springer-Verlag, Heidelberg.

Jeon, G.C., Park, M.S., Yoon, D.Y., Shin, C.H., Sin, H.S. and UM, S.J. (2005). *Exptl. Mol. Med.* **37(2):** 111-120.

Kantarjian, H.M., Talpaz, M., Santini, V., Murgo, A., Cheson, B. and O'Brian, S.M. (2001). *Cancer,* **92:** 1591-1603.

Kedei, N., Lundberg, D.J., Toth, A., Welburn, P., Garfield, S.H. and Blumberg, P.M. (2004). *Cancer Res.* **64:** 3243-3255.

Kessmann, H. and Schnurr, B. (2002). *In:* Plant biotechnology and transgenic plants, (Eds. Oksman-Caldentey, K.M. and Barz, W.H.) pp. Marcel Dekker, Inc, New York, Basel 45-57.

Keung, W.M. and Vallee, B.L. (1998). *Phytochem.* **47(4):** 499-506.

Kingston, D.G.I. (2001). *Chemical comm..* 867-880.

Knowles, J. and Gromo, G. (2003). *Nature Revs/Drug Disc* **2:** 63-69.

Kritz-Silverstein, D. and Goodman-Gruen, D.L. (2002). *J.Women's Health and Gender-Based Med.* **11:** 69.

Kurzer. M.S. and Xu, X. (1997). *Annu. Rev. Nutr.* **17:** 353.

Kumar, A. and Bhansali, R.R. (2006). *In:* Biotechnology secondary metabolites – plants and microbes , 2nd edition (Eds. Ramawat, K.G. and Merillon, J.M.) Sci. Pub. Inc. Enfield USA.

Kupachan, S.M., Court, W.A., Dailey, R.G., Gilmore, C.J. and Bryan, R.F. (1972). *J. Amer. Chem. Soc.* **94:** 7194-7195.

Lacey, J.V. Mink, Jr. P.J., Lubin, J.H., Sherman, M.E., Troisi, R., Hartge, P., Schatzkin, A., and Schairer, C. (2002). *JAMA.* **288:** 334-341.

Lancet, T. (1994). *The Lancet.* **343:** 1513-1515.

Lee, K.H. (2004). *J. Nat. Prod.* **67:** 273-283.

Lennon, G.G. (2005). *Drug Discovery Today,* **5:** 59-64.

Newman, D.J. and Cragg, G.M. (2005). *In:* Natural Products: Drug discovery and Therapeutic Medicine, (eds. Zhang, L. and Demain, A.L.) Humana Press Totowa New Jersey 129-168.

Newman, D.J., Cragg, G.M. and Snader, K.M. (2000). The *Nat. Prod. Rep.* **17:** 215-234.

Ondetti, M.A., Williams, N.J., Sabo, E.F., Pluscec, J., Weaver, E.R. and Kocy, O. (1971). *Biochemistry,* **10:** 4039.

Piver, B., M., Fer, X. Vitrac, Merillon, J.M., Dreano, Y. Berthou, F. and Lucas, D. (2004). *Biochem. Pharmacol.* **68(4):** 773-82.

Quilhot, W., Rubio, C. and Fernandez, E. (2006). *In:* Biotechnology secondary metabolites – plants and microbes, 2nd edition (Eds. Ramawat, K.G. and Merillon, J.M,) (Sci. Pub. Inc. Enfield USA).

Ramawat, K.G., Sonie, K.C. and Sharma, M.C. (2004). *In:* Biotechnology of medicial plants-Vitalizer and therapeutics, (ed. Ramawat, K.G.) Sci. Pub. Inc. Enfield USA, pp. 1-18.

Ramawat, K.G. (2006). *In:* Biotechnology secondary metabolites – plants and microbes, 2nd edition, (Eds. Ramawat, K.G. and Merillon, J.M.) Sci. Pub. Inc. Enfield USA.

Ramawat, K.G., Mathur, M., Tanwar, Y.S. and Goyal, S. (2006). Proc Inter Symp Herbal plant resources as Function of food, Univ Daegu, S. Korea, May 23-24, 2006 1-5.

Renaud, S. and De Lorgeril, M. (1992). *Lancet,* **339:** 1523-26.

Roessner, C.A. and Scott, A.I. (1996). *Ann. Rev. Microbiol.* **50:** 467-490.

Schippmann, U., Leaman, D.J. and Cunnningham, A.B. (2002). *In:* Biodiversity and the ecosystem Approach in agriculture, Forestry and Fisheries. Ninth Regular Session of the Commission on Genetic Resources for Food and agriculture. FAO, Rome, Italy 1-21.

Shapiro, S. (2003). *Mod. Drug Disc.* **12:** 9.

Sneader, W. (2005). Drug Discovery A history John Wiley & Sons, Hoboken, NJ.

Soleas, G.J., Diamadis, E.P. and Goldberg, D.M. (1997). *Clin. Biochem.* **30:** 91-113.

Strohl, W. (1997). *In:* (Ed. Strohl, W.) Biotechnology of Antibiotics, 2nd ed. Marcel Dekker, New York 1-47.

Torres, N., Torre-Villalvazo, I. and Tovar, A.R. (2006). *J. Nutr. Biochem.* **17:** 365-373.

Van Agtmael, M.A., Eggelte, T.A. and Van Boxtel, C.J. (1999). *Trends Pharma. Sci.* **20:** 199-205.

Verdine, G.L. (1996). *Nature,* **384:** Suppl: 11-13.

Vitrac, X., Krissa, S., Decendit, A., Deffieux, G. and Merillon, J.M. (2004). *In: Biotechnology of medicinal plants* (Ed. Ramawat, K.G.) Sci Pub. Enfield, USA. 33-75.

Yousef, M.I., Kamel, K.I., Esmail, A.M. and Baghdadi, H.H. (2004). *Food Chemical Toxicology,* **42:** 1497-1503.

Wall, M.E., Wani, M.C., Cook, C.E., Palmer, K.H., McPhail, A.T. and Sim, G.A. (1966). *J. Amer. Chem. Soc.* **88:** 3888-3890.

Wani, M.C., Taylor, H.L., Wall, M.E., Coggon, P. and McPhail, A.T. (1971). *J. Am. Chem.* **93:** 2325-27.

Wilson, E.K. (2002). *Chem. Eng. News.* **80(17):** 35-39.

Chapter 18

Agro-industrial Bioprocessing of Tropical Root and Tuber Crops—Current Research and Future Prospects

RAMESH C. RAY*, M. LATIFUL BARI[1] AND SEIICHI ISOBE[1]

Central Tuber Crops Research Institute (Regional Centre), Bhubaneswar–751 019, Orissa, India
[1]National Food Research Institute, 2-1-12, Kannondai, Tusukuba, 305-8642, (Japan)
E-mail : rc_ray@rediffmail.com

Tropical root and tuber crop are important staples for food security for about a fifth of the world population. The most important tropical root and tuber crops produced by the major world region in 2005 are cassava (*Manihot esculenta* crantz) sweet potato (*Ipomoea batatas* L; Lam, yams (*Dioscorea esculenta*, *D. rotundata*, *D.alata* and other species; colocasia (taro) (*Colocasia antiquorum* var.esculenta L, Schott) and potato (*Solanum tuberosum* L.). Potato is mainly grown in the developed nations of the temperate world, hence it is commonly considered as a temperate crop. Among these four tropical food crops the most important in terms of global production are cassava (48.8%) and sweet potato (39.8%), while yams (9.4%) and colocasia (2.0%) are less important. These root crops are very rich in starch and recognized as the most efficient converters of solar energy. For example, cassava and sweet potato produce 250103 kcal/ha, as compared to 176×10^3, 110×10^3 and 200×10^3 kcal/ha for rice, wheat and maize, respectively (Edision, 2004). Sweet potato, yams and colocasia are mainly processed as fresh vegetable, whereas 75% of the cassava in Africa, 75-80% in Asia and 65% in Latin America are processed either as fermented foods and feeds or as food additives such as starch, starch based sweeteners (glucose and fructose syrups), organic (acetic, lactic, glutamic and gluconic) acids, monosodium glutamate and microbial polysaccharides (santhan, pullulan. Sclerogjucan, etc.) and biofuel (ethyl alcohal). The fermented foods from cassava are *gari*, *fufu*, *lafun*, *chickwanghe*, *agbelima*, *attieke*, and *penjeum* in Africa, *gaplek*, *dosa* and *putto* in Asia and "cheese" *bread*, fermented starch based biscuits and coated peanut in Latin America. Similarly, sweet potatoes are either consumed as fresh vegetables or processed into starch, starch derived foods (*noodles*, *macroni*, *alcoholic* drink produced in China and Japan). The bagasse (residues after starch extraction) of these root crops serves as suitable substrate for production of microbial enzymes, organic acids, growth regulators, vitamins and low-calorific sweeteners in submerged and solid substrate fermentations.

INTRODUCTION

Tropical root and tuber crops are the third important food crops of humankind after cereals and grain legumes, and constitute either staple or subsidiary food for about a fifth of the world population (Chandra, 2003; Edison, 2004). These crops have their own role as an important staple in several countries in South America, Africa, South-East Asia, etc. Root and tuber crops have a higher biological efficiency as food producers and show the highest rate of dry matter production per day per unit area among all the crops. These crops are known to supply cheap source of energy especially for the weaker section of the population. Tuber crops can be broadly classified as temperate and tropical groups of crops. Potato is primarily a temperate crop, whereas all the other edible root and tuber crops, like cassava (*Manihot esculenta* Crantz; Family: *Euphorbiaceae*), sweet potato (*Ipomoea batatas* L.; Family: *Convolvulaceae*), yams (*Dioscorea esculenta, D. rotundata, D. alata* and other species; Family: *Dioscoreaceae*) and aroids [Colocasia (taro) (*Colocasia antiquorum* var. *esculenta* L. Schott and *C. esculenta* var. *antiquorum* L. Schott; Family: *Araceae* and elephant foot yam (*Amorphophallus paeoniifolius*; Family: *Araceae*] come under tropical root and tuber crops. The relevant characteristic of these crops is their potential to grow and yield even under low fertility conditions. With the ever increasing global population, the demand for food and feed will be tremendous and it may not be possible to meet the requirement with the present food crops such as cereals/millets and legumes. Further, with value addition, these root crops offer a tremendous scope in the industrial sector. Microbial bioprocessing is a premier technology in which these tuber crops can be value-added. In this chapter, bioprocessing of tuber crops with the aid of microorganisms into fermented food, food additives, feed and bioethanol has been described.

FERMENTATION PROCESS (BIOPROCESSING)

Many of the fermented food, feed and industrial enzymes and biochemicals are based on the *koji* process that belongs to a major technical fermentation category known as solid substrate (state) fermentation (SSF) differing from either submerged (SmF) or immobilization fermentations in the amount of free water that they contain.

Solid State Fermentation (SSF)

Solid state fermentation (also called solid state bioprocessing (SSB)) refers to the process where microbial growth and product formation occurs on the surface of solid materials. This process occurs in the absence of "free" water, where the moisture is absorbed to the solid matrix (Zheng and Shetty, 1999; Suryanarayan, 2003). Solid state fermentation (SSF) has a series of advantages over submerged fermentation (SmF) including lower cost, improved product characteristics, higher product yield, easiest product recovery and reduced energy requirement (Raimbault, 1998; Pandey *et al*., 2000a; Krishna, 2005; Ray *et al*., 2007). It is of special economic interest for countries with an abundance of agro-industrial residues that are inexpensive substrates for microbial growth (Krishna, 2005; Ray *et al*., 2006), besides preventing disposal-linked environmental problems. In recent years, SSF has shown much promise in the development of several bioprocesses and products (Pandey *et al.*, 2000a; Soccol and Vandenberghe, 2003).

Root crops, fruit, and vegetable wastes have been successfully converted into many value-added products via SSF strategies (Zheng and Shetty, 1998 a, b, 2000; Krishna, 2005; Ray and Ward, 2006; Ward *et al.*, 2006). These wastes contain mainly lignocelluloses, besides soluble sugars, starch (in case of root crops), fibres, phenols, and other hydrolysable materials (Tengerdy and Szakacs, 2003) that can be metabolized by a wide range of microorganisms into value-added products such as single-cell protein, microbial enzymes, etc.

Selection of Microorganisms in SSF

The ability of the microorganisms to grow on solid substrate is a function of their water activity requirements, their capacity to adhere to and penetrate into the substrate and their ability to assimilate mixtures of different polysaccharides from complex heterogenous substrates (Raimbault, 1998; Pêrez-Guerra *et al.*, 2001).

Thermophilic bacteria, fungi and yeasts can grow on solid substrates, and find applications in SSF processes. The filamentous fungi are the best adapted microorganisms for SSF owing to their physiological, enzymological and biochemical properties. Growth pattern of fungi in SSF has been studied in detail in many cases, and these can be summarized in two phases: (a) germination, germ tube elongation and mycelial branching to loosely cover most of the substrate, and (b) increase in mycelial density with aerial and penetrative hyphal development (Raimbault, 1998; Viniegra-González *et al.*, 2003). These features also give them a major advantage over unicellular microorganisms for their colonization of the substrate and the utilization of the available nutrients. In addition, their ability to grow at lower water activity (a_w) and under high osmotic pressure conditions (high nutrient conditions) makes fungi efficient and competitive in the natural microbial ecosystem for bioconversion of solid substrates (Viniegra-González *et al.*, 2003). From a practical application, vegetative growth is preferred over sporulation.

However, thermophilic bacteria and yeasts have also been used in traditional cultivation in SSF processes (Doelle *et al.*, 1992; Ray *et al.*, 2006). Bacteria have been used for enzyme and organic acid production, composting, ensiling and some food processing (e.g., *koji*, sausages, Japanese *natto*, etc.) (Weinberg and Ashbell, 2003; Ray *et al.*, 2006). Yeasts have been mainly used for ethanol production and protein enrichment of agricultural wastes (Ward and Singh, 2005; Ward *et al.*, 2006; Ray *et al.*, 2006).

Solid state fermentations have also been implemented using recombinant strains. For example, chymosin, the cheese-making enzyme from calf stomach was produced by recombinant *Aspergillus oryzae* in SSF media containing wheat bran (2%) (Tsushiya *et al.*, 1994).

Submerged Fermentation (SmF)

In contrast to SSF, SmF is the process of choice for industrial operations due to the very well-known engineering aspects such as fermentation modelling, bioreactor design and process control (Gutierrez-Correa and Villena, 2003). Food additives and enzyme production by SSF is not of wide use by commercial manufactures mainly due to difficulties in bioreactor design, although the advantages of this form of microbial cultivation as compared to SmF are well known.

In bioconversion of agricultural and food wastes to valuable chemicals, separation and purification of the products from the bulk liquid represented in highest percentages of the manufacturing cost. Therefore, the economic feasibility of reusing waste will strongly depend on the down streaming processing efficiency. In recent years, considerable advances have been achieved in biomolecular purification technologies; one of the key goals is to achieve more selective, more efficient, and safer separation routes (Angenent *et al.*, 2004). For example, a single-step direct lactic acid separation from fermentation broths has been successfully implemented using anionic fluidized-bed columns (Sosa, 2001). Superficial fluid technology (i.e., the use of fluid at a temperature and pressure that are greater than its critical levels) has also proven to be a useful tool for achieving separation of compounds directly from cultures, and is an attractive option because separation is usually performed at ambient temperatures using non-toxic, non-flammable solvents while achieving product crystallization (Angenent *et al.*, 2004).

BIOPROCESSING OF ROOT AND TUBER CROPS

Roots, tubers, and their residues after starch extraction (bagasse) serve as excellent economical substrate in bioprocesses. With the advent of biotechnological innovations, mainly in the area of enzyme and fermentation technology, many new avenues have opened for their utilization.

CASSAVA

Foods and Beverages

Fermentation is an important means of processing this crop to improve palatability, textural quality and to upgrade nutritive value by enrichment with proteins and for reduction of inherent toxic factors (cyanogenic glucosides: linamarin and lotaustralin) (Westby and Choo, 1994; Treche and Massamba, 1995; Zuleta *et al.*, 2004). A wide variety of fermented food items have been developed from cassava in Africa, Asia and Latin America. Solid state and submerged fermentations are the usual types of fermentation adapted before cassava is utilized for consumption, and the procedure adapted in different countries differs widely. Most of the traditionally fermented cassava foods are consumed immediately after fermentation, as it is produced in home and cottage industries for domestic consumption. Only very small proportions are produced on a commercial basis for later use. Microorganisms associated with fermented food and beverages from cassava are listed in Table 1. Some of the cassava-based fermented food products have been discussed below.

Gari

Cassava roots are harvested, peeled and washed, and grated and packed into coarsely knit bags. A heavy object is placed on top of the bag to express some of the juice. The bag contents are then left to undergo natural fermentation for several days. On *garifying*, the grated cassava is

Table 1: Microorganisms associated with fermented foods from tropical root and tuber crops*

Fermented foods	Microorganisms
Cassava	
Gari	*Lactobacillus plantarum, Geotrichum candidum, Leuconostoc, Corynebacterium*
Fufu	*Alcaligenes, Candida, Citrobacter freundii, Geotrichum* spp., *Candida* spp., *Streptococcus, Clostridium* spp. *Lactobacillus* spp., *Leuconostoc* spp
Lafun	*Bacillus subtilis, Lactobacillus, Leuconostoc, Streptococcus, Klebsiella, Corynebacterium, Candida*
Chickwanghe	*Streptococcus faecalis, Bacillus lichenoformis*
Agbelima	*Lactobacillus* spp. (*Lb. plantarum, Lb. brevis*); *Leuconostoc mesenteroides; Bacillus* spp. (*B. subtilis, B. lichenoformis, B. cereus, B. mycoides* and *B. polymixa); Penicillium* spp.; *Candida tropicalis, Candida krusei*; *Zygosaccharamyces*
Attieke and *Placali*	*Leuconostoc* spp., *Lactobacter* spp.
Kivunde	*Lactobacillus* spp.
Sweet potato foods and food additives	
Sour starch	*Lactobacillus* spp.; *Leuconostoc* spp. *Lactobacillus* spp.
Soya sauce	*Aspergillus oryzae, A. sojae*
Vinegar	*Acetobacter* sp.
Sochu	*Saccharomyces cerevisiae*
Sweet potato curd, yoghurt	*Lactobacillus bulgaricus, Streptococcus thermophilus, Leuconostoc* spp.
Lacto-fermented beverage	*Lb. plantarum*

*Source: Ray and Ward, 2006

dried to about 10% moisture content and the starch is probably partially dextrinized. At this stage, a little palm oil is added to give it colour; the final dry granular product is *gari*. Usually, *gari* is prepared from very fresh cassava roots as the roots are spoiled within 48 h by several physiological (vascular streaking and dehydration) and biological (fungi, bacteria and insects) factors. However, in a recent study, storage of cassava roots for 21 days in trench or sealed polyethylene bags did not significantly alter the overall acceptability of the *gari* from fresh roots (Akingbala *et al*., 2005).

Cassava fermentation to *gari* is associated with a community of microorganisms, which include yeasts (*Saccharomyces cerevisiae* and *Candida* sp.) and bacteria (*Lactobacillus, Leuconostoc, Alcaligenes, Corynebacterium,* etc.) (Ray *et al.,* 2007).

Fufu

Fufu is a fermented wet paste product processed from cassava. For production of *fufu*, the preliminary preparation of the cassava for fermentation is identical with the method for *gari* production. The roots are peeled, washed, cut into pieces and are then submerged in water in earthenware pots for five days. During this period, the cassava roots ferment and soften, releasing hydrogen cyanide into the soaked water and imparting the characteristics flavour of the retted cassava meal. The mash is sieved through small baskets to remove the ligneous central strands. The solid residue is pressed to drain off the water and formed into small balls for consumption. The balls are boiled in water and soft dough is produced. *Fufu* can be enriched with protein (from 1.85% to 5.5-8.2%) by co-fermentation with different proportion of cowpea and soya bean.

The involvement of more than one species of lactic acid bacteria has been reported in the production of *fufu*. In general, species of *Lactobacillus*, *Leuconostoc* and *Streptococcus* are the predominant microorganisms along with *Bacillus subtilis*, *Klebsiella*, *Candida krusei*, etc. (Oyewole, 1990, 2001).

Lafun

Lafun is a fine powdery cassava product that is prepared by fermentation and is commonly consumed in the western states of Nigeria. The whole or peeled roots are immersed in a stream, in stationary water, or in an earthenware vessel for 3-4 days and fermented until they become soft. The fermented roots are then taken out and the pulp broken into small crumbs and sun-dried on mats, racks and house roofs. Drying takes 1 to 3 days, depending on the weather. The dried crumbs are milled into flour. The flour is added into boiling water with constant stirring until a smooth thick paste is formed. The paste is cooled to about 35°C and is then served with soups. The fermented and dried cassava pulp, *lafun*, is similar to *cossettes* in Zaire and Rwanda, *kanyanga* and *mapanga* in Malawi and *makopa* in Tanzania.

Five major microbial groups, similar to the spectrum of those implicated to other fermented cassava foods, were consistently isolated during *lafun* preparation. These included species of *Bacillus*, *Klebsiella*, *Leuconostoc*, *Corynebacterium*, *Candida* and *Lactobacillus*.

Chickwanghe (kwanga)

Chickwanghe is the most popular processed food form of cassava in Zaire. *Myondo* and *bobolo* in Cameroon, *mboung* in Gabon and *mangbele* in Central African Republic also belong to this group.

Cassava roots are peeled, steeped in water and left for 3 to 5 days to ferment until they become soft. Fibres are removed from the pulp, which are heaped on a rack for further fermentation or covered with leaves and pressed using heavy objects to drain off excess liquid. The pulp is then ground on a stone, or pounded in a mortar. The fine pulp is steamed in pots. *Chickwanghe* is a very viscous paste, much thicker than *fufu*.

Fifteen microbial species, belonging to 10 microbial genera, have been isolated from the cassava dough to be used for *chickwanghe* preparation. When two of these species, *Streptococcus faecalis* and *Bacillus licheniformis*, were used as a starter culture to aseptically inoculate submerged cassava fermentation, a product having all the characteristics of chickwanghe was obtained (Nwankwo *et al.,* 1989).

Agbelima

Agbelima is a fermented product popular in the Ivory Coast. The roots are peeled, steeped in water for initial fermentation and then ground to paste like *fufu* and the paste is again left to ferment for two days in jute sacks and then pressed (Kofi *et al.*, 1996). Finally, the paste is removed from the sacks, crumbled or granulated, and then steamed. The difference from *fufu* preparation is the steaming step in the final stages. Microflora for fermenting cassava dough into *agbelima* showed a dominance of LAB consisting of species of *Lactobacillus*, i.e., *Lb. brevis*, *Lb. plantarum*, *Leuconostoc mesenteroides*.

Attieke and Placali

Attieke and *placali* are two other cassava-fermented products consumed in Ivory Coast and other parts of West Africa. *Attieke* is similar to *agbelima*, but has a slightly sour taste and is eaten with milk or meat or vegetables. The production involves fermentation, as is the case for the preparation of other cassava-based foods.

Kivunde

Kivunde is another traditional fermented food from cassava that is popular in Tanzania. Three cassava fermentation methods (spontaneous fermentation, back-slopping and the use of starter culture) for the production of *kivunde* were compared in terms of cyanide level reduction, microbiology and product quality improvement. *Lb. plantarum* strains were isolated from all types of fermentations based on their enzymatic activities and acid production (Kimaryo *et al.*, 2000).

Gaplek

The simplest and most common form of processing used widely in Indonesia is to peel the roots, wash and slice and then sun-dry for 2-3 days to produce dry cassava chips or chunks, known as *gaplek* in Indonesia. *Gaplek* can be stored and is traded in village markets. When needed, the dry root pieces are pounded into flour, which is shaken on a bamboo screen with some water to produce granules called *tiwal.* The size and shape of these granules are similar to rice grains and the *tiwal* is often cooked together with rice to extend the family's limited supply of rice. Presently, small processing plants in Indonesia buy fresh roots to be processed directly into various flour mixes (supplemented with vitamins and flavours) as well as semi-cooked instant *tiwal.* These are mainly destined for urban consumers (Howeler, 2004).

Putto

Putto is the most popular fermented rice preparation in Philippines similar to *idli* in India. Successful attempts were made to substitute rice flour with cassava flour in *putto* preparation and that at a 50% substitution; the volume of *putto* was nearly the same as that made with the flour (Yoshi and Garcia, 1984; Iyayi and Losel, 2001).

Fermented Sweet and Sour Flour in India

A technology has been developed at Central Tuber Crops Research Institute, India for extraction of fermented sweet and sour flour from cassava. In this process, a mixed starter culture of *Lactobacillus cellobiosus*, *Streptococus lactis*, *Corynebacterium* sp. and *Pichia membranaefaciens* was used to ferment cassava roots. It improved the extraction of starchy flour without liberating foul smells (George *et al.*, 1995). This approach beneficially modifies the functional properties of the starch while leaving the granular structure intact (Nanda and George, 1996). It also lowers the cyanide content (Padmaja *et al.*, 1993) and improves the shelf-life, raising property and cooking quality (Padmaja *et al.*, 1994; George *et al.*, 1995). Nutritional values and *in vitro* digestibility of the fermented cassava flour is much higher compared to non-fermented flour (Padmaja *et al.*, 1994). The starchy flour extraction process has been delineated as below.

Raw cassava flour possesses a cohesive texture on heating in the presence of water, and this imparts a rubbery consistency to the processed food products. The cohesive texture is due to the rapid breakdown of starch granules in raw cassava flour and adversely affects the substitution of cassava flour in place of cereal flour for preparation of food items. Fermented cassava flour offers additional stability to starch granules, thereby reducing the cohesive structures. The improvement in cooking quality of the fermented flour may be attributed to the enhanced quantity of fibrous residue and higher elasticity imparted to the starch granules, which prevent easy breakdown (Moorthy and George, 1998).

Fermented Cassava Starch in Brazil

The total consumption of fermented cassava starch in Brazil is around 50,000 tonnes per year (Cereda and Vilpoux, 2006). Some uses of the fermented starch in Latin America are described below.

Fermented cassava biscuits

Fermented starch allows biscuits to become light and voluminous during baking due to mass expansion. In some Brazilian rural regions, these biscuits are substitute for bread (Damasceno *et al.*, 2003). With suitable packaging, these biscuits can maintain good texture and flavour for several months.

"Cheese" bread

"Cheese" bread is made with fermented starch or with a mixture of 50/50 fermented and non-fermented starches supplemented with milk, fat, eggs, meat and cured cheese of the Parmesan. It is found in most of the bakeries and bars in south of Brazil. Just as with fermented starch biscuits, the main characteristic of this product is its expansion properties during baking without using specific agents such as yeast or baking powder. The total market for cheese bread in Brazil was estimated to be around 80,000 tonnes in 2002, with a total use of fermented starch of around 16,000 tonnes. Recently, the North American Company, Pillsbury, entered the cheese bread market in Brazil with an eye on the potential export markets in Japan and Europe (Cereda and Vilpoux, 2006).

Coated peanuts

In Brazil, peanuts coated with fermented starch called "Japanese peanuts" are common. Traditionally, the coating is made from half-natural cassava starch and half-fermented starch. This product is found in supermarkets all over Brazil.

In preparation of fermented food products from root crops, it can be concluded that lactic acid bacteria are of major importance in fermentation providing the taste, flavour and characteristic aroma (Achi and Akubor, 2000) as well as probiotic and prebiotic attributes (Holzapfel and Schillnger, 1992).

PROTEIN ENRICHED FEED

Silage

Ensiling or silage making is an age-old technique for the conservation of nutrients in green fodder grasses through fermentation and storing them for round-the-year feeding of animals (Lien *et al.*, 1994). Ensiling can also render some previously unpalatable products useful to live stocks by changing the chemical nature of the feed (Kayouli and Lee, 1998).

Cassava leaves are highly rich in proteins (16-18%, fresh weight basis) (Nassar and Marques, 2006). Ensiling is an economically viable alternative to sun drying of the cassava roots and leaves and further storage as dry chips, as the silage is totally free from insect attack. However, there are three specific problems encountered during ensiling of whole cassava roots:

(1) Due to high moisture content of roots, there is an excessive silage effluent production, which leads to nutrient leaching and subsequent quality deterioration of cassava silage.
(2) The acid pH attained within two days of ensiling tends to stabilize the acetone cyanohydrin, and the finished silage has often higher amounts of cyanogens than admissible for animal feeding
(3) During the utilization phase, the silage is subject to aerobic deterioration.

Nevertheless, successful ensiling technology has been developed for cassava roots and leaves overcoming the above problems (Padmaja *et al.*, 1995). Due to inherently high content of water-soluble carbohydrates in cassava there is no need to add fermentation stimulants like molasses although it consistently stimulated lactic acid production (Padmaja and Balagopalan, 1991). Silage absorbents, like dehydrated rice straw, cassava fibrous waste and wilted cassava leaves can completely control the effluent production in cassava silages.

Single-Cell Protein

The attractive characteristics in the use of microorganisms for single-cell protein (SCP) include their (1) fast growth rate in semi-solid and solid media; (2) high level of protein; (3) comparable good nutritional values and (4) easy genetically modification to growth under specific conditions on particular substrates (Adeyemi and Sipe, 2004; Ray *et al.*, 2006). Mostly fungi, *Aspergillus niger*, *Saccharomyces cerevisiae*, *Rhizopus* spp. and *Mucor* and yeasts (*S. cerevisiae* and *Kluyveromyces*) are employed for production of single-cell protein. Cassava roots with 25-30% carbohydrate and 0.17% protein content is a good source for production of SCP (Sriroth *et al.*, 2000). Further, in the manufacture of starch from cassava, four types of wastes are produced: outer skin (peel), inner rinds, wastewater and fibrous residues (Ray, 2004). These wastes (bagasse), constituting about 20-30% by weight of the cassava chips/roots processed, are good substrates for protein enrichment by cultivating microorganisms through solid substrate fermentation (Balagopalan, 1996; Belewu and Musa, 2003). In a recent study, growth of *Aspergillus niger* during treatment of cassava starch processing waste water with high contents of suspended solids was studied (Tung *et al.*, 2004).

Food and Feed Additives

Figure 1 shows in more details the various products, which can be made from cassava starch and dried chips, as well as the peels and fibrous residues—the by-products from the starch industry.

Microbial Enzymes

The production of microbial enzymes is an important and rapidly growing industry in many parts of the world. The enzymes produced find a wide range of applications in the feed, food industry, including the hydrolysis of starch, alcoholic fermentation, clarification of beers and fruit juices, in the manufacture of glucose and fructose syrups, de-hiding of leather, etc. Root crops as a whole, starch, flour or bagasse in particular could be used as cheap fermentation substrates for the production of microbial enzymes (Table 2) (Pandey *et al.*, 2000b; Ray *et al.*, 2006).

Organic Acids

Organic acid made from cassava starch and flour or bagasse include citric acid, acetic acid, fumaric acid, lactic acid and itaconic acid (Table 2), which are used in the food industry as well as for the production of plastics, synthetic resins, rubber products, etc. Lactic acid could be produced by *Rhizopus oryzae* from raw cassava starch (Wang Xiaodong *et al.*, 2000; Nar Anong *et al.*, 2001). Polylactic acid has become the main raw material for production of

Table 2: Bioprocessing of cassava and sweet potato roots and bagasse

Microorganisms	Substrate	Bioproducts	Process	References
Aspergillus flavus, A. niger	Cassava peel	Amylase	SSF	Sani *et al.*, 1992
A. niger	Raw cassava starch	Amylase	SmF	Okolo *et al.,* 1995
Rhizopus oryzae, Botryodiplodia theobromae	Cassava waste	Amylase	SmF	Ray, 2004
Rhizopus spp.	Cooked cassava roots	Amylase	SmF	Soccol *et al.*, 1994
Monascus purpureus	Cassava waste water	Glucoamylase	SmF	Timotius, 2005
Trichoderma sp.	Cassava peel	Cellulase	SSF	Ofuya and Nwajiuba, 1996
Rhizopus spp., *A. niger*	Cassava roots	Linamarase	SmF	Birk *et al.*, 1996
A. niger	Sweet potato bagasse	Protease	SSF	Guo *et al.*, 1986
R. oryzae	Cassava starch	Rhodanese	SmF	Ray, 1993
R. oryzae	Raw cassava starch	Lactic acid	SmF	Nar Anong *et al.*, 2001
A. niger	Sweet potato flour	Citric acid	SmF	Bindumole *et al.,* 2000
A. niger	Sweet potato starch	Citric acid	SmF	Lu *et al.*, 1995, 1997
A. niger LPB 21	Cassava bagasse	Citric acid	SSF	Kolicheski *et al.*, 1995
A. niger CFTRI 30	Cassava bagasse	Citric acid	SSF	Shankaranand and Lonsane, 1994
A. niger NRRL 2001	Cassava bagasse	Citric acid	SSF	Vandenberghe *et al.*, 1999
Candida lipolytica	Cassava bagasse	Citric acid	SmF	Vandenberghe *et al.*, 1998
Rhizopus arrahizus	Cassava bagasse	Fuamric acid	SmF	Carta *et al.,* 1999
R. oryzae	Cassava bagasse	Fumaric acid	SmF	Carta *et al.*, 1999
R. oligosporus	Cassava bagasse	Fumaric acid	SmF	Carta *et al.,* 1999
Brevibacterium glutamicu	Sweet potato starch, flour	Monosodium glutamate	SmF	Woolfe, 1992
Brevibacterium divaricatum	Cassava bagasse	Glutamic acid	SmF	Jyothi *et al.*, 2005b
Aureobasidium pullulans	Cassava bagasse	Pullulan	SSF	Ray and Moorthy, 2007
Ceratosystis fimbriata	Cassava bagasse	Aroma compounds	SSF	
Saccharomyces cerevisiae	Cassava starch, waste water	Baker's yeast	SmF	Ejiofor *et al.*, 1996
				Christen *et al.*, 1997;
				Bramorski *et al.*, 1998
Kluyveromyces marxianus		Aroma compounds	SSF	Medeiros, 1998
Pleurotus sajor-caju	Cassava bagasse	Mushrooms	SSF	Barbosa *et al.*,1995
Lentinus edodes	Cassava bagasse	Mushrooms	SSF	Beux *et al.*, 1995

SSF: Solid substrate fermentation; SmF: Submerged fermentation

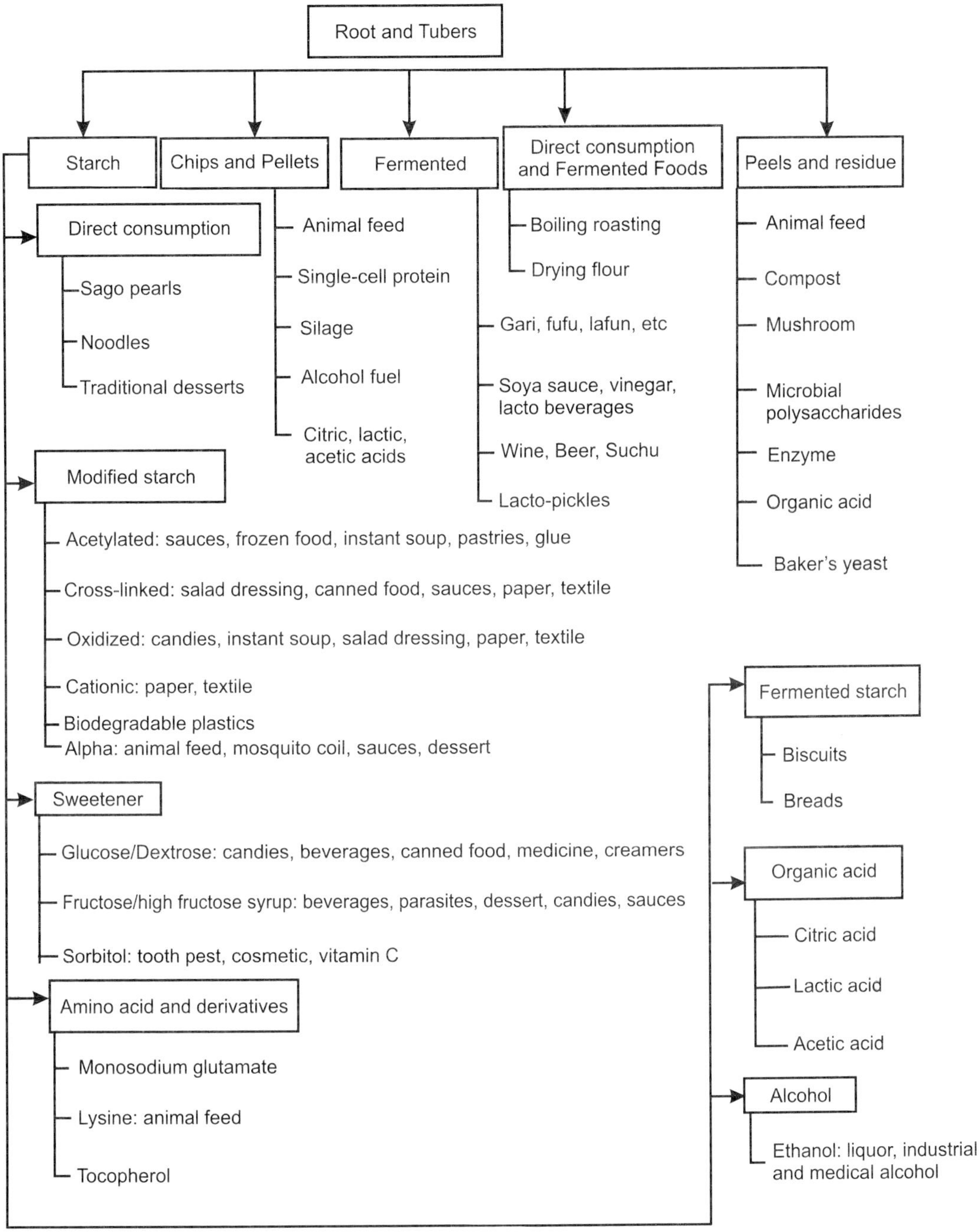

Fig 1. Processing of root crops into value-added products.

biodegradable plastics. Big American company (Cargill) produces polylactic acid from corn starch, however, cassava and sweet potato starch can be of equal potential for production of polylactic acid.

Jyothi *et al.* (2005b) optimized glutamic acid production from cassava starch factory residue hydrolysate using *Brevibacterium divaricatum*. The highest glutamate yield of 3.86% based on the weight of the bagasse was obtained at 30^{0}C at a pH of 7.0. Lysine is an important amino acid used as a supplement in animal feed, especially for pigs.

Sugar Syrups

Root crop starch, particularly from cassava can be used for the production of many types of sweeteners after hydrolyzation by either acids or enzymes, or both. These sweeteners include maltose, glucose syrups, glucose and fructose, which can be further processed into various oligosaccharides (Jin Shuren, 2001).

Mushroom Production

Bioconversion of crop residues for mushroom production is an area of profound economic importance. Cassava bagasse has been used for bioconversion mainly in solid substrate fermentation into mushrooms (Pandey and Soccol, 2000). Beux *et al.* (1995) compared the cultivation of *Lentinus edodes* on cassava bagasse and sugarcane bagasse, individually or in their mixture. Both the substrates were found suitable for mushroom production, but the best results were obtained when a mixture of cassava bagasse (80%) and sugarcane bagasse (20%) was used. Barbosa *et al.* (1995) also compared cassava bagasse and sugarcane bagasse for mushroom production. They used a different fungal culture, *Pleurotus sajor-caju*. Cassava bagasse showed better potential than sugarcane bagasse and the best results were obtained when cassava bagasse was used in a mixture with sugarcane bagasse (8:2 ratios, dry weight basis).

Polyunsaturated fatty acids were produced with *Mortierella alpina* by solid substrate fermentation. Rice bran was the most suitable substrate for polyunsaturated fatty acid production, followed by peanut meal residue, wheat bran, and sweet potato bagasse (Jang *et al.*, 2000).

Biodegradable Plastics

Various types of starches are being used for the production of biodegradable plastics. Starch is converted to lactic acid by microorganisms, and polymerization of lactic acid by thermo-chemical reaction leads to the production of polylactic acid, which is then blended with various other polymers to yield biodegradable plastics (Sriroth *et al.*, 2001). Further, fibres of cassava can be used as filler of bio-plastics. The use of cassava starch/fibre for these processes still requires much research.

Bioethanol

The use of ethanol as an alternative motor fuel has been steadily increasing around the world for a number of reasons. Domestic production and use of ethanol for fuel can decrease dependence on foreign oil, reduce trade deficits, create jobs in rural areas, reduce air pollution, and reduce carbon dioxide build-up. Ethanol, unlike gasoline, is an oxygenated fuel that contains 35% oxygen, which reduces particulate and NOx emissions from combustion (Sahoo *et al.*, 2005). Cane and beet molasses are the first choice for ethanol production all over the world; however, the gap between the demand and supply of molasses has been widened recently world over. Hence, demand for alternative sources of energy for bioethanol production from diverse raw materials (molasses, starchy crops like cassava and sweet potato) and sugar-rich crops like mahua (*Madhuca latifolia* L.) has increased (Swain *et al.*, 2006; Ward *et al.*, 2006; Ray *et al.*, 2007). Ethanol is an ideal fuel supplement or substitute for petrol, and starch from cassava and sweet potatoes can produce 400-450 litres (ethanol)/tonne of starch (equivalent to 160-180 litre. ethanol/tonne of raw cassava or sweet potato roots) (Ray *et al.*, 2004). The basic processes in the production of ethanol from cassava are somewhat cumbersome as compared to molasses, because of the requirement to hydrolyse the starch and include the following processes:

- Milling cassava chips or flour through sieve of 0.4 mm
- Liquefaction (using dilute inorganic acid or thermostable α-amylase)
- Saccharification (using amyloglucosidase)
- Fermentation, and
- Distillation

In some countries, cassava is used for the production of ethanol. In the late 1970s, several alcohol distilling factories were set up in Brazil using fresh cassava roots as raw material. The alcohol was used as automotive fuel, either mixed with gasoline (up to 20% alcohol) for which no motor modification is required, or as pure anhydrous ethanol, in which case the carburettor and some other parts need to be modified. Both result in less atmospheric pollution than the use of 100% gasoline.

The commercial production of ethanol from cassava is obviously not new in some parts of Asia like India and China. In China, several factories in Guangxi Province are now using the solid waste (bagasse) of the cassava starch industry for the production of ethanol (Gu Bi and Ye Gozhen, 2000). Other alcohol factories in China are switching from the use of molasses to that of cassava chips for alcohol production, because of strict pollution control requirements that makes the use of molasses uneconomical. In addition, it is expected that the Chinese government will support the use of "gasohol" (about 10-20% ethanol mixed with gasoline) as automotive fuel by 2008. This is expected to markedly increase demand for cassava chips, as this is probably cheapest raw materials. In Thailand, "gasohol" containing 10% ethanol is presently available in some gas stations in Bangkok. In India, too, "gasohol" contains 10% ethanol, but the ethanol is still produced from molasses although there are few factories in southern India producing ethanol from cassava (Edison, 2001).

SWEET POTATO

Sour Starch

Starch is the prime component of interest for food and industrial uses of sweet potatoes (Ray and Ravi, 2005). A traditional fermentation process involving the addition of liquid fermentate from peas and beans (called sour liquid containing LAB) at the separation stage speeds sedimentation and enhances separation and final product quality in China (Timmins *et al.*, 1991). Sour liquid is used even in enterprises with mechanized equipment.

Fermented Starchy Flour

The efficiency of starch extraction from sweet potato roots was improved by lactic acid fermentation using a mixed culture (*Lactobacillus cellobiosus, Streptococcus lactis* and *Corynebacterium* sp.) inoculum. Study of the properties of the starchy flour showed that there was a significant reduction in the starch content and consequently the soluble amylase contents of fermented samples from all six varieties of sweet potato used (Jyothi *et al.*, 2005a).

Lacto Pickles

Lactic acid bacteria influence the flavour of fermented foods in a variety of ways. In many cases, the most obvious change in lactic acid fermentation is the production of acid and lowering of pH, which results in an increase in sourness. Since most of the acid produced in fermentations results from the metabolism of sugars, sweetness is likely to decrease as sourness increases (McFeeters, 2004). This is the basic principle of the sauerkraut process in pickling of cabbage and other vegetables. Experimental work on pickling of sweet potato, taro and elephant foot yam by lactic acid fermentation (sauerkraut process) using 5-20% (w/v) brine solution has been carried out at Regional Centre of CTCRI, India (Panda *et al.*, 2007). Lactic acid fermentation not only produces lactic acid, which imparts taste and flavour, it also preserves ascorbic acid, phenols, and coloured pigments (β-carotene and anthocyanin): all these are considered as antioxidants (Shivashankara *et al.*, 2004).

Lactic Beverages

Non-alcoholic beverages from high anthocyanin/ β-carotene/anthocyanin sweet potato cultivars, possessing a nutritional value comparable with fruit drinks, have been formulated. The colour of the beverage varies according to the flesh colour of the roots, ranging from yellow to orange or pinkish purple.

In a modification of this process, the juices of β-carotene/anthocyanin sweet potato cultivars were fermented with *Lactobacillus plantarum*. The lactic beverages so obtained were slightly sour (lactic acid, 1.2 g/100 ml), sweetly flavoured and were bottled in 200 ml glass containers and pasteurized at 90-95°C. The lactic beverage had a pH of 3.3, total soluble solids of 10° Brix and insoluble solids of 9.0 mg/100 ml. A residual starch content of 8.4 g/100 g in the juice enhanced its resistance to separation. The product was not found to produce flatulence in consumers drinking more than 200 ml/day for a week (S.H. Panda and R.C. Ray, unpublished results).

Sweet Potato Curd and Yoghurt

Yoghurt and curd are reported to possess certain nutritional/dietary advantages over milk. It can be consumed by lactase-deficient individuals because much of the lactose in milk is converted to digestive lactic acid by curd- or yoghurt-producing bacteria during fermentation. While the starter culture for curd is a mixture of undefined cocktail of lactic acid producing microorganisms, i.e., *Lactobacillus bulgaricus*, *Streptococcus clemoris*, *St. thermophilus*, etc., the starter culture for yoghurt is the use of specific symbiotic/mixed culture of *Lb. bulgaricus* and *Streptococcus thermophilus*. Sweet potato yoghurt was prepared, which had high acceptability by consumers in USA (Collins *et al.*, 1991 a, b). In a recent study, a curd like product was prepared by co-fermenting boiled sweet potato pulp (8-16%) from β-carotene/anthocyanin-rich variety, sugar and curd inoculum. As the sweet potato concentration was increased in the fermented mash, some components (total and reducing sugars, amino acid, starch and anthocyanin) increased while others (titratable acidity, lactic acid and protein) gradually decreased (Panda *et al.*, 2006; Mohapatra *et al.*, 2006). Curd with 12-16% sweet potato pulp was most preferred by consumers' panelists (Ray *et al.*, 2005).

ALCOHOLIC BEVERAGES

Shochu

The production of *shochu*, a Japanese distilled liquor, which may also be made from rice or barley as well as sweet potato, has a long tradition (Yamakawa, 2000). During fermentation of this main mash, simultaneous starch conversion to sugar by the *koji* enzymes and fermentation of sugar to alcohol by the yeast *Saccharomyces cerevisiae* takes place. The final alcohol concentration of the mash is 13-15%. The mash is then pumped to the *still* and the alcohol is distilled off. Different batches of *shochu* may be blended to give a uniform product and the alcohol content is adjusted to 20-40% (v/v) before bottling (Woolfe, 1992).

Wine and Beer

Wine and beer are the other products which can be processed from coloured (anthocyanin/β-carotene) sweet potato cultivars (Yamakawa, 1997). Yellow, red and black coloured beverages like beer (sparkling liquor) and wine are being sold in the Kyushu Province in Japan (Yamakawa, 1997, 2000). A process of enzymatically saccharifying the sweet potato pulp (starch) and fermenting the mash into wine using wine yeast *S. cerevisiae* has been developed at Regional Centre, CTCRI, India. The wine prepared from an anthocyanin sweet potato cultivar is bright red in colour and fulfils all the quality attributes of fruit wines. Similarly, wine with medicinal properties against common ailments like cold, cough, muscle pain, stomachache, can be prepared by co-fermenting sweet potato and desired medicinal herbs, fruits or bark. Patents applications have been filed for the biotechnological processes developed for the preparation of sweet potato wines.

A rice beverage was prepared by adding barley sprouts, sweet potato, or a mixture of barley sprouts and sweet potato (1:1). The use of sweet potato resulted in an increase of sweetness, flavour and improved preference in rice beverage (Suh *et al.*, 2003).

FOOD ADDITIVES

Organic Acids

Sweet potato has been used a substrate for citric acid production using *Aspergillus niger* in solid substrate fermentation (Bindumole *et al.*, 2000). Several countries, notably China, Japan and Vietnam are now manufacturing citric acid from sweet potato starch or from its by-products. The process necessitates the initial breakdown of starch to sugars before these sugars are fermented by moulds, e.g., *A. niger* to citric acid. Lactic acid was produced from potato and sweet potato flour by *Lactobacillus* sp. in submerged fermentation (Ray *et al.,* 1991).

Monosodium Glutamate

Monosodium glutamate is a well-known flavour enhancing agent used in many Asian kitchens. China is the largest producer and consumer of monosodium glutamate in the world. It is made through the microbial fermentation of starch or sugar (molasses) in the presence of ammonium salts. The organism commonly used is *Brevibacterium glutamicum*, which converts sugar to glutamic acid.

Soya Sauce

Soya sauce, a popular condiment used everyday in Asian dishes is traditionally prepared from a mixture of soybeans and wheat, fermented by moulds, especially *Aspergillus oryzae* or *A.sojae*, to give a dark brown salty liquid used as a flavouring agent. The potential replacement of wheat flour with suitable substitutes in soya sauce manufacture, as in the case of bread, could mean a considerable utilization of available resources such as sweet potato flour (Data *et al.*, 1986).

Vinegar

Vinegar is a condiment made from sugar- or starch-containing materials by an alcoholic fermentation, followed by the microbial oxidation of alcohol to acetic acid. When starch-based material such as sweet potato is used as raw material, the starch requires initial hydrolysis to sugars before alcoholic fermentation by yeasts (*S. cerevisiae*) can take place. The next step, oxidation of the alcohol to acetic acid is carried out by acetic acid bacteria *(Acetobacter* spp.). The completed vinegar must contain a minimum of 4 g acetic acid/100 ml (Ward, 1989).

Sweet potato is rich in anti-oxidants such as high β-carotene, anthocyanin, caffeic acid and its derivatives, etc (Takenaka *et al.*, 2006). Vinegar has been processed from high β-carotene sweet potato cultivars (Ray, 2006). The appearance and taste is similar to that of white wine vinegar. Recently, a new red vinegar has been developed in Japan via fermentation with the storage root of purple fleshed sweet potato of Japanese variety (*Ipomoea batatas* L. cv. *Ayamurasaki*). The red vinegar had a higher antioxidative activity than white or black vinegars. The red vinegar contained some new components possibly derived from the original purple sweet potato. A major component identified was 6-O-(E) - caffeoyl-(2-O-beta-d-glucopyranosyl)-alpha-d-glucopyranose (caffeoylsophorose), a derivative of caffeic acid (Matsui *et al.*, 2003).

Silage

Sweet potato tops, especially leaves, are very rich in nutrients such as proteins, vitamins and minerals. Besides, they have amazing amounts of polyphenols, especially chlorogenic, isochlorogenic and caffeic acids (Takenaka *et al.*, 2006), which are antioxidants. However, one of the constraints to the use of the large quantities of sweet potato foliage for forage is that its availability is concentrated into one or two seasons of the year after harvest, which in general does not coincide with pasture scarcity. Preservation as hay (dehydration) or silage may therefore be needed. The latter method is more practical in the humid tropics where quick drying in the field is not possible. Simple small silos can be made with discarded gunny bags, polyethylene sugar bags and related materials, all of which can be afforded by rural farmers in the tropics (CIP, 2000). Tinh *et al.*, (2000) reported that sweet potato vines ensiled with chicken manure resulted in the highest quantity feed, increasing diet crude protein content and dry matter conservation rates. In a later study, Tinh *et al.* (2001) reported that cassava leaf meal, rice bran and chicken manure are good additives for fermenting sweet potato vines, in combination with salt, and the fermented sweet potato roots can be stored at least for 90 days without any significant reduction of quality. In a recent study, Giang *et al.* (2004) standardized the various proportions of sweet potato roots and vines to be mixed for silage preparation without any additives, which could be successfully stored for three months.

SCP

Sweet potato bagasse might also be a good candidate for SCP because of its abundant supply in several countries like China, Taiwan, and Japan at reasonable cost. The residue contains 2.32% protein and 65.4% total carbohydrates and is itself not a good source of protein for animal feeding (Ray, 2006). Fermented sweet potato feed stuffs are reported to be produced and utilized for livestock in China (Woolfe, 1992; Jiang *et al.*, 1993) and Korea (Ahn *et al.*, 1997). In a recent study, sweet potato roots and flour were fermented using pure strains of four species: *Aspergillus niger*, *Rhizopus stolonifer*, *Neurospora sitophila* and *Saccharomyces cerevisiae* using solid substrate fermentation at 30°C for 72 h. The protein content of the samples increased significantly from 2.34 in the non-fermented roots to values ranging from 11.50 to 12.62 in the fermented roots and 6.26 in the non-fermented flour to values ranging from 7.83 to 9.70 in the fermented flour. Besides, while the fermentation increased fat and ash content, crude fibre, carbohydrate, sugars and starch contents of both flours and roots substantially reduced (Onifade *et al.,* 2004). In yet another study in submerged fermentation, acid and gamma-irradiated pretreated sweet potato bagasse enhanced biomass yield and protein production when the bagasse was fermented with *Fusarium moniliforme* and *Saccharomyces cerevisiae* (Aziz and Mohseen, 2002).

Ethyl Alcohol

The manufacture of ethanol from sweet potato for human consumption as well as for chemical and pharmaceutical purposes is already established, especially in countries such as China, Japan and Korea. The process of manufacture of ethyl alcohol from sweet potato is essentially same as that of cassava. In China, alcohol is the most important product after starch and noodles; in Korea, however, alcohol production has been steadily rising, and in the 1990s represented more than 30% of total production, compared to only about 8% in the case of starch (Scott *et al.*, 2000). Japan uses only 6-8% of its production for alcohol (Yamakawa, 2000), and in India, technology has been standardized for alcohol production from sweet potato flour and roots (Ray *et al.*, 2004) but the process is yet to be commercialized.

Yams and Aroids

Yams and aroids are mostly consumed as fresh vegetables and rarely bioprocessed into fermented foods and industrial biochemicals. Only exceptions are 'poi' and 'amla', the fermented foods.

Poi and Amla

Poi is a fermented food product prepared from colocasia roots in Pacific islands (Matthews, 2000). Similarly, *amla* is a popular West African food item prepared from rehydrated yam flour.

SCP

Colocasia leaves could be included up to 10% as nitrogen source in SCP enrichment of taro roots for feed. No toxicity was observed during feeding the mixtures to weaned calves for 8 weeks as component of their crop diets (Anigbogu and Anigbogu, 1999).

CONCLUSION AND FUTURE PERSPECTIVES

For many years to come, root crops will continue to play a significant role in the diet of millions of people living in the tropics. For cassava, biotechnology research is needed to improve plant varieties and to introduce into its genome important characteristics for its value as food, which so far have been difficult to confer (e.g., absence of cyanogens; higher protein levels in the roots) (Fauquet and Taylor, 2001; Jin Shuren, 2001). For sweet potato, the major anti-nutrient factors discussed are α-amylase and trypsin inhibitors. Similarly, acidity, presence of mucilage, high tannins and polyphenols are the bitterness factors in yams and aroids.

Fermentation (bioprocessing) will probably remain the most important mode of processing of these root crops into edible foods and feeds (Fauquet and Taylor, 2001), because of their low cost and the desirable flavours they impart. The development of starter cultures would facilitate process standardization and quality control, and would set the stage for strain-improvement programmes through conventional microbiology and genetic engineering techniques, for better quality fermented products.

REFERENCES

Achi, O.K. and Akubor, P.I. (2000). *World J. Microbiol. Biotechnol.* **16(1):** 3-8.

Adeyemi, O.A. and Sipe, B.O. (2004). *Indian J. Animal Sci.* **74:** 321-323.

Ahn, Y.S., Jeong, B.C. and Oh, Y.B. (1997). *In: Proceedings of the International Workshop on Sweet potato Production System toward the 21st Century*, Dec 9-10, 1997, Miyakonojo, Miyazaki, Japan, pp. 137-147.

Akingbala, J.O., Oyewole, O.B., Uzo-Peters, P.I., Karim, R.O. and Baccus-Taylor, G.S.H. (2005). *J. Food Agric. Environ.* **3:** 75- 80.

Angenent, L.T., Karim, K., Al-Dahhan, M.H., Wrenn, B.A. and Dimiguez-Espinosa, R. (2004). *Trends Biotechnol.* **22:** 477-485.

Anigbogu, N.M. and Anigbogu, C.C. (1999). *J. Root Crop, India* **25(2):** 147-152.

Aziz, N.H. and Mohsen, G.I. (2002). *J. Ind. Microbiol. Biotechnol.* **29:** 264-267.

Balagopalan, C. (1996). *J. Sci. Ind. Res.* **55:** 479-482.

Barbosa, M.C.S., Soccol, C.R., Martin, B., Todeschini, M.L., Tonial, T. and Flores, V. (1995). *In:* Advances in Solid State Fermentation (Eds. Roussos, S., Lonsane, B.K., Raimbault, M. and Viniegra-Gonzalez, G.), Kluwer Academic Publishers, Dordecht, Netherlands, pp. 513-525.

Belewu, M.A. and Musa, A.K. (2003). *Trop. J. Anim. Sci.* **6:** 1-7.

Beux, M.R., Soccol, C.R., Martin, B., Tonial, T. and Roussos, S. (1995). *In:* Advances in Solid State Fermentation (eds. Roussos, S., Lonsane, B.K., Raimbault, M. and Viniegra-Gonzalez, G.), Kluwer Academic Publishers, Dordecht, Netherlands, pp. 499-511.

Bindumole, V.R., Sasikiran, K. and Balagopalan, C. (2000). *J. Root Crops, India* **26(1):** 38-42.

Birk, B., Bravdo, B. and Shoseyov, O. (1996). *Appl. Microb. Biotechnol.* **45:** 411-414.

Bramorski, A., Soccol, C.R., Christen, P. and Revah, S. (1998). *Rev. Microbiol.* **29:** 208-212.

Carta, F.S., Soccol, C.R., Ramos, L.P. and Fontana, J.D. (1999). *Biores. Technol.* **68:** 23-28.

Cereda, M.P. and Vilpoux, O.E. (2006). *In:* Microbial Biotechnology in Horticulture (Eds. Ray, R.C. and Ward, O.P.) Science Publishers, Enfield, New Hampshire, USA, pp. 397-420.

Chandra, S. (2003). *In:* Tropical Root and tuber crops: Opportunities for Poverty Alleviation and Sustainable Livelihoods in the Developing World. Thirteen Symposium of the International Society for Tropical Root Crops (ISTRC), November 10-14, 2003, AICC Arusha, Tanzania, pp. 19.

Christen, P., Meza, J.C. and Revah, S. (1997). *Mycol. Res.* **101:** 911-919.

CIP (International Potato Center) (2000). *CIP Program Report* 1999, pp. 312.

Collins, J.L., Ebah, C.B., Mount, J.R., Draughon, F.A. and Demott, B.J. (1991b). *J. Food Sci.* **56(3):** 682-684.

Collins, J.L., Ebah, C.B., Mount, J.R., Demott, B.J. and Draughon, F.A. (1991a). *J. Food Sci.* **56(3):** 685-688.

Damasceno, S., Cereda, M.P., Pastone, G.M. and Oliveira, J.G. (2003). *Process Biochem.* **39(3):** 411-415.

Data, E.S., Diamante, J.C. and Forio, E.E. (1986). *Ann. Trop. Res. (Phillip.)* **8(1):** 42-50.

Doelle, H.W., Mitchell, D.A. and Rolz, C.E. (1992). *Solid Substrate Cultivation.* Elsevier Science Publication, Essex, London, UK.

Edison, S. (2001). *In:* Casava's Potential in Asia in the 21st Century: Present Situation and Future Research and development Needs (Eds. Howeler, R.H. and Tan, S.L.), Proc. 6th Regional Workshop, held in Ho Chi Minh City, Vietnam, February 21-25, 2000, pp. 61-70.

Edison, S. (2004). *In:* Souvenir. National Seminar on Root and Tuber Crops for Nutrition, Food Security and Sustainable Environment, 29-31 October, 2004, Regional Centre of Central Tuber Crops Research Institute, Bhubaneswar, India, pp. 1-2.

Ejiofor, A.O., Chisti, Y. and Moo Young, M. (1996). *Enz. Microb. Technol.* **18(7):** 519-525.
Fauquet, C.M. and Taylor, N. (2001). *Food Nutr. Bull., Cancun, Mexico* **23(4):** 364-366.
George. M., Moorthy, S.N. and Padmaja, G. (1995). *J. Sci. Food Agric.* **69:** 367-371.
Giang, H.H., Ly, L.V. and Ogle, B. (2004). *Livestock Res. Rural Dev.* **16:** 12-21.
Gu, Bi. and Ye Guozhen. (2000). *In:* Cassava Starch and Starch Derivatives, Proc. Int. Symp. (eds. Howeler, R.H., Oates, C.G. and O'Brien, G.M.), held in Nannking, Guangxi, China, November 11-15, 1996, pp. 11-15.
Guo, Z.M., Sheng, W., Mian, W.J. and Zhi, Y.X.Z. (1986). http://www. Ncbi.nlm.nin.gov/entrez/guery.
Gutierrez-Correa, M. and Villena, G.K. (2003). *Rev. Peru Biol.* **10(2):** 113-124.
Holzapfel, W.H. and Schillinger, U. (1992). *In*: The Prokaryotes 2nd edn. (Ed. Balow, A.) Springer Verlag, N.W., pp. 1508-1534.
Howeler, R.H. (2004). *J. Root Crops* **30:** 81-92.
Jang, H-D., Lin, Y-Y. and Yang, S-S. (2000). *Bot. Bull. Acad. Sin.* **41:** 41-48.
Jin Shuren. (2000). *In:* Cassava Starch and Starch Derivatives, Proc. Int. Symp. (Eds. Howeler, R.H., Oates C.G. and O'Brien, G.M.), held in Nannking, Guangxi, China, November 11-15, 1996, pp. 32-36.
Jin Shuren. (2001). *In:* Casava's Potential in Asia in the 21st Century: Present Situation and Future Research and development Needs (eds. Howeler, R.H. and Tan, S.L.) Proc. 6th Regional Workshop, held in Ho Chi Minh City, Vietnam, February 21-25, 2000, pp. 553-563.
Iyayi, E.A. and Losel, D.M. (2001). *J. Food Technol. Afr.* **6:** 116-118.
Jiang, L., Rozelle, S. and Huang, J. (1993). Unpublished Report of CIP Sub-project 6211, pp. 63.
Jyothi, A.N. Wilson, B. Moorthy, S.N. and George, M. (2005). *J. Sci. Food Agric.* **85:** 1558-1563.
Jyothi, A.N., Sasikiran, K., Nambisan and Balagopalan, C. (2005). *Process Biochem.* **40:** 3576-3579.
Kayouli, C. and Lee, S. (1998). Silage from by-products for small-holders. http://www.fao.org/DOCREP/005/X8486E/x8486e01.htm
Kimaryo, V.M., Massawe, G.A., Olasupo, N.A. and Holzapfel, W.H. (2000). *Int. J. Food Microbiol.* **56:** 179-180.
Kofi, W., Amoa-Awua, A., Appok, F.E. and Jacobsen, M. (1996). *Int. J. Food Microbiol.* **96:** 87-98.
Kolicheski, M.B., Soccol, C.R., Marin, B., Medeiros, E. and Raimbault, M. (1995). *In:* Advances in Solid State Fermentation (Eds. Roussos, S., Lonsane, B.K., Raimbault, M. and Viniegra-Gonzalez, G.), Kluwer Academic Publishers, Dordrecht, The Netherlands, pp. 447-460.
Krishna, C. (2005). *Crit. Rev. Biotechnol.* **25:** 1-30.
Lien, L.V., Sansoucy, R. and Thien, N. (1994). *In:* Proceeding of National Seminar- Workshop on Sustainable Livestock production on Local Food Resources (Eds. Preston, T.R., Ogle, B., Ly, L.V. and Hieu, L.T.) Ho Chi Minh City, Vietnam, November 22-27, 1993, pp. 59-62.
Lu, M., Brooks, J.D. and Maddox, I.S. (1997). *Enz. Microb. Technol.* **21:** 392-397.
Lu, M., Maddox, I.S. and Brooks, J.D. (1995). *Bioresource Technol.* **54:** 235-239.
Matsui, T.N., Fukui, T., Sugita, K. and Matsumoto, K. (2003). *J. Agric. Food Chem.* **51:** 2539-2543.
Matthews, P.J. (2000). *In:* Twelfth Symposium of International Society of Tropical Root Crops (ISTRC), (Eds. Nakatani M. and Komaki, K.), 10-16 Sept., 2000, Tsukuba, Japan, pp. 484-511.
McFeeters, R.F. (2004). *J. Food Sci.* **69(1):** 35-37.
Medeiros, B.P.A. (1998). *Master's thesis*, Federal University of Parana, Curituba, Brazil.
Mohapatra, S., Panda, S.H., Sahoo, A.K., Sivakumar, P.S. and Ray, R.C. (2006). *Int. J. Food Sci. Technol.* **41:** *in press.*
Moorthy, S.N. and George, M. (1998). *Crit. Rev. Food Sci. Nutr.* **38:** 73-121.
Nanda, S.K. and George, M. (1996). *J. Food Eng.* **29:** 129-137.
Nar Anong, N., Poocharoean, D., Nuanphan, N. and Duangduan, P. (2001). *Kasetsart J. Natural Sci.* **19:** 1111-1122.
Nassar, M. and Marques, A. (2006). *J. Food Agric. Environ.* **4:** 187-188.
Nwankwo, D., Anadu, E. and Usoro, R. (1989). *MIRCEN J. Appl. Microbiol. Biotechnol.* **5:** 169-179.
Ofuya, C.O. and Nwajiuba, C.J. (1996). *J. Appl. Bacteriol.* **68:** 171-177.
Okolo, B.N., Ezeogu, L.I. and Mba, C.N. (1995). *J . Sci. Food Agric.* **69:** 109-115.
Onifade, A.K., Atum, H.N. and Adebolu, T.T. (2004). *Global J. Pure Appl. Sci.* **10:** 31-36.
Oyewole, O.B. (1990). *J. Appl. Biotechnol.* **68:** 49-54.
Oyewole, O.B. (2001). *Int. J. Food Microbiol.* **65(3):** 213-222.
Padmaja, G. and Balagopalan, C. (1991). *Trop. Sci.* **31:** 257-264.
Padmaja, G., George, M. and Moorthy, S.N. (1993). *J. Sci. Food Agri.* **63:** 473-476.
Padmaja, G., George, M. and Balagopalan, C. (1995). *In:* The Cassava Biotechnology Network : Proc. of the Second International Scientific Meeting, Bogor, Indonesia, 22-26 August, 1994, Cali, Colombia, CIAT, 1995 (Working document No.150) pp. 784-794.

Padmaja, G., George, M. and Moorthy, S.N. (1994). *J. Agric. Food Chem.* **42:** 766-770.

Panda, S.H., Parmanik, M., Sharma, P., Panda, S. and Ray, R.C. (2005). *In:* Microbes in Our Lives (eds. Mohanty, R.C. and Chand, P.K.), Department of Botany and Microbiology, Utkal University, Bhubaneswar, India, pp. 47-54.

Panda, S.H., Naskar, S.K. and Ray, R.C. (2006a). *J. Food Agric. Environ.* **4:** 124-127.

Panda, S.H., Paramanick, M. and Ray, R.C. (2007). *J. Food Process. Preserv.* **31(1):** *in press.*

Pandey, A. and Soccol, C.R. (2003). *J. Sci. Ind. Res.* **59:** 12-22.

Pandey, A., Soccol, C.R. and Mitchell, D. (2000a). *Process biochem.* **35:** 1153-1169.

Pandey, A., Soccol, C.R., Nigam, P., Soccol, V.T., Vandenberge, L.P.S. and Moham, R. (2000b). *Bioresource Technol.* **74(1):** 81-87.

Pêrez-Guerra, N., Torrado-Agrasar, A., López-Macias, C. and Pastrana, L. (2001). *Electronic J. Environ. Agric. Food Chem.* **1:** 1-20.

Raimbault, M. (1998). *Electron. J. Biotechnol.* **1:** 1-22.

Ray, L.G., Mukherjee, G. and Majumdar, S.K. (1991). *Indian J. Exptl. Biol.* **29:** 681-682.

Ray, R.C. (1993). *J. Environ. Biol.* **14(1):** 21-25.

Ray, R.C. (2004). *J. Environ. Biol.* **25(4):** 489-495.

Ray, R.C. (2006). *Food Rev. Int.* **22:** *in press.*

Ray, R.C., Naskar, S.K. and Sivakumar, P.S. (2005). *Technical Bulletin Series: **39***, Central Tuber Crops Research Institute, Thiruvanathapuram, India, 24 pp.

Ray, R.C. and Ravi, S. (2005). *Crit. Rev. Food Sci. Nutr.* **35:** 623-644.

Ray, R.C., Sahoo, A., Asano, K. and Tomita, F. (2006). *In:* Microbial Biotechnology in Agriculture and Aquaculture, Volume 2 (Ed. Ray, R.C.), Science Publishers, New Hampshire, New Hampshire, USA, pp. 511-552.

Ray, R.C., Shetty, K. and Ward, O.P. (2007). *In:* Microbial Biotechnology in Horticulture Volume 2 (Eds. Ray, R.C. and Ward, O.P.), Science Publishers, New Hampshire, New Hampshire, USA, *in press.*

Ray R.C. and Ward O.P. (2006). *In:* Microbial Biotechnology in Horticulture, Volume 1 (Eds. Ray, R.C. and Ward O.P.), Science Publishers, Enfield, New Hampshire, USA, pp. 345-396.

Ray, R.C., Kar, S. and Chaudhury, P. (2004). *In:* National Seminar on Root and Tuber Crops for Nutrition, Food Security and Sustainable Environment, 29-31 October, 2004, Regional Centre of Central Tuber Crops Research Institute, Bhubaneswar, India, Abstract, pp. 61-62.

Ray, R.C. and Moorthy, S.N. (2007). *J. Sci. Ind. Res.* **67:** *in press.*

Sahoo, A.K., Swain, M.R., Pattnaik, M.S., Ray, R.C. and Edison, S. (2005). *In:* Microbes in Our Lives (Eds. Mohanty, R.C. and Chand, P.K.), Department of Botany and Microbiology, Utkal University, Bhubaneswar, India, pp. 72-83.

Sani, A., Awe, F.A. and Akinyanju, J.A. (1992). *J. Ind. Microbiol.* **10:** 55-59.

Scott, G.H., Best, R., Rosegrant, M. and Bokanga, M. (2000). *A Co-Publication of the CIP/CIAT/IFPRI/IITA/IPGRI*, Printed in Lima, Peru, International Potato Centre.

Shankaranand, V.S. and Lonsane, B.K. (1994). *In:* Solid State fermentation (Ed. Pandey, A.), Wiley Eastern, New Delhi, India, pp. 149-152.

Shivashankara, K.S., Isobe, S., Al-Haq, M.J., Takenaka, M. and Shiina, T. (2004). *J. Agric. Food Chem.* **52(5):** 1281-1286.

Soccol, C.R., Iloki, I., Martin, B. and Raimbault, M. (1994). *J. Food Sci. Technol.* **31(4):** 320-323.

Soccol, C.R. and Vandenberghe, L.P.S. (2003). *Biochem. Eng. J.* **13:** 205-218.

Sosa, A.V. (2001). *Biotechnol. Prog.* **17:** 1079-1083.

Sriroth, K., Chollakup, R., Choineeranat, S., Piyachomkwan, K. and Oates, C.G. (2000). *Bioresource Technol.* **71(1):** 63-70.

Sriroth, K., Chollakup, R., Choineeranat, S., Piyachomkwan, K. and Oates, C.G. (2001). *In :* Casava's Potential in Asia in the 21st Century: Present Situation and Future Research and development Needs (Eds. Howeler, R.H. and Tan, S.L.), Proc. 6th Regional Workshop, held in Ho Chi Minh City, Vietnam, February 21-25, 2000, pp. 538-552.

Suh, H.J., Kim, J.M. and Choi, Y.M. (2003). *Int. J. Food Sci. Technol.* **38:** 145-151.

Suryanarayan, S. (2003). *Biochem. Eng. J.* (2003). **13:** 189-195.

Swain, M.R., Kar, S., Sahoo, A.K. and Ray, R.C. (2006). *Microbiol. Res: in press.*

Takenaka, M., Nanayama, K., Isobe, S. and Murata, M. (2006). *Biosci. Biotechnol. Biochem.* **70(1):** 172-177.

Tengerdy, R.P. and Szakacs, G. (2003). *Biochem. Eng. J.* **13:** 169-179.

Timmins, W.H., Marter, A.D., Westby, A. and Rickard, J.E. (1991). *In:* Product Development for Root and Tuber Crops. Volume I-*Asia* (Eds. Scott, G.G., Wiersema S. and Ferguson, P.I.). International Potato Centre, Lima, Peru, pp. 217-227.

Timotius, K.H. (2005). *World J. Microbiol. Biotechnol.* **21:** 615-617.

Tinh, N.T., Thuy, T.P.T., Thach, P.N.T. and Peters, D. (2000). *Information of Science and Technologies on Animal Production. No 4/2000. (In Vietnamese language).* Hanoi

Treche, S. and Massamba, J. (1995). *In:* Cassava Food Processing (Eds. Agbor, E., Brauman, A., Griffon, D. and Treche, S.), ORSTOM, Paris, pp. 133-150.

Tsuchiya, K., Nagasjhiam, T., Yamamoto, Y., Gomi, K., Kitamoto, K. and Umagai, C. (1994). *Biosci. Biotechnol. Biochem.* **58:** 895-899.

Tung, T.Q., Miyata, N. and Iwahori, K. (2004). *J. Biosci. Bioeng.* **97:** 329-335.

Vandenberghe, L.P.S., Soccol, C.R., Carta, F.S., Lebeault, J.-M., Milcent, P.F. and Machado, L. (1998). *Paper presented in Int. Congress Biotech' 1998*, Portugal.

Viniegra-González, G., Favela-Torres, E., Noe Aguilar, C., Rómero-Gomez, S.D.J., Díaz- Godínez, G. and Augur, C. (2003). *Biochem. Eng. J.* **13:** 157-167.

Wang Xioadong, G.X. and Rakshit, S.K. (2000). *In:* Casava's Potential in Asia in the 21st Century: Present Situation and Future Research and development Needs (Eds. Howeler, R.H. and Tan, S.L.), Proc. 6th Regional Workshop, held in Ho Chi Minh City, Vietnam, February 21-25, 2000, pp.198-201.

Ward, O.P. (1989). *Fermentation Biotechnology.* Open University Press. Biotechnology Series, Milton Keynes, U.K.

Ward, O.P. and Singh, A. (2005). *In:* Microbial Biotechnology in Agriculture and Aquaculture, Volume I (Ed. Ray, R.C.), Science Publishers, New Hampshire, USA, pp. 449-479.

Ward, O.P., Singh, A. and Ray, R.C. (2006). *In:* Microbial Biotechnology in Horticulture, Volume 1 (Eds. Ray, R.C. and Ward, O.P.), Science Publishers, Enfield, New Hampshire, USA, pp. 517-557.

Weinberg, Z.G. and Ashbell, G. (2003). *Biochem. Eng. J.* **13:** 181-188.

Westby, A. and Choo, B.K. (1994). *Acta Hort.* **375:** 209-215.

Woiciechowski, A.L., Soccol, C.R., Rocha, S.N. and Pandey, A. (2004). *Appl. Biochem. Biotechnol.* **118:** 305-312.

Woolfe, J.A. (1992). *Sweet potato an untapped food resource.* Cambridge University Press, New York, 643 pp.

Yamakawa, O. (1997). *In:* Proceedings of the International Workshop on Sweet potato Production System toward the 21st Century, Dec 9-10, 1997, Miyakonojo, Miyazaki, Japan, pp. 1-8.

Yamakawa, O. (2000). *In:* Potential of Root Crops for Food and Industrial Resources (Eds. Nakatani, M. and Komaki, K.), Twelfth Symposium of International Society of Tropical Root Crops (ISTRC), 10-16 Sept., 2000, Tsukuba, Japan, pp. 8-13.

Yoshi, H. and Garcia, V.V. (1984). *In: Tropical Root Crops: Post harvest Physiology and Processing* (Eds. Uritani, I. and Reyes, D.), Japan Scientific Society Press, Tokyo, 205.

Zheng, Z. and Shetty, K. (1998a). *J. Agric. Food Chem.* **46:** 783-787.

Zheng, Z. and Shetty, K. (1998b). *Process Biochem.* **33:** 323-329.

Zheng, Z. and Shetty, K. (1999). *In:* Encyclopaedia of Food Science and Technology, 2nd edition (Ed. Francis, F.J.), Wiley Publishers, New York, USA, pp. 2165-2174.

Zheng, Z. and Shetty, K. (2000a). *J. Agric. Food Chem.* **48:** 895-900.

Zuleta, A., Rio, M.E., Sambucetti, M.E., Mora, M., de Fabrizio, S.V. and Parada, J.L. (2004). *In:* Sixth International Scientific Meeting of the Cassava Biotechnological Network, 8-14 March, 2004, CIAT, Cali, Colombia, Abstract, p. 65.

Chapter 19

Pharma-active Compounds of Microbial Origin and their Diversity

RAGINI GOTHALWAL AND PRATIMA GUPTA

School of Biotechnology, Guru Ghasidas University, Bilaspur–495 009 (CG)
E-mail: ragini_gothalwal@yahoo.com

Pharmacologically active microbial products are generally classified as those useful for treating human diseases and disorders. The introduction of penicillin in the 1940s, which opened up the era of antibiotics, has been recognized as one of the greatest advances in the therapeutics. Since this dramatic medical breakthrough, the exciting prospects of microbial metabolites as therapeutic agents are of active areas of scientific interest. Pharmaceutical industries too depend on the discovery of new therapeutic metabolites from microbial origin where the drug discovery process relies on successful exploitation of microbial diversity to find new compounds by selective isolation technique, high throuhput screening and assay methods.

The present article is an effort to review the pharmacologically active metabolites from microbial origin for various infectious and non-infectious diseases. The review covers the microbial products used as antibiotics against bacterial, fungal, viral, protozoan diseases and also about antitumor antibiotics. Microbial metabolites have also provided enzyme inhibitors with therapeutic activity in treatment of various metabolic disorders such as for hypercholestrolemia, hyperglycemia, obesity, etc. Vasoactive substances have also been isolated as drugs effective against circulatory blood diseases. Discovery of metabolites acting as immunosuppressors have widespread use in prevention of organ transplant rejection. Screening and discovery of successful class of new compounds which are of paramount importance such as against resistant pathogens and for some diseases like cancer and neural diseases are still continued.

INTRODUCTION

Microorganisms are important natural sources of chemical diversity for drug discovery programs. The strength of these natural products is their structural diversity and complexity and their high incidence of biological activity. A statistical comparison of structural features of natural product and synthetic compound found that natural products are sterically more complex and around 40% of natural products have no synthetic equivalents (Henkel *et al.*, 1999). The selectivity of reactive microbial metabolites, which bind irreversibly to their targets with high affinity, can provide useful leads of finding. Identifying bioactive compounds from microorganisms and establishing their health effects are active areas of scientific interest. There are exciting prospects that select such compounds to reduce the risk of many diseases.

Pharmacologically active natural products are generally classified as those useful for treating human disease and disorders including hyperlipidemia, cancer, immunoregulatory disorders, inflammation, neurological and metabolic disease. Approaches to the discovery of pharmacologically active natural products and the developments status of leads arising from various discovery programs have been reviewed over the past decade or so (Nisbet and Portes 1989; Franco and Coutinho 1991; Shu, 1998; Hill *et al.*, 1998; Newman *et al.*, 2000). Screening and discovery of new pharma active compounds are of paramount importance to fight against resistant pathogens, newly emergent disease and some of the diseases like cancer, neural disorders, etc. for which the search is going on. Microbial products have a rich heritage in this area and they are like miniature chemical factories as they have the capacity to convert a variety of new materials used as substrate into a series of value-added products. There is the greatest possibility of finding many more substances with interesting bioactivities (such as immunomodulative, vasoactive, antiulcer and hypotensive) among microbial metabolites.

A very large number of biologically active compounds from microorganisms have been discovered so far. The search for new pharma active compounds of microbial origin continues by application of new screening methods, advancement of experimental techniques, and discovery of new sources. Dr. H. Umezawa and his coworkers in 1972, initiated research and development of microbial enzyme inhibitors as therapeutic agents for the control of abnormalities of homeostasis. This approach led his and other research groups to discover a number of interesting pharmacologically active microbial metabolites such as bestatin (aminopeptidase inhibitor, immunostimulant, FK-506 (immunosuppressant), mutastein (glycosyltransferase inhibitor) and pravastatin (HMG-CoA reductase inhibitor). The success and effective exploitation of microbes to a large extent rely on the diversity of microorganisms in the environment. Particularly pharmaceutical industry recognizes that diversity is essential and potentially important in the development of new drugs for health and disease control. The notable great success in search of microbial products as therapeutic agents include the indole alkaloids from ergot (*Claviceps purpurea*) and their derivatives, agonist and antagonists of adregenic, dopaminergic and seratoninergic receptors with utility in the treatment of migraine, Parkinson's disease amd postparum hemorrhage (Shu, 1998), antitumor agents such as the anthracyclins, daunomycin and doxorubicin, bleomycins and mitomycins from various *Streptomyces* (Newmann *et al.*, 2000; Cragg *et al.*, 2000) from fungi (Cardenas *et al.*, 1998, Duncan *et al.*, 2003) from cyanobacteria (Reddy *et al.*, 2000) and telomerase inhibitor from *Streptomyces annulatus* (Shin-ya *et al.*, 2001). Cyclosporin A from *Tolypocladium inflatum*, the first of new class of immunosuppressants that is widely used for prevention of organ transplant rejection (Borel *et al.*, 1976; Mann, 2001) and the hydroxymethylglutaryl CoA reductase inhibitors, mevinolin (statins) from *Aspergillus terreus*, the first in very successful class of drugs lowering the level of cholesterol and low density lipoprotein with hypercholesterolemia (Alberts *et al.*, 1980; Endo and Hasumi, 1993). Novel squalene synthase inhibitors of microbial origin continue to report such as bisabosquals from

Stachybotrys sp. (Minagawa *et al.*, 2001), CJ-15,183 from *Aspergillus aculateus* (Watanabe *et al.*, 2001).

The potential for drug discovery is being further enhanced by recent advances in procedures for microbial cultivation, nucleic acid extraction from environmental samples, resulting in the identification of novel microbes that provide a vast untapped reservoir of genetic and metabolic diversity.

ANTIBACTERIAL AGENTS

Penicillin, discovered by Fleming in 1928, was rediscovered by Chain *et al.*, (1940) as a chemotherapeutic agent. After this first breakthrough of discovery of antibiotic, many clinically useful antibacterial antibiotics such as streptomycin, chloramphenicol, chlorotetrayclin (aureomycin), neomycin, oxytetracyclin, erythromycin, etc. were discovered and most bacterial infection seemed to be conquered. The clinically useful antibacterial antibiotics can be classified into groups according to their mechanism and site of action:

1. Folic acid synthesis inhibitors (Sulphonamides and trimethoprim)
2. Cell wall inhibitors (Penicillins, cephalosporins, vancomycin, etc.)
3. Cell membrane inhibitors (Polymyxin)
4. Protein synthesis inhibitors (aminoglycosides and spectinomycin, tetracyclin, erythromycin, chloramphenicol, etc.)
5. Nucleic acid inhibitors (nalidixic acid and rifampicin)

Bacterial cell wall received great emphasis as the target of antibiotic action, and many inhibitors have been identified from microorganisms. Various types of antibacterial antibiotics with their source are listed in Table 1.

A large number of antibacterial compounds have been discovered now, but the search is still going on for the discovery of novel microorganism for new metabolites as antibacterial agent. Maskey *et al.* (2004) isolated antibacterial trioxacarcins from ethyl acetate extract of the *Streptomyces sp.* isolate B8652. In the year 2006, Maskey and his group again isolated Kettapeptin, a new hexadepsipeptide antibiotic of the azinothricin type from *Streptomyces sp.* isolate GW99/1572 against Gram-positive bacteria. Recent challenges in antibacterial discovery is to search for new useful antibiotic against antibiotic and multidrug resistant strains of bacteria after the spread of antibiotic therapy such as emergence of multidrug resistant strains of *Mycobacterium tuberculosis,* methicillin resistant *Staphylococcus aureus* (MRSA), vancomycin-resistant enterococci (VRE) and ß-lactam-resistant *Streptococcus pneumoniae* among the most serious problems. Actinomycetes still appear to remain the best potential source of antibacterial compounds (Watve *et al.*, 2001; Lazzarini *et al.*, 2001). Litosterol is a C-19 hydroxy steroid isolated from *Litophyton viridis* and inhibits 90% growth of *M. tuberculosis* at a concentration of 3.13μg/ml (Donia and Hamann, 2003). Lall *et al.* (2006) reported the antituberculosis activity from *Helichrysum melanacme.* Pyocyanine, a metabolite from *Pseudomonas aeruginosa* exhibited antistaphylococcal activity even against methicillin resistant *Staphylococcus aureus* (MRSA), and coagulation negative *Staphylococcus cocci* (CONS) (Arunkumar *et al.*, 1997). *Actinopolyspora sp.* AH1 showed good antibacterial activity against *Staphylococcus aureus*, *S. epidermis* and *Bacillus subtilis* (Kokare *et al.*, 2003). Figure 1 (a) and (b) shows the structure of antibacterial agents.

Table 1: Major groups of antibacterial and antifungal antibiotics, their source, and the year their discovery was publicly disclosed in the literature or the patent literature

Compound	Activity	Scaffold	Original microorganism producing the scaffold	Year of discovery
Penicillins	AB	ß-Lactam (penam)	*Penicillium notatum*	1928
Tyrothricin	AB	Peptides mixture	*Bacillus sp.*	1936
Gramicidin	AB	Cyclic peptide	*Bacillus brevis*	1941
Strepromycin	AB	Aminoglycoside	*Streptomyces griseus*	1944
Nisin	AB	Lantibiotic	*Lactococcus lactis*	1944
Bacitracin	AB	Peptide	*Bacillus licheniformis*	1945
Chloramphenicol	AB	Phenylpropanoid	*Streptomyces venezuelae*	1948
Cephalosporins	AB	ß-Lactam (ceph-3-em)	*Acremonium chrysogenum*	1948
Polymyxin B	AB	Cyclic peptide	*Bacillus polymyxa*	1948
Chlorotetracycline	AB	Tetracycline	*Streptomyces aureofaciens*	1948
Neomycin	AB	Aminoglycoside	*Streptomyces sp.*	1949
Oxytetracycline	AB	Tetracycline	*Streptomyces rimosus*	1950
Nystatin	AF	Polyene	*Streptomyces noursei*	1950
Erythromycin	AB	14-Membered macrolide	*Saccharopolyspora erythrea*	1953
Synercid	AB	Streprogramins	*Streptomyces pristinaespiralis*	1953
Tetracycline	AB	Tetracycline	*Streptomyces viridofaciens*	1953
Vancomycin	AB	Glycopeptide	*Streptomyces orientalis*	1955
Novobiocin	AB	Coumarin	*Streptomyces spheroides*	1955
Cycloserine	AB	Amino acid analog	*Streptomyces orchidaceus*	1955
Amphorericin B	AF	Polyene	*Streptomyces nodosus*	1955
Rifampin	AB	Rifamycin	*Streptomyces mediterranei*	1959
Kanamycin	AB	Aminocyclitol	*Streptomyces kanamycetus*	1960
Spectinomycin	AB	Aminocyclirol	*Streptomyces spectabilis*	1961
Lincomycin	AB	Lincomycin	*Streptomyces lincolnensis*	1962
Fusidic acid	AB	Sterol	*Fusidium coccineum*	1962
Everninomycin	AB	Orrhosomyen	*Micromonospora carbonacea*	1964
Fosfomycin	AB	Phosphonic acid	*Streptomyces sp.*	1969
Cefoxitin	AB	ß-Lactam (cephamycin)	*Nocardia lactamdurans*	1971
Mupirocin	AB	Pseudomonic acid	*Pseudomonas fluorescens*	1971
Cuspofungin	AF	Echinocandin	*Glarea lozoyensis*	1974
Clavulanic acid	AB	ß-Lactam (clavam)	*Streptomyces clavuligerus*	1975
Imipenem	AB	ß-Lactam (carbapenem)	*Streptomyces cattleya*	1976
Aztreonam	AB	ß-Lactam (monobacram)	*Chromobacterium violaceum*	1981

Source : Bull, A.T. (2004). Antimicrobials in *Microbial Diversity and Bioprospecting.*

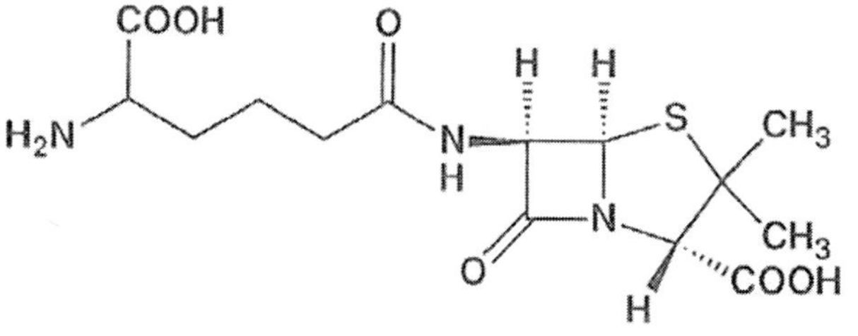

(a)

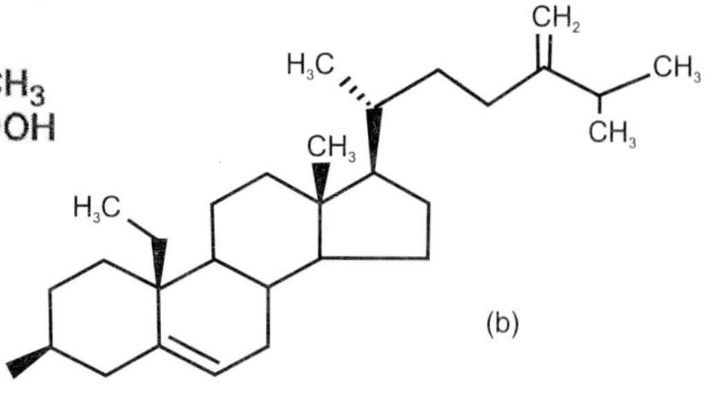

(b)

Fig.1. Antibacterial agent (a) Penicillin (b) Litosterol.

ANTIFUNGAL SUBSTANCES

In 1958, the antidermatophytic effect of griseofulvin was discovered by Gentile. Griseofulvin was followed by other antifungal agents of microbial origin such as polyene macrolide antibiotics (nystatin, amphoteracin and pyrolonitrin) (Arima *et al.*, 1964). Some major groups of antifungal antibiotics of clinical use have been given in Table 1. The polyene antibiotics are potent antifungal agents with broad antifungal spectra. Among these amphoteracin B is commonly used in systemic fungal diseases. The other narrow spectrum antifungal agents are cycloheximides, active against *Saccharomyces*, actimycin active against *Saccharomyces*, *Candida* and highly active against *Pyricularia*, polyoxins are active against *Pyricularia*, nikkomycins against *Candida*, *Mucor* and *Pyricularia* (Omura, 1992). A very high proportions of compounds isolated from myxobacteria have antifungal activity (Reichenbach, 2001), including sorangicin, epolithone, ambruticin, haliangicin, myxin, stigmatellin, myxalamide, cystahiazole, glidobactins and ratzadon. Although marine sources are a prolific source of new antifungal compounds (Mohapatra *et al.*, 2002), many marine compounds with antifungal may be non-specific cytotoxins. The newest antifungal drug in the market is Cancidas (caspofungin), a semi-synthetic echinocandin that inhibits the synthesis of ß-1,3–glucans, major component of fungal cell wall (Powels *et al.*, 1998; Tkacz and DiDomenico, 2001), isolated from filamentous fungi.

Antifungal activities against *Aspergillus niger*, *A. fumigatus A. flavus*, *Fusarium oxysporum*, *Penicillium* sp. and *Trichoderma sp.* is exhibited by *Actinopolyspora sp.* AH 1 (Kokare *et al.*, 2003). A new 5-alkelenyl resorcinol sch 725681 is identified and isolated from the culture of an *Aspergillus* sp. showing the inhibitory activity against *S. cerevisae* and *Candida albicans* (Yang *et al.*, 2006). Novel antifungal lipopeptides FR 227673 and FR 190293 are isolated from the fermentation broths of fungal strain *Chalara sp.* no. 22210 and *Tolypocladium parasitium* no. 166216 respectively against *A. fumigatus* and *Candida albicans* (Kanasaki *et al.*, 2006). Gilled mushrooms of order Agaricales have more species with antifungal activities (Suay *et al.*, 2000). Recently progress has been made in the area of antifungal chemotherapy, which is expected to facilitate discovery of newer and safer antifungal antibiotics. Fig. 2 (a) and (b) show the structure of antifungal agents.

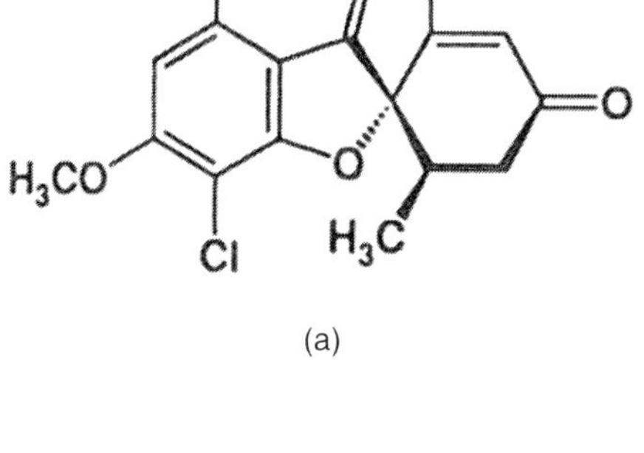

(a)

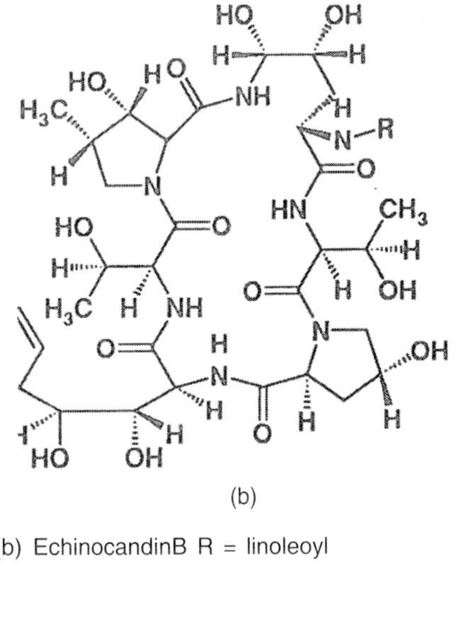

(b)

Fig. 2. Antifungal agent (a) Griseofulvin (b) EchinocandinB R = linoleoyl

ANTIVIRAL SUBSTANCES

The pace of the discovery of clinically useful antiviral drugs has become markedly accelerated in contrast to the slow progress at the beginning of search. Today, the major disease targets of antiviral chemotherapy are influenza, rhinovirus, cold, herpes and human immunodeficiency virus (HIV), the causative agent of AIDS. HIV inhibitors called sulfonic acid, containing glycolipid, isolated from extract of blue green algae, *Lyngbya lagerheimeii* and *Phorimidium tenue,* has been reported by Gustafson *et al.* (1989). Cyanoviridin-N isolated from blue cyanobacterium inactivates the strains of HIV virus and inhibits cell to cell and virus to cell fusion (Yang et al., 1997). Calcium spirulan (Ca-Sp), a sulfated polysaccharide, is an antiviral agent, which selectively inhibits the entry of enveloped virus (herpes simplex, human cytomegalo virus, measles virus) into the cell (Hayashi *et al.*, 1996; Hayashi and Hayashi, 1996; Ayhunie *et al.*, 1998). Sato *et al.* (2006) reported the modified quinolone antibiotic to produce novel HIV integrase inhibitor, JTK-303 as antiviral substances.

Arginine deiminase of *Mycoplasma argini* suppresses the growth of HIV type I replication and this enzyme is known to be a chemotherapeutic agent against arginine requiring tumors and potential application in antiviral therapy (Kubo *et al.*, 2006). Various semisynthetic derivative of glycopeptide antibiotics including vancomycin, eremomycin, teihoplanin, ristocetin A and DA-40296 showed inhibitory activity against feline infectious peritonitis virus (FIPV) and human (SARS) coronavirus (Balazarini *et al.*, 2006). Acetoin extract containing the active constituents 2, 4',6'- trihydroxy-3'prenyl chalcone and 4',6',5'-trihydroxy-6'-6'- dimethyldihydropyrano [2",3"-2',3'] chalcone of *Helichrysum melanacme* are active against Human influenza virus (Lall *et al.*, 2006).

ANTICANCER AGENTS

Antitumor antibiotics have played an important role in the development of anticancer agents. These agents can only identify and differentiate tumor cells with normal cells on the basis of rapidly dividing nature of tumor cells. Antitumor antibiotics isolated from various *Streptomyces spp.* - namely, the anthracyclins, daunomycin, doxorubicin and their semisynthetic derivatives, the glycopeptide bleomycins, the peptolide actinomycins, the mitosanes such as mitomycin have all proved to be clinically important cancer therapeutic agents (Cragg *et al.*, 2000). Table 2 summarizes some of the antitumor antibiotics from microbial sources. Cell and mechanism based antitumor discovery approaches continue to be productive in terms of new compounds identified. Epothilones A and B from myxobateria (Gerth *et al.*, 1994) and its lactam anolog, BMS-24755 (Lee *et al.,* 2001) were found to be and developed as an anticancer agent. The DNA cleaving capability of the calicheamicin enediyene-containing antitumor antibiotic from *Micromonospora echinospora* (Lee *et al.,* 1993) and angiogenesis inhibitory properties of fumagilin from *Aspergillus fumigatus* and its analog (Ingber *et al.*, 1990) are among the potent antitumor agents.

Novel antitumor substances possessing more specific activity or new type of structures are always being looked for. Discovery of compounds with anticancer agent potential to act via screening programs is based on cytotoxicity and target-direct. The National Cancer Institute operates a panel of 60 human cancer cell lines for this purpose (Cragg *et al.*, 2000). The active component,

Table 2: Clinical and lead experimental microorganism-derived anticancer drugs.

Anticancer drug	Type	Comments
Adacinomycin A	Anthracycline	Produced by *Streptomyces galiaells*
Bleomycin	Glycopeptide	Produced by *Streptomyces verticillus*
Carminomycin	Anthracycline	Produced by *Actinomadllra carminata*
Dactinomycin	Acylpeptidolactone	Produced by *Streptomyces parvbulus*
Daunorubicin	Anthracycline	Produced by *Streptomyces pelucetius*
Doxorubicin	Anthracycline	Produced by *S. pellcetills* subsp. *caesilus*
Epothilones B, D.	Substituted macrolides	Produced by *Sorangillm celllllosllm;* in clinical trials
Irinotecan	Camptothecin	Approved for metastatic colon and rectal cancers
Mitomycin C	Cytotoxic drug	Produced by *Streptomyces caespitosus*
Pentosta tin	Nucleoside analog	Produced by *Streptomyces antibioticus*
Plicamycin.	Polyketide	Mithramycin; produced by *Streptomyces argilaceus*

Source : Bull, A.T.(2004) Antimicrobials in *Microbial Diversity and Bioprospecting.*

XR842 from fermentation broth of *Streptomyces cyaneus* has been reported with potent cytotoxic properties (Menendez *et al.*, 2002). Targets of interest are several enzyme inhibitors such as inhibitors of protein kinase C (PKC), farnesyl protein transferase (FTPase), proteasome, etc. An early target of interest was PKC. Overactivation of this key signaling enzyme is postulated to be important in cancer. Staurosporine from *Saccharothrix sp.* (Omura *et al.*, 1995a) UCN-01, 7-hydoxystaurosporine from *Streptomyces sp.* (Newman *et al.*, 2000) and balanol from *Verticillium balantoides* are identified as potent inhibitors of PKC in screening programs from microbial metabolites. FTPase inhibitor are able to block the ras dependent tumorigenesis, and the ras gene with point mutations are associated with 30% of all human tumors. Erbstatin from *Streptomyces sp.* (Umezawa *et al.*, 1986) and clavaric acid from *Clavariciadelphus truncatus* (Lingham *et al.*, 1998) are therapeutically effective FTPase inhibitors. Histone deacetylase inhibitor trichostatin A from actinomycete extract, caused the arrest at both G1 and G2 phases (Ueki *et al.*, 2001), and a novel fungal metabolite, lucilactaene, caused arrest at G1 phase of cell cycle progression of tumorous growth (Kakeya *et al.*, 2001)

Dysfunction of the *p53* gene is often associated with the development of cancers, and over 50% of all human tumors contained a mutated form. Compounds that disturb the interaction between MDM2 (protein that prevent the transcription *p53*) and *p53* could potentially treat tumor expressing high levels of MDM2 such as human sarcoma. Chlorofusin, a novel metabolite of the fungus *Microdochium caespitosum* is reversible MDM2 binder and inhibitor of this protein-protein interaction (Duncan, 2003). The *Streptomyces* sp. metabolite lactacystin (Omura *et al.*, 1991) was one of the first proteasome inhibitors to be identified. Two recently reported proteasome inhibitors are tyropeptins A and B, peptidyl aldehydes produced by *Kitasatosporia* sp. (Momose *et al.*, 2001) and the cyclic hexapeptide phepropeptins from a *Streptomyces sp.* (Sekizawa *et al.*, 2001).

Tumor cells acquire proliferative immortality by activating telomerase that maintains telomere length. Telomere inhibitors therefore represent a potentially useful chemotherapeutic agent. Telemostatin, identified from *Streptomyces annulatus*, is the most potent and selective telomerase inhibitor (Shin-ya *et al.*, 2001). Jose and Janardhanan (2002) reported the antioxidant activities of ethylacetate and methanol extract of *Plerotus florida* with remarkable capacity to inhibit the solid tumors. A new cytotoxic compound, streptokordin of methylpyridine class and four known

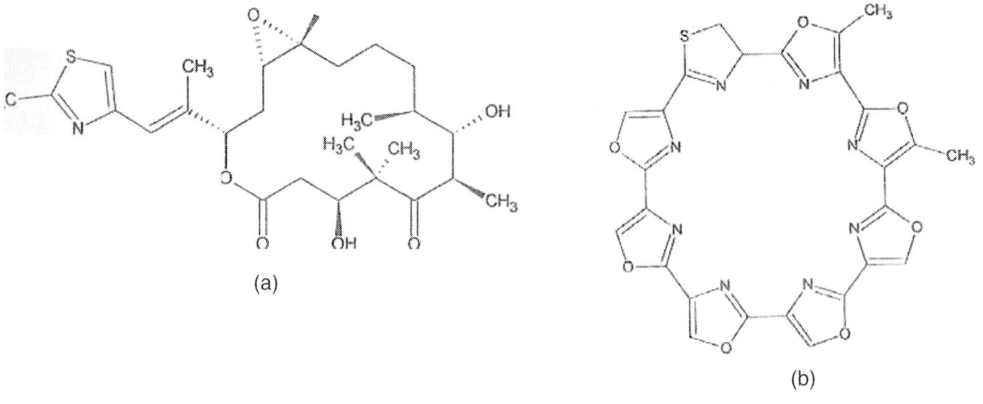

Fig. 3. Anticancer agent (a) Epothilone A (b) Telemostatin

compounds, nonactic acid, dilactone, trilactone and nonactin isolated from marine actinomycete exhibited significant cytotoxocity against human cancer cell lines (Jeong *et al.*, 2006). Figure 3 (a) and (b) shows the structure of anticancer agents.

ANTIPARASITIC AGENTS

In recent year, new types of parasitic diseases have appeared due to opportunistic infection. Antibiotics of practical use include the macrocyclic lactones avermectins (broad spectrum nematode insecticidal activities), tetracyclines (antiprotozoal activities against *Plasmodium*, *Entamoeba* and *Anaplama*), the macrolide spriromycin (antiprotozoal activity against *Cryptosporida* and *Toxoplama*), the polyene macrolide amphoteracin B (against *Neagleria* and *Leishmania*) and trichomycin (anti trichomonas), the aminoglycosides paramomycin (antiprotozoal and cestocidal activities against *Entamoeba*, *Giardia*, *Balantida*, *Leishmania* and *Taenia*) hygromycin B and destomycin (antinematodal acitivity against *Ascaris*), clindamycin (against Plasmodium, Babesia and Toxoplasma) and the polyether ionophores monensin and salinomycin (anticoccidal ativity) (Bosche *et al.,* 1986; Campbell and Rew 1986) Brown et al., (2005) reported the systemic treatment of new world cutaneus leishmaniasis caused by *Leishmania braziliencis* with liposomal amphoteracin B. Trioxacarcins present in ethyl acetate extract from *Streptomyces sp.* isolated B8652 showed higher antimalarial action (Maskey *et al.*, 2004).

Among them the avermectin complex is practically useful as an antihelminthic chemotherapeutic agent. Dihydroxy tetrahydrofuran from marine brown algae *Notheia anomala* exhibits the selective nematicidal activity against *Trichostrorgyrus* and also geodin A macrocyclic polyketide lactam tetramic aid (Donia and Hamann, 2003), Jasplakinolide—a potent antiparasitic agent against *Nippo-stronglylus braziliensis* from blue-green algae (Singh *et al.,* 2005). Fig. 6(b) shows the structure of antiparasitic agents.

IMMUNOMODULATOR SUBSTANCES

The Immunomodulating substances that attack specific target sites of cells are thought to be potent tool for the cellular and biochemical events of the immune responses and also provide drugs for immunotherapy. Many immunomodulators have also been discovered among the primary metabolites of microorganisms, such as cell wall components. These substances can be further categorized as immunopotentiators and immunosuppressors depending upon the activation and suppression of immune response.

Swainsonine has potential immunomodulator for the immunocompromised hosts isolated from *Swainsona sp.* (Colegate *et al.*, 1979). Some other immunopotentiators such as kifunensine (FR-900494) and FR 900483 have also been found in fermentation broth of *Kitasatosporia kifunense* sp. no. 9482 and *Nectria lucida* F-4490, respectively (Iwami *et al.*, 1987; Shibata *et al.*, 1988).

Immunosuppresants have achieved the widespread use in the prevention of organ transplant rejection. It may well prove to have wider utility against autoimmune disease including rheumatoid arthritis, systemic lupus erythematosus, and Crohn's disease (Mann, 2001). Cyclosporin A from *Tolypocladium inflatum*, FK 506 from *Streptomyces tsukubaensis* and rapamycin from *Streptomyces hygroscopicus* are the three metobolites that have had the most impact to date as immunosuppressant. Two other notable examples of compound of microbial origin that have reached clinical use are the fungal metabolites, mycophenolic acid, causing selective inhibition of DNA synthesis in lymphocytes as the drug mycophenolate mofetil (Sievers *et al.,* 1997), and synthetic derivative of 15–deoxyspergualin isolated from *Bacillus lateosporus* in prevention of organ transplant rejection (Mann, 2001).The myriocins, cyclophillin A– binding compounds and cytokine production inhibitors provide more recent examples of interesting microbial immunosuppressant. Myriocin A, a sphingosine like compound isolated from the fungus *Isaria sinclairii* is now known to be highly potent inhibitor of serine palmitoyl transferase involved in biosynthesis of sphingolipids of cellular proliferation and programmed cell death (Chen *et al.,* 1999). The sanglifehrins from a *Streptomyces* sp. (Sanglier *et al.,* 1999) and cymbimicins A and B from *Micromonospora* sp. (Fehr *et al.,* 1997) are cyclophilin binding compounds that interferes with T cell signaling pathways. There have been number of other recent reports of cytokine production inhibitors such as the pyridine divertive CJ 14,897 from *Marasmiellus sp.* (Ichikawa *et al.,* 2001a), a novel benzo fluoranthen 3-one XR774 from *Cladosporium cf. cladosporoides* 3-one XR774 from *Cladosporium cf cladosporioides* (Wriggly *et al.,* 2001) and series of diterpenes from *Oidiodendron griseum* (Ichikawa *et al.,* 2001b). The potent inhibitors of TNF-α production are 5-7 oxo zeanol from the fungus *Cochliobolus lunatus* (Rawlins *et al.,* 1999) and a closely related resorcylic acid lactone, L-783, 277 from a *Phoma sp.* (Zhao *et al.,* 1999) and a novel (6S)- 4,6 dimethyl dodeca 2E, 4E dienoylester of a *Phomopsis* sp. (Wrigley *et al.,* 1999). Figs. 4 (a) and (b) show the structure of immunosuppressive agents.

CARDIOVASCULAR AND METABOLIC DISORDERS

Many enzymes have important roles in the maintenance of homeostasis. Some inhibitors of enzymes in which the correlation with disease process has been established are very important as pharmacologically active drugs. One of the most successful class of drugs over the past decade has been the antihypercholesterolemic agent termed as statin. Statins are hydroxymethylglutaryl-CoA reductase (HMG-COA) inhibitors, the enzyme involved in the rate limiting step in human cholesterol biosynthesis. The first statin discovered was mevastatin from *Penicillium citrinum*

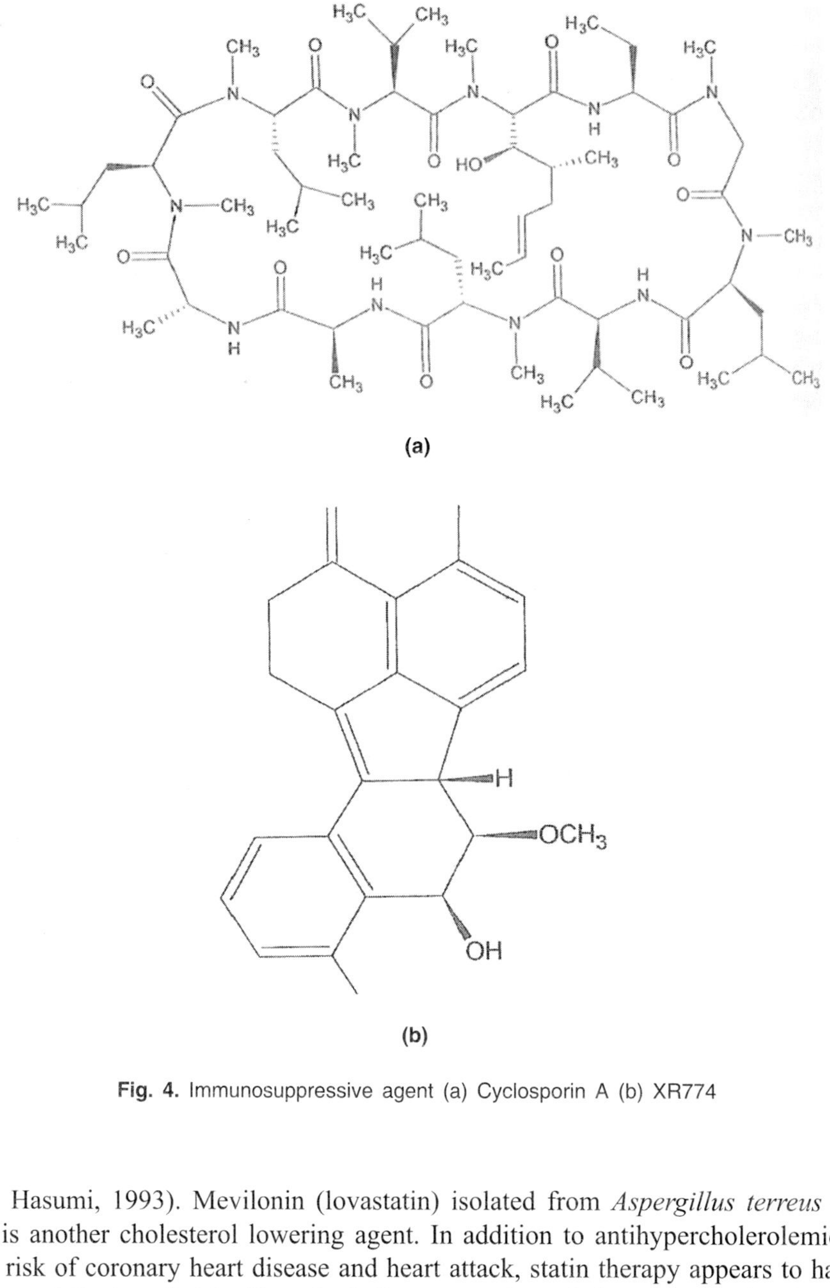

Fig. 4. Immunosuppressive agent (a) Cyclosporin A (b) XR774

(Endo and Hasumi, 1993). Mevilonin (lovastatin) isolated from *Aspergillus terreus* (Alberts *et al.,* 1980) is another cholesterol lowering agent. In addition to antihypercholerolemic effects to reduce the risk of coronary heart disease and heart attack, statin therapy appears to have number of other potentially beneficial effects including promoting nitric oxide mediated angiogenesis (Kureishi *et al.,* 2000), stimulation of bore formation (Mundy *et al.,* 1999), and antioxidant and anti-inflammatory properties (Simons, 2000). The inhibitors of squaline synthase and acylCoA: cholesterol acyltransferase (ACAT) could be used to treat hypercholesterolemia and atherosclerosis, as these are involved in cholesterol biosynthesis. Zaragozic acid A, B, and C is isolated from unidentified fungus, *Sporomiella intermedia* and *Leptodontium elatius*, respectively. (Bergstrom *et al.,* 1993). Sequalestatin from *Phoma sp.* (Dawason *et al.,* 1992), bisabosquals from *Stachybotrys sp.* (Minagawa *et al.*, 2001) and CJ-15,183 from *Aspergillus aculateus* (Watanabe *et al.,* 2001) are potent inhibitors of squalene synthase. The most potent ACAT inhibitor, pyripyropene A, is isolated from *Aspergillus fumigatus* (Omura *et al.,* 1995b). Similarily, microbial metabolites have also provided enzyme inhibitors with therapeutic activity in the treatment of obesity and diabetes, as they reduce the gastrointestinal absorption of dietary factors important for these diseases. Lipstatin, isolated from *Streptomyces toxytricini* is a potent inhibitor of pancreatic lipase, a key enzyme involved in intestinal fat digestion. Its semi-synthetic derivative orlistat is used in the treatment of obesity because of hypolipidemic properties (Guerciolini, 1997). Acarbose is a pseudo tetrasaccharide isolated from *Actinoplanes sp.* is reversible and competitive inhibitor of pancreatic β-amylase and intestinal enzymes such as glucoamylase, sucrose and maltase. It is used in postprandial hyperglycemia and hyper insulinemia in type II diabetes patients where dietary treatment alone provides glycemic control (Laube, 2002). Vasoactive substances from microorganism have also been isolated, which act as important drugs and effective against circulatory blood disease. Antithrombocitic drugs such as aggreticin from *Streptomyces sp.* OM-484 (Omura *et al.*, 1988), vasodialators, amauromine from *Amauroascus*

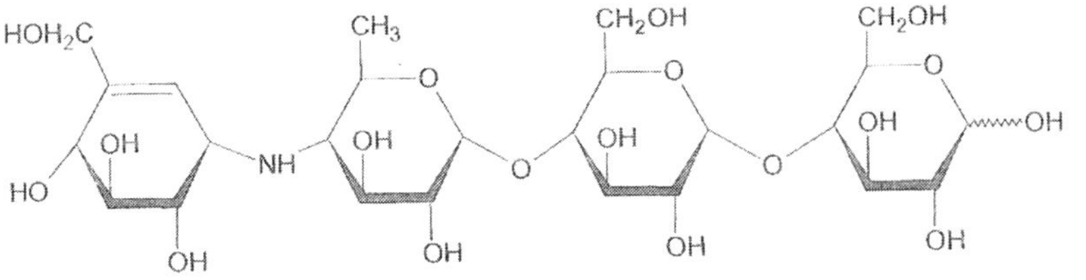

(a)

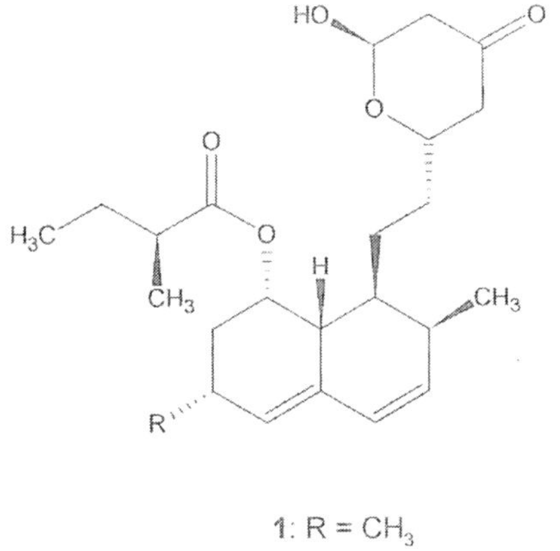

1: R = CH_3

2: R = H

(b)

Fig. 5. (a) Acarbose (antihyperglycemic) (b) Mevastatin (antihypercholestrolemic agent)

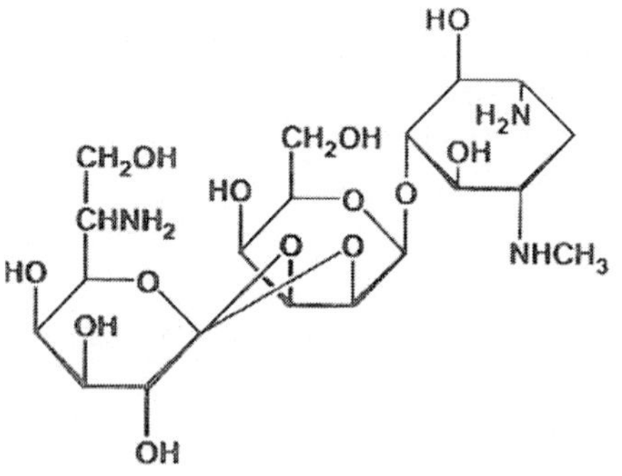

(a)

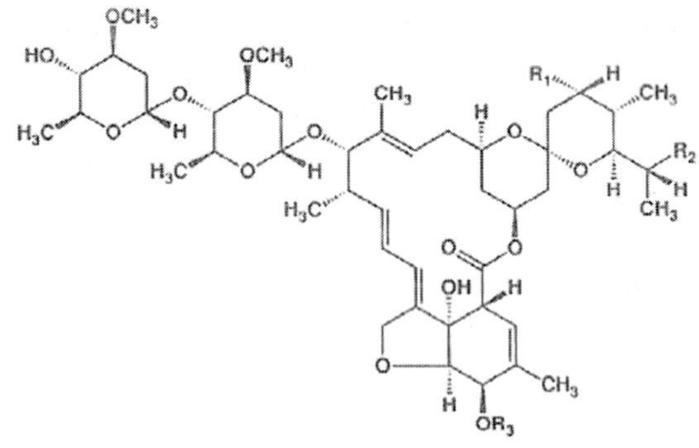

(b)

Fig. 6. (a) Hygromycin B (antiviral agent) (b) Avermectin (antiparasitic agent)

sp. no. 6327 (Takese *et al.*, 1984a) and fibrinolytic agent, WB3559 A, B, C and D from *Flavobacterium* sp. No. 3559 (Yoshida, 1985) have been isolated. A more recent approach to final compounds with the potential to treat diabetes resulted in the identification of non-peptidyl quinone-containing metabolite from *Pseudomassaria sp.*, L-783-281, which acted as activators of insulin receptor tyrosine kinase. This compound shows the promise for the development of new therapies for diabetes (Zhang *et al.,* 1999). Figure 5 (a) and (b) shows the structure of antihyperglycemic and antihypercholesterolemic agents, respectively.

CONCLUSION

It is becoming more apparent with each year that microorganisms offer potentially new and excellent sources of antibiotics and other nonantibiotic phamacophores of clinical importance. Microbial products in general offer a large pool of complex and highly interesting chemically diverse natural product having the role in drug discovery process. These clinically relevant marketed drugs are either unmodified natural products or semi-synthetic derivatives of its parental compound. Since past decades the wide and incredible microbial diversity has been explored and exploited for the production of antimetabolites. Berdy (1995) showed that out of 11,9000 natural products that had been discovered in 1994, approx. 6,600 (55%) are produced by *Streptomyces spp.* The remainder of producing organisms included the filamentous fungi, which produced 2,600 (22%), non-actinomycetes bacteria, which produced 1,400 (12%) and non-streptomyces strain of actinomycetes, which produced 1,300 (11%). When broken down by pharmacological activity, however, the *Streptomyces* and other actinomycetes produced 66% of the compounds which demonstrate antibacterial, antifungal and antitumor activity. About 63% of the compounds isolated from extract of filamentous fungi are bioactive pharmacological targets other than antibiotics. As per the basic need to address the problems such as antibiotic and multidrug resistant pathogen and newly emerging infectious disease as well as to discover broad spectrum drug, the best way is to search many new targets and new compounds that can discriminate between the new and already discovered function. Several such approaches have been explored and exploited in the primitive prokaryote upto fungi for product of clinical importance. This continuous drug discovery process has given a enormous spectrum to bioactive compounds of clinical use. For example, antibacterial agents penicillin to kettapeptin, and antifungals from griseofulvin to caspofungin.

There is no question that the future of pharmacological industry depends on the discovery of new agents from microbial origin. The Indian biopharma industry is the single largest contributor to biotech industry of India. Indian manufacturer account for 62% of the total biopharma business. The primary issue for the drug discovery process today will be selective isolation technique to discover uncultured diversity, ability to find new compounds out of the extract quickly and efficiently, high throughput screening and developing assay that will differentiate between known compounds and potentially new, exciting product leads.

REFERENCES

Alberts, A.W., Chen, J., Kuron, G., Hunt, V., Huff, J., Hoffman, C., Rothrock, J., Lopez, M., Jashua, H., Harris, E., Patchett, A., Monaghan, R., Currie, S., Stapley, E., Albers-Schongberg, G., Hensens, O., Hirshfield, J., Hoog Steen, K., Liesch, J. and Springer, J. (1980). *Proc. Natl. Acad. Sci.* USA. **77:** 3957-3961.

Arun Kumar, G., Gowrish Rao, S. and Shivananda, P.G. (1997). *Curr. Sci.*, **72:** 581-583.

Ayhunie, S., Belay, A., Baba, T.W. and Ruprecht, R.M. (1998). *Immuno. Defic. Syndr. Hum. Retrovirol.* **18:** 7-12.

Balzarini, J., Keyaerts, E., Vijgen, L., Egberink, H., De Clercq, E., Van Ranst, M., Printsevskaya, S.S., Olsufyeva, E.N., Solovieva, S.E. and Preobrazhenskaya, M.N. (2006). *Antiviral Res.* **72:** 20-23.

Berdy, J. (1995). *In:* Proceedings of ninth International Symposium on the Biology of Actinomycetes. (Eds. Debabov, V.G., Dudnik, Y.V. and Denilenko, V.N.) All-Russia Scientific Research Institute for Genetics and Selection of Industrial Microorganisms, Moscow, Russia. pp. 13-34.

Bergstrom, J.D., Kurtz, M.M., Rew, D.J., Amend, A.M., Karkas, J.D., Bostedor, R.G., Bansal, V.S., Dufresne, C., Van- Middlesworth, F.L., Hensens, O.D., Liesch, J.M., Zink, D.J., Wilson, K.E., Onishi, J., Milligan, J.A, Bills, G., Kaplan, L., Nallin Omstead, M., Jenkins, R.G., Huang, L., Meinz, M.S., Quinn, L., Burg, R.W., Kong, Y.L., Mochales, S., Mojena, M., Martin, I., Pelaez, F., Diez, M.T. and Alberts, A.W. (1993). *Proc. Natl. Acad. Sci.* USA. **90:** 80-84.

Borel, J.F., Feurer, C., Gubler, H.U. and Stahelin, H. (1976). *Agents Actions.* **6:** 468-475.

Bossche, V.H., Thienpont, D. and Jansens, G.P. (1986). *In:* Handbook of Experiment Pharmocology. pp. 77 Springer Verlag New York.

Brown, M., Nourasadeghi, M., Boyle, J. and Davidson, R.N. (2005). *Br. J. Dermatol*, **153:** 203-5.

Bull, A.T. (2004). *In:* Microbial Diversity and Bioprospecting. ASM press, Washington. pp. 337 and 349.

Campbell, W.C. and Rew, R.S. (1986). *In:* Chemotherapy of parasitic disease. (Eds. Campbell, W.C. and Rew, R.S.) Plenum Press, New York.

Cardenas, M.E., Sanfridson, A., Cutler, N.S. and Heitman, J. (1998). *Trends Biotechbnology.* **16:** 427-433.

Chain, E.B., Florey, H.W., Gardner, A.D., Hetley, N.G., Jerrings, M.A. Orr-Ewing, J. and Sanders, A.G. (1940), *Lancet* **II:** 226-228.

Chen, J.K., Lane, W.S. and Schreiber, S.L. (1999). *Chem. Biol.* **6:** 221-235.

Colegate, S.M., Dorling, P.R. and Huxtable, C.R. (1979). A spectroscopic investigation of swansonine: An α-mannosidase inhibitor isolated from *Swainsona canescens*. 2257-2264.

Cragg. G.M., Boyd, M.R., Hallock, Y.F., Newman, D.J., Sausville, E.A. and Wolpert, M.K. (2000). *In:* Biodiversity: New leads for the Pharmaceutical and Agrochemical industries (Ed. Wrigley, S.K., Hayes, M.A., Thomas, R., Crystal E.J.T. and Nicholson, N.) Royal Society of Chemistry, Cambridge, United Kingdom. pp. 22-24.

Dawson, M.J., Farthing, J.E., Marshall, P.S., Middleton, R.F., O'Neill, M.J., Shuttleworth, A., Stylli, C., Tait, R.M., Taylor, P.M., Wildman, H.G., Buss, A.D., Langley, D. and Hayes, M.V. (1992). *J. Antibiot.* **45:** 639-647.

Donia, M. and Hamann, M.T. (2003). *The Lancet,* **3:** 338-348.

Duncan, S.J., Cooper, M.A. and Williams, D.H. (2003). *Chem. Commun.* **2003:** 316-317.

Endo, A. and Hasumi, K. (1993). *Nat. Prod. Rep.* **10:** 541-550.

Fehr, T., Quesniaux, V.F.J., Sanglier, J.J., Oberer, L., Gschwind, L., Ponelle, M., Schilling, W., Wehrli, S., Enz, A., Zenke, G. and Schuler, W. (1997). *J. Antibiot.* **50:** 893-899.

Franco, C.M.M. and Coutinho, L.E.L. (1991). *Crit. Rev. Biotechnol.* **11:** 193-276.

Gerth, K.N., Bedorf, G., Hofle, H., Irschik, H. and Reichenbach, H. (1996). *J. Antibiot* **49:** 560-563.
Guerciolini, R. (1997). *Int. J. Obesity,* **21:** S12-S23.
Gustafson, K., Cardellina, J., Fullar, R.W., Weislow, O.S., Kiser, R.F. Snader, K.M., Patterson, G.M.L. and Boyd, R.M. (1989). *J. National Cancer Institute.* **81:** 1254-1258.
Hayashi, K. Hayashi, T. and Kojima, I. (1996). *AIDS Res. Hum. Retrovir.* **12:** 1463-1471.
Hayashi, T. and Hayashi, K. (1996). *J. Nat. Prod.* **59:** 83-87.
Henkel, T., Brunne, R.M., Muller, H. and Reichel, F. (1999). *Angew. Chem. Int. Ed.* **38:** 643-647.
Hill, D.C., Wrigley, S.K and Nisbet, L.J. (1998). *Adv. Biochem Eng. Biotechnol.* **59:** 73-121.
Ichikawa, K., Hirai, H., Ishiguro, M., Kambara, T., Kato, Y., Kim, Y.J., Kojima, Y., Matsunaga, V., Nishida, H., Shiomi, Y., Yoshikawa, N. and Kojima, N. (2001a). *J. Antibiot.* **54:** 703-709.
Ichikawa, K., Hirai, H., Ishigaro, M., Kambara, T., Kato, Y., Kim, J., Kojima, Y., Matsunaga, Y., Nishida, H., Shiomi Y., Yoshikawa, N., Huang, L.H. and Kojima, N. (2001b). *J. Antibiot.* **54:** 697-702.
Ingber, D., Fujita, T., Kishimoto, S., Sudo, K., Kanamaru, T., Brem, H. and Folkman, J. (1990). *Nature.* **348:** 555-557.
Iwami, M., Nakayama, O., Terano, H., Kohsaka, M., Aoki, H. and Imanaka, H. (1987). *J. Antibiot.* (Tokyo) **40:** 612-622.
Jeong, S.Y., Shin, H.J., Kim, T.S., Lee, H.S., Park, S.K. and Kim, H.M. (2006). *J. Anitibiot.* (Tokyo) **59:** 234-40.
Jose, N. and Janardhanan, K.K. (2000). *Current Science.* **79:** 941-943.
Kakeya, H., Kageyama, S.I., Nie, L., Onose, R., Okada, G., Beppu, T., Norbury, C.J. and Osada, H. (2001). *J. Antibiot.* **54:** 850-54.
Kanaski, R., Kobavashi, M., Fujine, K., Sato, I., Hashimoto, M., Takase, S., Tsurumi, Y., Fujie, A., Hino, M., Hashimoto, S. and Hori, Y. (2006). *J. Antibiot.* (Tokyo). **59:** 158-67.
Kokare, C.R., Mahadiek, K.R., Kadam, S.S. and Chopade, B.A. (2004). *Current science,* **86:** 593-597.
Kubo, M., Nishitsuji H., Kurihara, K., Hayasi, T., Masuda J. and Kannagi, M. (2006). *J. Gen. Virol.* **87:** 1589-93.
Kureishi, Y., Luo, Z., Shiojima, I., Bialik, A., Fulton, D., Lefer, D.J., Sessa, W.C. and Walsh, K. (2000). *Nat. Med.* **6:** 1004-1010.
Lall, N., Hussein, A.A. and Meyer, J.J. (2006). *Fitoterapia.* **77:** 230-2.
Laube, H. *Clin. Drug Invest,* **22:** 141-156.
Lazzirini, A., Cavaletti, Toppo, G. and Marinelli, F. (2001). *Antonie Leeuwenhoek.* **79:** 399-405.
Lee, M.D., Durr, F.E., Hinman, L.M., Hamann, P.R. and Ellestad, G.A. (1993). *Adv. Med. Chem.* **2:** 31-66.
Lee, Y.F.Y., Borzilleri, R., Fairchild, C.R., Kim, S.H., Long, B.H., Reventos-Suarez, C., Vite, G.D., Rose, W.C. and Kramer, R.A. (2001). *Clin. Cancer Res.* **7:** 1429-1437.
Lingham, R.B., Silverman, K.C., Jayasuriya, H., Kim, B.M., Amo, S.E., Wilson, F.R., Raw, D.J., Schaber, M.D., Bergstrom, J.D., Koblan, K.S., Graham, S.L., Kohl, N.E., Gibbs, J.B. and Singh, S.B. (1998). *J. Med. Chem.* **41:** 4492-4501.
Mann, J. (2001). *Nat. Prod. Rep.* **18:** 417-430.
Maskey R.P., Fotsos, S., Sevvana, M., Uson, I., Grunwollnv, I. and Laatsch, H. (2006). *J. Antibiot.* (Tokyo). **59:** 309-314.
Maskey, R.P., Helmke, E., Kayser, O., Fiebig, H.H., Maier, A., Busche, A., Laatch, H. (2004). *J. Antibiot,* (Tokyo). **57:** 771-9.
Menendez, A. *et al.,* (2002). Discovery and characterization of XR842, an apoptotic natural product with novel structure and mechanism of action, Abstr. 93rd Annu. Meet. Am. Assoc. cancer Res. Abstr. 4656.
Minagawa K., Kouzuki, S., Nomura, K., Yamaguchi, T., Kawamura, Y., Matsushima, K., Tani, H., Ishii, K., Tanimoto, T. and Kamigauchi, T. (2001). *J. Antibiot.* **54:** 890-895.
Mohapatra, B.R., Bapuji, M. and Sree, A. (2002). *Folia Microbiol.* **47:** 51-55.
Momose, I., Sekizawa, R., Hashizume, H., Kinoshita, N., Homma, Y., Hamada, M., Iinuma, H. and Takeuchi, T. (2001). *J. Antibiot.* **54:** 997-103.
Mundy, G., Garerett, R., Harris, S., Chan, J., Chen, D., Rossini, G., Boyce, B., Zhao, M. and Gutierrez, G. (1999). *Science.* **286:** 1946-1949.
Newman A.D.J., Gragg, G. M. and Snader, K.M. (2000). *Nat. Prod. Rep.* **17:** 215-234.
Nisbet, L.J. and Portes N. (1989). *In:* (Eds. Banmberg, S., Hunter, I. and Rhodes, M.) Society for General Microbiology symposium 44. Cambridge University Press, Cambridge. United Kingdom. pp. 309-342.
Omura, S. (1992). *In:* The Search for Bioactive Compounds from Microorganisms. Springer Verlag. New York. pp. 30-44.
Omura, S., Nakagawa, A., Fukamachi, N., Miura, S., Takahashi, H., Komiyama, K. and Kabayashi, B. (1988). *J. Antibiot.* (Tokyo). **30:** 275-282.
Omura, S., Sasaki, Y., Iwai, Y. and Takeshima, H. (1995a). *J. Antibiot.* **48:** 535-548.

Omura, S., Tomoda, H. and Sunazuka, T. (1995b). *In: Sekundarmetabolimus bei Mikroorganismen*, (Eds. Kuhn, W. and Fiedler, H.P.) Attempto Verlag Tubifgen Germany. pp 37-49.

Rawlins, P., Mander, T., Sandhegi, R., Hills, S., Gammon, G., Foxwell, B., Wrigley, S. and Moore, M. (1999). *Int. J. Immunopharmacol.* **21:** 799-814.

Reddy, C.M., Bhat, V.B. Kiranmai, G., Reddy, M.N., Reddanna, P. and Madyastha, K.M. (2000). *Res. Commun.* **3:** 599-603.

Sanglier, J.J., Quesniaux, V., Fehr, T., Hofmann, H., Mahnke, M., Memmert, K., Schuler, W., Zenke, G., Gschwind, L., Maurer, C. and Schilling, W. (1999). *J. Antibiot.* **52:** 466-473.

Sato, M., Motomura, T., Aramaki. H., Matsuda, T., Yamashita, M., Ito, Y., Kawakami, H., Matsazaki, Y., Watanabe, W., Yamataka, K., Ikeda, S., Kodama, E., Matsuoka, M. and Sinkai, H. (2006). *J. Med. Chem.* **9:** 1506-8.

Sekizawa, R., Momose, I., Kinoshita, N., Naganawa, H., Hamada, M., Muraoka, Y., Iinuma, H. and Takeuchi, T. (2001). *J. Antibiot.* **54:** 874-881.

Shibata, T., Nakayama, O., Tsurumi, Y., Okuhara, M., Terano, H. and Kohsaka, M. (1988). *J. Antibiot.* (Tokyo). **41:** 296-301.

Shin–ya, K., Wierzba, K., Matsco, K., Ohtani, T., Yamada, Y., Funihata, K., Hayakawa, Y. and Seto, H. (2001). *J. Am. Chem. Soc.* **123:** 1262-1263.

Shu Y-Z. (1998). *J. Nat. Prod.* **61:** 1053-1071.

Sievers, T.M., Rossi, S.J., Ghobrial, R.M., Arriola, E., Nishimura, P., Kawano, M. and Holt, C.D. (1997). *Pharmacotherapy.* **17:** 1178-1197.

Simons, M. (2000). *Nat. Med.* **6:** 965-966.

Singh, S., Kate, B.N. and Banerjee, U.C. (2005). *Critical Reviews in Biotechnology.* **25:** 73-93.

Suay, I., Arenol, F., Asenio, F., Basilio, A., Cabello, M. and Diez, M.T. (2000). *Antonie van Leeuwenhoek.* **78:** 129-39.

Takese, S., Iwami, M., Ando, T., Okamoto, M., Yoshida, K., Horiai, H., Kohsaka, M., Aoki, H. and Imanaka, H. (1984a). *J. Antibiot.* (Tokyo). **37:** 1320-1323.

Tkacz, J. and DiDomenico, B. (2001). *Curr. Opin. Microbiol.* **4:** 540-546.

Ueki, M., Teruya, T., Nie, L., Usami, R., Yoshida, M. and Osada, H. (2001). *J. Antibiot.* **54:** 1093-1095.

Umezawa, H. (1972). *Enzyme inhibitors of microbial origin.* University of Tokyo Press, Tokyo. pp. 1-114.

Umezawa, H., Imoto, M., Sawa, T., Isshikoi, N., Matsuda, N., Uchida, H., Iinuma, M., Hamada, M. and Takeuchi, T. (1986). *J. Antibiot.* **39:** 170-173.

Watanabe, S., Hirai, H., Ishiguro, M., Kamabara, T., Kojima, Y., Matsunaga, T., Nishida, H., Suzuki, Y., Sugiura, A., Harwood, H.J., Huang, L.H. and Kojima, N. (2001). *J. Antibiot.* **54:** 904-910.

Watve, M.G. Tickoo, R., Jog, M.M. and Bhole, B.D. (2001). *Arch. Microbiol.* **176:** 386-390.

Wrigley, S.K., Ainsworth, A.M., Kau, D.A., Martin, S.M., Bahl, S., Tang, J.S., Hardick, D.J., Rawlins, P., Sadhegi, R. and Moore, M. (2001). *J. Antibiot.* **54:** 479-488.

Wrigley, S.K., Sadhegi, R., Bahl, S., Whiting, A.J., Ainsworth, A.M., Martin, S.M., Katzer, W., Ford, R., Kou, D.A., Robinson, N., Hayes, M.A., Elcock, C., Mander, T. and Moore, M. (1999). *J. Antibiot.* **52:** 862-872.

Yang, H., Lee, E. and Kim, H. (1997). *Life Sci.* **61:** 1237-1244.

Yang, S.W., Chan, T.M., Terracciano, J., Loebenberg, D., Patel, M., Gullo, V. and Chu, M. (2006). *J. Antibiot,* (Tokyo). **59:** 190-2.

Yoshida, K., Iwami, M., Umehara, Y., Nishikawa, M., Uchida, I., Kohsaka, M., Aoki, H. and Imanaka, H. (1985). *J. Antibiot.* (Tokyo). **38:** 1469-1475.

Zhang, L.H. and Liu, J.O. (2001). *J. Immunol.* **166:** 5611-5618.

Zhao, A., Lee, S.H., Mojena, M., Jenkins, R.G., Patrick, D.R., Huber, H.E., Goetz, M.A., Hensens, O.D., Zink, D.L., Vilella, D., Dombrowski, A.W., Lingham, R.B. and Huang, L. (1999). *J. Antibiot.* **52:** 1086-1094.

Chapter 20

Industrially Useful Microbial Bioresources

Alok Satlewal[1], Reeta Goel[1]* and Govind K. Garg[2]

[1]Department of Microbiology,
[2]Department of Molecular Biology and Genetic Engineering,
G.B. Pant University of Agriculture and Technology, Pantnagar–263 145 (India)
E-mail: rg55@rediffmail.com

Microbes are ancient forms of life on earth that have continued to evolve in response to the environment so as to create an unparalleled diversity. Over the years man has skilfully exploited them for domestic and commercial uses for example, for food processing, production of antimicrobial agents, enzymes, bioinoculants for soil fertility, disease control, antifouling and antibiofilm agents and pollutant biodegradation. The current review attempts to summarize the current status of microbial diversity with special emphasis on the industrially useful microbes isolated from the soils of Uttarakhand state.

INTRODUCTION

The earth remains unique, at least in perception, as much as in its diversity, structure and function of the inhabitants and in its support to various life forms. Biodiversity is a broad term that refers to the diversity amongst and within biological species independent of the ecosystems in which these organisms thrive. The creation of biodiversity is thought to be the result of their responses to the challenges offered by the change in their habitat and environment, hence microbial diversity is not an exception. Microbial diversity is therefore defined and classified according to the observed variations in morphological, physiological and genetic characteristics apart from their geographical locations. Tremendous efforts have been made to explore, characterize and exploit the vastness of microbial diversity.

The Himalayas are the youngest and the highest mountain ecosystem in the world. These are spread across eight countries of Asia including India. In India, the Himalayas cover 14 states (fully or partially) and 90 districts in two distinct geographical flanks of the western and eastern region; Uttarakhand belongs to the western Himalayan region. This region covers a large area (62.4% of total Himalayas) having steep slopes. It has severe winters and is covered by predominantly coniferous forest vegetation (Samra *et al.*, 1999). The wide range of variations in altitude, latitude and longitude generate large variations in micro-climate and soils, consequently it becomes habitat of unusually diverse flora and fauna. This enormous wealth of natural resources comprising of land, water and life forms is under severe threat from the ever-increasing human and livestock population.

NATIONAL SCENARIO

India is among the important mega diversity centres of the world, with an immense contribution from the Himalayan ecosystem. The primary measure of biodiversity is in the richness of species, that is the number of species present in a given area or habitat. Generally speaking, more the number of species in the area, the more biodiverse it is. Therefore, a comparative look at the species documented in India and the world (Table 1) clearly establishes the richness of microbial diversity in India vis-a-vis the world.

Table 1: Comparative statement of recorded number of species in India and the world.

Taxa	Species		Percentage of India to the world
	India	World	
Bacteria	850	4000	21.25%
Viruses	Unknown	4000	—
Algae	6500	40000	16.25
Fungi	14,500	72000	20.14
Lichens	2000	17000	11.80
Bryophytes	2850	16000	17.80
Pteridophytes	1100	13000	8.46
Gymnosperms	64	750	8.53
Angiosperms	17500	250000	7

Source: MoEF 1999.

The Indian concern for proper management of its biodiversity found expression in devising a National Biodiversity Strategy and Action Plan (NBSAP), which gives equal emphasis to microbial diversity. It has been developed with the support of UNDP and Global Environmental Facility (GEF) and the involvement of a Technical and Policy core Group (TPCG). Its formulation, was coordinated by a non-government organization. Kalpavriksh (Pune) and the administrative support was provided by Biotech Consortium India Limited (BCIL, New Delhi). The National Action Plan (NAP) was synthesized from the plans made at the local, eco-regional, state and thematic levels. A detailed format for the National Action Plan was formulated after a series of discussions, and work on the National Action Plan was started by March 2002 (www.kalpavriksh.org).

The national plan has recommended a series of measures to ensure that the country's ecological security is assured, and that the livelihoods of the millions of people dependent on biological resources are also secured. These measures call for a major reorientation of the process of economic development and of governance of natural resources, such that the health of the environment especially critical ecosystems, wildlife habitats, and livelihoods of biomass dependent communities, become central to all planning, and decision-making.

Moreover, the Department of Biotechnology has begun a programme on digitization of the available data for cataloguing and conservation of the existing gene pool, and this baseline information should be ready shortly. In this regard microbial diversity task force constituted by the Department of Biotechnology has taken several initiatives to facilitate management of microbial diversity in its entity. It will need active participation of microbiologists for the continued and sustained exploration and screening of microbes.

In the present scenario of IPR and patent issues, a need has been felt for the establishment of international culture collection centre and bureau. Therefore, an internationally recognized Microbial Type Collection Centre (MTCC) has been developed at the Institute of Microbial Technology (IMTech); Chandigarh with the support from the Department of Biotechnology (DBT), Govt. of India. Similarly, a National Bureau of Agriculturally Important Microorganisms (NBAIM), Mau has been set up by Indian Council of Agriculture Research (ICAR), Govt. of India to look after the management of agriculturally important microorganisms.

A PERSPECTIVE OF MICROBIAL DIVERSITY IN UTTARAKHAND

Uttarakhand is a bioresource rich state, spread over an area of 51,082 sq km, wherein only 13% land is cultivated. The rest 87% area represents several unique ecological niches, which harbor a great variety of flora and microbes that live in synergy. Further, microbial diversity of this "Hot spot" (Central Himalayan region) is unique by way of geographical locales including variation in vegetation, temperature regime, cropping practices and other physical parameters of soil and water. It is situated at 79.32° to 86.10° in longitude and N 29.29° to N 30.31° in latitude with varies altitude from 1000 m to 5000 m. The pH range of soil varies from 6.8 to 5.8 and can be even lower than this; however, the average temperate varies from 28° to 0°C to –20°C and subzero. Four regions are represented completely (subtropical warm temperate, cold temperate, and cold zone) and one partly (>3500 m snow coverage) in the hill and mountain agro-ecosystem of this state (Fig. 1).

Soils from Uttarakhand have been a very potent source of microorganisms that have been used in delineating metabolic diversity, understanding mechanism of environmental stress specially temperature, nutrient recycling, symbiosis, parasitism, etc. Organisms that could degrade camphor, variety of other cyclic and aromatic compounds have been isolated from these soils, although, records or documentations are scanty. The folklore in Dr Gunsalus' lab stated that camphor degrading *Pseudomonas putida* had originated from the soils under the camphor tree. Late Dr. O.P. Shukla, who worked at Pantnagar and Central Drug Research Institute, at Lucknow often referred to how he succeded in isolating *Bacillus Sp.* from Pantnagar soil that could degrade a variety of pyridine based compounds that led him to understand their metabolism (Shukla, 1984). Similarly, several other strains of *Bacillus brevis* were isolated by our lab from Pantnagar's soil that helped us in understanding the role of glutamine synthetase in sporulation (Tiwari *et al.*, 1989). In view of the fact that G. B. Pant University has an active interdisciplinary group of

Fig. 1. Uttarakhand Map.

scientists working on microbial bioresources for various applications, we are summarizing the major findings of this group.

STRATEGIC ROLE OF GB PANT UNIVERSITY, PANTNAGAR IN BIOPROSPECTING OF MICROBIAL DIVERSIY

The university is represented by the Krishi Vigyan Kendra (KVK) or the Research Stations in each of the 13 districts of Uttarakhand, covering Kumaon and Garhwal regions, both at temperate and subtropical zone covering tarai and Bhabhar ecozones. The group of scientists, working in the departments of microbiology, soil science, molecular biology and genetic engineering, chemistry, biochemistry and plant pathology, etc. have worked relentlessly to isolate various bacterial and fungal strains for their unique genes, properties and uses. The isolated organisms have been conserved and field tested for their activities as required.

Centre for Bacteria and Archea

This university with the support from the ministry of environment has established a "Centre for Bacteria and Archea", and under this programme the scientists have collected more then 700 microbial strains from different parts of the country. Two hundred bacterial isolates of these are from Kumaon and Garhwal hills located in a range from 1000 to 2000 mts. About 30 isolates were characterized at biochemical and molecular level such as *nif*$^{+}$, 'P' solubilizer, heavy metal transformers, (As, Ni, Co, Cu and Pb), and low/high density polyethylene degraders, and their 16S RNA sequences were submitted to the National Centre of Biotechnology Information (NCBI) database (Table 2).

Table 2: Properties of bacterial strains whose 16S RNA sequences are deposited to NCBI database.

S. No.	Accession No.	Definition	Strain	Journal	Authors	Title/Property
1.	AY970345	*BacillusCereus* 16S r RNA	AG27	Unpublished	Gupta, A.; Goel, R.	Arsenic resistant
2.	AY970346	*Bacillus cereus* 16S r RNA	AG24	Unpublished	Gupta,A.; Goel, R.	Arsenic resistant
3.	DQ118019	*Acinetobacter sp.*	GB1	Unpublished	Pratap, K.C.; Chaturvedi, P.; Goel, R.	Polysulfone degrading
4.	DQ118020	*Sphingobacterium sp.*	GB2	Unpublished	Pratap, K.C.; Chaturvedi, P.; Goel, R.	Polysulfone degrading
5.	DQ205427	*Pesudomonas putida* 16S r RNA	knp9	Curr. Microbiol.	Tripathi, M.; Goel, R	Siderophore producing
6.	DQ205428	*Acinetobacter sp.* 16S r RNA	cn4	Unpublished	Satlewal, A.; Goel, R	Polycarbonate degrading
7.	DQ205429	*Arthrobacter sp* 16S r RNA.	cna2	Unpublished	Satlewal, A.; Goel, R	Polycarbonate degrading
8.	DQ205430	*Proteobacterium sp.* 16S r RNA	cp1	Unpublished	Satlewal, A.; Goel, R	Polycarbonate & LDPE degrading
9.	DQ205431	*Enterobacter sp.* 16S r RNA	lna3	Unpublished	Satlewal, A.; Goel, R	Polycarbonate & LDPE degrading
10.	DQ209209	*Bacillus pumilus* 16S r RNA	S2	Unpublished	Soni, R; Goel, R.	LDPE degrading
11.	DQ209210	*Bacillus cereus* 16S r RNA	PN15	Unpublished	Soni, R; Goel, R.	LDPE degrading
12.	DQ118015	*Bacillus firmus* 16S r RNA	PSR4	Unpublished	Chaturvedi, P.; Goel, R.	Isolated from Uttarakhand soil
13.	DQ118016	*Bacillus pumilus* 16S r RNA	RGR7	Unpublished	Chaturvedi, P.; Goel, R.	Polyester, polyamide & LDPE degrading
14.	DQ118017	*Escherichia coli* 16S r RNA	RGR13	Unpublished	Chaturvedi, P.; Goel, R.	P' solubilization
15.	DQ118018	*Chryseobacterium sp.* 16S r RNA	PSR10	Unpublished	Chaturvedi, P.; Goel, R.	'P' solubilization
16.	AY952928	*Pseudomonas fluorescens* 16S r RNA	CRC1	Unpublished	Khan, M.; Goel, R.	cold shock protein gene
17.	DQ111780	*Pseudomonas fluorescens* 16S r RNA	CRC2	Unpublished	Khan, M.; Goel, R.	cold shock protein gene
18.	DQ205432	*Proteus vulgaris* 16S r RNA	KNP3	Unpublished	Tripathi, M.; Goel, R	Pb & Cd resistant
19.	DQ423486	*Bacillus cereus* 16S r RNA	RGR10	Unpublished	Soni, R; Goel, R.	LDPE degrading
20.	DQ423487	*BacteriumTe68R* 16S r RNA	PN12	Unpublished	Soni, R.; Goel, R. and Souche, Y.	LDPE degrading
21.	DQ423485	*Bacillus cereus* 16S r RNA	PN24	Unpublished	Soni, R.; Goel, R.	LDPE degrading
22.	DQ523735	*Bacillus sp.* 16S r RNA	PN13	Unpublished	Soni, R.; Goel, R. and Souche, Y.	LDPE degrading

Table 2: Cotnd.

S. No.	Accession No.	Definition	Strain	Journal	Authors	Title/Property
23.	DQ486130	*Arthrobacter luteolus partia* 16S r RNA	LNR3	Unpublished	Soni, R.; Goel, R. and Souche, Y.	LDPE degrading
24.	DQ318884	*Mycobacterium sp.* 16S r RNA	MK3	Unpublished	Khan, M. and Goel, R.	HDPE degrading
25.	DQ318885	*Pseudomonas putida* 16S r RNA	MK4	Unpublished	Khan, M. and Goel, R	HDPE degrading
26.	DQ318886	*Agromyces sp.* 16S r RNA	MK5	Unpublished	Khan, M. and Goel, R	HDPE degrading
27.	DQ318887	*Pseudomonas sp.* 16S r RNA	MK7	Unpublished	Khan, M. and Goel, R	HDPE degrading
28.	DQ517938	*B.cereus* Arsenate reductase gene	AG27	Unpublished	Gupta, A.; Goel, R.	Arsenic resistant
29.	DQ517939	*B.cereus* Arsenate reductase-like gene	AG24	Unpublished	Gupta, A.; Goel, R.	Arsenic resistant
30.	DQ517500	*Bacillus sp.* 16S rRNA	BG1	Unpublished	Singh, S.; Goel, R	Rhizospheric diversity
31.	DQ517501	*Bacillus sp.* 16S rRNA	RG1	Unpublished	Singh, S.; Goel, R	Rhizospheric diversity
32.	EF092429	*Bacillus sp.* 16S rRNA	RG4	Unpublished	Singh, S.; Goel, R	Rhizospheric diversity
33.	EF092430	*Bacillus sp.* 16S rRNA	S1	Unpublished	Singh, S.; Goel, R	Rhizospheric diversity
34.	EF092431	*Bacillus sp.* 16S rRNA	BG1	Unpublished	Singh, S.; Goel, R	Rhizospheric diversity
35.	EF092432	*Bacillus sp.* 16S rRNA	BG2	Unpublished	Singh, S.; Goel, R	Rhizospheric diversity
36.	EF092433	*Bacillus sp.* 16S rRNA	BG3	Unpublished	Singh, S.; Goel, R	Rhizospheric diversity
37.	EF092434	*Bacillus sp.* 16S rRNA	BG4	Unpublished	Singh, S.; Goel, R	Rhizospheric diversity
38.	EF207715	*Pseudomonas putida*	710A	Unpublished	Satlewal, A.; Goel. R.	HDPE degrading psychrotolerant bacteria
39.	EF207716	*Comamonas aquatica*	710B	Unpublished	Satlewal, A.; Goel. R.	HDPE degrading psychrotolerant bacteria

Moreover, as part of the ongoing taxonomy initiative, strains have been collected from rhizosphere and nodules of cereals and legume plants from different geographical locations of Uttarakhand. About 60 strains have been characterized functionally as well as by partial 16S rRNA sequencing (Table 3). This study shows that the population associated with wheat rhizosphere is too diverse (Shilpi, 2004). Out of 49 strains isolated from wheat rhizosphere, about 59% strains belonged to the genus *Pseudomonas,* and remaining 41% of the population constituted of strains belonging to *Proteus, Bacillus, Stenotrophomonas* and *Serratia*. Similarity, searches through NCBI database revealed that majority of these strains were more than 80% similar with those of already described forms. Similarity level for strains ranged between 65.8–53.6%. However, bacteria associated with leguminous host *Dalbergia sisoo* belonged to the group *rhizobia*; with strains belonging to the different species of genera *Rhizobium* and *Mesorhizobium* (Sahgal *et al*., 2004, 2006).

Table 3. Microbial diversity characterized in relation to host plant.

S.No.	Name	Host plant	Identification
1.	SLHRE9	Wheat rhizosphere	*Proteus vulgaris*
2.	SHHRH77	-do-	*Pseudomonas sp.*
3.	SLLRE82	-do-	*P. fluorescens* strain WRind 4510
4.	SLLRE88	-do-	*P. fluorescens*
5.	456	-do-	*P. fluorescens*
6.	902	-do-	*Serretia marcescens*
7.	4513	-do-	*B. megaterium*
8.	4510	-do-	*P. fluorescens*
9.	4516	-do-	*Enterobacter* sp.
10.	454	-do-	*B. cereus*
11.	4514	-do-	*B. megaterium*
12.	WP-S-17	-do-	*B. megaterium*
13.	Pb-2	-do-	*P. fluorescens*
14.	PP12	-do-	*Pseudomonas* sp.
15.	PS6	-do-	*Pseudomonas* sp.
16.	PS7	-do-	*Pseudomonas fluorescens*
17.	RS7	-do-	*Pseudomonas stutzeri*
18.	RS16	-do-	*Pseudomonas* sp.
19.	RS17	-do-	*Pseudomonas fulva*
20.	RP17	-do-	*Pseudomonas putida*
21.	RP13	-do-	*Pseudomonas* sp.
22.	PP12	-do-	*Pseudomonas* sp.
23.	1S16	-do-	*Pseudomonas filiscindens*
24.	1P20	-do-	*Pseudomonas* sp.
25.	2S1	-do-	*Pseudomonas syringae*
26.	2S3	-do-	*Pseudoxanthomonas* sp.
27.	2S16-F	-do-	*Stenotrophomonas maltophilia*
28.	2S17	-do-	*Pseudomonas putida*
29.	3S19	-do-	*Pseudomonas straminea*
30.	3S2-F	-do-	*Pseudomonas* sp.
31.	3S13	-do-	*Pseudomonas* sp.
32.	3S23	-do-	*Pseudomonas* sp.
33.	3P11	-do-	*Pseudoxanthomonas mexicana*
34.	3S7	-do-	*Pseudomonas* sp.
35.	SHHRE344	-do-	*P. fluorescens*
36.	1P4	-do-	*Stenotrophomonas maltophilia*
37.	PP6	-do-	*Pseudoxanthomonas* sp.
38.	PS8	-do-	*Stenotrophomonas maltophilia*
39.	RS10-F	-do-	*Stenotrophomonas maltophilia*
40.	3S12	-do-	*Pseudomonas* sp.
41.	2S13	-do-	*Pseudomonas anguilliseptica*
42.	SHHRH380	-do	*Proteus vulgaris*
43.	SLHRE425	-do	*A. faecalis*
44.	SHHRH508	-do	*P. fluorescens*
45.	SHHRE635	-do	*P. putida*
46.	SLHRH642	-do	*Bacillus subtilis*
47.	SHHRE645	-do-	*Serratia ficaria*

Table 3. Contd.

S.No.	Name	Host plant	Identification
48.	MB4	Mandua rhizosphere	*B.pumilus*
49.	MB2	-do-	*P.putida*
50.	MTCC 4186	*Dalbergia sissoo*	*Rhizobium* sp.
51.	MTCC 4187	-do-	*Mesorhizobium* sp.
52.	MTCC 4188	-do-	*Mesorhizobium ciceri*
53.	MTCC 4189	-do-	Rhizobium sp.
54.	MTCC 4190	-do-	*Rhizobium* sp.
55.	MTCC 4191	-do-	*Mesorhizobium* sp.
56.	MTCC 4192	-do-	*Rhizobium* sp.
57.	MTCC 4193	-do-	*Rhizobium* sp.
58.	MTCC 4194	-do-	*Rhizobium* sp.

Source: Shilpi Mittal, 2004 and Sahgal *et al.*, 2004, 2006

The Prospective Bioinoculants

Plants and microbes live in close association and it is also known that plants influence the microbial community quantitatively and qualitatively. Plants secrete a variety of root exudates, and microbes utilize these as their energy source (Das *et al.*, 2003; Katiyar and Goel, 2003). Some of them survive as free-living bacteria, whereas others take their nutrients from root exudates (bacterial community dependent on rhizosphere and rhizoplane). Bacteria may show a harmful or beneficial effect on plant growth. Those which are plant growth promotory in nature (PGPR) may fix atmospheric nitrogen, solubilize minerals such as phosphorus, produce siderophores that solubilize and sequester iron, produce plant growth regulators such as hormones, produce antagonistic substances, or induce resistance against plant pathogens (Das *et al.*, 2003; Katiyar and Goel, 2003). In nature one or more than one properties are found associated with one strain.

Diazotrophs

Biological nitrogen fixation is estimated to contribute 180×10^6 metric tonnes/year globally, of which eighty per cent comes from symbiotic associations and the rest from free-living or associative systems of bacteria and archaea (Graham, 1988). The symbiotic nitrogen-fixing forms include *Rhizobium* (obligate symbionts in leguminous plants) and *Frankia* (in non-leguminous plants), whereas non-symbiotic N_2-fixing forms now include cyanobacteria, *Azospirillum*, *Azotobacter*, *Acetobacter diazotrophicus*, *Azoarcus*, etc. These free-living and associative microorganisms are becoming more popular day by day in various agronomical practices. Each laboratory now tends to isolate location specific non-symbiotic microorganisms, which could be used as biofertilizers (Tilak *et al.*, 2005; Thajuddin N. and Subramanian, 2005).

During the last few years there have been considerable changes in the context of taxonomic status of rhizobia. Sahgal and Johri (2006) have outlined the current status of rhizobial taxonomy and enlisted 36 species distributed among seven genera (*Allorhizobium*, *Azorhizobium*,

Bradyrhizobium, *Mesorhizobium*, *Methylobacterium*, *Rhizobium* and *Sinorhizobium*), based on the polyphasic taxonomic approach. They suggested that in view of the fact that taxonomy of rhizobia is in a continuous state of flux, the only stability factor could be 16S rRNA sequences, which have become comparatively easy with modern molecular tools and machines (Sahgal and Johri, 2006).

Phosphatic Fertilizer

Phosphorus is the second most important macronutrient for plant growth after nitrogen. Availability of phosphorus is widely influenced by pH, salinity, temperature, etc. These factors render a large proportion of the inorganic 'P' insoluble that is unavailable to plants. Thus, 'P' acquisition is not the problem of total supply but of unavailability caused by extreme insolubility of phosphorus at both acidic and alkaline pH. The rhizospheric microbes, especially *Pseudomonas*, *Bacillus* and *Azospirillum* species have the potential to release phosphate from sparingly soluble minerals and are hence important in plant productivity and sustainability. One does not have to depend only on the natural biodiversity, even mutagenesis can enhance the 'P' solubilizing ability. The data presented in Fig. 2 shows that chemically mutagenised cultures of fluorescent *Pseudomonads* can give rise to mutants with better 'p' solubilizing ability (Table 4).

Other Plant Growth Promoting Rhizobacteria

Some soil microorganisms produce a variety of secondary metabolites, which directly or indirectly affect plant growth. Many species of bacteria and fungi are known to produce indole acetic acid, gibberellins, cytokinins, ethylene, etc., which act as supplements for growth promotion, thereby improving the yields of crop plants. Some of these include *Agrobacterium, Pseudomonas, Arthrobacter* and *Bacillus.* Thus, microbial bioinoculants promote plant growth in more than

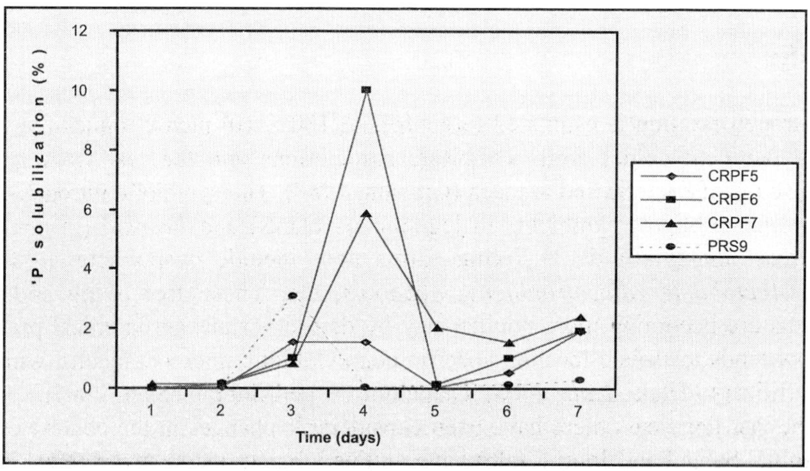

Fig. 2. 'P' solubilization in Mung bean plants by fluorescent *Pseudomonads*
Source : Das *et al.,* 2004

one way. A list of important bioinoculants is provided here, which have been isolated from different locations of Uttarakhand and are available in gene pool of GBPUA&T (Table 4). Several of these strains have been found to have multiple activities and are very popular with farmers, more so with those who are practicing organic farming.

Table 4. Microbial bioresources available at G.B. P. U & T., Pantnagar

Isolates\Mutants	Properties	Crop Tested	Reference
Bioinoculants			
KNP9	Sid+, Cd^r, Cu^r, Pb^r, Growth Promotion	Mung Bean	Tripathi, 2005
PRS 9	Hg^r, Growth Promotion	Soybean	Gupta *et al.*, 2005
CRPF5,CRPF8	'P' Dynamics of soil	Mung Bean	Das *et al.*, 2003
NBRI 4014	Sid+ ,P+, IAA+	Soybean	Gupta *et al.*, 2002 and 2004
CRPF8	Sid+, Growth Promotion	Mung Bean	Katiyar and Goel, 2004
TH18	Cu^r ,'P' Solubilizer	Black Gram	Gupta *et al.*, 2004
CRPF1	Cold Resistant, Growth Promotion	Mung Bean	Katiyar and Goel, 2003
CRPF7	Cold Resistant, 'P' Solubilizers	Mung Bean	Das *et al.*, 2003
CD7	Metal Resistant, Osmophilic	Pulses	Mittal *et al.*, 2003
CG1	Cu^r ,'P' Solubilizer	Black Gram	Prameela *et al.*, 2002
GRS1	'P' Solubilizer & Sid+	Soybean	Mishra and Goel, 1999
PRS1	Sid+, Biocontrol	Wheat	Saxena *et al.*, 1999

MICROBES IN ENVIRONMENTAL MANAGEMENT

Isolation of Cold Tolerance Gene(s)

A large area of the Himalayan region in Uttarakhand has temperate climate, thus providing us with an opportunity to isolate, identify and screen bacteria showing cold tolerance. Considering the unique features and biology of psychrophiles, studies were carried out where twenty-five of such strains were isolated from heights of 2000 mts and above. These were characterized in lab for their ability to grow at 4°C or below. These remain uncharacterized except for two strains, which were found to be growth promoting in wheat and lentil (Katiyar, 1999).

In another strategy a strain of *Pseudomonas* sp. was selected and subjected to mutation with nitrosoguanidine (Mishra and goel, 1999). When mutated cultures were grown at 4°C, eight mutants could be isolated with their ability to grow efficiently at this temperature as well as solubilizing phosphate at 10°C (Fig. 3).

These mutants were useful as bioinoculants at low temperature for mung, soybean and wheat. Interestingly the primers based on database of cold shock protein(s)/gene(s) of *Pseudomonas*, when used for PCR based cloning of cold shock proteins at last one full length gene could be isolated (Khan, 2005). Its transfer to a mesophilic *E. coli* host made it to grow at 4°C. The gene was later expressed at transcriptional and translational level. It has been sequenced and deposited in NCBI database (DQ111780, Table 2).

These psychrophilic and psychrotolerant bacteria would be helpful in understanding the mechanism of cold shock tolerance. There have been reports that at molecular level their

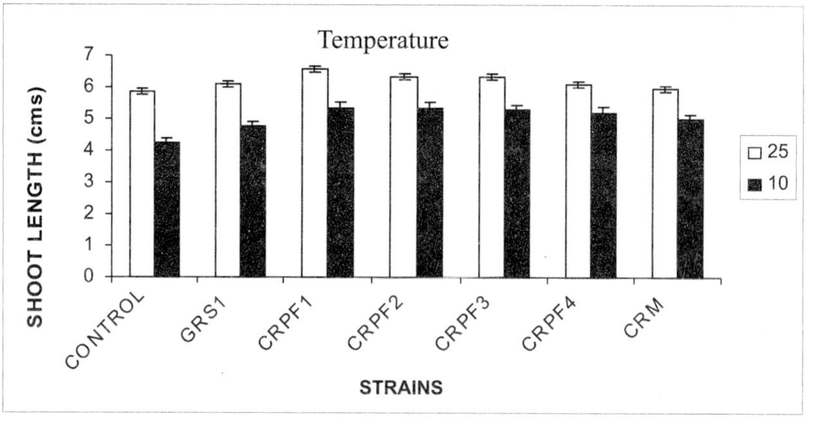

Fig. 3. Fluorescent *Pseudomonads* as diazotrophs both at 25 and 10°C.

adaptation at low temperature is dependent on a number of survival strategies such as modulated membranes fluidity, biochemical reaction at low temperatures, and ability to regulate gene expression at low temperature (Chintalapati, 2004; Alberts, *et al.,* 1994; Russell, 2002; Neidhart, *et al.,* 1990). The presence of cold active enzymes/proteins (Wunderlich, *et al.* 2005; Gerdy, *et al.,* 1997), and the ability to support transcription and translation at low temperature is also documented (Yamanka, *et al.,* 1999; Berry and Foegeding, 1997; Jones, *et al.,* 1987). In addition, genes essential for survival at low temperature, and genes that only express at low temperature are identified. Moreover, polymeric substances, which could serve as cryoprotectant, and cold shock and cold acclimation proteins, which function as transcriptional enhancers, are also characterized (Schindler, 1999; Jones and Inouye, 1994; Jones, *et al.,* 1987). This understanding of cold resistant biology may lead to the development of plant varieties that are able to tolerate greater cold stress, which is indispensable in view of the cold climate.

Use of Microorganisms in Bioremediation

The industrial revolution has become a cause for widespread contamination of land and water. Effluents released from industries include a wide range of synthetic chemicals such as plasticizers, dyes, pigments, agrochemicals, solvents, pharmaceuticals, hydraulics, fire retardants, halogenated compounds, etc. Also the modern agricultural practices utilize large amounts of chemicals such as fertilizers, herbicides and insecticides. Presence of these xenobiotic compounds, pollutants along with heavy metals has severely affected the productivity of soil. They do have the potential to disturb the population dynamics and growth of plants in metal contaminated sites. In India about 174 million hectare area is considered as degraded land and is not fit for agricultural purposes. Since land is a limiting resource in agriculture, one of the alternatives to improve the crop production in these lands is through selected bioinoculants (mutants, Table 4), which can tolerate or degrade these soil contaminants effectively, thereby facilitating utilization of degraded polluted or unused lands, which are otherwise unfit for agricultural practices (Tripathi, 2005).

Bioremediation of Heavy Metals

Few major industrial sites are located in Uttarakhand too (Table 5), which show metal pollution.

Table 5. Major industrial sites and pollutants in Uttarakhand

Industries	Metal Pollutants
Paper Industry (Century Pulp and Paper Mill, Lalkuan, Nainital)	Pb, Cu
Electronic Industry, BHEL, Haridwar	Pb, Cu, Cd, Ni
Dyeing Industries, Almora, Bageshwar	Pb, Cr
Indian Drugs and Pharmaceuticals, Rishikesh	As, Ni

The management of the environment is a key and controversial issue in modern agriculture, and the role of microorganisms in environmental management holds significance. Considering the toxic effects of heavy metals in Uttarakhand, various strains have been isolated from industrial and groundwater sources, which have high arsenic resistance property. The molecular characterization of these potent strains has also been (Gupta, 2006) documented under the new strain improvement programme (Table 6).

Table 6. Microbes in bioremediation at GBPUA & T

Isolates	Properties	Metal/Polymer	Reference
AGM 6	Arsenic Resistant Bacteria	Ar	Gupta, 2006
AGM 16	Arsenic Resistant Bacteria	Ar	Gupta, 2006
AGM 32	Arsenic Resistant Bacteria	Ar	Gupta, 2006
AGM 13	Arsenic Resistant Bacteria	Ar	Gupta, 2006
LNA 3	HDPE/LDPE/Polycarbonate degrading bacteria	HDPE/LDPE/Polycarbonate	Soni, 2006
PN 12	HDPE/LDPE degrading bacteria	HDPE/LDPE	Soni, 2006
CNA 2	LDPE degrading bacteria	LDPE	Soni, 2006
PN 13	LDPE degrading bacteria	LDPE	Soni, 2006
CN 4	LDPE degrading bacteria	LDPE	Soni, 2006

Biodegradation of Plastic Polymers

The ever-increasing plastic waste has posed a great threat to the environment all around, but more so in the state of Uttarakhand, which is a premier spot for tourism. Work has been in progress for biodegradation of polymers from plastic waste. These studies have been carried out on polycarbonate, polyester (Kusumlata, 2004), polysulfone (Pratap K.C., 2004) and polyethylene. The strains that could degrade both polymers of polyethylene (high density and low density polyethylene, HDPE/LDPE) were isolated by enrichment selection from waste disposal sites (Kumari S., 2005; Soni R., 2006). The strains used in this study were collected from different agroclimatic zones of Uttarakhand, and after their rigorous screening and selection, some important ones were characterized at molecular level and their sequences were submitted to NCBI database along with those of heavy metal resistant strains (Table 1). A consortium was also developed with better efficiency and performance.

The polythene samples were taken in powder form and incubated for ten days with cultures isolated from enrichment sites in minimal media without carbon source. The residual powder was recovered by evaporation of the filtrate after centrifugation at 5000 rpm for 10 min. This sample was analyzed by FTIR spectroscopy along with what remained as residue over the filter paper. The results presented here (Fig. 4) clearly show the sign of degradation as LDPE and its degraded product has common absorption frequencies corresponding to CH stretching 3039.4 - 3514.5, CH_2 deformations 1595.1-1598.8 (Ferguson, 2004), CH bending (asymmetrical) 1383.7-1465.2, CH_2 bending (symmetrical) 1351-1384 and CH_2 rocking 718–769 cm^{-1}. A significant shifting in CH_2 stretching and C-O stretching frequencies was observed owing to microbial degradation of LDPE. LDPE did not show any C-O stretching frequency, whereas its biodegradation in the presence of consortium showed C-O stretching frequency ranging from 1056-1164.8 cm^{-1}. These observations clearly indicate that all consortia have significantly affected the 'C' chain of LDPE and induced bioactive hydrolysis. Prolonged adaptation and molecular characterization are needed before definite mechanism can be delineated. However, these studies provide hope of attaining microbial degradation of these recalcitrant materials.

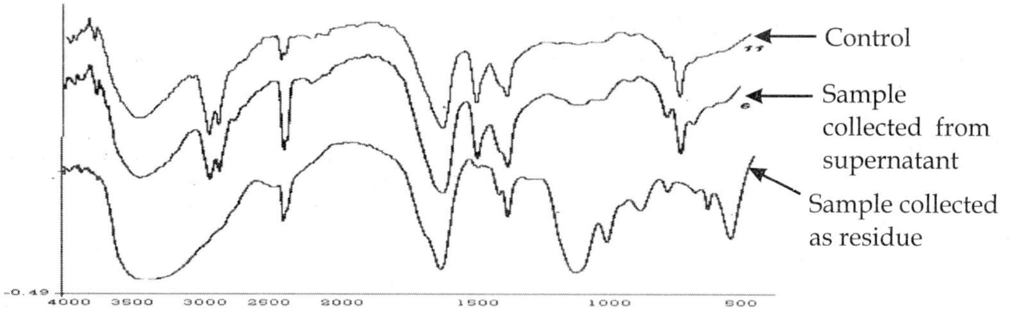

Fig. 4. Comparative FTIR spectra of control vs supernatant and residue.

OTHER IMPORTANT CONTRIBUTIONS

In addition to GBPUAT; Gurukul Kangri University (GKU) has also played a pivotal role in exploring the novel microbial diversity of this region. This work is significant in view of understanding of plant-microbe interactions (Pandey and Maheshwari *et al.*, 2007), PGPR activity (Pandey *et al.*, 2005, 2005), biocontrol activity of different strains (Gupta *et al.*, 2006) and exploration of rhizobial genetic diversity of sub-Himalayan region of Uttarakhand (Pandey *et al.*, 2004).

Pandey and Maheshwari (2007) have studied the interactions between two plant growth promoting rhizospheric isolates, *Burkholderia* sp. MSSP and *Sinorhizobium meliloti* PP3. The microorganisms were grown as monospecies or mixed species culture and studied for growth profile, IAA production and phosphate solubilization. They have found that IAA production was increased upto 50% in mixed species culture and there was an exceptional increase in seedling growth in mixed species, coinoculated consortium (Table 7).

Table 7. Effect of Burkholderia sp. MSSP and *Sinorhizobium meliloti* PP3 on vegetative growth of *Cajanus cajan* after 40 days of seed germination
Growth measurement*

Strain	Seed germination (%)	Total plant length (mm)	Shoot length (mg)	Fresh root weight (mg)	Fresh shoot weight (g)
Control	70	125a	101a	401a	2.15a
MSSP	100	188b	150b	835b	3.18b
PP3	90	142a,b	111a	560a	3.47b
MSSP+PP3	100	312c	242c	750c	6.06c

*Mean value was calculated from three replicates of each treatment. Results were analyzed by Student's *t* test and there was no significant difference between treatments followed by the same letters in each column at *P*-0.05.
Source: Pandey and Maheshwari 2007

Arora *et al.* (2007) studied the role of chitinase and β-1, 3-glucanase activities produced by fluorescent pseudomonads and *in vitro* inhibition of phytopathogenic fungi, *Phytophthora capsici* and *Rhizoctonia solani*. The results have shown the involvement of chitinases, β-1, 3-glucanases and antifungal activity against *Sclerotonia sclerotiorum* (Gupta *et al.*, 2006). The scanning electron microscopic (SEM) studies have shown morphological abnormalities such as perforation, lysis and fragmentation of hyphae of *S. sclerotiorum* caused by *P. aeruginosa* GRC1. Few important bacterial strains isolated and functionally characterized by Department of Botany and Microbiology, GKU are summarized in Table 8.

Table 8. Important microbes functionally characterized at GKU

Strain	Species	Significance	Reference
PP1	*Rhizobacteria sp.*	Rhizobacterial diversity	Pandey *et al.*, 2004
PP2	*Rhizobacteria sp.*	-do-	Pandey *et al.*, 2004
PP3	*Sinorhizobium meliloti*	-do- PGPR activity	Pandey *et al.*, 2004; Pandey and Maheshwari, 2007
PP4	*Rhizobacteria sp.*	-do-	Pandey *et al.*, 2004
PP5	*Rhizobacteria sp.*	-do-	Pandey *et al.*, 2004
PP6	*Rhizobacteria sp.*	-do-	Pandey *et al.*, 2004
PP7	*Rhizobacteria sp.*	-do-	Pandey *et al.*, 2004
PP8	*Rhizobacteria sp.*	-do-	Pandey *et al.*, 2004
PP9	*Rhizobacteria sp.*	-do-	Pandey *et al.*, 2004
MSSP	*Burkholderia sp.*	Antifungal activity, PGPR activity	Kang *et al.*, 2004; Pandey *et al.*, 2005; Pandey and Maheshwari, 2007
GRC1	*Pseudomonas aeruginosa*	Antifungal activity, PGPR activity,	Pandey *et al.*, 2005; Gupta *et al.*, 2006
GRC3	*Pseudomonas sp.*	Antifungal activity	Arora *et al.*, 2007
GRC4	*Pseudomonas sp.*	Antifungal activity	Arora *et al.*, 2007

CONCLUSION

In view of the Himalayan region as a "Hot spot" for biodiversity, it is imperative for us to explore the unique microflora of Uttarakhand. Thus with the support of Ministry of Environment and Forests (MOEF), a "Centre for Bacteria and Archea" has been established and under this programme scientists have collected more than 200 microbial strains from Kumaon and Garhwal hills. More than 80 strains of these have been characterized at biochemical and molecular level. Studies have been conducted in context of their association with host plants such as wheat, mandua, and *Dalbergia sissoo* also. They have been observed for their ability to fix nitrogen, solubilizing phosphate, heavy metal bioremediation and biodegradation of xenobiotic and recalcitrant compounds. The 16S rRNA sequences of potent strains have been submitted to NCBI database. Since agro-climatic and physiological conditions of this state are unique and fragile, thus for the socio-economic development of local population, better agricultural practices are required. Now, considering the fact that only 1% of microbial diversity could be explored through culture dependent methods, much more efforts are needed for work on culture dependent and independent methodologies. This would be helpful in developing a database of novel genes and genomics for sustainable development in the fields of agriculture, environment management and commercial applications. Moreover, this large bacterial gene pool would also be helpful in the development of Biotech Park in Uttarakhand.

ACKNOWLEDGEMENTS

The work presented here has been supported by DST, DBT, UGC, and CSIR grants to identified departments at different points of time. We thank Dr. (Ms.) B. Saini, Associate Professor for carefully reading the manuscript.

REFERENCES

Alberts, B., Bray, D., Lewis, J., Raff, M., Roberts K. and Watson J.D. (1994). *In:* Molecular biology of the cell. pp. 477-506.
Arora, N.K., Kim, M.J., Kang, S.C. and Maheshwari, D.K. (2007). *Can. J. Microbiol* **53(2):** 207-212.
Berry, E.D. and Foegeding, P.M. (1997). *J. Food. Prot.* **60(12):** 1583-1594.
Chintalapati, S., Kiran, M.D. and Shivaji S. (2004). *Cell. Mol. Biol.* **5:** 631-642.
Das, K., Katiyar, V. and Goel R. (2003). *Microbiol. Res.* **158:** 1-4.
Ferguson J.D., Weimer A.W. and George S.M. (2004). *Chem. Mater.* **(16):** 5602-5609.
Gerdy, C., Aittalab, M., Arpingly, J.L., Baise, E., Chessa, J.P., Grasoux, G., Petrescu, I. and Feller, G. (1997). *Biochim. Biophys. Acta.* **1342:** 119-131.
Graham, P.H. (1988). *Principles and Application of Soil Microbiology.* 322-345.
Gupta, A., Kumar, M. and Goel, R. (2004). *Biol. Trace. Ele. Res.* **98:** 1-9.
Gupta, A., Meyer, J.M. and Goel, R. (2002). *Curr. Microbiol.* **45:** 323-327.
Gupta, A., Rai, V., Bagdwal, N. and Goel, R. (2005). *Microbiol. Res.* **160(4):** 385-388.
Gupta, C.P., Kumar, B., Dubey, R.C. and Maheshwari, D.K. (2006). *Biocontrol* **51:** 821-835.
Gupta, A. (2006). Ph.D Thesis, GB Pant University of Agriculture and Technology, Pantnagar.
http://www.kalpavriksh.org/kalpavriksh/kalpavriksh/f1/f1.1/index_html
Jones, P.G. and Inouye, M. (1994). *Mol. Microbiol.* **1(5):** 811-8.
Jones, P.G., Van Bogelen, R.A. and Neidhardt, F.C. (1987). *J. Bacteriol.* **169:** 2092-2095.
Kang, J.G., Shin, S.Y., Kim, M.J., Bajpai, V., Maheshwari, D.K. and Kang, S.C. (2004). *J. Antibiot,* **57:** 726-731.
Katiyar, V. and Goel, R. (2003). *Microbial. Res.* **158:** 163-168.
Katiyar, V. and Goel, R. (2004). *Plant Growth Regulation.* **42(3):** 239-244.
Katiyar, V. and Goel, R. (2004). *J. Microbio. & Biotechnol.* **14(4):** 653-657.
Katiyar, V. (1999). M.Sc. Thesis, GB Pant University of Agriculture and Technology, Pantnagar.
Khan, M. (2005) Ph.D Thesis, GB Pant University of Agriculture and Technology, Pantnagar.
Kumar, C.P. (2004). M.Sc. Thesis, GB Pant University of Agriculture and Technology, Pantnagar.
Kusumlata. (2004). Thesis, GB Pant University of Agriculture and Technology, Pantnagar.
Mishra, M. and Goel, R. (1999). *J. Biotechnol.,* **75:** 71-75.

Mittal, S., Meyer, J.M. and Goel, R. (2003). *Ind. J. Biotechnol.*, **2(4):** 00.
Mittal, S. (2004). Ph.D. thesis. G.B. Pant University of Agriculture & Technology, Pantnagar.
Neidhart, F.C., Ingraham, J.L. and Schaeter, M. (1990). Physiology of the bacterial cell: a molecular approach. Sunderland, Mass.: Sinauer Associates Inc.
Pandey, P. and Maheshwari, D.K. (2007). *Curr. Sci.,* **92:** 1137-1142.
Pandey, P., Kang, S.C., Gupta, C.P. and Maheshwari, D.K. (2005). *Curr. Microbiol,* **51:** 303-309.
Pandey, P., Sahgal, M., Maheshwari, D.K. and Johri, B.N. (2003). *Current Science,* **86:** 202-207.
Prameela, M.S. and Goel, R. (2002). *Asian J. Microbiol. Biotech. Env. Sci.* **4(3):** 103-105.
Report of National Forest Commission (annual report 2004-2005). Ministry of Environment and Forests (Govt. of INDIA). http://www.envfor.nic.in
Russell, N.J. (2002). *Int. J. Food. Microbiol.*, **79:** 27-34.
Sahgal, M. and Johri, B.N. (2006). *Curr. Sci.*, **90(4):** 486-487.
Sahgal, M., Sharma A., Johri, B.N. and Prakash, A. (2004). *Symbiosis.* **36:** 83-96.
Samra, J.S., Dhyani, B.L. and Sharma, A.R. (1999). Problems and prospects of natural resource management in Indian Himalayas–A Base paper. Dehradun: Allied Printers pp. 1-32.
Sarita, K. (2005). M.Sc. Thesis, GB Pant University of Agriculture and Technology, Pantnagar.
Saxena A. and Goel R. (1999). *Physiol. Mol. Biol. Plants.*, **5:** 181-184.
Schindler, T., Graumann, P.L., Perl, D., Ma, S., Schmid, F.X. and Marahiel, A. (1999). *J. Biol. Chem.* **274:** 3407-3413.
Shukla, O.P. (1984). *J. Sci. Ind. Res.,* **43:** 98-116.
Soni, R. (2006). M.Sc. Thesis, GB Pant University of Agriculture and Technology, Pantnagar.
Thajuddin, N. and Subramanian, G. (2005). *Curr. Sci.*, **89:** 47-57.
Tilak, K.V.B.R., Ranganayaki, N., Pal, K.K., De, R., Saxena, A.K., Nautiyal, C.S., Mittal, S., Tripathi, A.K. and Johri, B.N. (2005). *Curr. Sci.*, **89:** 136-150.
Tiwari, R.P., Gaur, N.K. and Garg, G.K. (1989). *Indian J. Biochem. Biophys.*, **26:** 43-47.
Tripathi, M. (2005). Ph.D Thesis, GB Pant University of Agriculture and Technology, Pantnagar.
Unesco's man and the biosphere programme (MAB), http://www.unesco.org/mab/mabProg.shtml
Wunderlich, M., Martin, A. and Schmid, F.X. (2005). *J. Mol. Biol.,* **347:** 1063-1076.
Yamanka, K. (1999). *J. Mol. Microbiol. Biotechnol.*, **1:** 193-202.

Chapter 21

Antimicrobial Properties of Essential Oils and their Potential Applications in Pharmaceutical Industries

R.C. DUBEY*, K.K. GUPTA AND R.R. PANDEY**

Department of Botany and Microbiology,
Gurukul Kangri University, Haridwar–249 404 (Uttarakhand)
***Department of Life Sciences, Manipur University–Canchipur, Imphal–795 003 (India)*
E-mail: profrcdubey@gmail.com

Essential oils are commonly a mixture of lipophilic compounds mainly belonging to the class of terpenes, aliphatic and aromatic compounds. Since middle ages, essential oils have been widely used for bactericidal, virucidal, fungicidal, insecticidal, antiparasitical, medicinal and cosmetic applications. Nowadays, pharmaceutical, sanitary, cosmetic, agricultural and food industries are involved in the production of wide ranges of products based on essential oils. Phytochemical analysis indicates that many of them contain more than 100 individual compounds, and therefore, it is difficult to attribute specific properties of the entire oils to the presence or absence of a single compound. Antimicrobial studies of essential oils provides a great deal of knowledge on their possible use in therapeutics. The aim of this review article is to discuss comprehensive information on various aspects of plant essential oils ranging from their traditional uses in India to mode of action, extraction procedures, antimicrobial bioassays and quality control.

INTRODUCTION

The plants used for medicinal purposes throughout the world are clubbed under the category of Medicinal Plants. According to World Health Organization (WHO, 1978), *'a medicinal plant is any plant which, in one or more of its organs, contains substances that can be used for therapeutic purposes,* or *which are precursors for chemo-pharmaceutical semi-synthesis'*. The World Health Organization has defined the Traditional Medicine as 'the sum total of all the knowledge and practices, whether explicable or not, used in diagnosis, prevention and elimination of physical, mental or social imbalance and relying exclusively on practical experience and observation handed down from generation to generation, whether verbally or in writing' (WHO, 1978). Traditional medicine is, therefore, used mainly to distinguish the ancient and culture-bound healthcare practices, which existed before the application of science to health matters in official modern scientific medicine or allopathy. Rawat (2006) suggested some common features of traditional medicines as below:

- Herbal medicines are different from clinically defined medicines in their character as well as in their medicinal value.
- They are based mostly on herbal products.
- Usually, they are multi-drug formulations including animal and mineral products as essential components or additives.
- In herbal therapy, data on pre-clinical investigations are often incomplete, although in majority of cases the therapeutic experiences have been accumulated over centuries.
- Some of them follow practices based on, for example, mistaken beliefs, faulty experimentation, or inaccurate information that can be dangerous.
- They mostly include empirically defined doses and course of medication.
- The identity of plant species used is often controversial.
- Safety measures are poorly adopted, in most cases.
- Additives are frequently used; many of them also have therapeutic actions.

In recent years, the study of medicinal plants has assumed great importance in India and abroad. The less studied Indian system of medicine is now being considered a field of national concern. Medicinal plants-based drug industry has four major segments: (i) plant drugs for WHO recognized Indian systems of medicine, (ii) over the counter non-prescription items involving plant parts, extracts and galenicals, (iii) phyto-pharmaceuticals, and (iv) essential oils.

The purpose of this paper is to provide an over view of the published data on the antimicrobial properties of essential oils and its components that may be considered suitable for potential application in pharmaceutical industries.

WHAT IS ESSENTIAL OIL?

Aromatic plants are generally referred to as essential oil yielding plants. They have volatile, odoriferous oils in special cells, glands or ducts located in different parts of a plant, such as, the leaves, barks, roots, flowers and fruits and sometimes in just one or two parts. The oils are usually present in very small amounts and comprise only a small fraction of the entire plant material. The oils are produced during some metabolic processes of the plant and are secreted or excreted as odoriferous by-products. The fragrant oils may not necessarily be present as such in the living plants, but may occur as odourless compounds termed as glycosides. An enzyme reaction occurs after macerating the tissues, which in turn causes the glycosides to undergo a chemical change. This action in turn liberates the distinctive essential oil (Anonymous, 2004). Essential oils are also known as volatile or ethereal oils. These are aromatic liquids obtained from plant material i.e., flowers, buds, seeds, leaves, twigs, bark, herbs, wood, fruits and roots.

The term essential oil is thought to derive from the name coined in 16th century by the Swiss reformer of medicine, Paracelsus von Hohenheim; he named the effective component of a drug *Quinta essentia* (Guenther, 1948).

USE OF AROMATIC PLANTS IN TRADITIONAL MEDICINE IN INDIA

In India, the tradition of using plants for healthcare goes back several thousand years. Vedic texts deal extensively with medicinal properties of plants. In Rigveda (4500-9000 B.C.), which is an old repository of human knowledge, about 67 plants of medicinal value have been described. The *Atharvaveda* and the *Yajurveda* mention 200 and 81 species of plants of therapeutic value. Charak (1000 B.C.) and Sushruta (800 B.C.) quoted 341 and 395 medicinal plants, respectively in their *Samhitas*. Out of 17,500 flowering plant species known from India, more than 4000 species are used as medicinal plants, of which 300 species yield gum and dyes and about 100 species yield essential oils used as raw materials in drug industry. Table 1 shows some commonly used selected essential oils of some plants.

Table 1: Common uses of selected essential oils.

Essential oil	Common uses
Citrus	Industrial solvents, fragrance for cleaning products, flavouring
Spearmint	Toothpaste, mouthwash, confectionery flavouring
Peppermint	Toothpaste, mouthwash, chewing gum, food flavouring, cosmetics, tobacco flavouring
Lavender/lavandin	Fragrances, toiletries
Eucalyptus	Cough/cold remedies, solvents, cleaning agents, flavouring
Tea tree	Toiletries, insect repellents, germicides, cosmetics
Boronia	Food flavouring, fragrance
Blackcurrant bud	Food and beverage flavouring

Indian system of medicine (ISM) exists at two levels. The *classical system* encompassing the well documented, codified systems like Ayurveda, Siddha, Unani, which have textbooks, is taught in colleges, which grant degrees and the practitioners of which are incorporated into the official health care system. The name ayurveda comes from two Sanskrit words: *ayur* meaning 'life' and *veda* meaning 'knowledge'. According to *ayurvedic* teaching, everyone and everything in the universe consists of three basic forces or elements. In Sanskrit they are called *vata, pitta* and *kapha* (often pronounced *vat, pit* and *kaph*). They together control all physical and mental processes and are compared to the workings of the wind, the sun and the moon. *Vata* is linked to the wind, which is constantly on the move, and controls the central nervous system. *Pitta* is like the sun, a source of energy. It controls the digestive system and all biochemical processes. *Kapha* governs the balance of tissue fluid, controlling cell growth and the firmness of the body, as the moon governs the tides. The other is the *informal system* of folk medicine, the *Lok Parampara* (tradition) which exists in communities, passed orally from generation to generation, for which there are no books and little documentation and which is not part of the official system. These folk traditions are rich and diverse knowledge base complex, and with the their ability to heal a wide range of ailments, these (often tribal) healers or vaidyas are the backbone of healthcare for 80% of India (Sahai, 2000).

Curcuma longa (= *haridra* or *haldi*) has been intimately associated with customs and traditions of India since the time immemorial. 'Haridra' indicates that it is dear to Hari (Lord Vishnu); it is also believed that the cloths worn by Lord Krishna (*Peethambaram*) were dyed with turmeric. In many Indian villages during ancient period, turmeric was used for the treatment of jaundice. Turmeric paste was applied over the body of patient suffering from jaundice and after that the turmeric was washed off. It was believed that the disease also got washed off together with the turmeric (Remadevi and Ravindran, 2005). In *ayurveda*, turmeric is used to treat diabetes, eye disease, ulcers, oedema, anaemia, anorexia, leprosy, scrofula, allergy and to boost immunity and enhance complexion. Its paste combined with that of neem leaves is used to cure ringworm, itching, eczema and other parasitic skin diseases (Thankamani *et al.*, 2005). Yield of essential oils in different parts of plants vary. Essential oil yield from different parts of *Curcuma longa* is given in Table 2.

Table 2: Essential oil yield from *Curcuma longa* L. (Leela *et al.*, 2002)

Plant part	Essential oil yield (%)
Flowers[a]	0.3
Leaves[b]	1.3
Rhizomes[b]	3.8
Roots[b]	4.3

[a]Fresh mass basis; [b]Dry mass basis

Ayurvedic practitioners refer ginger (*Zingiber officinale*) as medicine for stomach. Ginger juice reduces motion sickness, aids digestion and absorption. Ginger eases the transport of substances through digestive tracts, thus decreasing the irritation in intestinal wall. The triad scholars of ayurveda viz., Charak, Sushruta and Vagbhata mentioned the use of ginger in Udara roga (disease of abdomen), Shotha (Oedema), Karnasula (Earaches) and disease of cough and Vatha roga (Thankamani *et al.*, 2005). Ayurvedic medicine uses coriander (*Coriandrum sativum*) for numerous purposes. The crushed and roasted seeds mixed with warm water are used as a gargle for ulcers, or mixed with cold water as an eye lotion in cases of conjunctivitis. Coriander leaves and seeds were employed as an aid to digestion. The pharmacological value of 'elaichi' (*Elettaria cardamomum*) belongs to seed, having importance against urine retention and stomach disorders. Seeds are used in ayurveda as a tonic for heart and as an excellent expectorant for treating respiratory diseases.

In the first century A.D., Charaka was prescribing dried leaves and bark of 'tejpatta' (*Cinnamomum tamala*) for fever, anaemia, and body odour. Its seeds were crushed and mixed with honey or sugar, and administered to children for dysentery or coughs. Ayurveda uses garlic (*Allium sativum*) to treat a wide range of diseases, e.g., rheumatism for the lung as an expectorant and powerful decongestant, for heart to lower blood pressure and reduce high cholesterol, for the stomach as an anti-helminthic that expels roundworms from the system, for gynaecological problems to regulate or bring on menstruation and for improving the libido. Ayurveda also prescribes garlic for anorexia and for diseases of the vocal cords. Externally, garlic is applied for deafness and pain; crushed garlic fried in mustard or coconut oil is used as an antiseptic liniment, especially in the relief of scabies sores and ulcerating wounds.

The oil from bark of 'dalchini' (*Cinnamomum zeylanicum*) is useful for toothaches and painful gums. The cinnamon bark is also administered by ayurvedic doctors for anorexia, bladder disorders, and as a tonic for the heart. Its most common use is as an expectorant and decongestant, where cinnamon bark is boiled in water and inhaled as a vapour for coughs, colds, and sore throats. Ayurveda considers lime juice (*Citrus medica*) boiled with water as particularly useful against vomiting attacks and dehydration. It is also widely used to help cure coughs, and in patients suffering from anorexia, as its vitamin content makes it both a tonic and an appetizer. Charaka recommended the use of 'jaiphal' (*Myristica fragrans*) as a medicine for external applications only. However, when used in small amounts, powdered jaiphal forms one of the constituents of many ayurvedic prescriptions for nausea, dysentery, insomnia, as well as smoker's cough and other bronchial irritations. The pungent and aromatic clove (*Synzygium aromaticum*) is widely used by Indians as a mouth freshener.

Charaka noted the value of clove as an antiseptic for inflammations of the mouth and throat, as well as its capacity to cure digestive action, since bad breath is often caused by poor digestion. Later, ayurvedic physicians called the clove the most stimulating and carminative or all aromatics. The oil is a strong antiseptic and germicide, although its potency as a local anesthetic, as in cases of toothache, is quite feeble. Most common use of camphor or kapur (*Cinnamomum camphora*) is in vapours and liniments for colds, chills, and inflammatory conditions. It is used in ayurveda locally to numb the peripheral sensory nerves and as a counter-irritant in rheumatisms, sprains and inflammatory conditions. Physicians prescribe camphor internally for its calming influence in hysteria, neuralgia, and other nervous ailments, and it is sometimes used in cases of heart failure. But ayurvedic science stresses that camphor is very acrid, and if taken in large doses, it may be highly poisonous. *Eucalyptus globulus* oil is used locally as an antiseptic, especially in the treatment of the infections of upper respiratory tract and in certain skin diseases (Saini, 2004).

EXTRACTION OF ESSENTIAL OILS

Essential oils are usually obtained by physical means such as steam distillation (SD) or hydrodistillation (HD) (Whish, 1996; Ramanadhan, 2005).

In steam distillation, the steam is passed through crushed plant material that is placed on a mesh between the steam inlet and the condenser (Whish, 1996; Houghton and Raman, 1998). The process results in the rupture of gland cells and release of the substances therein. The steam plus volatile oil passes through a cold water condenser allowing the volatile oil fraction to float on the top of water. Essential oil is drawn from the mixture of water and oil. Steam distillation makes use of the principle that two immiscible liquids, when mixed, each exert a vapour pressure, as if each liquid were pure (Houghton and Raman, 1998). The total vapour pressure of the boiling mixture is, therefore, equal to the sum of the partial pressures exerted by each of the individual components of the mixture. When the total vapour pressure reaches atmospheric pressure, the mixture starts to boil. This implies that the boiling point of the mixture is reached at a lower temperature than the boiling points of the individual components. Steam distillation is, therefore, able to separate volatile from nonvolatile components with a reduction in boiling point, thereby avoiding extreme temperatures. An added advantage is the displacement of atmospheric oxygen by the steam, which protects the compounds from oxidation (Krell, 1982). However, hydrolysable compounds such as esters as well as thermolabile components may be decomposed during the distillation process (Houghton and Raman, 1998). In addition, partial loss of more polar constituents of the oil, due to their affinity for water, may also occur (Masango, 2004; Baker *et al.*, 2000).

In hydrodistillation procedures, the material is immersed in water, which is heated to boiling point using an external heat source. Once the vapours get condensed, the liquid mixture is poured in a separator where the oil and water are separated by density difference (Denny, 1991). The water sometimes is also used and is marketed as floral waters (also called hydrosol or sweet water), such as rosewater, lavender water and orange water. Some disadvantages are the variability on the distillation rate due to the uncontrolled heat rate input, possible overheated and eventually the aromatic materials can get burned. Improved distillation control can be obtained by using steam from a separate boiler. A further disadvantage of this system is that it requires the heating of a large quantity of water and the associated energy demand adds to the costs and time needed for each distillation (Eikani and Rowshanzamir, 2004). In both hydro- and steam-distillation techniques, the vapours are allowed to condense and the oil is then separated from the aqueous phase. Care must be taken to ensure efficient condensation of steam, thereby preventing the loss of the more volatile oil components (Houghton and Raman, 1998).

Water/steam distillation is very similar to the hydrodistillation, the difference is that in this method the still contains a grid which keeps the plant material above the water level. This arrangement can be compared to a kitchen steamer basket. The plant material is supported in a basket over boiling water, thus exposing the plant material only to the rising steam vapours. The water is boiled below the charge and wet steam passes through the plant material. It is important in both water/steam and steam distillation that the still is packed evenly and not too tightly, so that steam can extract from the complete charge efficiently. Overpacking of the still can cause the steam to force rat holes through the charge and leave other parts of the charge unextracted (Reverchon 1994).

Solvent extraction is a different technique which gives different results from steam distillation (Scheffer, 1993). The choice of organic solvent for extracting volatile oils is usually restricted to alcohols (methanol and ethanol). Solvent extraction can be used to extract at lower temperatures, which is an advantage where composition of oil could be changed by heat. Difficulties may arise in separating essential oils from the solvent and this could involve an extra stage in the extraction process. Hydrodistillation apparatus of European Pharmacopoeia is standard research apparatus and this is useful for comparisons. However, it may not be most suitable for thermally labile components, which may rearrange if temperature exceeds 40°C during extraction. For these, a solvent mix (e.g., pentane + diethyl ether) may be used for the extraction. This may lead to contamination and difficulty, or expense, of separation by chromatography.

The other method of extraction is the solvent free microwave extraction (SFME), which is a combination of microwave heating and distillation at atmospheric pressure. Microwaves cause excitation of water molecules in the plant tissues causing plant cells to rupture and release the essential oils trapped in the extracellular tissues of the plant (Bélanger *et al.*, 1991). The benefit of this new alternative is mainly the reduction of extraction time which leads to a big opportunity in energy savings and improving the product quality (Michael Spiro 1995). Microwaves cause the physical rupture of the cells and the glands more rapidly than in conventional hydrodistillation or steam distillation (Lucchesi *et al.*, 2007). Lucchesi *et al.* (2004) conducted experiments using Ajowan and Cumin showing that the composition of essential oils extracted by SFME and hydrodistillation has major differences in their aromatic profiles. Larger amount of desired compounds were found in the oils extracted by SFME due to the diminution of thermal and hydrolytic effects, since this method uses the *in situ* water of the plant. Kwon *et al.* (2003) compared the potential of SFME with the conventional hydrodistillation method for the extraction of essential oil from cardamom seeds. It was found that extraction time, irradiation power or moisture content can be optimized to obtain a high yield of essential oil, or to obtain essential oils of differing composition.

Microwave ovens use radio waves to convey energy at a frequency which is approximately 2500 megahertz. Waves in this frequency range are absorbed by water, fats and sugars and other molecules with a sufficiently polar oxygen group. In the specific case of water, which is an electric dipole (positive charge on one end and negative charge on the other), once the electric field is induced, the molecules try to align within it. This motion and the collision with other molecules produce heat. In other words, when microwaves excite the molecules producing rotational and translational motion and this motion generates heat (Buffler, 1993). Heat is generated everywhere at once because the microwave energy applied excites molecules throughout the material. However, radio waves penetrate unevenly in thick pieces, therefore depending on the sample, the microwaves sometimes do not penetrate all the way to the middle, and there is also localized heat caused by wave interference (Buffler, 1993). Microwave energy is an advantageous alternative to several thermal applications due to its efficient volumetric heat production (Ramanadhan, 2005). The volumetric heating (heating of the bulk) as opposed to transferring heat from the surface inwards, is more efficient and uniform (Sau Soon Chen, 1995). The ability to control heat and the rapid heat transfer are probably the greatest advantages of microwaves over conventional thermal technologies. In processing applications, the ability to instantaneously stop the heat source makes enormous difference to the product quality and hence the production economics (Wang *et al.*, 2006).

Some other techniques consist of extracting oils using a mechanical and thermochemical reaction (Bouzid *et al.*, 1997). Yield data of these oils is therefore determined on a fresh weight basis (w/v).

CHEMICAL COMPOSITION OF ESSENTIAL OIL

Volatile oils are very complex mixtures of compounds. The constituents of the oils are mainly monoterpenes and sesquiterpines, which are hydrocarbons with the general formula $(C_5H_8)n$. Oxygenated compounds derived from these hydrocarbons include alcohols, aldehydes, esters, ethers, ketones, phenols and oxides. It is estimated that there are more than 1000 monoterpene and 3000 sesquiterpene structures. Other compounds include phenylpropenes and specific compounds containing sulphur or nitrogen. Numerous publications have presented data on the composition of the various essential oils. The major components of the economically interesting essential oils are summarised by Bauer *et al.* (2001). Detailed compositional analysis of essential oils is done by gas chromatography and mass spectrometry (Salzer, 1977; Scheffer and Baerheim Svendsen, 1981; Wilkins and Madsen, 1991; Daferera *et al.*, 2000; Juliano *et al.*, 2000; Jerkovic *et al.*, 2001; Delaquis *et al.*, 2002).

Essential oils may contain more than sixty individual components (Senatore, 1996; Russo *et al.*, 1998). Table 3 shows the composition of essential oils of *Curcuma longa.* Major components can constitute up to 85% of the essential oil, whereas other components are present only as a trace (Senatore, 1996; Bauer *et al.*, 2001). The phenolic components are mostly responsible for the antibacterial properties of essential oils (Cosentino *et al.*, 1999). Minor components have a critical part to play in antibacterial activity, possibly by producing a synergistic effect between other components for instance sage (Marino *et al.*, 2001), certain species of *Thymus* (Lattaoui and Tantaoui-Elaraki, 1994; Paster *et al.*, 1995; Marino *et al.*, 1999) and oregano (Paster *et al.*, 1995).

Table 3: Composition of essential oils of *Curcuma longa* L. (Leeal *et al.*, 2002)

Component	Concentration (%)			
	Leaf	Flower	Root	Rhizome
α-Pinene	2.1	0.4	0.1	0.1
β-Pinene	2.8	0.1	0.1	tr
Myrcene	2.3	0.2	tr	0.1
α-Phellandrene	32.6	–	0.1	0.1
δ-3-Carene	1.1–	0.6	–	–
α-Terpinene	1.3	0.1	–	–
p-Cymene	5.9	1.6	3.3	3.0
β-Phellandrene	3.2	tr	–	tr
1,8-Cineole	6.5	4.1	0.7	2.4
Z-β-Ocimene	0.2	–	–	–
E-β-Ocimene	0.4	–	–	–
γ-Terpinene	1.5	–	–	–
Terpinolene	26.0	7.4	0.1	0.3
Linalool	0.7	1.1	0.1	–
1,3,8-Paramenthatriene	0.2	0.3	–	–
p-Methylacetophenone	0.1	0.3	tr	tr
p-Cymen-8-ol	0.8	26.0	1.5	0.3
α-Terpineol	0.4	1.1	0.1	0.2
Thymol	0.3	–	0.1	–
Carvacrol	0.1	–	0.3	0.1
γ-Curcumene	0.1	tr	0.4	0.1
ar-Curcumene	0.2	1.9	7.0	6.3
α-Zingiberene	0.5	0.8	tr	tr
β-Bisabolene	–	0.9	2.3	1.3
β-Sesquiphellandrene	0.3	1.1	tr	2.6
E-Nerolidal	0.1	1.1	–	–
Dehydrocurcumene	–	–	4.3	2.2
ar-Turmerone	0.1	1.2	46.8	31.1
Turmerone	0.9	1.0	–	10.0
Curlone	0.2	0.3	0.6	10.6
Curcuphenol	tr	tr	0.6	0.5
6S,7R bisabolene	0.1	0.4	1.2	0.9
Other	9.0	48.0	30.3	27.8

Factors Affecting Chemical Composition of Essential Oils

The composition of essential oils from a particular plant species can differ between harvesting seasons and between geographical sources (Arras and Grella, 1992; Marotti *et al.*, 1994; McGimpsey *et al.*, 1994; Cosentino *et al.*, 1999; Marino *et al.*, 1999; Juliano *et al.*, 2000; Faleiro *et al.*, 2003). Generally, oils produced from herbs harvested during or immediately after flowering possess the strongest antimicrobial activity (McGimpsey *et al.*, 1994; Marino*et al.*, 1999). Enantiomers of essential oil components have been shown to exhibit antimicrobial activity to different extents (Lis-Balchin *et al.*, 1999). The composition of essential oils from different parts of the same plant can also differ widely. For example, oil obtained from the seeds of coriander (*Coriandrum sativum*) has a quite different composition as compared to oil of cilantro, which is obtained from the immature leaves of the same plant (Delaquis *et al.*, 2002).

The soil type could be statistically linked to certain components in the essential oil of *Tagetes minuta* (Zygadlo *et al.*, 1990). This finding was confirmed by Chalchat *et al.* (1995), who compared essential oil components of *Tagetes minuta* harvested in different regions of Rwanda and France. In addition, it was demonstrated that the addition of mineral fertilizers significantly affected the yields of specific components in the essential oil of *Salvia officinalis* (Piccaglia *et al.*, 1989). Results obtained by Graven *et al.* (1991) indicated that inherent N, P and S deficiencies cause variation in oil composition and yield of *Tagetes minuta* oil. Zheljazkov and Jekocov (1996) investigated the heavy metal content of essential oils of ten plant species grown on soil contaminated with hazardous metals, including Cd and Pb. Metal concentrations in the oils were lower than the detection limits of the graphite furnace atomic absorption and inductively coupled plasma techniques used. These findings were extremely important, since they show the potential of growing essential oil crops, as substitutes for edible crops, on soil heavily polluted with heavy metals, or under atmospheric pollution, without adverse effects on the product.

Postharvest drying of material is an accepted practice in the production of essential oils (Whish and Williams, 1996). Drying methods include exposure to natural air in the shade, sundrying, as well as drying by blowing warm air over the material (Soysal and Öztekin, 2001). Postharvest drying is thought to improve oil yield and accelerate distillation by improving heat transfer, in addition to providing increased loading capacity due to loss of plant moisture. Further advantages are the reduction of microbial growth and the inhibition of some biochemical reactions in dried material (Baritaux *et al.*, 1992; Soysal and Öztekin, 2001). However, oil may be lost due to volatilization and mechanical damage to oil glands. Essential oil components (including terpenoids) are usually present in the free form, but may also be present bound to sugar moieties, usually mono- or disaccharides (Skoula and Harborne, 2002). Enzymatic or chemical reactions which take place during maturation, drying or storage of plants may result in hydrolysis of the sugars, thereby facilitating the release of volatile compounds.

Light-induced photolysis, oxidation, hydrolysis, polymerization and other chemical reactions affect the quality of essential oils during storage (Kaul *et al.*, 1997). These reactions may produce undesirable compounds, which result in off-flavours and off-aromas in the product (Luque de Castro *et al.*, 1999). Incorrect storage conditions of rose-scented geranium oil (*Pelargonium* spp.) resulted in changes in the oil composition, with some compounds increasing in concentration to the detriment of other sample components. Correct storage conditions were described as oil that had been dried with anhydrous sodium sulphate, transferred to fully filled aluminium or amber glass containers, which were then tightly sealed and stored under ambient temperature conditions. The essential oil of *Lippia* species contains the same constituents as other essential oils: monterpenes, oxygenated monoterpenes and sesquiterpenes. The same storage regimes should, therefore, be adequate in preventing decomposition of the oil components. Terblanché (1995) used the same storage conditions as described here and concluded that if any oxidation of components had taken place under these conditions, it was unavoidable. Other factors that affect the growing plants, thus leading to variations in oil yield and composition, include length of exposure to sunlight (Burbott and Loomis, 1957), growing temperature (Lincoln and Langenheim, 1978), availability of water, height above sea level (Galambosi and Peura, 1996), plant density (Graven *et al.*, 1990; Clark and Menary, 1979), time of sowing (Galambosi and Peura, 1996) and the presence of fungal diseases and insects (Graven *et al.*, 1990; Margina and Zheljazkov, 1994). The oil composition and yield may also change as a result of the harvesting methods used (Bonnardeaux, 1992).

METHODS FOR TESTING *IN VITRO* ANTIMICROBIAL ACTIVITY OF ESSENTIAL OILS

Tests of antimicrobial activity can be classified as diffusion, dilution or bioautographic methods (Rios *et al.*, 1988). The principles and practice of these test methods are explained in the literature (Barry, 1976; Davidson and Parish, 1989; Hodges and Hanlon, 1991). The NCCLS method for antibacterial susceptibility testing, which is principally aimed at the testing of antibiotics, has been modified for testing essential oils (Hammer *et al.*, 1999; NCCLS, 2000). A fourth and upcoming test method is the conductimetric assay, which detects microbial growth as a change in the electrical conductivity or impedance of the growth medium (Sawai *et al.*, 2002). In relation to essential oils, only diffusion and dilution methods are discussed here.

Agar-Diffusion Methods (Cos *et al.*, 2006)

In diffusion technique, a reservoir containing the test compound at a known concentration is brought into contact with an inoculated medium and the diameter of the clear zone around the reservoir (inhibition zone in mm) is measured at the end of the incubation period. In order to enhance the detection limit, the inoculated system is kept at lower temperature (25°C–30°C) for several hours before incubation to favour compound diffusion over microbial growth, thereby increasing the inhibition diameter. Different types of reservoirs can be used, such as filter paper discs (agar disc diffusion), holes punched in the agar medium (agar well diffusion) and stainless steel cylinders placed on the surface. The fixed agar is left on the bottom of the hole to ensure that the sample does not leak under the agar layer (Cole, 1994). Figure 1 shows the effect of different concentrations of essential oil of *Spillanthes acmella* on growth of *Streptococcus pyogenes* by agar well diffusion method.

In filter paper disc method, care should be taken to ensure that the contact of the disc with the agar is even. The diffusion method is not appropriate for testing non-polar samples or samples that do not diffuse easily into the agar. In general, the relative antimicrobial potency of different samples may not always be compared mainly because of differences in physical properties such as solubility, volatility and diffusion characteristics in agar. Dubey *et al.* (2005) evaluated antimicrobial activity of essential oils of some medicinal plants viz., *Trachyspormum ammi, Cinnamomum zelyanicum, Syzygium aromaticum, Zingiber officinale, Curcuma longa* and *Acorus calamus* against certain human pathogens such as *Salmonella typhi, Klebsiella pneumoniae, Escherichia coli, Staphylococcus aureus* and *Bacillus subtilis* by filter paper disc diffusion method.

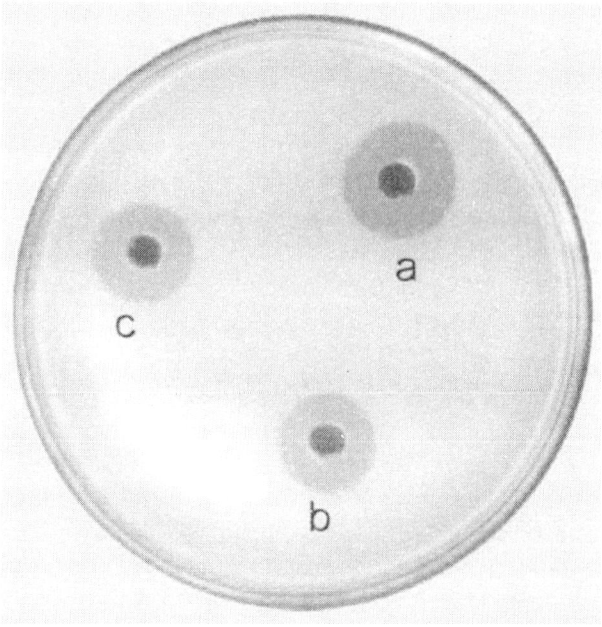

Fig. 1. Effect of different concentrations of essential oil *Spillanthes acmella* on *Streptococcus pyogenes*; a- 1,000 ppm, b- 500 ppm; c- 200 ppm oil (Source: Rana, 2005).

The oils of *T. ammi* and *C. zelanicum* exhibited the maximum inhibitory effect as compared to the oils of the other medicinal plants.

Dilution Methods (Cos *et al.*, 2006)

The test compounds are mixed with suitable medium that has previously been inoculated with the test organism. It can be carried out in liquid as well as in solid media and the growth of the microorganisms can be measured in a number of ways. In agar dilution method, the minimal inhibitory concentration (MIC) is defined as the lowest concentration able to inhibit visible microbial growth. In liquid or broth dilution methods, turbidity is most frequently used. Turbidity can be estimated visually or obtained more accurately by measuring the optical density at 405 nm. In this case, the test samples not fully soluble may interfere with turbidity readings, thereby emphasizing the need for a negative control, i.e., extract dissolved in blank medium without microorganisms. The liquid-dilution method also allows determination whether a compound or extract has cidal or static action at a particular concentration. The minimum bactericidal or fungicidal concentration (MBC or MFC) is determined by plating out samples of completely inhibited dilution cultures and assessing growth (static) or no-growth (cidal) after incubation. In general, dilution methods are considered to be appropriate for assaying polar or non-polar extracts or compounds for determining MIC and MBC/MFC values.

Selection of Test Organisms for Antimicrobial Testing

Selection of test organism usually depends on the aim of investigation. If the purpose is human health management, the organism should be selected on the basis of their clinical and medical importance. For example, multi-drug resistance (MDR) strains of *Mycobacterium* spp. may be used including extremely drug resistant (XDR) strains of the organism for which currently there is no cure. In case of bacteria, both Gram-positive and Gram-negative should be included. In primary screening, drug-sensitive strains are preferably used and should represent common pathogenic species of different classes. Extracts with prominent activity against Gram-positive cocci should also be tested against methicillin-resistant *Staphylococcus aureus* (MRSA) and vancomycin-resistant *Enterococci* (VRE), as they represent the greatest current medical need (Cos *et al.*, 2006). Choice of reference fungi includes *Epidermophyton floccosum* and *Trichophyton mentagrophytes* as representatives of the dermatophytes. Opportunistic filamentous fungi such as *Aspergillus niger*, *Fusarium solani* yeast *Candida albicans* can also be listed for the said purpose.

Factors Affecting the Results of Agar Diffusion and Dilution Methods

The outcome of a test can be affected by factors such as the method used to extract the essential oil from plant material, choice of a particular solvent, the level of inoculum, growth phase, culture medium used, pH of the media and incubation time and temperature (Rios *et al.*, 1988). Several workers have surveyed the methods used for antibacterial activity studies carried out with essential oils (Koedam, 1977a,b; Shelef, 1983; Janssen *et al.*, 1987; Rios *et al.*, 1988; Nychas, 1995) and comparison of published data on these factors is complicated (Janssen *et al.*, 1987; Friedman *et al.*, 2002). Factors such as the volume of essential oil placed on the paper disks, the thickness of the agar layer and whether a solvent is used vary considerably between studies.

The antibacterial activity of essential oils is also influenced by the degree to which oxygen is available. This could be due to the fact that when little oxygen is present, fewer oxidative changes can take place in the essential oils, and/or that cells obtaining energy via anaerobic metabolism are more sensitive to the toxic action of essential oils (Paster *et al.*, 1990).

The study of antimicrobial activity of essential oils presents some difficulty due to poor solubility in water-like constituents. The use of tween 80 (detergent) or ethanol (alcohol) in order to facilitate the dispersion of essential oils in the aqueous medium/agar is reported to interfere with the estimation of antimicrobial activity of essential oils (Remmal *et al.*, 1993). Furthermore, MIC (minimal inhibitory concentrations) for different bacterial species in the presence of agar appears to be significantly lower than those observed in the presence of Tween 80 or ethanol (Remmal *et al.*, 1993).

Use of a particular solvent for extraction can affect the result. For example, herb essential oils extracted by hexane have been shown to exhibit greater antimicrobial activity than the corresponding steam distilled essential oils (Packiyasothy and Kyle, 2002). The solvents used to make test compound solutions include dimethyl sulphoxide (DMSO), methanol and ethanol. The latter two have the disadvantage of rapid evaporation, whereby the stated concentration of stock solutions cannot be maintained. On the other hand, the added advantages of stock solutions in 100% DMSO include good solubility during serial dilution procedures and elimination of microbial contamination, thereby reducing the need for sterilization by autoclaving or other strenuous methods, which affects the quality of test samples. DMSO is potentially toxic for cells and many microorganisms, therefore, to avoid later interference in the biological test systems, the in-test concentration of DMSO should not exceed 1% (Cos *et al.*, 2006). Because of the variability of individual compounds, there are no general storage conditions that guarantee the sample integrity (Verkman, 2004). A practical recommendation for storage of compounds or extracts is either without solvent for long term storage or at 100% DMSO at –20°C with minimum exposures to freeze-thaw cycles and humidity (Cos *et al.*, 2006).

Slight differences in the composition of growth medium can greatly affect the antibacterial activity of a compound or extract. For example, the addition of sheep blood to Mueller-Hinton medium increases the MIC of flavomycin from 0.12 to 256 mg/l (Butaye *et al.*, 2000). Mueller-Hinton (MH) agar or broth and tryptic soy agar (TSA) or broth (TSB) is general growth media for bacteria, while sabouraud (SAB) agar or broth is used for fungi. Growth of fastidious microorganisms such as *Streptococcus pneumoniae* and *Legionella pneumophila* may requires more complex media, enrichment of incubation atmosphere with 5% CO_2 and/or extension of incubation time. Mueller Hinton medium allows good growth of non-fastidious bacteria and is generally poor for antagonists. It also meets the requirements of the NCCLS standard and is recommended as reference medium for agar- and broth-dilution tests (Anon, 2000; 2003).

The level of inoculum or inoculum concentration may also affect the antibacterial or antifungal potency of a sample. Therefore, inoculum also needs standardization (Anon, 2003). In dilution methods, 10^5 CFU/ml is adequate for most bacterial species, while for yeast and fungi, it ranges between 10^3 and 10^4 CFU/ml (Hadacek and Greger, 2000). A too low (10^2 CFU/ml) or too high (10^7 CFU/ml) inoculum size may lead to the interpretation of false negative results. Bacteria or yeast inoculum should be prepared from overnight cultures or existing biofreeze stocks. It is generally recommended to collect from cultures during logarithmic growth phase and always to take four or five colonies of a pure culture on agar to avoid selecting a typical variant (Anon 2003; Cos *et al.*, 2006).

MODE OF ANTIMICROBIAL ACTION OF ESSENTIAL OILS (BURT, 2004)

Important characteristics of essential oils and their components are their hydrophobicity, which enable them to partition in the lipids of the bacterial cell membrane and mitochondria, disturbing the structures and rendering them more permeable (Knobloch *et al.*, 1986; Sikkema *et al.*, 1994). Leakage of ions and other cell contents can then occur (Oosterhaven *et al.*, 1995; Gustafson *et al.*, 1998; Helander *et al.*, 1998; Cox *et al.*, 2000; Lambert *et al.*, 2001; Skandamis *et al.*, 2001; Carson *et al.*, 2002; Ultee *et al.*, 2002). Although a certain amount of leakage from bacterial cells may be tolerated without loss of viability, extensive loss of cell contents or the exit of critical molecules and ions will lead to death (Denyer and Hugo, 1991a). There is some evidence from studies with tea tree oil and *E. coli* that cell death may occur before lysis (Gustafson *et al.*, 1998). Generally, the essential oils possessing the strongest antibacterial properties against food-borne pathogens contain a high percentage of phenolic compounds such as carvacrol, eugenol and thymol (Farag *et al.*, 1989; Thoroski *et al.*, 1989; Cosentino *et al.*, 1999; Dorman and

Deans, 2000; Juliano *et al.*, 2000; Lambert *et al.*, 2001). Mechanism of action is considered to be the disturbance of the cytoplasmic membrane, disrupting the proton motive force, electron flow, active transport and coagulation of cell contents (Denyer and Hugo, 1991b; Sikkema *et al.*, 1995; Davidson, 1997).

The chemical structure of the individual essential oil components affects their precise mode of action and antibacterial activity (Dorman and Deans, 2000). The importance of the presence of the hydroxyl group in phenolic compounds such as carvacrol and thymol in the antimicrobial action has been confirmed by several workers (Knobloch *et al.*, 1986; Dorman and Deans, 2000; Ultee *et al.*, 2002). The position of the hydroxyl group on the phenolic ring does not seems to influence the degree of antibacterial activity. For example, the action of thymol against *B. cereus*, *Staphylococcus aureus* and *Pseudomonas aeruginosa* appears to be comparable to that of carvacrol (Lambert *et al.*, 2001; Ultee *et al.*, 2002). However, in one study, carvacrol and thymol were found to act differently against Gram-positive and Gram-negative species (Dorman and Deans, 2000). Ultee *et al.* (2002) demonstrated the lack of activity of menthol as compared to carvacrol indicating the significance of the phenolic ring (destabilised electrons) in carvacrol. Dorman and Deans (2000) reported that the addition of an acetate moiety to the molecule appeared to increase the antibacterial activity; geranyl acetate was more active against a range of Gram-positive and negative species than geraniol. As far as non-phenolic components of essential oils are concerned, the type of alkyl group has been found to influence activity (alkenyl > alkyl). For example, limonene is more active than p-cymene (Dorman and Deans, 2000).

Components of essential oil are supposed to act on cell proteins embedded in the cytoplasmic membrane (Knobloch *et al.*, 1989). Enzymes such as ATPases are known to be located in the cytoplasmic membrane and bordered by lipid molecules. Two possible mechanisms have been suggested whereby cyclic hydrocarbons could act on these. Lipophilic hydrocarbon molecules could accumulate in the lipid bilayer and distort the lipid–protein interaction; alternatively, direct interaction of the lipophilic compounds with hydrophobic parts of the protein is possible (Juven *et al.*, 1994; Sikkema *et al.*, 1995).

Some essential oils have been found to stimulate the growth of pseudomycelia (a series of cells adhering end-to-end as a result of incomplete separation of newly formed cells) in certain yeasts. This could be an indication that essential oils act on the enzymes involved in the energy regulation or synthesis of structural components (Conner and Beuchat, 1984). Cinnamon oil and its components have been shown to inhibit amino acid decarboxylases in *Enterobacter aerogenes*. The mechanism of action was thought to be the binding of proteins (Wendakoon and Sakaguchi, 1995). Indications that essential oil components may act on proteins were also obtained from studies using milk containing different protein levels (Pol *et al.*, 2001).

Carvacrol and thymol appear to make the cell membrane permeable (Lambert *et al.*, 2001). Both are able to disintegrate the outer membrane of Gram-negative bacteria, releasing lipopolysaccharides (LPS) and increasing the permeability of the cytoplasmic membrane to ATP. The presence of magnesium chloride has been shown to have no influence on this action, suggesting a mechanism other than chelation of cations in the outer membrane (Helander *et al.*, 1998). Studies with *B. cereus* have shown that carvacrol interacts with the cell membrane, where it dissolves in the phospholipid bilayer and is assumed to align between the fatty acid chains (Ultee *et al.*, 2000a). This distortion of the physical structure would cause expansion and destabilisation of the membrane, increasing membrane fluidity, which in turn would increase passive permeability (Ultee *et al.*, 2002). Carvacrol forms channels through the membrane by pushing apart the fatty acid chains of the phospholipids, allowing ions to leave the cytoplasm (Ultee, 2000). Oregano essential oil, containing carvacrol as a major component, causes leakage of phosphate ions from *S. aureus* and *P. aeruginosa* (Lambert *et al.*, 2001). Juven *et al.* (1994) examined the working of thymol against *S. typhimurium* and *S. aureus* and hypothesised that thymol binds to membrane-proteins hydrophobically and by means of hydrogen bonding, thereby changing the permeability characteristics of the membrane. Thymol was found to be more inhibitive at pH 5.5 than 6.5. At low pH the thymol molecule would be undissociated and therefore more hydrophobic, and so may bind better to the hydrophobic areas of proteins and dissolve better in the lipid phase (Juven *et al.*, 1994).

Eugenol is a major component (approximately 85%) of clove oil (Farag *et al.*, 1989). Sublethal concentrations of eugenol have been found to inhibit production of amylase and proteases by *B. cereus.* Cell wall deterioration and a high degree of cell lysis were also noted (Thoroski *et al.*, 1989). The hydroxyl group on eugenol is thought to bind proteins, preventing enzyme action in *E. aerogenes* (Wendakoon and Sakaguchi, 1995).

The biological precursor of carvacrol, *p*-cymene is hydrophobic and causes swelling of the cytoplasmic membrane to a greater extent than does carvacrol (Ultee *et al.*, 2002). *p*-Cymene is not an effective antibacterial when used alone (Juven *et al.*, 1994; Dorman and Deans, 2000; Juliano *et al.*, 2000; Ultee *et al.*, 2000a), but when combined with carvacrol, synergism has been observed against *B. cereus in vitro* and in rice (Ultee *et al.*, 2000b). The greater efficiency of *p-cymene* at being incorporated in the lipid bilayer of *B. cereus* very likely facilitates transport of carvacrol across the cytoplasmic membrane (Ultee *et al.*, 2002).

When tested in a liposome model system at concentrations above the MIC, carvone (2-methyl-5-(1-methylethenyl)-2-cyclohexen-1-one) dissipated the pH gradient and membrane potential of cells. The specific growth rate of *E. coli, Streptococcus thermophilus* and *L. lactis* decreased with increasing concentrations of carvone, which suggests that it acts by disturbing the metabolic energy status of cells (Oosterhaven *et al.*, 1995). In contrast, another study found that carvone was ineffective on the outer membrane of *E. coli* and *S. typhimurium* and did not affect the intracellular ATP pool (Helander *et al.*, 1998).

Although cinnamaldehyde (3-phenyl-2-propenal) is known to be inhibitive to growth of *E. coli* O157:H7 and *S. typhimurium* at similar concentrations to carvacrol and thymol, it did not disintegrate the outer membrane or deplete the intracellular ATP pool (Helander *et al.*, 1998). The carbonyl group is thought to bind with proteins, preventing the action of amino acid decarboxylases in *E. aerogenes* (Wendakoon and Sakaguchi, 1995).

γ-Terpinene did not antagonize growth of *S. Typhimurium* (Juven *et al.*, 1994), whereas α-terpinene inhibited 11 of the 25 bacterial species screened (Dorman and Deans, 2000).

QUALITY CONTROL OF HERBAL DRUGS

Microbial Contamination of Herbal Drugs

Although modern analytical tools have facilitated indepth studies of medically relevant compounds contained in various plants, the field of microbiological parameters of medicinal plants and/or herbal drugs has not been extensively considered so far (Czech *et al.*, 2001). Hygienic situation of medicinal drugs is a result of an array of positive and negative effects, which are determined by extrinsic and/or intrinsic factors (Basilico and Basilico, 1999; Cosentino *et al.*, 1999; Arici, 1999; Kerr, 1999). Intrinsic factors include nature of the plant and natural barrier, structure of the plant, plant composition (antimicrobial compounds and agents) and intracellular microbial contaminations. Extrinsic factors include climate, humidity, location/position, harvesting method, post harvesting, physical state, technological treatment, packaging and storage conditions, and exogenous microbial contaminations (Kneifel *et al.*, 2002). In principle, the microbial load of plants is the result of the series of influences caused by animate and inanimate sources. The microbial contaminants are easily transferred via air-borne (fungal spores, fungi, bacteria, bacterial spores, viruses) and soil-borne (fungal spores, bacterial spores, viruses) vectors. The bacterial endospores and fungal spores can be regarded as the two dominant groups of contaminants associated with medicinal plants, however, broad diversity of bacterial and fungal cells can be found in/on the plant material (Frank, 1989; Wichtl, 1997; Schilcher, 1982; McKee, 1995).

Among these microorganisms, pathogens may also occur, and this fact particularly limits the utilization of these plants (Frank, 1989; Czech *et al.*, 2001), besides quality reduction caused by microbiologically induced spoilage. Apart from these, extraneous matter and filth material originating from rodents, insects and inorganic sources, for e.g., stones, may be present in some herbal preparations (Schilcher, 1982; Kalinovic and Rozman, 2000; Kneifel and Berger, 1994; Farrell, 1990). The degree of contamination usually depends on the distance from the soil bottom the plant has been grown (Alonzo *et al.*, 1994). Czech *et al.* (2001) have introduced an array of methods that may be used to assess the microbiological quality of medicinal plants. According to European Pharmacopoeia, these drugs should meet general hygiene standards which aim at low viable microbial counts and absence of pathogenic microorganisms. In addition they must not contain insects and have to be reasonably free from extraneous matter and defective material. Drugs used for topic applications must not contain aerobic mesophilic bacteria, yeast and mould counts exceeding 100 CFU/g, enterobacteria should be lower than 10 CFU/g (Czech *et al.*, 2001). According to Kniefel *et al.* (2002), the absence of pathogens in at least 10 grams of product and a low level of microbial contamination (i.e., low total aerobic mesophilic, endospore and enterobacterial count) should be the main goal for each producer of medicinal plants. This also forms the optimum prerequisite for providing herbal plants of high quality and with desired therapeutic benefits.

Standardisation of Herbal Medicines

According to WHO, Herbal Medicines should be regarded as *finished, labelled medicinal products that contain as active ingredients aerial or underground parts of plants, or other plant material, or combinations there of, whether in the crude state or as plant preparations*. Plant material includes juices, gums, fatty oils, essential oils, and any other substance of this nature. Herbal medicines may contain excipients in addition to the active ingredients. Medicines containing plant material combined with chemically defined active substances, including chemically defined, isolated constituents of plants are not considered to be herbal medicines. Exceptionally, in some countries herbal medicines also contain natural organic or inorganic active ingredients which are not of plant origin (Anonymous, 1991).

Herbal medicines are prepared from materials of plant origin which are prone to contamination, deterioration and variation in composition. It is, therefore, necessary to maintain phytochemical profile of a herbal drug used in crude forms. Several manufacturers are now producing standardized herbal extracts to attain better batch to batch consistency and, therefore, to achieve more consistent pharmaceutical quality (De Los Reyes and Koda, 2002). Such extracts are processed to contain a specific quantity of the active constituent(s), and in some cases the unwanted or toxic constituents are removed.

For herbs, where identification of active constituents is not possible, the products may be standardized on their content of certain marker compounds, i.e., chemical characteristic of herb or chemicals present in large quantities (Barnes, 2002) or chromatographic fingerprinting. A marker can be defined as a chemical entity in the plant material which may or may not be chemically defined and serves as a characteristic fingerprint for that plant (Shrikumar and Ravi, 2007). Various analytical techniques like TLC, HPLC and HPTLC can be used to detect presence of marker compound in the plant and also to quantify it to ascertain the limits. A biomarker on the other hands is a group of chemical compounds, which are in addition to being unique for that plant material also correlate with biological efficacy (Shrikumar *et al.*, 2005). Batch to batch variations start from the collection of raw material itself in the absence of any reference standards for proper identification, and multiply during storage and further processing. Chromatographic fingerprinting emphasizes an integral formulation of pharmacologically active and phytopharmaceutically characteristic components of samples with similar or different attributions. This technique can be used for the assessment of quality consistency and stability of herbal extracts or products by visible observation and comparison of the standardized fingerprint pattern (Li *et al.*, 2006). Fingerprinting of the herbals are done by high performance liquid chromatography (HPLC), high-performance thin layer chromatography (HPTLC) and thin layer chromatography (TLC), etc.

Similarly, extracts obtained from each batch are loaded onto a TLC plate and are developed in suitable solvent system. Different solvent systems may be used to develop a better TLC pattern. This developed plated is then studied under different conditions in the scanner, which utilizes a laser beam and detects and quantifies distinct spots due to different compounds on the plate. The different conditions, which can be used for detecting spots on plate, include UV at different wavelength, derivitisation of plate to obtain distinct coloured spots which are then detected by the software programme and a characteristic graph obtained. There is a very distinct graphic pattern for each plant material and the extracts from different batches usually give overlapping graphic patterns. So basically this pattern can be optimized to give a distinct HPTLC profile for each fractionated plant part (Bhutani, 2000; Shrikumar and Ravi, 2007).

Ecotype Pharmacological Evaluation of Medicinal Plants

The relative proportions of phytochemicals present in medicinal plants can vary in different ecotypes. Therefore, the major problem in quality control is batch to batch variations in the quality of medicinal plants. The pharmacological activities of certain compounds can be influenced by their interaction with other compounds. For example, there might be traces of certain toxic chemicals in a plant and relative proportions of these toxins in the plant body can increase or decrease under certain ecological conditions. Further, the toxicity of certain chemicals can be reduced or increased by the presence of certain other chemicals. These effects could not be determined by phytochemical analysis alone, but an overall pharmacological evaluation of each ecotype is essential even if major chemical constituents are known (Subramoniam, 2001). Precautions have to be taken to overcome certain problems in the evaluation of ecotypes of medicinal plants. In some cases genotype variations could be confused with ecotype variations. The ecotype variations among medicinal plants cannot be generalized into specific patterns. A particular soil or climatic condition can increase the medicinal value of one plant species (or genotype), whereas the same condition can, possibly, decrease the medicinal value of another species (or genotype). Man-made interferences in the ecosystem such as widespread use of pesticides, heavy metals and other environmental toxins lead to their accumulation in the medicinal plants. If adequate measures are not taken to detect such toxins, they could interfere with ecotype pharmacological evaluation (Subramoniam, 2001).

Safety and Toxicity of Essential Oils

The use of animal testing to establish safety and toxicity is not so important in case of herbal drugs as they are time tested and used widely in traditional forms, and also they are recommended for oral administration by the ayurvedic practioners. On the other hand, synthetic molecules drug development requires about 12-15 years for the drug to come into the market with an expense of US $ 900 million (Verdine, 1996; Chauthaiwale, 2003). But the knowledge and experimental data base of ayurvedic medicine can provide new functional leads to reduce time, money and toxicity—the three main hurdles in drug development (Shrikumar and Ravi, 2007). These medicines are effectively tested for thousand of years on people (Patwardhan and Hooper, 1992).

However, it is dangerous to assume, just because a tea or alcoholic extract of a plant used as a herbal medicine is harmless, that the essential oil derived from that plant is also safe. The drammatic increase in concentration of the essential oil compared with that in the whole plant (often the yield is 0.01%) demonstrates that essential oils are not equivalent to the whole herb. Essential oils are also volatile and fat-soluble and therefore differ from the mainly water-soluble whole herb extracts used in herbal medicine. The toxicity of essential oils can also be entirely different from that of the herb, not only because of their high concentration, but also because of their ability to pass across membranes very efficiently due to their lipophilicity (Anomymous, 2006).

Most essential oils have GRAS (generally recognized as safe) status granted by the Flavour and Extract Manufacturers Association (FEMA) and approved by the U.S. Food and Drug Administration (FDA) for food use, and many appear in the Food Chemical Codex (Anonymous, 2006). Essential oils are used in the food industry usually in very minute amounts of 10 ppm, and even up to 1000 ppm and above in the case of mint confectionery or chewing gum. This does not make it safe to use drops of undiluted essential oils on sugar lumps for oral application or on suppositories for anal or vaginal application. Severe damage to mucous membranes could result from such practices, and the dangers are magnified if cheap, adulterated oils are used

(Anonymous, 2006). There are also reports indicating irritation and toxicity of essential oils. For example some essential oils such as eugenol, menthol and thymol cause irritation of mouth tissues, when applied in root canal treatments (Burt, 2004).

Cytotoxicity study on these compounds refers to conclusion that gum irritation may be related to membrane lysis and the tissue penetration may be due to membrane affinity and lipid solubility (Manabe *et al.*, 1987). Also other essential oils, for example, cinnamaldehyde, carvacrol, carvone and thymol exhibit toxic effects at the cellular level (Stammati *et al.*, 1999). Some essential oils and their components have been known to cause allergic contact dermatitis in people who use them frequently. Preventive measures may be needed to ensure the well-being of workers if these substances were to be used on a larger scale (Bleasel *et al.*, 2002; Boyle, 1955; Board and Gould, 1991; Brocklehurst *et al.*, 1995; Burt and Reinders, 2003; Canillac and Mourey, 2001; Carson and Riley, 1993, 1995, 2001; Carson *et al.*, 1995a, 1995b, 2002). Various oils used in medicine, paramedicine and aromatherapy have been shown to exhibit spasmolytic or spasmogenic properties, although these are difficult to associate with a particular component (Lis-Balchin *et al.*, 1996; Madeira *et al.*, 2002). Enantiomers of α-pinene have been shown to have very different spasmogenic effects (Lis-Balchin *et al.*, 1999).

The reported adverse effects of herbal remedies include allergic reactions, toxic reactions, adverse effects related to a herb's desired pharmacological actions, possible mutagenic effects, drug interactions, drug contamination and even mistaken plant identities. The use of herbal products containing *Aristolochia* species has been associated with serious renal toxicity (Anonymous, 2005). All these aspects make it necessary to retrieve and evaluate scientific evidence on the toxicity of herbal medicine used in traditional medicinal system in India and to grade the toxicity of individual herbs. The resulting database should assist in promoting the safe and effective use of Indian ayurvedic herbal medicines. In India, moves are being made to establish more effective regulation of quality, safety and efficacy (QSE) of ayurvedic medicine (Anonymous, 2007 in Wan, 2007).

ESSENTIAL OIL-BASED PRODUCTS

In most of the pharmaceutical formulations, essential oils are used directly. Formulations contain single essential oils or mixtures of several oils to enhance their efficacy. A list of some of the manufacturers and exporters of essential oil-based products in India is given in Table 4. Popular formulations like Gripewater and digestive/carminative preparations are used by people without going in for any prescription. Gripewater is mainly constituted by dill, anise and caraway oils, while Pudinhara, the most popular digestive preparations, contains peppermint, spearmint and cinnamon oils. Most of the cough syrups/expectorants like Glycodin and Cadistin contain the oils of eucalyptus, peppermint, pine, cubeb or tolu balsam. Eucalyptus and cinnamon oils alongwith clove oils are also ingredients of many formulations used in dental pain/infections. A number of essential oils like those of eucalyptus, terpentine, cinnamon, camphor, and lemon grass are used as ingredients of many popular pain balms. Oils of anise and eucalyptus along with extracts of lobelia and datura are common ingredients of many formulations providing relief in asthma (Rana, 2005). A commercial product Vicks vaporub by Procter and Gamble also contain flowers of Ajowan and *Eucalyptus* oil.

Table 4: Manufacturers and exporters of some essential oil-based products in India[a]

Essential oil	Name of the Company	Address	Website
Sandalwood Oil, Sandalwood Chips, Sandalwood Powder	Crown Essential Oil Co.	15-6-399, Begum Bazar, Ramveer Plaza, Plot No.121, Hyderabad – 500012, Andhra Pradesh	–
Essential oils blend with oils and ointments	LDG International	A-42, Badli Extension, Village Badli, Delhi – 110042, India	http://www.ldg-intl.com
Neem oil	Shiv Sales Corporation	252, Kotla, Mayur Vihar, Phase-I, Near Gujar Bhawan, Delhi – 110091	http://www.shivorganics.com
Oils of Ajowan, Clove, Black pepper, Turmeric, Cardamom, *Curcuma aromatica*	Parekh Enterprise	A-409, Fairdeal house, opp. St. Xavier's ladies hostel, Navrangpura, Ahmedabad – 382009	–
Sandalwood Oil (Indian), Kapoor Kachari Oil (*Hydechim* Root Oil), etc.	Lala Pragdutt Jagdish Prasad and Company	7/110-B, Swaroop Nagar, Kanpur – 208002, Uttar Pradesh, India	http://www.naturalfragrances.net, www.indianattars.com
Natural Menthol, Spearmint Oil, Melted Menthol	CLG Impex	CLG Impex Br-64a, West Shalimar Bagh, Delhi – 110088, India	–
Stevia oils, cultivation of aromatic herbs, purification of aromatic oils, Phenyl blended with essential oil, etc.	Arjun Agro Industries	Gut #76, Near Bombay Highway, Sharnapur, Aurangabad – 431001, Maharashtra, India	http://www.arjunagro.com
Aroma Oils, Aromatic Chemicals, Herbal Extracts	Bansal Aroma	E-1/6/40, sec-15, Rohini, Delhi – 110085, India	http://www.bansalfab.8m.com /bansalaroma.com
Menthal oil (*Mentha arvensis*), Spearmint oil (BP/USP), Peppermint oil (*Mentha piperata*)	Silverline Chemicals	Silverline chemicals Shingla Carpets Building., GT road, Near tehsil, Panipat–132103, Haryana, India	http://www.silverlinechemicals.com
Cinamon Extract, *Curcuma longa* Extract, Garlic Oil, Ginger Extract	Best Exports	Flat # 36/37/38, 1st floor, 'c' block, Kushal Garden Arcade, Plot No. 1A, II Phase, Peenya Industrial Area, Bangalore–560058, Karnataka, India	http://www.bestexports.net

[a]All information has been taken from the links of web page: http://www.indiachemicalexporters.com/TradeLeads/sell/Health_and_Beauty/Essential_Oils_and_Aromatics on May 22, 2008.

REFERENCES

Alonzo, V., Monforte, M.T., Tumino, G., Ragusa, S. and Bisignano, G. (1994). *Pharmeuropa,* **6:** 47-55.

Annon. (2000). *Clinical Microbiology and Infection,* **6:** 509-515.

Annon. (2003). *Clinical Microbiology and Infection,* **9:** 1-7.

Anomymous. (2005). EMEA/HMPC/139215: "Public statement on the risks associated with the use of herbal products containing *Aristolchia species*", Adopted November 2005.

Anonymous. (2004). *In:* Final report of research study on "Mechanism for sustainable development & promotion of herbal & medicinal plants in the state of Uttaranchal (India)", forwarded to SER Division, Planning Commission, Govt. of India, Parliament Street, New Delhi. Natural Resources India Foundation (NRIF), 93, GH-9, Pocket, Sunder Vihar, New Delhi – 110087.

Anonymous. (2006). *In:* Aromatherapy Science. Pharmaceutical Press, pp. 75-92.

Anonymous. (1991). WHO Consultation in Munich, Germany, June 1991. pp. 19-21.

Anonymous. (2007). Website of Department of Ayurveda, Yoga, & Naturopathy, Unani, Siddha, and Homeopathy (AYUSH), Ministry of Health and Family Wealthfare, Government of India.

http://indianmedicine.nic.in/html/ayurveda/ayurveda.htm#de

Arici, M. (1999). *Ernahrung/Nutrition,* **23:** 298-301.

Arras G. and Grella, G.E. (1992). *J. Hort. Sci.*, **67(2):** 197-202.

Baker, G.R., Lowe, R.F. and Southwell, I.A. (2000). *J. Agri. Food Chemistry,* **48:** 4041-4043.

Baritaux, O., Richard, H., Touche, J. and Derbesy, M. (1992). *Flavour and Fragrance J.,* **7:** 267-271.

Barnes, J. (2002). *Herbal The Pharmaceutical J.,* **268:** 804-806.

Barry A.L. (1976). The Antimicrobic Susceptibility Test: Principles and Practices. Lea and Febiger, Philadelphia.

Basilico, M.Z. and Basilico, J.C. (1999). *Letters in Applied Microbiology,* **29:** 238-241.

Bauer K., Garbe, D. and Suburg, H. (2001). Common Fragrance and Flavor Materials: Preparation, Properties and Uses. Wiley-VCH, Weinheim, p. 293.

Bélanger, A., Landry, B. and Dextraze, L., *et al.* (1991). *Rivista Italiana* EPPOS **2:** 455.

Bhutani, K.K. (2000). *The Eastern Pharmacist,* 21-26.

Bleasel, N., Tate, B. and Rademaker, M. (2002). *Australasian J. Dermatology* **43:** 211-213.

Board, R.G. and Gould, G.W. (1991). p. 267-284. *In:* (N.J.R. a. G.W. Gould Ed.), Food preservatives. Blackie, Glasgow.

Bonnardeaux, J. (1992). *J. Essential Oil Research,* **4:** 65-69.

Bouzid, N., Vilarem, G. and Gaset, A. (1997). *In:* Proceedings of the Intern. Congr. Arom. Medicinal Plants & Essential Oils, (Benjilali, B., Ettalibi, M., Ismaili-Alaoui, M. and Zrira, S. Eds.). *Actes Editions, Rabat, Morocco,* 115-120.

Boyle, W. (1955). *The American Perfumer and Essential Oil Review.* **66:** 25-28.

Brocklehurst, T.F., Parker, M.L., Gunning, P.A., Coleman, H.P. and Robins, M.M. (1995). *J. Appl. Bacteriology,* **78:** 609-615.

Buffler, C.R. (1993). Microwave cooking and processing: engineering fundamentals for the food scientist.

Burbott, A.J. and Loomis, W.D. (1957). *Plant Physiology,* **42:** 20-28.

Burt, S. (2004). *Int. J. Food Microbiol,* **94:** 223-253.

Burt, S.A. and Reinders, R.D. (2003). *Letters in Applied Microbiology,* **36:** 162-167.

Butaye, P., Devriese, L.A. and Haesebrouck, F. (2000). *J. Antimicrobial Chemotherapy,* **46:** 713-716.

Canillac, N. and Mourey, A. (2001). *Food Microbiology,* **18:** 261-268.

Carson, C.F., Cookson, B.D., Farrelly, H.D. and Riley, T.V. (1995a). *J. Antimicrobial Chemotherapy,* **35:** 421-424.

Carson, C.F., Hammer, K.A. and Riley, T.V. (1995b). *Microbios,* **82:** 181-185.

Carson, C.F., Mee, B.J. and Riley, T.V. (2002). *Antimicrobial Agents and Chemotherapy,* **46:** 1914-1920.

Carson, C.F. and Riley, T.V. (1993). *Letters in Applied Microbiology,* **16:**49-55.

Carson, C.F. and Riley, T.V. (1995). *J. Appl. Bacteriology,* **78:** 264-269.

Carson, C.F. and Riley, T.V. (2001). *Contact Dermatitis,* **45:** 65-67.

Chalchat, J.C., Garry, R.P. and Muhayimana, A. (1995). *J. Essential Oil Research,* **7:** 375-386.

Chauthaiwale, V. (2003). Rediscovering drug discovery. Indian Pharmaceutical Congress, Pune. pp. 176.

Clark, R.J. and Menary, R.C. (1979). *J. American Soc. for Horticultural Sci.,* **104(5):** 702-706.

Cole, M.D. (1994). *Biochemical Systematics and Ecology,* **22:** 837-856.

Conner, D.E. and Beuchat, L.R. (1984). *J. Food Sci.,* **49:** 429-434.

Cos, P., Vlietinck, A.J., Berghe, D.V. and Maes, L. (2006). *J. Ethnopharmacology,* **106:** 290-302.

Cosentino S., Tuberoso, C.I.G., Pisano, B., Satta, M., Mascia, V., Arredi, E. and Palmas, F. (1999). *Lett. Appl. Microbiol.,* **29:** 130-135.

Cox, S.D., Mann, C.M., Markham, J.L., Bell, H.C., Gustafson, J.E., Warmington, J.R. and Wyllie, S.G. (2000). *J. Appl. Microbiology,* **88:** 170-175.

Czech, E., Kneifel, W. and Kopp, B. (2001). *Planta Medica,* **67:** 263-269.

Daferera D.J., Biogas, B.N. and Polissiou, M.G. (2000). *J. Agric. Food Chem.,* **48:** 2576-2581.

Davidson P.M. and Parish, M.E. (1989). *Food Technol.,* **43:** 148-155.

De Los Reyes, G.C. and Koda, R.T. (2002). *Am. J. Health-Syst. Pharm.*, **59:** 545-547.

Delaquis P.J., Stanich, K., Girare, B. and Mazza, G. (2002). *Int. J. Food Microbiol.*, **74:** 101-109.

Denny, E.F.K. (1991). Field Distillation for Herbaceous Oils. Denny, McKenzie Associates.

Denyer, S.P. and Hugo, W.B. (1991b). pp. 331-334. *In:* Mechanisms of action of chemical biocides. (Eds. Denyer, S.P. and Hugo W.B.) Blackwell Scientific Publications, Oxford.

Denyer, S.P. and Hugo, W.B. (1991a). *In:* Mechanisms of action of chemical biocides. (Ed. Denyer, S.P. and Hugo, W.B). The Society for Applied Bacteriology, Technical Series No 27. Oxford Blackwell Scientific Publication, Oxford.

Dorman, H.J.D. and Deans, S.G. (2000). *J. Appl. Microbiol.,* **88:** 308-316.

Dubey, R.C., Rana, A. and Shukla, R.K. (2005). *Indian Drugs,* **42(7):** 443-446.

Eikani, M.H., Rowshanzamir, S. (2004). Modeling and simulation of superheated water extraction of essential oils, Prague, Czech Republic, Czech Society of Chemical Engineering, Prague **1:** 116-68, Czech Republic.

Faleiro M.L., Miguel, M.G., Ladeiro, F., Venâncio, F., Tavares, R., Brito, J.C., Figueiredo, A.C., Barroso, J.G. and Pedro, L.G. (2003). *Lett. Appl. Microbiol.,* **36:** 35-40.

Farag, R.S., Daw, Z.Y., Hewedi, F.M. and El-Baroty, G.S.A. (1989). *J. Food Protection,* **52:** 665-667.

Farrell, K.T. (1990). Spices, Condiments and Seasonings, second edition. Chapman and Hall, New York.

Frank, B. (1989). Mikroorganismen in Drogen. Der mikrobiologische Status von Drogen und Drogenzubereitungen und seine Beurteilung. Deutsche Apotheker Zeitung **129:** 617-623.

Friedman M., Henika, P.R. and Mandrell, R.E. (2002). *J. Food Prot.,* **65:** 1545-1560.

Galambosi, B. and Peura, P. (1996). *J. Essential Oil Research,* **8:** 389-397.

Graven, E.H., Webber, L., Benians, G., Venter, M. and Gardner, J.B. (1991). *J. Essential Oil Research,* **3:** 303-307.

Graven, E.H., Webber, L., Venter, M. and Gardner, J.B. (1990). *J. Essential Oil Research,* **2:** 215-220.

Guenther, E. (1948). The Essential Oils. D. Van Nostrand, New York.

Gustafson, J.E., Liew, Y.C., Chew, S., Markham, J.L., Bell, H.C., Wyllie, S.G. and Warmington, J.R. (1998). *Letters in Appl. Microbiol.,* **26:** 194-198.

Hadacek, F. and Greger, H. (2000). *Phytochemical Analysis,* **11:** 137-147.

Hammer, K.A., Carson, C.F. and Riley, T.V. (1999). *J. Appl. Microbiol.,* **86:** 985-990.

Helander, I.M., Alakomi, H.L., Latva-Kala, K., Mattila-Sandholm, T., Pol, I., Smid, E.J., Gorris, L.G.M. and Von Wright, A. (1998). *J. Agricultural and Food Chemistry,* **46:** 3590-3595.

Hodges, N.A. and Hanlon, G.W. (1991). *In:* Mechanism of action of chemical biocides (Eds. Denyer, S.P. and Hugo, W.B.) Blackwell, Oxford, pp 297-310.

Houghton, P.J. and Raman, A. (1998). 1st ed. Laboratory handbook for the fractionation of natural extracts. London: Chapman and Hall: 32.

Janssen, A.M., Scheffer, J.J.C. and Baerheim Svendsen, A. (1987). *Planta Med.,* **53:** 396-398.

Jerkovic, I., Mastelic, J. and Milos, M. (2001). *International J. Food Sci. Technol.,* **36:** 649-654.

Juliano, C., Mattana, A. and Usai, M. (2000). *J. Ess. Oil Res.,* **12:** 516-522.

Juliano, C., Mattana, A. and Usai, M. (2000). *J. Essential Oil Research,* **12:** 516-522.

Juven, B.J., Kanner, J., Schved, F. and Weisslowicz, H. (1994). *J. Appl. Bacteriology,* **76:** 626-631.

Kalinovic, I. and Rozman, V. (2000). *Plant Protection Science,* **36:** 21-22.

Kaul, P.N., Rao, B.R.R., Bhattacharya, A.K., Mallavarapu, G.R. and Ramesh, S.I. (1997). *J. Essential Oil Research,* **9:** 115-117.

Kerr, J.R. (1999). *Microbial Ecology in Health and Disease,* **11:**129-142.

Kneifel, W. and Berger, E. (1994). *J. Food Protection,* **57:** 893-901.

Kneifel, W., Czech, E. and Kopp, B. (2002). *Planta Medica,* **68:** 5-15.

Knobloch, K., Pauli, A., Iberl, B., Weigand, H. and Weis, N. (1989). *J. Essential Oil Research,* **1:** 119-128.

Knobloch, K., Weigand, H., Weis, N., Schwarm, H.M. and Vigenschow, H. (1986). *In:* Progress in Essential Oil Research: (ed. Brunke, E.J.), 16th International Symposium on Essential Oils. De Gruyter, Berlin.

Koedam A. (1977a). *Riechs. Arom. Kosmet.,* **27(2):** 36-41.

Koedam A. (1977b). *Riechs. Arom. Kosmet.,* **27(1):** 8-11.

Krell, E. (1982). Techniques and instrumentation in Analytical Chemistry, Vol. 2: Handbook of laboratory distillation. 2nd ed. Amsterdam: Elsevier Scientific Publishing Company.

Kwon, J.H. and Belanger, J.M.R. *et al.* (2003). *Food Research International,* **36(5):** 491-498.

Lambert, R.J.W., Skandamis, P.N., Coote, P. and Nychas, G.J.E. (2001). *J. Appl. Microbiology,* **91:** 453-462.

Lattaoui N. and Tantaoui-Elaraki, A. (1994). *Rivista Italiana* EPPOS, **13:** 13-19.

Leela, N.K., Tava, A., Shafi, P.M., John, S.P. and Chempakam, B. (2002). *Acta Pharm.,* **52:** 137-141.

Li, B.Y., Hu, Y., Liang, Y.Z., Xice, P.S. and Ozaki, Y. (2006). *The Analyst,* **131**: 538-46.

Lincoln, D.E. and Langenheim, J.H. (1978). *Biochemical Systematics and Ecology,* **6(1):** 21-32.
Lis-Balchin M., Ochoka, R.J., Deans, S.G., Asztemborska, M. and Hart, S. (1999). *J. Ess. Oil Res.* **11:** 393-397.
Lis-Balchin, M., Hart, S., Deans, S.G. and Eaglesham, E. (1996). *Spices and Medicinal Plants,* **4:** 69-86.
Lis-Balchin, M., Ochoka, R.J., Deans, S.G., Asztemborska, M. and Hart, S. (1999). *J. Essential Oil Research,* **11:** 393-397.
Lucchesi, M.E. and Chemat, F. *et al.* (2004). *J. Chromatography A,* **1043(2):** 323-327.
Lucchesi, M.E. and Smadja, J. *et al.* (2007). *J. Food Engineering,* **79(3):** 1079-1086.
Luque De Castro, M.D., Jiménez-Carmona, M.M. and Fernández-Peréz, V. (1999). *Trends in Analytical Chemistry,* **18(11):** 708-716.
Madeira, S.V.F., Matos, F.J.A., Leal-Cardoso, J.H., and Criddle, D.N. (2002). *J. Ethnopharmacology,* **81:** 1-4.
Manabe, A., Nakayama, S. and Sakamoto, K. (1987). *Japanese J. Pharmacology,* **44:** 77-84.
Margina, A. and Zheljazkov, V. (1994). *J. Essential Oil Research,* **6:** 607-615.
Marino M., Bersani, C. and Comi, G. (2001). *Int. J. Food Microbiol.,* **67:** 187-195.
Marino M., Bersani, C. and Comi, G. (1999). *J. Food Prot.* **62(9):** 1017-1023.
Marotti M., Piccaglia, R. and Giovanelli, E. (1994). *Flav. Frag. J.,* **9:** 125-129.
Masango, P. (2004). *J. Cleaner Production,* **13(8):** 833-839.
McGimpsey J.A., Douglas, M.H., Van Klink, J.L., Beauregard, D.A. and Perry, N.B. (1994). *Flav. Frag. J.,* **9:** 347-352.
McKee, L.H. (1995). *Lebensmittel-Wissenschaft und Technologie,* **28:** 1-11.
Michael Spiro, S.S.C. (1995). *Flavour and fragrance J.,* **10:** 259-272.
NCCLS, (2000). Methods for dilution antimicrobial susceptibility tests for bacteria that grow aerobically; approved standard— Fifth edition. NCCLS document M7-A5. [ISBN 1-56238-394- 9] NCCLS, 940 West Valley Road, Suite 1400, Wayne, PA 19087-1898. USA.
Nychas, G.J.E. (1995). *In:* New Methods of Food Preservation. (Ed. Gould, G.W.). Blackie Academic and Professional, London, pp. 58-89.
Oosterhaven, K., Poolman, B. and Smid, E.J. (1995). *Industrial Crops and Products,* **4:** 23-31.
Packiyasothy E.V. and Kyle, S. (2002). *Food Australia,* **54(9):** 384-387.
Paster N., Menasherov, M., Ravid, U. and Juven, B. (1995). *J. Food Prot.,* **58(1):** 81-85.
Paster, N., Juven, B.J., Shaaya, E., Menasherov, M., Nitzan, R., Weisslowicz, H. and Ravid, U. (1990). *Letters in Appl. Microbiology,* **11:** 33-37.
Patwardhan, B. and Hooper, M. (1992). *Int. J. Altrenative Complement. Med.,* **10:** 9-11.
Piccaglia, R., Marotti, M. and Galletti, G.C. (1989). *J. Essential Oil Research,* **2:** 73-83.
Pol, I.E., Mastwijk, H.C., Slump, R.A., Popa, M.E. and Smid, E.J. (2001). *J. Food Protection,* **64:** 1012-1018.
Ramanadhan, B. (2005). Microwave extraction of essential oils (from black pepper and coriander) at 2.46 GHz. 169.
Rana, A. (2005). *Ph.D. Thesis,* Gurukula Kangri University, Haridwar, Uttarakhand (India).
Rawat, R.B.S. (2006). *In:* Proceedings of the workshop – "Approaches towards evaluation of medicinal plants prior to clinical trials" organized by The Foundation For Medical Research, 84-A, R.G. Thandani Marg, Worli, Mumbai-400018. November 08, 2006, Yashada, Pune, India. pp. 16.
Remadevi, R. and Ravindran, P.N. (2005). *Spice India,* **18(8)**: 11-17.
Reverchon, A.A.F.S. (1994). *Flavour and Fragrance J.,* **9:** 19-23.
Rios J.L., Recio, M.C. and Villar, A. (1988). *J. Ethnopharmacol.,* **23:** 127-149.
Russo, M., Galletti, G.C., Bocchini, P. and Carnicini, A. (1998). *J. Agric. Food Chem.* **46:** 3741-3746.
Sahai, S. (2000). *In:* UNCTAD Expert Meeting on Systems and National Experiences for Protecting Traditional Knowledge, Innovations and Practices, Geneva 30 October – 1 November 2000. Gene Campaign, J- 235/ A Sainik Farms, Khanpur, New Delhi – 110 062.
Saini, S. (2004). Ph.D. thesis Gurukula Kangri University, Haridwar–249404, Uttarakhand (India).
Salzer, U.J. (1977). *Crit. Rev. Food Sci. Nut.,* **9(4):** 345-373.
Sau Soon Chen, M.S. (1995). *Flavour and fragrance J.,* **10:** 101-112.
Sawai, J., Doi, R., Maekawa, Y., Yoshikawa, T. and Kojima, H. (2002). *J. Industrial Microbiology and Technology,* **29:** 296-298.
Scheffe, J.J.C. and Baerheim Svendsen, A. (1981). *Riech. Arom. Kosmetic.,* **31(2):** 39-41.
Scheffer, J.J.C. (1993). *Acta Horticulturae,* **344:** 1-8.
Schilcher, H. (1982). *Planta Medica,* **44:** 65-77.
Senatore, F. (1996). *J. Agric. Food Chem.,* **44:** 1327-1332.
Shelef, L.A. (1983). *J. Food Saf.,* **6:** 29-44.
Shrikumar, S., Athem, M., Sukumar, M. and Ravi, T.K. (2005). *Indian J. of Pharm. Sci.,* **67(6):** 721-24.

Shrikumar, S. and Ravi, T.K. (2007). *Pharmacognosy Reviews,* **1(1):**180-184.

Sikkema, J., De Bont, J.A.M. and Poolman, B. (1994). *J. Biological Chemistry,* **269:** 8022-8028.

Sikkema, J., De Bont, J.A.M. and Poolman, B. (1995). Mechanisms of membrane toxicity of hydrocarbons. *Microbiological Reviews,* **59:** 201-222.

Skandamis, P., Koutsoumanis, K., Fasseas, K. and Nychas, G.J.E. (2001). *Italian J. Food Science,* **13:** 65-75.

Skoula, M. and Harborne, J.B. (2002). *In: Oregano: The genera Origanum and Lippia.* (Ed. Kintzios, S.E.) 1st ed. London: Taylor & Francis. pp. 91.

Soysal, Y. and Öztekin, S. (2001). *J. Agricultural Engineering Research,* **79(1):** 73-79.

Stammati, A., Bonsi, P., Zucco, F., Moezelaar, R., Alakomi, H.I. and Von Wright, A. (1999). *Food and Chemical Toxicology,* **37:** 813-823.

Subramoniam, A. (2001). *Indian J. Pharmacology,* **33 :** 145-146.

Terblanché, F.C. (1995). The extraction and medicinal activity of essential oil components of Lippia scaberrima Sond. M.Sc. dissertation, Pretoria, University of Pretoria.

Thankamani, C.K., Kandiannan, K. and Sheriff, P.A. (2005). *Spice India,* **18(8):** 22-28.

Thoroski, J., Blank, G. and Biliaderis, C. (1989). *J. Food Protection,* **52:** 399-403.

Ultee, A., Bennink, M.H.J. and Moezelaar, R. (2002). *Appl. and Environmental Microbiology,* **68:** 1561-1568.

Ultee, A., Kets, E.P.W., Alberda, M., Hoekstra, F.A. and Smid, E.J. (2000a). *Archives of Microbiology,* **174:** 233-238.

Ultee, A., Slump, R.A., Steging, G. and Smid, E.J. (2000b). *J. Food Protection,* **63:** 620-624.

Verdine, G.L. (1996). *Nature,* **11(3):** 384.

Verkman, A.S. (2004). *American J. Physiology-Cell Physiology,* **286:** 465-474.

Wan, J.W. (2007). Herbal Medicinal Products: Quality, safety and efficacy considerations. Pharmalink Consulting Ltd., UK, pp. 7.

Wang, Z. and Ding, L. *et al.* (2006). *J. Chromatography* A, **1102(1-2):** 11-17.

Wendakoon, C.N. and Sakaguchi, M. (1995). *J. Food Protection,* **58:** 280-283.

Whish, J.P.M. (1996). *J. Essential Oil Research,* **8:** 405-410.

Whish, J.P.M. and Williams, R.R. (1996). *J. Essential Oil Research,* **8:** 47-51.

WHO. (1978). The Promotion and Development of Traditional Medicine. WHO Technical Report Series, No. 622: 8, Geneva, Switzerland.

Wichtl, M. (1997). *In:* Teedrogen und Phytopharmaka. Ein Handbuch fur die Praxis auf wissenschaftlicher Grundlage. 3. Auflage Stuttgart: Wissenschaftliche Verlags-GesmbH, 5-31.

Wilkins C. and Madsen, J.O. (1991). *Zeitschrift für Lebensmittel-Untersuchung und-Forschung,* **192:** 214-219.

Zheljazkov, V.D. and Jekov, D. (1996). *In:* Proceedings International Symposium for Medicinal and Aromatic Plants, (Eds. Craker, L.E., Nolan, L. and Shetty, K.) 427-433.

Zygadlo, J.A., Grosso, N.R., Aburra, R.E. and Guzman, C.A. (1990). *Biochemical Systematics and Ecology,* **18(6):** 405-407.

Index